AF441963

European Climate Vulnerabilities and Adaptation

European Climate Vulnerabilities and Adaptation

A Spatial Planning Perspective

Edited by

Philipp Schmidt-Thomé
Geological Survey of Finland (GTK), Espoo, Finland

Stefan Greiving
Institute of Spatial Planning (IRPUD), TU Dortmund University, Germany

WILEY Blackwell

Texts and maps stemming from research projects under the ESPON programme presented in this book do not necessarily reflect the opinion of the ESPON Monitoring Committee.

Registered office: John Wiley & Sons, Ltd, The Atrium, Southern Gate, Chichester, West Sussex, PO19 8SQ, UK

Editorial offices: 9600 Garsington Road, Oxford, OX4 2DQ, UK
The Atrium, Southern Gate, Chichester, West Sussex, PO19 8SQ, UK
111 River Street, Hoboken, NJ 07030-5774, USA

For details of our global editorial offices, for customer services and for information about how to apply for permission to reuse the copyright material in this book please see our website at www.wiley.com/wiley-blackwell.

Library of Congress Cataloging-in-Publication Data

Schmidt-Thomé, Philipp.
  European climate vulnerabilities and adaptation : a spatial planning perspective / editors, Philipp Schmidt-Thomé, Stefan Greiving.
    pages cm
  Includes bibliographical references and index.
  ISBN 978-0-470-97741-5 (hardback)
  1. Climatic changes – Europe. 2. Climatic changes – Government policy – Europe. 3. Climatic changes – Economic aspects – Europe. 4. Europe – Climate. I. Greiving, Stefan. II. Title.
  QC903.2.E85S36 2013
  363.738'742094 – dc23

                                                                                        2013008530

A catalogue record for this book is available from the British Library.

Wiley also publishes its books in a variety of electronic formats. Some content that appears in print may not be available in electronic books.

Cover image: Supplied by iStock.
Cover design by Dan Jubb.

Typeset in 10.5/12.5pt Minion by Laserwords Private Limited, Chennai, India.
Printed and bound in Singapore by Markono Print Media Pte Ltd.

1 2013

Contents

Biographies

Philipp Schmidt-Thomé is a senior scientist and project manager at the Geological Survey of Finland (GTK) and an Adjunct Professor at the University of Helsinki. He trained as a geographer (M.Sc.) and holds a Ph.D. in geology. He leads the Working Group on Climate Change Adaptation under the International Union of Geosciences Commission on Geo-Environment. His scientific focus is on geoscience communication and interdisciplinary cooperation. His recent project work has focused on integrating natural hazards, climate change and risks into land-use planning practices. He is a regular lecturer at several universities and a visiting fellow to the South East Asia Disaster Prevention Institute (SEADPRI).

Stefan Greiving is Executive Director of the Institute of Spatial Planning at TU Dortmund University, Germany. He holds a diploma in spatial planning, a Ph.D. in urban planning and a habilitation in planning and administration. He was coordinator of the ESPON Climate project. Professor Greiving is author of about 140 publications. His main research focus is on assessment and management of spatially relevant risks and effects of climate change. He is member of a UN expert working group on measuring disaster vulnerability and full member of the German Academy for Spatial Planning and Research.

Acknowledgements

The Editors acknowledge the significant contribution of Anika Nockert who was largely responsible for the technical and administrative revision process of this book. **Anika Nockert** has a B.Sc. in Geography and is an M.Sc. student in 'Physical geography of human environment-systems' at the Humboldt University in Berlin. She worked as a research assistant in the 'Climate Impacts and Vulnerability' Research Domain at the Potsdam Institute for Climate Impact Research (PIK) and at the Geological Survey of Finland (GTK).

Philipp Schmidt-Thomé would like to thank the Internal Union of Geological Sciences' Comission on Geo-Science for Environmental Management (IUGS-GEM) for its endorsement in the preparation of this book.

List of Contributors

Gábor Bálint
VITUKI Environmental and Water
   Management Research Institute
Kvassay 1
1095 Budapest
Hungary

Daniel Baumgartner
Swiss Federal Institute for Forest, Snow and
   Landscape Research WSL
Zürcherstrasse 111
8903 Birmensdorf
Switzerland

Arno Bouwman
PBL Netherlands Environmental
   Assessment Agency
A. van Leeuwenhoeklaan 9
3721 MA Bilthoven
The Netherlands

Alina Chicoş
VITUKI Environmental and Water
   Management Research Institute
Kvassay 1
1095 Budapest
Hungary

Mária Csete
BME (Budapest University of Technology and Eco-
   nomics)
Műegyetem rkp.
1111 Budapest
Hungary

Simin Davoudi
Newcastle University
School of Architecture, Planning and
   Landscape
Claremont Towe
Newcastle upon Tyne, NE1 7RU
United Kingdom

Jan Dzurdzenik
Agency for the Support of Regional
   Development Košice
n.o., Strojárenská 3
040 01 Košice
Slovakia

Florian Flex
TU Dortmund University
Institute of Spatial Planning (IRPUD)
August-Schmidt-Strasse 10
44227 Dortmund
Germany

Annamária Göncz
VÁTI Nonprofit Ltd.
Spatial Planning Department
Gellérthegy u. 30–32
1016 Budapest
Hungary

Stefan Greiving
TU Dortmund University
Institute of Spatial Planning (IRPUD)
August-Schmidt-Strasse 10
44227 Dortmund
Germany

Anne Holsten
Potsdam Institute for Climate Impact Research
P.O. Box 60 12 03
14412 Potsdam
Germany

Tesliar Jaroslav
Agency for the Support of Regional Development
  Košice
n.o., Strojárenská 3
040 01 Košice
Slovakia

Sirkku Juhola
Aalto University
Department of Real Estate
Planning and Geoinformatics
P.O. Box 12200
00076 Aalto, Espoo
Finland
and
University of Helsinki
Department of Environmental Sciences
P.O. Box 65
00014 Helsinki
Finland

Joost M. Knoop
PBL Netherlands Environmental
  Assessment Agency
A. van Leeuwenhoeklaan 9
3721 MA Bilthoven
The Netherlands

Jürgen P. Kropp
Potsdam Institute for Climate Impact Research
P.O. Box 60 12 03
14412 Potsdam
Germany

Sylvia Kruse
Swiss Federal Institute for Forest, Snow and
  Landscape Research WSL
Zürcherstrasse 111
8903 Birmensdorf
Switzerland

Ove Langeland
Norwegian Institute for Urban and Regional
  Research (NIBR)
Gaustadalléen 21
0349 Oslo
Norway

Bjørg Langset
Norwegian Institute for Urban and Regional
  Research (NIBR)
Gaustadalléen 21
0349 Oslo
Norway

Christian Lindner
TU Dortmund University
Institute of Spatial Planning (IRPUD)
August-Schmidt-Strasse 10
44227 Dortmund
Germany

Johannes Lückenkötter
TU Dortmund University
Institute of Spatial Planning (IRPUD)
August-Schmidt-Strasse 10
44227 Dortmund
Germany

Hug March
Autonomous University of Barcelona
Geography Department
08193 Bellaterra
Spain

Javier Martín-Vide
University of Barcelona
Department of Physical Geography
08001 Barcelona
Spain

Petteri Niemi
Aalto University
Department of Real Estate
Planning and Geoinformatics
P.O. Box 12200
00076 Aalto, Espoo
Finland

Jorge Olcina
University of Alicante
Department of Regional Geographical Analysis
03080 Alicante
Spain

Emilio Padilla
Autonomous University of Barcelona
Department of Applied Economics
08193 Bellaterra
Spain

Tamás Pálvölgyi
BME (Budapest University of Technology
   and Economics)
Műegyetem rkp.
1111 Budapest
Hungary

Lasse Peltonen
Aalto University
Department of Real Estate
Planning and Geoinformatics
P.O. Box 12200
00076 Espoo
Finland
and
Finnish Environment Institute
P.O. Box 140
00251 Helsinki
Finland

Alexandru-Ionut Petrisor
URBAN-INCERC
Soseaua Pantelimon, nr. 266
Sector 2
021652 Bucharest
Romania

Marco Pütz
Swiss Federal Institute for Forest, Snow and
   Landscape Research WSL
Zürcherstrasse 111
8903 Birmensdorf
Switzerland

Olivia Roithmeier
Potsdam Institute for Climate Impact Research
P.O. Box 60 12 03
14412 Potsdam
Germany

David Saurí
Autonomous University of Barcelona
Geography Department
08193 Bellaterra
Spain

Philipp Schmidt-Thomé
Geological Survey of Finland (GTK)
P.O. Box 96
02151 Espoo
Finland

Krisztián Schneller
VÁTI Nonprofit Ltd.
Spatial Planning Department
Gellérthegy u. 30–32
1016 Budapest
Hungary

Anna Serra-Llobet
Autonomous University of Barcelona
Geography Department
08193 Bellaterra
Spain

Teresa Sprague
TU Dortmund University
Institute of Spatial Planning (IRPUD)
August-Schmidt-Strasse 10
44227 Dortmund
Germany

Manuela Stiffler
Swiss Federal Institute for Forest, Snow and
   Landscape Research WSL
Zürcherstrasse 111
8903 Birmensdorf
Switzerland

Emmanouil Tranos
VU University Amsterdam
Faculty of Economics and Business
  Administration
De Boelelaan 1105
1081 HV Amsterdam
The Netherlands

Jarmo Vehmas
Finland Futures Research Centre
University of Turku
20014 Turku
Finland

José Fernando Vera
University Alicante
Institute for Tourism Research
03080 Alicante
Spain

Hans Visser
PBL Netherlands Environmental
  Assessment Agency
A. van Leeuwenhoeklaan 9
3721 MA Bilthoven
The Netherlands

Carsten Walther
Potsdam Institute for Climate Impact Research
P.O. Box 60 12 03
14412 Potsdam
Germany

Chapter 1

Introducing the pan-European approach to integration on climate change impacts and vulnerabilities into regional development perspectives

Philipp Schmidt-Thomé[1] and Stefan Greiving[2]

[1]*Geological Survey of Finland (GTK), P.O. Box 96, 02151 Espoo, Finland*
[2]*TU Dortmund University, Institute of Spatial Planning (IRPUD),*
August-Schmidt-Strasse 10, 44227 Dortmund, Germany

Abstract

There is a political demand towards a territorial response to climate change. Since the development of territorially differentiated adaptation strategies calls for an evidence basis, a cohesive approach to developing an integrated vulnerability assessment is introduced. Although the European Observation Network for Territorial Development and Cohesion (ESPON) Climate project was the first attempt at a pan-European and cross-sectorial climate change vulnerability assessment, the further research that is needed in just about every aspect of climate change that the project touched upon is discussed. The three parts of the book are then outlined.

1.1   Introduction

Territorial development is generally considered to be very important when dealing with climate change. For example, it is regarded as being responsible for and capable of reducing regional vulnerabilities to climate change as well as developing climate mitigation and adaptation capacities against the impacts of climate change (Stern, 2007; IPCC, 2007). The World Bank Report '*The Global Monitoring Report 2008*', which deals with climate change and the Millennium Development Goals, concludes that the advancement of adaptive urban development strategies is a fundamental field of action for dealing with the challenges of climate change (World Bank, 2008). The European Union (EU) White Paper '*Adapting to Climate Change: Towards a European Framework for Action*', explicitly relates to spatial planning and territorial development, respectively, stating that 'a more strategic and long-term approach to spatial planning will be necessary, both on land and on marine areas, including in transport, regional development, industry, tourism and energy policies' (Commission of the European Communities, 2009, p. 4). In the EU Territorial Agenda it is stipulated under Priority 5 that '. . . joint trans-regional and integrated approaches and strategies should be further developed in order to face natural hazards, reduce and mitigate greenhouse gas emissions and adapt to climate change. Further work is required to develop and intensify territorial cohesion policy, particularly with respect to the consequences of territorially differentiated adaptation strategies' (BMVBS (Federal Ministry of Transport, Building and Urban Development), 2010, p. 7).

The above-mentioned quotes show that there is a political demand towards a territorial response to climate change. Since the development of territorially differentiated adaptation strategies calls for an evidence basis, this book presents a cohesive approach to developing an integrated vulnerability assessment. The methodology was developed under the European Observation Network for Territorial Development and Cohesion (ESPON) 'Climate' project. The ESPON Climate project was given the task of developing a pan-European vulnerability assessment as a basis of identifying regional typologies of climate change exposure, sensitivity, impact and vulnerability. On this basis, tailor-made adaptation options were derived to cope with regionally specific patterns of climate change. In the ESPON Climate project, this regional specificity

was addressed by several case studies from the trans-national to the very local level.

This book summarises the results achieved by the ESPON Climate project. It is structured into several chapters that display the development of the methodology, the selection, evaluation and assessment of data sets, towards the production of indicators and maps. Following the European overview, there are applications of the approach for local case studies to test and approve the methodology.

The territorial perspective and dimension on climate change vulnerabilities displayed in this book are somehow unique, because so far most of the existing vulnerability studies have a clear sectorial focus, that is, addressing very specific impacts of climate change on single elements of a particular sector. To date, such a comprehensive methodological approach, especially one covering almost an entire continent, has not available. Specialised research is sensible and necessary, but the findings of such focused studies are not easily transferable between sectors or between regions. Research results are often not comparable due to methodological differences. This is particularly troublesome in an international policy context such as the European Union, when it needs to be determined what the consequences of climate change are on the competiveness of Europe as a whole, or on the territorial cohesion of European regions. This book therefore shows the development of a new comprehensive vulnerability assessment methodology, applying it to all regions belonging to 'ESPON space'. The methodology may be applied to develop a response to climate change from the perspective of a European territorial development policy.

Any climate change vulnerability assessment will definitely be confronted with uncertainties, which are based on the uncertainties of the underlying models and emissions scenarios. The vulnerability assessment methodology presented in this book used the COSMO model in Climate Mode (CCLM) as a regional climate model that covers almost the entire ESPON space. The forcing scenario used was the SRES A1B scenario of the Intergovernmental Panel on Climate Change (IPCC). It is important to underline that the methodology is not tied to any specific models, emissions scenarios or indicator data sets. Therefore, the results displayed here may be improved at some point in the future if better input data becomes available, both at this scale and in a comparable format. The developed methodology is scientifically acknowledged, and may thus be used for other similar assessments on entire continents or specific regions.

This book thus displays one possible vulnerability scenario that shows what Europe's future in the wake of climate change might look like. The results are not a forecast, but they give some evidence-based hints as to what European adaptation should address in view of the identified regional typologies of climate change, from a regional development perspective. For example, the book shows that key patterns of regional climate change vulnerability run counter to a major pillar of European policy: territorial cohesion. Several regions in the South and East of the continent are highly vulnerable to climate change. Simultaneously, the current economic performance of those vulnerable regions is weak, as compared with other European regions. This underlines the need for a tailor-made adaptation policy at the European level.

1.2   Further research

The ESPON Climate project was the first attempt at a pan-European and cross-sectorial climate change vulnerability assessment. The project succeeded in developing and implementing a comprehensive methodology that integrates data and interrelations across a vast range of relevant fields. For each indicator a detailed methodology was developed, building on existing research findings, establishing causal relations to other indicators and utilising most appropriate and up-to-date data. Through this course, the project developed several advanced methods for assessing climate change impacts for the pan-European study on a very fine-grained scale. For example, the assessment of many indicators was performed on a 100×100 meter grid cell basis, for example to identify exactly those parts of a region's

population that are sensitive to river flooding inundation or which live in urban heat islands and are especially sensitive to heat waves.

Further research is needed in just about every aspect of climate change that the project touched upon. This includes research on second-order and indirect effects of climatic changes. For example, the project estimated the potential effects of a changing climate on the tourism sector of each NUTS 3 (Nomenclature of Territorial Units for Statistic) region. Through backward and forward linkages, these direct effects have multiplier effects on other (sub-) sectors. Such further analysis is certainly possible and would allow a more complete assessment of the economic impacts of climate change. Relevant economic linkages are likely to, for example; also reach into adjoining regions, thus adding an additional layer of complexity. This would require more economic modelling, which was clearly beyond the scope of this project.

Besides a deeper understanding of detailed mechanisms of climate change, what are needed are pan-European methodologies and comparative research. There are many studies that have been conducted at a national or a regional level, which should be scaled up to a European level. An expert-based, multi-criteria classification of all 231 habitat types of the NATURA 2000 directive in regard to their climate change sensitivity is one example, as so far only about 80 of the central European habitat types have been classified accordingly.

Besides expanding, up-scaling and integrating existing research approaches, this book identifies a great need to make qualitative and institutional aspects of climate change, as well as adaptation and mitigation, compatible with the quantitative assessments conducted. The Alpine space study charted a way forward in this regard, but systematic, pan-European methodologies, including reviews and classifications are needed to integrate institutional aspects into pan-European studies.

Current climate models differ greatly in their projections of future climatic conditions. It should be important that future research projects on climate change vulnerability are resourceful enough to use of all, or at least the major, climate model data. This would, first of all, decrease the uncertainty, which is very high when using only one climate model and one emission scenario, as done exemplarily here. Using more models and scenarios would also lead to a more robust database upon which to perform sensitivity, impact and vulnerability analyses.

Most importantly, further research is needed with respect to projecting sensitivity indicators into the future. ESPON's DEMIFER project broke new ground in projecting demographic trends up to the year 2100. However, what about other social and economic trends? Of course it is difficult, some may say impossible, to make such long-term projections for issues and variables that are volatile and constantly shaped by human intervention. Thus the challenge of climate change and the advances made in modelling future climates puts pressure on other disciplines to also develop sophisticated models or scenarios. Without such research, any climate change impact or vulnerability assessment is fraught with the great weakness that one can only relate dynamic, future-oriented climate data to static sensitivity data.

1.3   Structure of the book

This book is structured into three parts, each of which starts with introductory chapters. The first part starts with the methodological framework and approach and explains the selection of the forcing scenario and the climate model. The following chapters then analyse the climatic stimuli and the climatic exposure of Europe towards selected climate change parameters. Two chapters assessing economic impacts and an integrated impact assessment to determine regional vulnerability patterns follow this. European adaptive and mitigative capacities, respectively, are subsequently analysed. The adaptive capacity is then integrated into the climate change impacts to determine European regional vulnerabilities.

The second part of the book describes how the methodological approach of the project was both applied and further developed in several case studies. These case studies represent different scales, starting from multi-national river regions through national scales towards a federal state. The case studies also represent different geologic, climatic and socio-economic settings.

The book concludes with future challenges for Europe in integrating climate change vulnerabilities into regional development, for example, cohesion funds.

References

BMVBS (Federal Ministry of Transport, Building and Urban Development) (ed.) (2010) National Strategies of European Countries for Climate Change adaptation: A Review from a Spatial Planning and Territorial Development Perspective. (pdf) Available at: <http://www.google.de/url?sa=t&rct=j&q=&esrc=s&source=web&cd=1&ved=0CCoQFjAA&url=http%3A%2F%2Fwww.bbsr.bund.de%2Fnn_629248%2FBBSR%2FEN%2FPublications%2FBMVBS%2FOnline%2F2010%2FDL__ON212010%2CtemplateId%3Draw%2Cproperty%3DpublicationFile.pdf%2FDL_ON212010.pdf&ei=Udh2UK66NMjmtQaj2oH4Dw&usg=AFQjCNFE-iqs3El57AyBbLL8JWsjpispSA> (accessed 10 October 2012).

Commission of the European Communities (2009) Impact Assessment. Commission Staff Working Document accompanying the White Paper Adapting to Climate Change: Towards a European Framework for Action. Commission of the European Communities, Brussels.

IPCC (2007) Climate Change 2007 – Contribution of Working Group II to the Fourth Assessment Report of the Intergovernmental Panel on Climate Change, Cambridge University Press, Cambridge.

Stern, N. (2007) The Economics of Climate Change: The Stern Review (online), Cambridge University Press, Cambridge. Available at: <http://www.hmtreasury.gov.uk/independent_reviews/stern_review_economics_climate_change/sternreview_index.cfm> (accessed 26 September 2012).

World Bank (2008) The Global Monitoring Report 2008, World Bank, Washington.

Chapter 2

Methodology for an integrated climate change vulnerability assessment

Johannes Lückenkötter, Christian Lindner and Stefan Greiving
TU Dortmund University, Institute of Spatial Planning (IRPUD), August-Schmidt-Strasse 10, 44227 Dortmund, Germany

Abstract

The ESPON Climate project is based on an IPCC conceptual framework that is widely used in the climate change and impact research community. According to this framework, rising anthropogenic greenhouse gas emissions contribute to global warming and thus to climate change. This anthropogenic contribution runs parallel to natural climate variability. The resulting climate changes differ between regions, that is, each region has a different *exposure* to climate change. In addition, each region has distinct physical, environmental, social, cultural and economic characteristics that result in different *sensitivities* to climate change. Exposure and sensitivity together determine the possible *impact* that climatic changes may have on a region. However, a region might in the long run be able to adjust, for example, by increasing its dikes. This *adaptive capacity* enhances or counteracts the climate change impacts and thus leads to a region's overall *vulnerability* to climate change.

2.1 Introduction

This chapter describes a methodology that addresses the major climate change vulnerability. After a short overview of the main phases of the methodology, each step of the assessment is defined in detail. The chapter closes with methodological reflections on strengths and weaknesses of the described method and what challenges are ahead for climate change vulnerability assessments in the coming years.

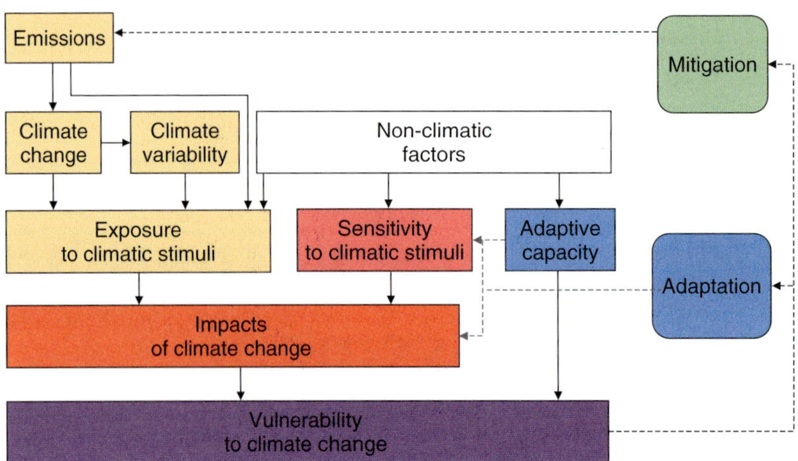

Figure 2.1   Climate change research framework. Adapted from Füssel & Klein, 2002 and 2006.

2.2   Overview of the methodology

Following the conceptual framework given in Figure 2.1, the ESPON project's methodology (see McCarthy *et al.*, 2001) consisted of five main components (see Figure 2.2 for a graphic overview).

The *exposure analysis* focused on the climatic changes as such. It made use of existing projections on climate change and climate variability from the CCLM climate model, whose results have been used, among others, by the 4th IPCC assessment report on climate change. Using the IPCC climate scenario A1B (Nakicenovic *et al.*, 2000), the ESPON Climate project aggregated data for two time periods (1961–1990 and 2071–2100) for eight climate stimuli.[1] River flooding and sea level rise were added as two immediate 'triggered effects' of these climate stimuli.

Each region was then assessed with respect to its climate change *sensitivity*. For each sensitivity dimension (physical, environmental, social, economic and cultural) several sensitivity indicators were developed. Each of the 24 sensitivity indicators[2] were calculated in absolute and relative terms and then combined. This integrates two equally valid perspectives on sensitivity: while relative sensitivity (e.g. density of sensitive population) is advantageous from a comparative point of view, the absolute sensitivity (e.g. absolute number of sensitive inhabitants) is more relevant from a policy/action point of view.

Exposure and sensitivity were then combined to determine the potential *impacts* of climate change. The analysis thus focused on what would be the consequences on human and natural systems if climate change took place unrestrictedly and impacted on the regions without further preparation. To determine impacts, each sensitivity indicator was related to one or more specific exposure indicator(s). For example, heat sensitive population (persons older than 65 years living in urban heat islands) was related to changes in the number of summer days (above 25 °C), while forests sensitive to fire were related to summer days and summer precipitation. After determining the individual impacts,

[1] Exposure indicators used by ESPON Climate related to changes in annual mean temperature, frost days, summer days, winter precipitation, summer precipitation, heavy rainfall days, snow cover days and evaporation.

[2] Sensitivity indicators used by ESPON Climate related to roads, railways, airports, harbours, thermal power stations, refineries, settlements, coastal population, population in river valleys, heat sensitive population in urban heat islands, NATURA 2000 protected areas, occurrence of forest fires, soil organic carbon, soil erosion, museums, cultural World Heritage Sites, energy supply and demand, agriculture and forestry employment and GDP, tourism comfort index and tourist beds.

Table 2.1 Weights resulting from the Delphi survey.

Impacts		Adaptive capacity	
Environmental impacts	0.31	Knowledge and awareness	0.23
Economic impacts	0.24	Technology	0.23
Physical impacts	0.19	Economic resources	0.21
Social impacts	0.16	Institutions	0.17
Cultural impacts	0.1	Infrastructure	0.16

all impacts of one dimension were aggregated. The impact values of the five sensitivity dimensions were finally combined to give one overall impact value.

This aggregation of the various impact dimensions (and later the integration of exposure, sensitivity and adaptive capacity) raises normative issues induced by the theoretical framework. At these stages of the assessment process weighting takes place, even if no weighting is deliberately performed: no weighting amounts to giving equal weights to each dimension. The weighting ultimately refers to cultural beliefs and political preferences, for example, how one values human lives in comparison with economic damage. The ESPON Climate project decided to address these normative issues openly and conducted a survey based on the *Delphi* method among the members of the ESPON Monitoring Committee, which represented the European Commission, 27 European countries and four Partner States. Committee members were asked to propose individual weights for the major phases and dimensions of the assessment (see results in Table 2.1).[3]

A third major component of the project was the assessment of *adaptive capacity* in regard to climate change, that is, the economic, socio-cultural, institutional and technological ability of a region to adapt to the impacts of a changing regional climate. This could mean preventing or moderating potential damage, but also taking advantage of new opportunities. In total 15 adaptive capacity indicators were developed, grouped into the five

adaptive capacity dimensions: economic resources, knowledge and awareness, infrastructure, institutions and technology. The indicators were combined for each dimension and finally aggregated into an overall adaptive capacity. This aggregation was again conducted on the basis of the Delphi survey results.

To determine the overall *vulnerability* of regions to climate change, the impacts and the adaptive capacity to climate change were combined for each region. The underlying rationale is that a region with a high climate change impact may still be moderately vulnerable if it is well adapted to the anticipated climate changes. On the other hand, high impacts would result in high vulnerability to climate change if a region has a low adaptive capacity.

Mitigation of climate change refers to actions that are aimed at reducing concentrations of greenhouse gases and thus global warming. Mitigation is highly relevant for territorial development and cohesion since climate policy implementation and the transition to a low-carbon society will have differential effects on sectors and regions. Mitigation measures, even implemented at the regional level, will not have significant effects on regional climate but only contribute to an overall reduction of global climate change. Therefore, the project's mitigation analysis could only determine the mitigation capacity of each region but could not determine what effect this would have locally or regionally.

Finally, seven *case studies* at the trans-national, regional and local level cross-checked and deepened the findings of the pan-European assessment and explored the diversity of response approaches to climate change. Basically, the same methodology was applied in the case studies as in the pan-European analysis. However, additional methods as well as data sources were utilised in order to explore special regional aspects of climate change impacts, adaptive capacity and vulnerability (Chapters 10–16).

2.3　Methodology in detail

The following section describes in detail the individual steps that needed to be performed within

---

[3]The survey yielded equal weights for exposure versus sensitivity in addition to impact versus adaptive capacity.

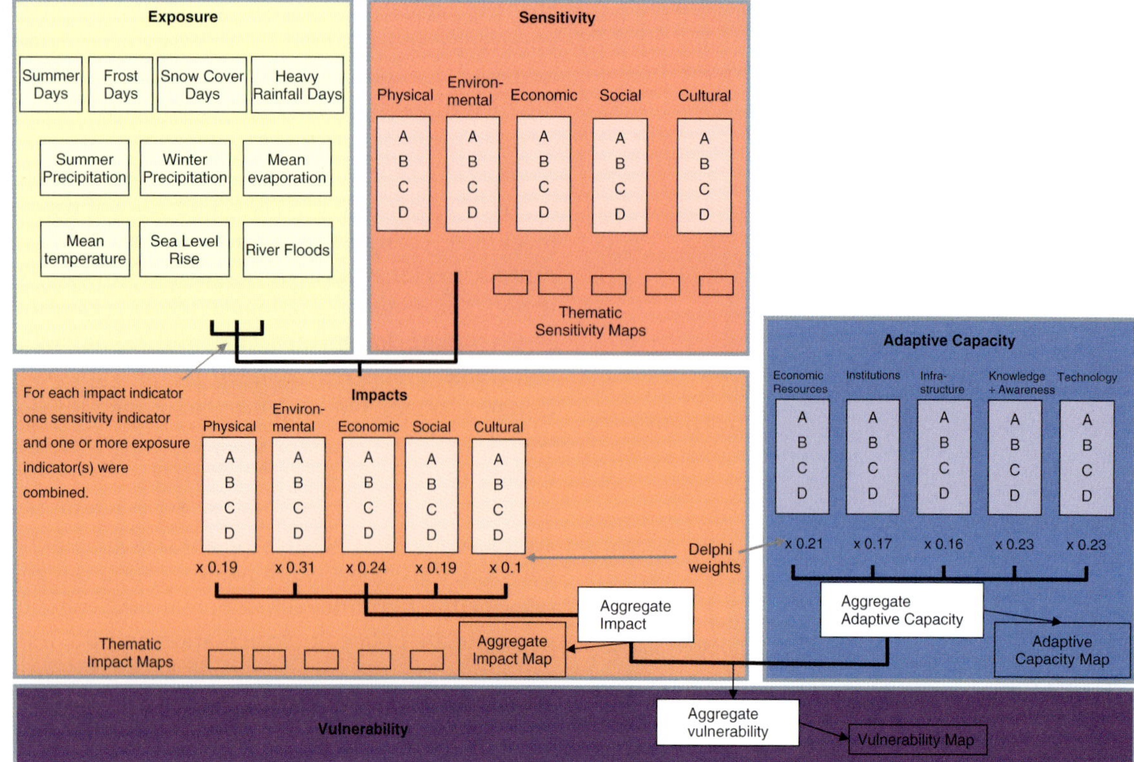

Figure 2.2   Overview of the ESPON Climate vulnerability assessment.

each component of the climate change vulnerability assessment. Figure 2.2 summarises the various steps and may serve as an orientation for the textual explanations.

2.3.1 *Exposure assessment*

2.3.1.1 Aggregation of exposure data

The exposure analysis, based on the CCLM climate model, yielded data for each NUTS 3 region for each of the eight exposure indicators (Chapter 3). For further analysis these exposure variables E_i were normalised. In order to account for the direction of change (decreasing or increasing climatic stimulus), the maximum absolute change in either direction (E_{max}) was used as the reference point for

the normalisation.[4] Thus, the exposure variables are defined by:

$$E_{norm} = E_i / |E_{max}| \qquad (2.1)$$

A special type of exposure indicators need to be highlighted, which were termed 'triggered climate effects' as they are directly triggered by other climatic stimuli. For example, globally rising mean temperatures lead to rising mean sea levels, or the amount of winter precipitation in a river catchment area determines the likelihood and extent

---

[4]In the impact analysis where exposure indicators were related to sensitivity indicators it was sometimes necessary to reverse the mathematical sign of some exposure variables. For example, increased forest sensitivity has to be related to *decreased* (not increased) summer precipitation.

of river flooding in downstream areas. These two triggered climate effects are therefore dependent on global climate changes or on the accumulated effects of climate changes in larger regions. The data for these two triggered climate effects are therefore not taken from the CCLM climate data for a particular raster cell, but are derived from global climate change projections and special hydrological models, respectively.

A cluster analysis was then performed using all eight exposure variables as an informative overview (Chapter 3). The subsequent impact and vulnerability assessment maintained and used only the individual exposure indicators.

2.3.2 *Sensitivity assessment*

2.3.2.1 Identification of sensitivity indicators

To assess the sensitivity of regions to climate change, five sensitivity dimensions were identified, namely physical, environmental, economic, social and cultural sensitivity. For each of these dimensions indicators were identified based on current literature in order to capture the most important regional sensitivities to the climatic changes projected in the exposure analysis (see Chapters 5 and 6 for detailed discussions of each dimension).

2.3.2.2 Determining individual sensitivities

Each sensitivity indicator was calculated individually, that is different data were used and possibly combined to develop meaningful indicators. For some indicators this was relatively straight forward, for example, calculating the relative p of senior citizens in a NUTS 3 region. For other indicators it was necessary to use additional data and perform more complex calculations, such as, when determining the settlement area sensitive to heavy rainfall flash floods (see details in Chapters 5 and 6).

For each sensitivity indicator one absolute and one relative indicator was calculated. For example,

for roads sensitive to river flooding the percentage of the region's road network and the total length of roads sensitive to river flooding were calculated for each NUTS 3 region. Both of these aspects are important, because a sparsely developed region might only have a few kilometres of flood sensitive transport infrastructure, but in relation to the total transport infrastructure of that region this is quite relevant. On the other hand, a more densely developed region might have many kilometres of flood sensitive transport infrastructure, which might nevertheless only account for a small fraction of the total infrastructure of that region. But for e.g. policy-making or disaster management it is still quite relevant that in absolute terms one region only has a few kilometres and the other many kilometres of flood sensitive infrastructure. Thus, absolute and relative indicators used in combination yield a more comprehensive measure of a region's sensitivity.

2.3.2.3 Normalisation and aggregation of sensitivity data

The sensitivity data for all indicators were transformed to be able to first aggregate and later relate them to the exposure indicators. In a first step, the absolute and the relative values for a particular sensitivity indicator were normalised separately using the MinMax normalisation method, that is, the sensitivity values S_j are based on the minimum (S_{min}) and maximum (S_{max}) values within the data range. Thus, the sensitivity values were defined by

$$S_{norm} = (S_j - S_{min})/(S_{max} - S_{min}). \qquad (2.2)$$

This normalisation procedure yields values ranging from 0 (low sensitivity) to 1 (high sensitivity). On this basis the arithmetic mean of the relative and absolute value of each sensitivity indicator was calculated and afterwards normalised again as described above.

2.3.3 *Impact assessment*

2.3.3.1 Combination of exposure and sensitivity

The combining of climate change exposure with the climate change sensitivity results in the (potential) impact of climate change. This process of relating exposure to sensitivity is not performed at the aggregate level but at the indicator level, taking into account that for each sensitivity indicator a different combination of exposure indicators is relevant (see Table 2.2 for an overview). In order to ensure that in cases of no exposure to climate change ($E_i = 0$) the calculated impact I_i would also be zero, multiplication of exposure and sensitivity values was chosen as the most suitable aggregation method. Thus, for each region the value of a particular sensitivity indicator was multiplied with the arithmetic mean of the particular exposure indicators considered relevant for this sensitivity indicator. For example, the climate change impact value for airports was defined by:

$$I_{phys\_air} = S_{norm\_air} \times (E_{norm\_river} + E_{norm\_coast})/2$$

$$(2.3)$$

where E_{norm_river} and E_{norm_coast} refer to the normalised exposure values for river flooding and sea-level rise adjusted coastal flooding, respectively. Afterwards each impact value was normalised following the procedures described above.

2.3.3.2 Aggregating impact scores based on a Delphi survey

In a next step the normalised values of all indicators belonging to one dimension (e.g. social impacts) were combined. Usually this was done by calculating the arithmetic mean of all indicators of a particular dimension; but within the physical and the economic impact dimensions some closely related indicators were first grouped.[5] As an example

____

[5] In the physical impact dimension, arithmetic means were first calculated for the indicator pairs: road and rail transport, airports and harbours, thermal power stations and refineries. Also in the economic impact dimension, the indicators relating to one sector were first grouped before calculating the overall economic impact across sectors.

of the standard procedure, the total social impact (I_{soc}) was defined by:

$$I_{soc} = (I_{soc\_river} + I_{soc\_flash} + I_{soc\_coast} + I_{soc\_heat})/4$$

$$(2.4)$$

where the individual social impact indicators relate to impacts on population due to river flooding (I_{soc_river}), flash floods (I_{soc_flash}), coastal flooding (I_{soc_coast}) and heat days (I_{soc_heat}). The combined average for each impact dimension was subsequently normalised as described above, resulting in one impact value for each impact dimension for each region. On this basis, comparable summary maps were created for each impact dimension (Figure 2.2, Chapters 5 and 6).

Then all dimensions' impact values were aggregated once again to yield one overall impact score. However, averaging the values of the five dimensions would have implied that all dimensions are equally important, that is, that the sensitivity of humans to climate change is as important as, for example, the sensitivity of cultural monuments to climate change. In order to make such normative assumptions transparent and allow the perspectives and preferences from various ESPON countries to enter into the assessment, an internet-based Delphi survey was conducted.

The Delphi method is based on a structured process for collecting and synthesizing knowledge from a group of experts. The aim is to achieve a maximum level of agreement among the participants through several rounds of anonymous opinion surveys that are, nevertheless, informed by the summary results of the preceding round(s) (Helmer, 1966; Linstone and Turoff, 1975; Cooke, 1991). The principle advantages of this approach are that it: (i) avoids key persons exerting a higher influence on a group's responses, (ii) overcomes the geographical constraints and costs of bringing together a group of experts and (iii) allows Delphi participants to express their personal views freely due to the anonymity of answers.

Furthermore, by design the Delphi method is particularly useful for a topic where strong differences

Table 2.2 Relating exposure to sensitivity indicators.

	Δ Annual mean temperature	Δ Frost days	Δ Summer days	Δ Mean winter precipitation	Δ Mean summer precipitation	Δ Heavy rainfall days	Δ Annual mean evaporation	Δ Snow cover days	Triggered climate effects Δ River flooding	Triggered climate effects Δ Mean sea level
Physical sensitivity										
Settlements sensitive to flash, river and coastal floods						•			•	•
Roads sensitive to flash, river and coastal floods						•			•	•
Railways sensitive to flash, river and coastal flooding						•			•	•
Airports sensitive to river and coastal flooding									•	•
Harbours sensitive to river and coastal flooding									•	•
Refineries sensitive to river and coastal flooding									•	•
Thermal power plants sensitive to river and coastal flooding									•	•
Environmental sensitivity										
Forests sensitive to forest fires			•		•					
Protected natural areas	•	•	•	•	•	•	•	•		
Areas prone to soil erosion						•				
Soil organic carbon	•			•	•		•			
Social sensitivity										
Population sensitive to summer heat			•							
Population sensitive to coastal flooding										•
Population sensitive to river flooding									•	
Population sensitive to flash floods						•				
Cultural sensitivity										
UNESCO Cultural World Heritage Sites sensitive to river and coastal flooding									•	
Museums sensitive to river and coastal flooding									•	
Economic sensitivity										
Agriculture sensitive to water availability							•			
Forestry sensitive to water availability							•			
Summer tourism sensitive to high summer temperatures					•					
Winter tourism sensitive to snow cover changes								•		
Energy demand sensitive to summer heat			•							
Energy demand sensitive to winter frost		•								
Energy supply sensitive to changing river water levels					•					

of opinion or high levels of uncertainty exist. As an example, it has been applied for the successful identification of adaptation measures to climate change (Doria *et al.*, 2009).

As participants of the Delphi survey, the members of the ESPON Monitoring Committee were chosen. This committee was considered the most relevant community to be surveyed as it represents all ESPON member states and also accounts for the final ESPON policy recommendations to the EU institutions and member states, respectively. In the first round, 25 of the 47 members of the Monitoring Committee participated and 27 in the second round of weighting. Correct understanding of the concepts and methods used in the survey were ensured by detailed explanations on the ESPON Climate Delphi survey website as well as follow-up phone calls.

The survey itself was conducted in two rounds.

In a first round, all members of the ESPON Monitoring Committee were asked for their initial opinions. Using a survey website they had to allocate percentages for each sensitivity/impact dimension as well as for each component of the two 'pairs' exposure versus sensitivity and impacts versus adaptive capacity. Each of these three estimations added up to a sum of 100%.

Before the second round, all participants were informed about the results of the first round and were then asked to again distribute percentage scores. Typically, those participants, whose opinions differed significantly from the average scores of the first round, often gave scores that are more moderate in the second round.

Usually a third round is conducted in a Delphi survey. However, after the second round the scores of the participants had already converged to such a degree that it was considered unnecessary to conduct yet another round of weighting. Hence, the weights, that is, preferences, expressed by the participants after the second round were used as the relative weights for the various components of the vulnerability assessment.

On this basis, the various impact values of the individual impact dimensions could be aggregated in a way that reflected the preferences of the ESPON Monitoring Committee. Each dimension's impact value was multiplied with the respective weight before an overall arithmetic mean was calculated:

$$I_{\text{total}} = (a \times I_{\text{soc}} + b \times I_{\text{phys}} + c \times I_{\text{env}} + d \times I_{\text{cult}} + e \times I_{\text{econ}})/5 \qquad (2.5)$$

where a, b, c, d and e are the respective weights derived from the Delphi survey. The resulting aggregate impact I_{total} was then normalised as described above.

The resulting overall impact value incorporates three 'dimensions': a relative, dynamic dimension (exposure measured as projected *changes* of climate), an absolute, static dimension (sensitivity measured as relevant regional *conditions vis-à-vis* climate change) and a normative dimension (relative importance of impact dimensions on the basis of expressed *preferences* of survey participants).

2.3.4 Adaptive capacity assessment

2.3.4.1 Adaptive capacity calculation and aggregation

The assessment of the adaptive capacity to climate change was also divided into five dimensions: economic resources, institutions, infrastructure, knowledge and awareness as well as technology were considered the most relevant assets a region has for adapting to climate change. For each dimension several indicators were identified and then aggregated as described for the impact indicators. On this basis the arithmetic mean was calculated for each dimension. Using the results from the Delphi survey the weighted scores of the five dimensions were added up, resulting in an aggregate adaptive capacity value for each NUTS 3 region. This was finally normalised again – following the same procedure as for determining the aggregate impact value of each region. Maps of each dimension's average and of the aggregate adaptive capacity were produced for pan-European comparison.

2.3.5　*Vulnerability assessment*

2.3.5.1　Vulnerability calculation

The results of the impact assessment were multiplied with the aggregate adaptive capacity values ($V = I \times AC$) and then normalised as described in the preceding sections to calculate the aggregate vulnerability score for each region. A final vulnerability map concluded the pan-European assessment.

2.4　Methodological reflections

Reflecting on the project's methodology a number of key features and challenges are apparent. First of all the project used a generally accepted conceptual framework and on this basis was able to build a coherent vulnerability assessment methodology. Nevertheless, the selection, calculation and aggregation of the individual indicators involved not only scientific knowledge, but also normative decisions on what aspects of such concepts as climate change, sensitivity or adaptive capacity are to be captured and assessed. In addition, the choices of indicators are also shaped by the availability and quality of statistical data. Each selected indicator was carefully assessed and revised concerning its relevance for climate change impacts on European Regions, any comparable existing studies, as well as data sources and, finally, its applicability respective to indicator methodology employed in the ESPON Climate project (Chapters 5 and 6). The last was necessary because many indicators used in the project are made up of several input variables. The construction of such composite indicators is especially challenging as it involves complex choices regarding the selection of data, normalisation procedures, weighting schemes and aggregation methods (OECD/JRC 2008).

Challenges of the methodology relate especially to the various mathematical procedures, that is, calculating averages, multiplying and normalising data sets. While the sequence and logic of these operations serve the purpose of combining a great number of very different indicators, the dimensions of the indicators are lost. Scores of different indicators needed to be made compatible by means of normalising their values before calculating arithmetic means or multiplying them. It was not possible to retain the magnitude of the individual indicators because the extreme value of each indicator is by definition set to 1 (or -1). This also means that all aggregated values are inherently relative. A regional impact or vulnerability score is only 'high' or 'low' in relation to all other European regions.

The normalisation method employed in this research places high emphasis on extreme values. Because the regional values of the various individual impact indicators (and likewise the impact dimensions) have different statistical distributions, this can lead to one impact indicator having many regions with high impacts, whereas another indicator has only one region with high impacts. When later combining these two indicators, the indicator with higher impact values has a greater weight in the summary indicator. As a remedy, one could normalise all indicators in a way that they have the same mean absolute value before combining them. Alternatively, area weighted means could be calculated for percentiles before aggregation of indicators. On the other hand, one would in this way 'harmonise' and smooth out certain differences between indicators, which from another perspective might also be objectionable. For example, the thus normalised values would not necessarily reflect the existing relative distances between all data points anymore.

The participants of the Delphi were chosen because of their particular skills with regional development perspectives in Europe – and from a pan-European perspective, which was the addressee of the ESPON Climate project. Had the survey been conducted e.g. only among climatologists, the results would certainly have been different. Therefore, the vulnerability assessment method presented in this chapter and in this book cannot and does not claim 'objectivity'. Exemplarily, Chapter 9 displays the results of using equal weights among the impact dimensions in comparison of the weights derived from the Delphi survey.

A second Delphi survey related point of criticism could be that the participants actually indicated weights with average impacts in mind, whereas the impact assessment (as discussed above) used extreme values as the main reference points. This could well be, and this issue was not explicitly clarified when the Delphi survey was conducted. However, it seems more likely that intuitively participants oriented their ratings more on impacts from extreme events than on average impacts or impacts of small, creeping changes, because most people's perception is tuned to (highly publicised) extremes and not average or gradual changes.

At a more fundamental level, the overall project methodology can be criticised for being rather crude. Using only 8 exposure variables, 24 sensitivity variables and 15 adaptive capacity variables is certainly an oversimplification of the reality of climate change impacts. This is compounded by the fact that only climate data from one climate model (CCLM) based on one forcing scenario (A1B) of the IPCC's Special Report on Emission Scenarios (SRES) was used. More robust results could be achieved using more indicators, several climate models and SRES scenarios (or better IPCC Representative Concentration Pathway (RCP), which will be published in the IPCC 5th Assessment Report). Scientists specialising in one of the chosen indicators could also justifiably demand more complex impact modelling. However, all this was beyond the capacity of a small two-year research project. ESPON Climate's main goal was to make indicators and data compatible and combine them in an overarching, coherent methodological framework. As shown by the project's case studies, this methodology can even be employed at various spatial levels.

Finally, the project's methodology admittedly includes a major challenge: like most other comparable studies ESPON Climate related projections of future climate to current socio-economical and environmental sensitivity conditions. In effect this assumes that the climatic changes that are modelled to occur between 2071 and 2100 would happen all at once and at present. The correct way would be to relate future climate to future sensitivity. However, there are hardly any sensitivity projections for such a long-term perspective. In addition, given the current structural economic crisis in Europe it does not seem likely that such projections (especially for socio-economic indicators) will be published soon and be available for indicator-based and regionally specific impact research. Nevertheless, the parallel modelling approach that underlies the (still unpublished) IPCC's Fifth Assessment Report (to be finalised in 2014) addresses these challenges: research groups throughout the world are busy developing innovative methods that try to cope with the limitations concerning sensitivity projections.

2.5   Conclusion

The main advantage of the ESPON Climate's assessment methodology is its transparency and flexibility. Underlying normative decisions have been made explicit and subjected to normative input from the ESPON Monitoring Committee. The resulting weights of the various dimensions of the methodology can easily be changed to produce respective information (maps). In addition, individual indicators (or new data for a particular indicator) can easily be updated or replaced or new indicators be added without needing to change the methodology of the assessment. This applies to exposure, sensitivity and adaptive capacity indicators. Even the causal relations between particular exposure and sensitivity indicators can easily be modified on the basis of new research findings. This flexibility makes the project's methodology capable of incorporating new findings and data from various research fields.

References

Cooke, R.M. (1991) *Experts in Uncertainty: Opinion and Subjective Probability in Science*, Oxford University Press, New York, Oxford.

Doria, M., Boyd, E., Tompkins, E., Adger, W.N. (2009) Using expert elicitation to define successful adaptation to climate change. *Environmental Science and Policy*, **12**, 810–819.

Füssel, H.-M. and Klein, R.J.T. (2002) Vulnerability and adaptation assessments to climate change: An evolution of conceptual thinking, in UNDP Expert Group Meeting 'Integrating Disaster Reduction and Adaptation to Climate Change'. Havana, Cuba.

Füssel, H.-M. and Klein, R.J.T. (2006) Climate change vulnerability assessments: An evolution of conceptual thinking. *Climate Change*, **75**, 301–329.

Helmer, O. (1966) *Social Technology*, Basic Books, New York.

Linstone, H.A., and Turoff, M. (eds) (1975) *The Delphi Method. Techniques and Applications*, Reading/Mass.

McCarthy, J.J., Canziani, O.F., Leary, N.A., Dokken, D.J., White, K.S. (eds.) (2001) Climate Change 2001: Impacts, Adaptation, and Vulnerability: Contribution of Working Group II to the Third Assessment Report of the Intergovernmental Panel on Climate Change, Cambridge University Press, Cambridge.

Nakicenovic, N., Alcamo, J., Davis, G., *et al.* (2000) IPCC Special Report on Emissions Scenarios, Cambridge University Press, Cambridge and New York.

OECD/JRC (2008) *Handbook on Constructing Composite Indicators: Methodology and User Guide*, OECD Publications, Paris.

Chapter 3

Identifying a typology of climate change in Europe

Carsten Walther, Anne Holsten and Jürgen P. Kropp

Potsdam Institute for Climate Impact Research, P.O. Box 60 12 03, 14412 Potsdam, Germany

Abstract

We distinguish climate change regions of Europe in a spatially explicit way. For this we apply a cluster analysis of impact related climatic indicators for the years 1961–1990 to 2071–2100 under scenario A1B. A high spatial resolution is achieved by using the regional climate model CCLM with a grid edge length of around 20 km. We overcome the challenge of identifying the optimal number of clusters by means of a stability-based approach and obtain five climate change clusters. To analyse the sensitivity of the discovered typology we compare the outcomes of the clustering of two different climate model runs and of two different methods of calculating the climate change indicators. We find the typology to be stable for most of the study area. Our results show European regions that will exhibit similar climatic changes in the future. This can provide a basis for a multisectoral impact analysis assessing the effects of climate change until the end of the century.

3.1 Introduction

Society is facing major challenges through a changing climate. A great deal of effort has to be put into mitigation to prevent further climate change. At the same time, society has to adapt to the already ongoing and therefore unavoidable changes. Knowledge about typologies within the spatially diverse changes in climate could, for example, help to optimise and transfer methods of adaptation for regions facing a comparable thread.

Climate itself was first categorised by the global climate classifications of Köppen (1936). He distinguished the climate on the basis of five vegetation groups as well as precipitation and air temperature. Updated versions based on recent data sets exist (i.e. Kottek *et al.*, 2006). In Gerstengarbe, Werner and Fraedrich (1999), a cluster analysis to distinguish regions with similar climate was applied to precipitation and temperature data. The spatial extent of the analysis was the Northern half of Europe.

We want to distinguish not only climate but climate change regions. Each of the revealed clusters then consists of regions that are similar with regard to the changes of the climatic stimuli in these areas. Gerstengarbe and Werner (2007) applied a typology of climate change on the basis of empirical data from 1901 to 2000. Mahlstein and Knutti (2009) moved a step further and additionally used projected climate data in their cluster analysis. For the exploration of typologies of climate change they implemented the state and the change of climate via the variables precipitation and temperature over the whole world. Owing to the large spatial extent, the resolution was relatively coarse, with grid cells of 200–300 km.

Our aim is to distinguish climate change regions of Europe by using impact related climate change indicators that will give a deeper insight into the projected climate and its influence on society (see Figure 3.1). The identified regions then provide a basis for a multisectoral impact analysis of case studies. We achieve a high spatial resolution by using a regional climate model with grid edge length of around 20 km. To gain an insight into the sensitivity of the discovered typology of five clusters, we compare the outcomes of the clustering of two different model runs and of two different methods to calculate the climate change indicators.

3.2 Methods applied

3.2.1 *Climate change indicators*

The climate data that we applied is generated with the regional climate model CCLM (Lautenschlager *et al.*, 2009) averaged over different model runs within the validation and projection period in the scenario A1B. The climate change indicators are calculated as the difference between the validation

period 1961–1990 and the projected period of 2071–2100. The spatial extent of the analysed data (see Figure 3.1) covers the European continent, excluding parts of Eastern and Southern Europe (Ukraine, Belarus, Turkey and most parts of the Balkan States) (comprising 15 061 grid cells in total). To analyse the changing climate we selected the eight indicators given in Table 3.1.

Beside the indicators *change in summer precipitation*, *change in winter precipitation* and *change in evaporation*, where the difference between the two time periods is measured in relative terms, we use absolute values of change. The calculation of the relative change ΔX_i^{rel} and the absolute change ΔX_i^{abs} is done in the following way:

$$\Delta X_i^{abs} = X_i^F - X_i^P \quad \Delta X_i^{rel} = (X_i^F - X_i^P / X_i^P) \cdot 100$$

$$(3.1)$$

where X_i^F is the value of the ith indicator averaged over the projected period 2071–2100 and X_i^P is the ith indicator averaged over the reference period 1961–1990. We chose a relative definition of change in order to take into account the state from which the change occurs. Thereby it is possible to differentiate between cases such as a region in the Alps with 470 mm precipitation in the summer season and a place in the North of Portugal with a summer precipitation of 170 mm. Here both regions had the same absolute change of around 100 mm between the reference and the projection period, but in relative terms they differ between −20% and −60%. With absolute changes it is not possible to reflect these disparities.

The spatial distributions of the projected changes in the climate indicators within the raster cells are summarised in Figure 3.1.

The spatial distribution of *change in mean annual temperature* shows the highest values in the most Northern and the Mediterranean regions. The three variables *summer precipitation, winter precipitation* and *evaporation* show increasing values in the North, whereas the South is mostly affected by a decline. The centre of Europe exhibits an increase in *winter precipitation* but only small change rates in *summer*

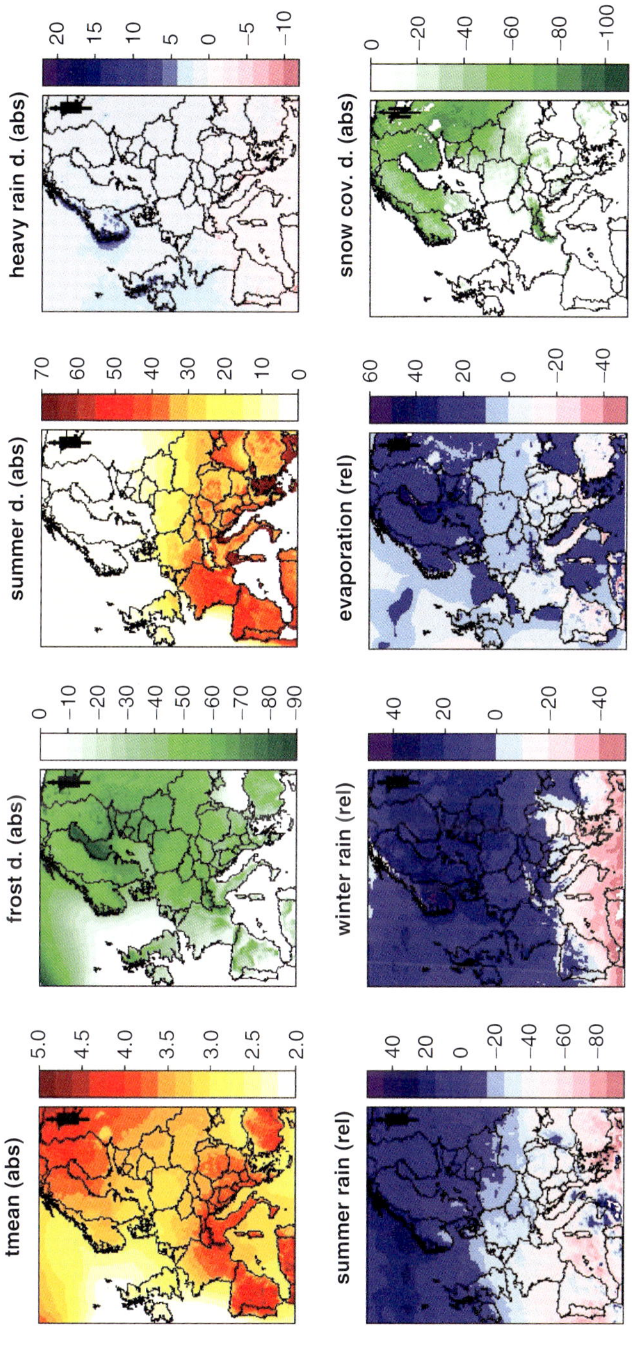

Figure 3.1   Changes of the eight considered climate variables of the model CCLM between the time periods 1961–1990 and 2071–2100 (for abbreviations see Table 3.1). The African continent and parts of Eastern and Southern Europe are not part of the analysis.

Table 3.1 Titles and abbreviations of climate change indicators.

Title of climate change indicator	Abbreviation
Absolute change in mean annual temperature	tmean ($^{\circ}$C) (abs)
Absolute change in number of frost days	frost days (abs)
Absolute change in number of summer days	summer days (abs)
Absolute change in number of days with heavy precipitation	heavy prec. days (abs)
Relative change in amount of summer precipitation	summer prec. (%) (rel)
Relative change in amount of winter precipitation	winter prec. (%) (rel)
Relative change in evaporation	evaporation (%) (rel)
Absolute change in number of days with snow cover	snow cover days (abs)

precipitation and *evaporation*. The *change in the number of summer days* is very small in the Northeast and becomes larger further Southwest. More or less the opposite behaviour is exhibited for the *change in frost days* with almost no change in the Southwest and the strongest decreasing rates in the Northeast. The Alpine regions and the Northwest of Europe have the most pronounced decrease in *days with snow cover*. It is apparent that the characteristics of the last three indicators strongly depend on the pattern of the underlying climate variable. Where the number of days in a climate variable is small, the change in absolute terms is mostly low. The *change in days with heavy precipitation* is dominated strongly by very high increasing rates at the Atlantic coast in Scotland and Norway (see Figure 3.1).

The variables *change in frost days* and *change in days with snow cover* show only negative values (thus a decreasing number of days) for all cells, whereas the variables *temperature change* and *change in summer days* show only positive values (and thus increasing temperature or days). For the other variables, both increases and decreases are projected for Europe.

Before using the data for the cluster analysis we had to apply the following pre-treatment: owing to the fact that outliers can heavily influence the

distribution of an indicator we decided to apply a 99.8% winsorisation of our data set. That means that all data points with a projected change below the 0.1st percentile were set to the 0.1st percentile and all values above the 99.9th percentile were substituted by this threshold.

The variable *change in days with heavy precipitation* is treated in a particular way due to the fact that the influence of extreme values in this indicator is stronger than in the remaining variables. So for most cells only slight changes are projected, whereas strong changes are projected for only a small number of cells (see also Figure 3.1). These extreme values narrow the main part of the data set, such that cluster centres would be restricted to a small value range. Hence, we apply an even stronger winsorisation of this indicator with the 99.9th and the 0.1st percentiles. The effect of this winsorisation is exemplarily shown for this variable in Figure 3.2.

After the winsorisation, the whole data set is standardised by its range to values between 0 and 1 (Milligan and Cooper, 1988). Figure 3.3 gives an overview of the frequency distribution of the normalised and winsorised values of the climate variables for the considered cells.

3.2.2 Technique of the cluster analysis

A cluster analysis categorises the dimensions of a data set by allocating the objects into groups in such a way that the objects within each group are more similar to each other than to objects in other groups. The cluster mechanisms can be distinguished in hierarchical, partitioning and density-based methods (Handl, Knowles and Kell, 2005). Our analysis is mainly based on the second method. A hierarchical clustering is only used to initialise the partitioning method.

In the hierarchical part the data set is transformed into a distance matrix containing all pair wise distances between the objects in the data set. Using specific amalgamation rules, at first the objects and later on the accumulated groups are merged. The 'ward'-method is applied, which merges the

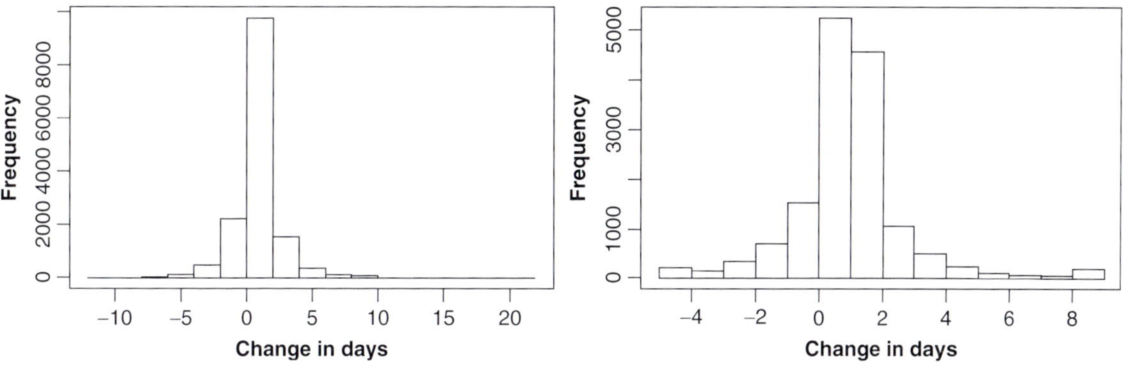

Figure 3.2   Particular treatment of the variable *changes in days with heavy precipitation* – untreated (left) and winsorised with 1st and 99th percentile (right) ($n = 15\,061$ cells).

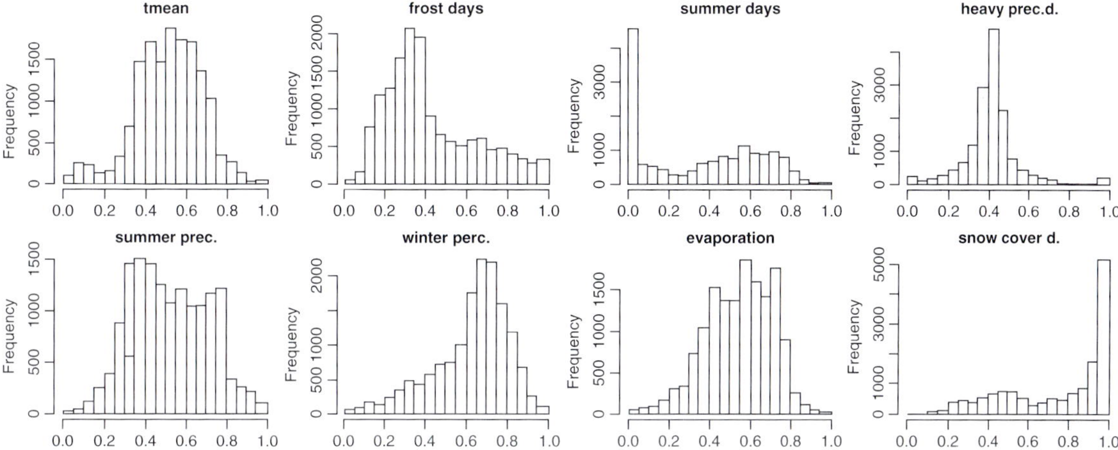

Figure 3.3   Frequency distributions of the climate variables for the considered cells ($n = 15\,061$ cells) after winsorisation and standardisation.

pair of groups that contributes least to the within cluster variance of the whole partition (Ward, 1963). As mentioned above, the hierarchical clustering is used to cluster a small subset of objects to create a starting partition for the subsequent partitioning method – what has been demonstrated to be a valid initialising method (Peterson, Ghosh and Maitra, 2010). For discovering the structure in the data set the widely known partitioning method of K-means is applied (Macqueen, 1967; Hartigan and Wong, 1979). This algorithm minimises the total within-cluster sum-of-squares (TSS) criterion (Steinley, 2006). If the data set consists of P variables and the number of groups is chosen to be K, the criterion is defined by:

$$TSS = \sum_{j=1}^{P} \sum_{k=1}^{K} \sum_{i \in C_k} (x_{ij} - \bar{x}_j^{(k)})^2. \qquad (3.2)$$

where x_{ij} is the value of the indicator j of object i and $\bar{x}_j^{(k)}$ is the mean value of indicator j in cluster k. The objects are assigned to the K given initial cluster centres. Then the new centre is calculated as the average of all objects within the cluster and again all objects are assigned to their nearest cluster centre.

This procedure is repeated until an abort criterion is reached (e.g. the points no longer change position or maximum number of loops). The greatest advantages of K-means are the high calculation speed and the applicability for very large datasets. On the other hand, there is a risk of local minima in the optimisation process and the user has to choose in advance the expected number of clusters. To overcome this disadvantage, we introduce a method to find an appropriate number of clusters in the next part.

3.2.3 Determination of the number of clusters

For identifying the most robust and therefore most representative number of clusters we use a consistency measure, which belongs to the group of stability based methods (see also Ben-Hur, Elisseeff and Guyon, 2002; Roth et $al.$, 2002). It is based on the idea that if the pre-given number of clusters does not fit the underlying structure of the data, a stochastically initialised cluster algorithm generates indefinite and different results.

The procedure of the chosen method is to generate pairs of cluster results, that is, run K-means twice, for a pre-given cluster number K. The size of the overlap of these pairs is assigned as a measure for the consistency, showing how much the two cluster results vary (see Figure 3.4).

A lower variability and a higher value for the consistency measure imply a higher similarity between the pre-given number of clusters and the underlying structure in the analysed data. This pair-wise matching is replicated several times to achieve a robust mean value for the consistency measure. The overall procedure is repeated for different cluster numbers K.

Additionally, the silhouette width as a measure for the confidence of the observations assignment to the clusters is calculated (Kaufman and Rousseeuw, 1990). The silhouette value $S(i)$ for every single observation is defined as:

$$S(i) = \frac{b_i - a_i}{\max(b_i, a_i)}, \qquad (3.3)$$

with b_i as the mean distance to data in the nearest neighbouring cluster and a_i as the mean distance between an object and all other data in the same cluster. By averaging over all the silhouette values we obtain the silhouette width of a cluster result.

A third method for analysing the most appropriate number of clusters is the variance ratio (VRC), which was first proposed by Calinski and Harabasz (1974) and is defined as:

$$VRC = \frac{BCS/(K-1)}{WCS/(n-K)}, \qquad (3.4)$$

where BCS is the between cluster sum of squares, WCS the within cluster sum of squares, K the number of clusters and n the number of objects. It measures the ratio of the between-cluster isolation and the within-cluster compactness. The optimal number of clusters is determined by maximising the VRC.

The statistical software R (R Development Core Team, 2010) has been used to apply the above mentioned methods.

3.2.4 Uncertainty analysis

In order to gain more insight into the stability of our results, we apply an uncertainty analysis in a twofold manner.

First, we compare the achieved results with another way of calculating the difference between the climate indicators of the two time intervals. As shown in Table 3.1, we calculate the change in the indicators with two different methods. Five indicators have been processed with an absolute difference and three with a relative one. As described in Section 3.2.1, this is done for the reason of comparability of regions with differing levels of precipitation. Beside the calculation of the difference, the data set is treated in the same way as the data in Section 3.2.1.

Furthermore, we examine the difference between the clustering of two model runs of CCLM-A1B with the fixed cluster number of five to test the uncertainty of the typology of Europe's climate change regions.

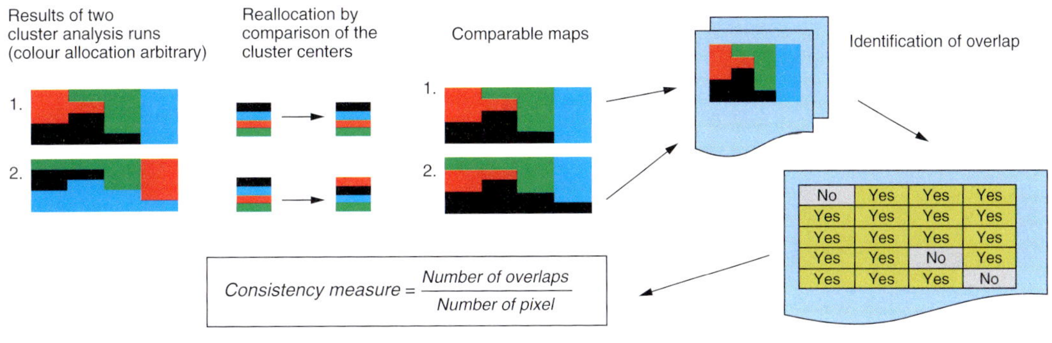

Figure 3.4 Determination of the number of clusters by means of measuring their consistency. Reprinted from Sietz, Lüdeke and Walther, Copyright 2011, with permission from Elsevier.

To illustrate the differences obtained from both comparisons we map the areas that show different and those that show equal allocations to the clusters in the particular typology.

3.3 Results and discussion

3.3.1 *Typologies of climate change regions*

The cluster analysis of European typologies of climate change is based on the averaged CCLM data. As shown in Figure 3.5, the silhouette plot (middle) and variance ratio (right) do not give a clear result (very weak local maximum at $K = 8$ in the silhouette plot). In the consistency graph (left) a small local maximum at nine clusters is apparent. The highest values of consistency are reached in the case of two and three clusters. Partitions with four and five clusters exhibit a stability that is still on a significantly higher level than for the remaining cluster partitions from six to ten. Since we want to show a diverse picture of the typology of climate change in the study area, we choose cluster number five as a compromise between heterogeneity and stability. The geographical distribution of the typology of climate change is shown in Figure 3.6.

According to their spatial appearance the clusters are named as *Northern Europe (NE;*

red), which includes Scandinavia, parts of the Baltic States and the Alps, *Southern central Europe (ScE; orange)*, which covers a band from Eastern France to Romania and some scattered parts in the Southern countries, *Northern central Europe (NcE; green)*, the countries on the coast of the Southern Baltic Sea down to the Czech Republic and parts of Scotland, *Northern western Europe (NwE; purple)*, which comprises Ireland, most parts of Britain and the continental coast along the Channel and finally the *Mediterranean region (Mr; yellow)*.

In Figure 3.6 we further included the case studies discussed in Chapters 11–15 and it appears that the selection of these studies fits well to the revealed typology of climate change in Europe. So the *NwE*-cluster, the *NE*-cluster, Mr-cluster and *ScE*-cluster are each covered with at least one case study. Additionally the North-Rhine Westphalia study covers the three clusters *ScE*, *NwE* and *NcE*. Furthermore, two case studies are located in an area assigned to two different climate change clusters: North-Rhine Westphalia in the *ScE*- and the *NcE*-cluster, as well as the coastal aquifers of Romania, which are situated in *ScE*- and *Mc*-cluster.

The strongest increase in temperature between the time intervals of 2071–2100 and 1961–1990 under CCLM-A1B has to be expected in the region of the *Mr*-cluster ($\sim$3.8 °C) (see Figure 3.7). The lowest increase is projected for the *NwE*-cluster.

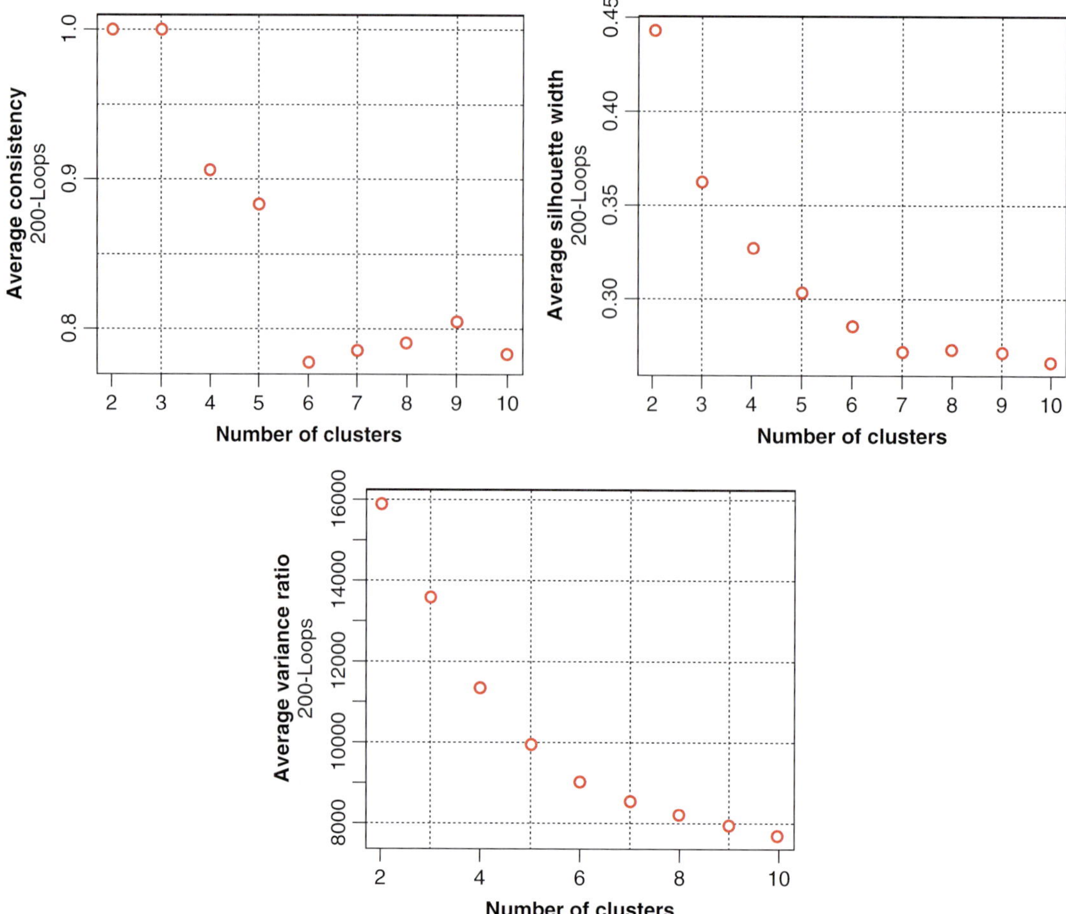

Figure 3.5 Results of consistency measure (left), silhouette index (middle) and variance ratio (right) for the average of the model runs of CCLM.

Here the change in the mean temperature amounts to around 2.7 °C. The change in number of frost days is negative for all clusters and weakest in the *Mr*-cluster, respectively, strongest in both the *Northern* and *Northern central* cluster (∼−57 and ∼−58 days, respectively). In the two southern clusters *ScE* and *Mr* the increases in summer days are highest. Concerning change in precipitation in the winter months, the *Northern* and *Northern central Europe* clusters show particularly strong increases, while for summer months most significant changes in terms of strong decrease can be observed in *Southern central Europe* and *Mediterranean region* clusters. The variable *heavy precipitation* does not show very

strong changes for any of the clusters, although the development in the *Mr*-cluster is in contrast to a small increase in the rest of the clusters slightly decreasing. The same holds for the indicator *evaporation*. The *NE*-cluster has an outstanding position in the variable *days with snow cover*, since the projected decrease is much stronger here than in any other region (∼−61 days).

We further analyse the quality of the cluster representation of each cell (expressed by the distance between the data point and the cluster centre) (Figure 3.8). Pixels with a relatively high distance to their cluster centre are the Alpine region, the Norwegian coast, the Atlantic coast, Scotland and Greece.

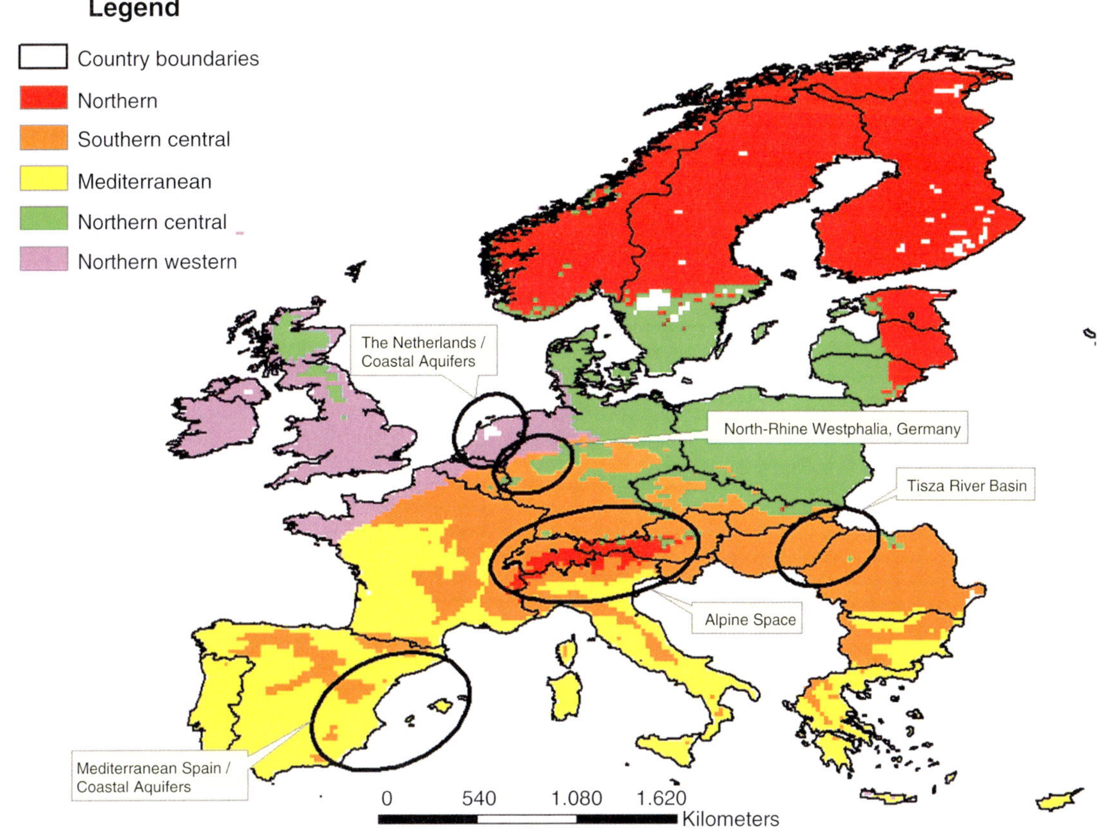

Figure 3.6 European climate change regions analysed by cluster analyses of the average CCLM data. Additionally the five case studies for the impact analysis of Chapters 11–15 are implemented.

This can be traced back to various constellations in the data set. For example, the outstanding high values for the change in heavy precipitation along the Norwegian coast could contribute to this high distance to the corresponding cluster centre.

The different ranges of the cluster centres in Figure 3.7 suggest that each of the eight indicators of climate change has a different influence on the separation of the clusters. We therefore analyse the importance of the indicators for the achieved typology of climate change.

Figure 3.9 shows the result of an ANOVA and the disaggregation-index according to Fraiman, Justel and Svarc (2006). The ANOVA displays the percentage of explained variance by the cluster centres (between cluster sums over total residuals). These workers applied the blinding of an indicator by its mean value to examine the amount of change for the cluster result.

The left graph in Figure 3.9 shows that the highest ratio for the analysis of variance is found for the variable *change in summer days* whereas only a small ratio can be found for the variable *change in days with heavy precipitation*. For the former case this implies that cluster centres represent the distribution of the variable better and that in the latter case the representation is worse. Small values of the Fraiman-Index denote indicators that contribute more to the cluster result. Here the variable *change in days with snow cover* denotes the smallest value.

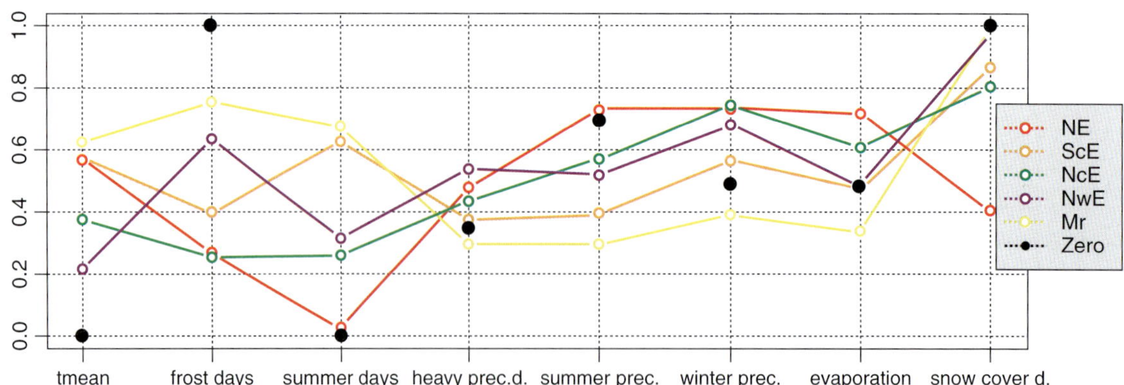

Figure 3.7 Detailed information about the eight climate change indicators (averaged CCLM data) for the five typical regions (description of abbreviations see Table 3.1). The average values for every cluster are plotted and the location of the value zero (no change) is shown by black dots.

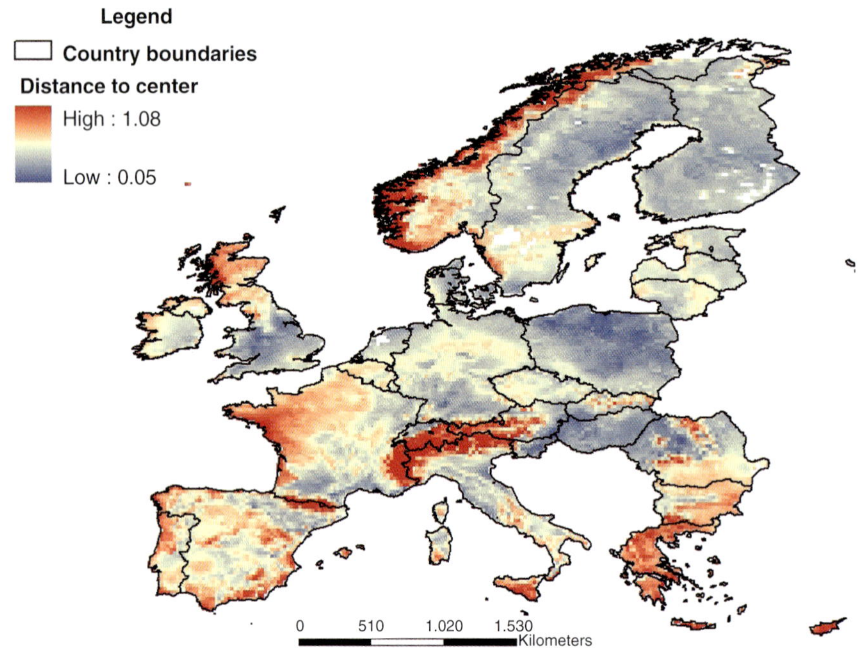

Figure 3.8 Spatial distribution of the distance of the properties of each data point to the corresponding cluster centre. The blue pixels are well represented by their cluster centre. In contrast the red pixels exhibit a relatively high distance to their cluster centroids.

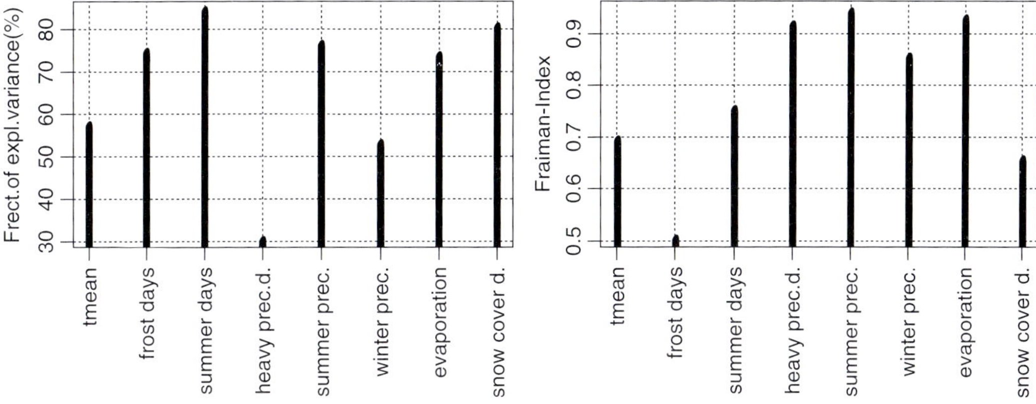

Figure 3.9   Analysis of the importance of the indicators to the cluster result by applying an ANOVA (fraction of explained variance) and calculating a Fraiman-Index (Fraiman, Justel and Svarc, 2006) (for descriptions of abbreviations see Table 3.1).

At this point, two shortcomings of the method have to be mentioned. Firstly, the method is sensitive to correlations between the variables, because in the presence of correlations the blinding of an indicator could be masked by a correlated variable.

Owing to the fact that there is correlation between the indicators, the Fraiman-Index can only give a lower bound for the importance of each indicator. The second weakness is the necessity of a repeating of the clustering, which could lead to a local minimum. We therefore repeated this clustering several times and took the result with the lowest value for the objective function (see Equation 3.2). This procedure reduces the risk of 'getting caught' in a local minimum.

3.3.2 Uncertainty analysis

The results of the uncertainty analysis regarding different ways of calculating the difference between the climate indicators of the two time intervals is shown in Figure 3.10. The left figure shows the typology based solely on absolute indicators, and

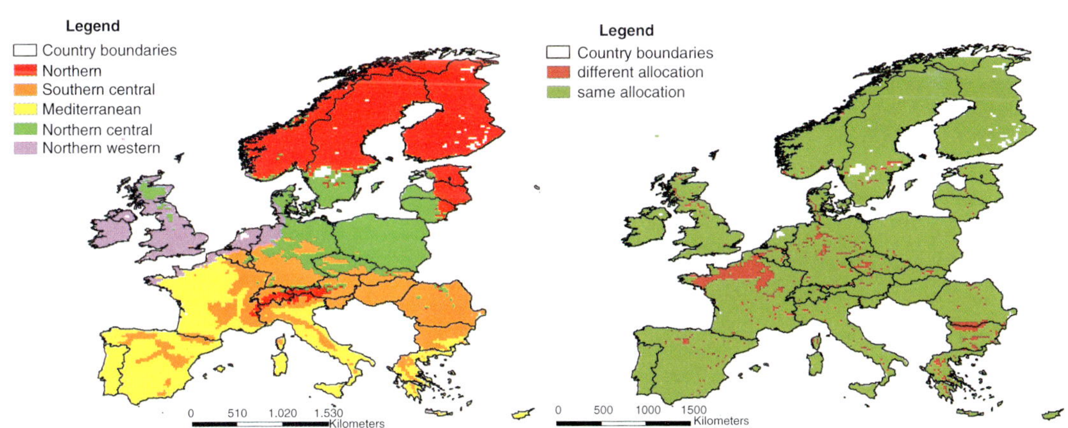

Figure 3.10   European climate change regions achieved by clustering of data set calculated only with absolute differences between the validation and the projected time period (left). Further, the comparison of the cluster allocations between the typology of the left figure and the typology of Figure 3.6 is shown (right).

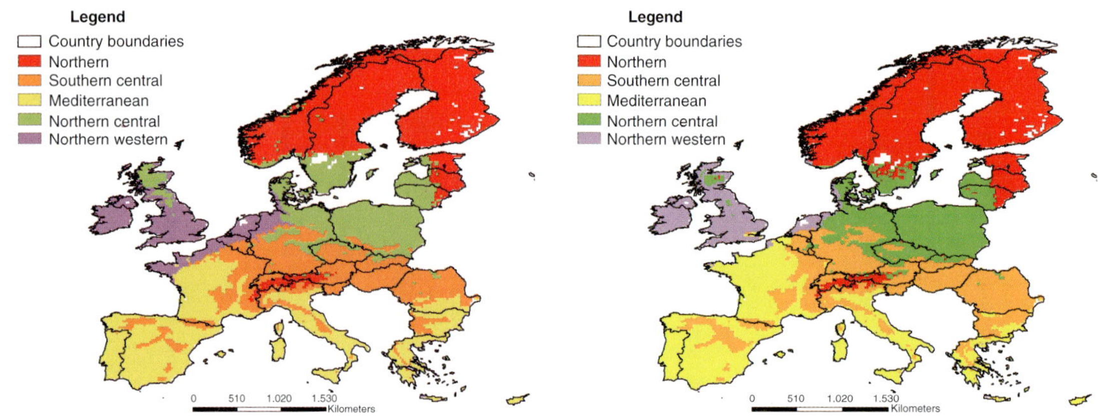

Figure 3.11   European climate change regions achieved by clustering of CCLM model run 1 (left) and 2 (right).

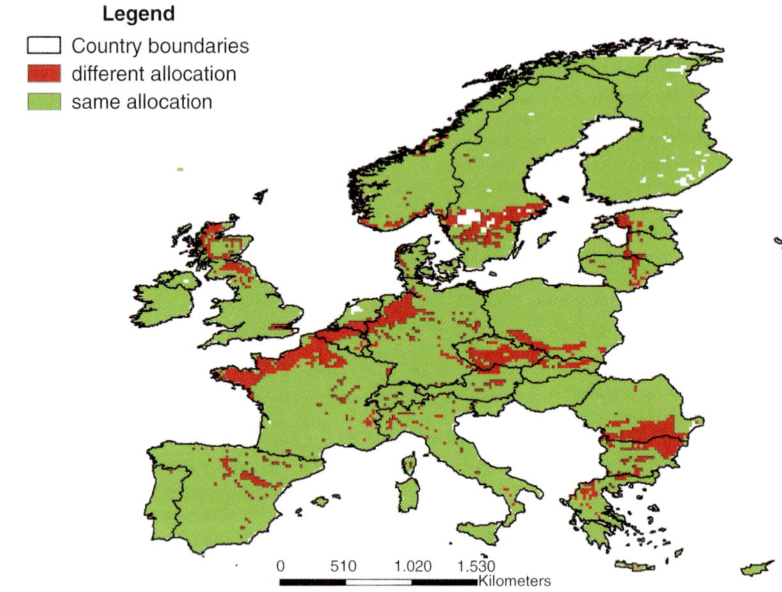

Figure 3.12   Uncertainty analysis of the cluster allocation against the two different model runs of CCLM.

on the right are pictures of grid cells with differing allocations in comparison with the original typology of Figure 3.6. Wide areas of Europe could be found in the same cluster. Only parts of Northwestern France and the border region of Romania and Bulgaria changed their cluster. It thus can be seen that the decision to include relative values for some of the variables only slightly changes the typology of European climate change.

By looking at the stability of the cluster results of different numbers of clusters in the previous section, we found the group number of five to be a stable typology of the European climate change variables. In Figure 3.11, the five cluster solution for the first and the second model run is shown. Both typologies have a strong similarity in the spatial extent of the five clusters. The grid cells with a similar (green) and those with an unequal (red) allocation are displayed in Figure 3.12. The stable allocations cover approximately 88% of the area of the analysed countries. Unstable regions with regard to the clustering are Southern parts of Sweden, parts of the Czech Republic, an area on the border between Romania and Bulgaria as well as a band of grid cells along the Channel coast.

For these pixels, specific indicators tend to have values that are located at the intersection between two clusters. So, due to slightly changing characteristics, they are allocated to one or the other cluster. However, this analysis shows that the qualitative structure represented by the five clusters is relatively stable against the usage of another CCLM model run.

3.4   Conclusion

The resulting spatial typologies of the applied cluster analysis (see Figure 3.6) divide the study area into five different climate change regions. It has to be emphasised that these clusters do not constitute 'climate clusters', but 'climate change clusters', that is, each cluster consists of regions that are similar in regard to the *changes* of the climatic stimuli as presented in the previous sections. Furthermore, the names of these clusters only serve the purpose of providing easy to understand and easily distinguishable labels. As such they should not be considered as completely accurate in a geographical sense.

Within the process of the analysis we had to overcome the problem of the unknown number of clusters. Here we used measures that account for separation and stability of the different clusters. The results showed that only the signal in the stability measure, namely the consistency measure, provided sufficient information.

To analyse the stability of the typology we applied an uncertainty analysis in a twofold way. First we compared two different ways of indicator generation and found that a data set solely based on absolute indicators only differs in a few areas from our typology generated with a mix of relative and absolute indicators. To deal with the above described problem of comparability of various types

of changes, it will be worthwhile to use the state *and* change of climate variables combined in a continuative analysis. The second part of the uncertainty analysis was to compare two different model runs of CCLM-A1B. Here we also found that the typology of climate change is stable for most parts of the study area.

Finally, we showed that the choice of the case study areas made in Chapters 11–15 satisfactorily covers the five European climate change regions.

References

Ben-Hur, A., Elisseeff, A. and Guyon, I. (2002). A stability based method for discovering structure in clustered data. *Pacific Symposium On Biocomputing*, **17** (7), 6–17. Retrieved from http://www.ncbi.nlm.nih.gov/pubmed/11928511 (accessed 1 June 2011).

Calinski, T., and Harabasz, J. (1974) A dendrite method for cluster analysis. *Communications in Statistics Theory and Methods*, **3** (1), 1–27. doi: 10.1080/03610927408827101.

Fraiman, R., Justel, A. and Svarc, M. (2006) Selection of variables for cluster analysis and classification rules. *Journal of the American Statistical Association*, **103** (483), 28. Retrieved from http://arxiv.org/abs/math/0610757 (accessed 1 June 2011).

Gerstengarbe, F.W. and Werner, P.C. (2007) Der rezente klimawandel, in (eds W. Endlicher and F.-W. Gerstengarbe), *Der Klimawandel – Einblicke, Rückblicke und Ausblicke*, Springer, Berlin, pp. 34–43.

Gerstengarbe, F.W., Werner, P.C. and Fraedrich, K. (1999) Applying non-hierarchical cluster analysis algorithms to climate classification: some problems and their solution. *Theoretical and Applied Climatology*, **64** (3–4), 143–150. doi: 10.1007/s007040050118.

Handl, J., Knowles, J. and Kell, D. B. (2005) Computational cluster validation in post-genomic data analysis. *Bioinformatics (Oxford, England)*, **21** (15), 3201–3012. doi: 10.1093/bioinformatics/bti517.

Hartigan, J.A. and Wong, M.A. (1979) A K-means clustering algorithm. *Applied Statistics*, **28** (1), 100–108. doi: 10.1016/S0167-8655(03)00146-6.

Kaufman, L. and Rousseeuw, P.J. (1990) *Finding Groups in Data.* (eds L. Kaufman and P.J. Rousseeuw), John Wiley & Sons, Inc., New York (p. 368). doi: 10.1002/9780470316801.

Kottek, M., *et al.* (2006) World map of the Köppen-Geiger climate classification updated. *Meteorologische Zeitschrift*, **15** (3), 259–263. doi: 10.1127/0941-2948/2006/0130.

Köppen, W. (1936) Das geographische System der Klimate, in (eds W. Köppen and R. Geiger), *Handbuch der Klimatologie*, vol. 1, Part C, Gebrüder Bornträger, pp. 1–44.

Lautenschlager, M., Keuler, K., Wunram, C., *et al.* (2009) *Climate Simulation with CLM, Climate of the 20th Century run no. 1, 2 and 3 and Scenario A1B runs no. 1, 2, Data Stream 3*. Data set available at: www.cera.wdc-climate.de (accessed 1 June 2011).

Macqueen, J. (1967) Some methods of classification and analysis of multivariate observations, 281–297. Retrieved from http://www.citeulike.org/user/filippone/article/903715 (accessed 1 June 2011).

Mahlstein, I. and Knutti, R. (2009) Regional climate change patterns identified by cluster analysis. *Climate Dynamics*, **35** (4), 587–600. doi: 10.1007/s00382-009-0654-0.

Milligan, G.W. and Cooper, M.C. (1988) A study of standardization of variables in cluster analysis. *Journal of Classification*, **5** (2), 181–204.

Peterson, A.D., Ghosh, A.P. and Maitra, R. (2010) A systematic evaluation of different methods for initializing the K-means clustering algorithm. *Knowledge Creation Diffusion Utilization*, 1–11.

R Development Core Team. (2010) R: A Language and Environment for Statistical Computing. *R Foundation for Statistical Computing Vienna Austria*, R Foundation for Statistical Computing, Vienna, Austria. Retrieved from http://www.r-project.org (accessed 1 June 2011).

Roth, V., Lange, T., Braun, M., and Buhmann, J. (2002) A resampling approach to cluster validation, in (eds W. Härdle and B. Rönz), *Proceedings Computation Statistics 15th Symposium Held in Berlin*, Physika-Verlag, (pp. 123–128). Retrieved from citeseer.ist.psu.edu/article/roth02resampling.html (accessed 1 June 2011).

Sietz, D., Lüdeke, M.K.B. and Walther, C. (2011) Categorisation of typical vulnerability patterns in global drylands. *Global Environmental Change*, **21** (2), 431–440. doi: 10.1016/j.gloenvcha.2010.11.005.

Steinley, D. (2006) K-means clustering: a half-century synthesis. *The British Journal of Mathematical and Statistical Psychology*, **59** (Pt 1), 1–34. Retrieved from http://www.ncbi.nlm.nih.gov/pubmed/16709277 (accessed 1 June 2011).

Ward, J.H. (1963) Hierachical grouping to optimize an objective function. *Journal of American Statistical Associaton*, **58**, 236–244.

Chapter 4

Climate change exposure assessment of European regions

Christian Lindner[1], Stefan Greiving[1] and Anne Holsten[2]

[1]*TU Dortmund University, Institute of Spatial Planning (IRPUD), August-Schmidt-Strasse 10, 44227 Dortmund, Germany*
[2]*Potsdam Institute for Climate Impact Research (PIK), P.O. Box 60 12 03, 14412 Potsdam, Germany*

Abstract

Climate change exposure refers to the nature and degree to which a system is exposed to climatic variations. This chapter describes the exposure analysis carried out by the ESPON Climate project in order to determine patterns of climate change across Europe and its regional variations. The analyses are based on the outputs of a regional climate model and serve as an input to further analysis of climate change impacts combining the model output's together with sensitivity.

4.1 Introduction

Climate change exposure depends on global trends of climate change and – due to spatial variations – on the system's location (compare with Füssel and Klein, 2006, p. 313). The exposure analysis of the ESPON Climate project is based on the results of the regional climate model CCLM (see below). Taken together with sensitivity[1] to climate change as well as adaptive capacity, exposure becomes a component of impacts of climate change (potential as well as residual). The climate exposure values used in the ESPON climate project are based on the Intergovernmental Panel on Climate Change (IPCC) emission scenarios published in 2000 (IPCC, 2000) and employed within the IPCC

[1]'The distinction between changes in sensitivity and changes in exposure is not always straightforward for processes that affect the extent or spatial structure of the exposure unit. Consider the vulnerability to flooding of a country that experiences significant internal migration from the highlands into the flood plains. This migration changes the exposure of certain population groups to flooding events. Aggregated to the country level, however, the effects of migration represent changes in the sensitivity of the population to flooding events' (Füssel and Klein, 2006, p. 317).

European Climate Vulnerabilities and Adaptation: A Spatial Planning Perspective, First Edition.
Edited by Philipp Schmidt-Thomé and Stefan Greiving.
© 2013 John Wiley & Sons, Ltd. Published 2013 by John Wiley & Sons, Ltd.

Fourth Assessment Report in 2007. Based on these scenarios the CCLM model has been run simulating future climate change for almost the whole European territory (Lautenschlager *et al.*, 2009). Subsequently, the IPCC scenarios and the CCLM projections will be elaborated, followed by the results from the analysis of different climatic parameters derived from CCLM data for the European territory.

4.2 Future climate projections: the CCLM model

The COSMO-CLM (or CCLM) model is a non-hydrostatic unified weather forecast and regional climate model developed by the COnsortium for Small scale MOdelling (COSMO) and the Climate Limited-area Modelling Community (CLM). It is run with the boundary conditions of the General Circulation Model ECHAM5/MPI OM. The CCLM model was selected due to its fine spatial resolution ($\sim$20 km), an extended and transient simulation period until 2100, spatial coverage of Europe and its state-of-the-art climate module offering a large output of climate variables. In contrast to the ENSEMBLES (compare with Van der Linden and Mitchell, 2009, p. 160) database of regional models, CCLM provides aggregated information on variables representing extreme events such as days with heavy rainfall, frost days, summer days and days with snow cover, which are of particular importance within the case studies of this project. Moreover, at the starting time of the ESPON Climate project, the CCLM simulation runs were the most up to date (December 2008), whereas in the ENSEMBLES database of regional models, older versions of climate models were used.[2]

The CCLM model leans on the emission scenarios as defined by the IPCC in its 2000 special report on emissions scenarios (IPCC, 2000) in order

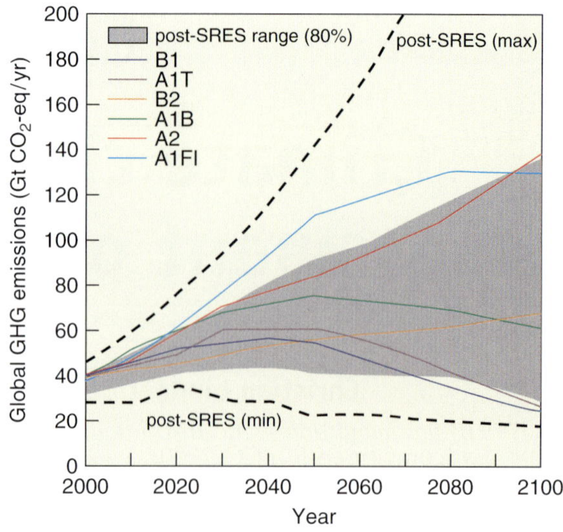

Figure 4.1 IPCC scenarios of global GHG emissions until 2100. *Source*: Climate Change 2007: Synthesis Report. Contribution of Working Groups I, II and III to the Fourth Assessment Report of the Intergovernmental Panel on Climate Change, Figure 3.1. IPCC, Geneva, Switzerland.

to produce future climate projections. IPCC has presented six scenarios on the development of greenhouse gas emissions (GHG) from 2000 to 2100 (SRES scenarios) (see Figure 4.1). These scenarios presume the absence of additional climate policies, which may affect GHG emissions and cover a wide range of GHG emission drivers in the fields of demography, economy and technology. Divided into four scenario families (A1, A2, B1 and B2) they explore alternative development pathways with respect to the evolution of future GHG emissions (IPCC, 2007a, p. 44).[3]

---

[2] The use of a single climate model of course represents a certain shortcoming, which has to be considered along with the results. However, projections of the CCLM model are compared with other models within the case studies of the project. Further projects should aim at comparing the European wide results of this project with a larger range of global and regional climate models and scenarios.

[3] "The A1 storyline assumes a world of very rapid economic growth, a global population that peaks in mid-century and rapid introduction of new and more efficient technologies. A1 is divided into three groups that describe alternative directions of technological change: fossil intensive (A1FI), non-fossil energy resources (A1T) and a balance across all sources (A1B). B1 describes a convergent world, with the same global population as A1, but with more rapid changes in economic structures toward a service and information economy. B2 describes a world with intermediate population and economic growth, emphasising local solutions to economic, social, and environmental sustainability. A2 describes a very heterogeneous world with high population growth, slow economic development and slow technological change' (IPCC, 2007a, p. 44).

The A1 storyline is a case of rapid and successful economic development, in which regional average income per capita converge – current distinctions between 'poor' and 'rich' countries eventually dissolve. The primary dynamics are:

- strong commitment to market-based solutions,
- high savings and commitment to education at the household level,
- high rates of investment and innovation in education, technology, and institutions at the national and international levels,
- international mobility of people, ideas, and technology (compare with IPCC, 2000).

There are subsets to the A1 scenario family based on their technological emphasis: The chosen A1B subset is based on a balanced use of all energy sources. Thus, A1B is used for almost all vulnerability assessments as a moderate scenario.

Scenario B1, which assumes more global (environmental) sustainability, is not realistic anymore due to the fact that human GHG emissions have already reached the high-end of the IPCC scenarios, that is, A1FI.[4] It was thus decided to only continue with the A1B scenario as it displays a reasonable average (in case emissions would in fact decrease).

4.3   Indicators on exposure to climatic stimuli

The CCLM model has been used to produce climate change runs with three realisations for the time period 1961–1990 and two realisations for each scenario for the time frame 2001–2100. Regional climate models are better able to reproduce small-scale climate features and certain weather extremes due to their higher resolution. However, they inherit errors related to the simulation of large-scale circulation patterns from the global models that are driving them. For European-wide data the spatial resolution available is approximately 20 km. Based on these model projections different climate change indicators have been calculated constituting the basis for the current analysis of exposure to climatic stimuli.[5]

In principle, the CCLM model delivers a wide range of climate-related output parameters (compare with Wunram, 2007). These parameters relate to many different fields relevant within meteorology and climate research. For almost all output parameters, data are provided on an hourly to daily basis. Thus, for the purpose of this research, selected parameters have originally been aggregated for the time frames 1961–1990 and 2071–2100 for the A1B scenario in order to attain mean values exhibiting projected mean changes for the European territory.

The focus on central climate parameters is crucial since the CCLM model delivers a broad range of parameters (also varying by so-called 'data streams'). A larger range of output data is available for data stream 3 of the model outputs, compared with data stream 2, including aggregated data on 'extreme' events, such as days with heavy rainfall, summer days or frost days. To represent these events within the study, climate information from data stream 3 were used, which covers a large part of Europe, but excludes some countries such as Iceland, which are part of the ESPON space but not covered by data stream 3.

---

[4]The worst-case scenario A1FI may become more and more likely due the recently observed CO_2 emissions, which in 2010 were the highest in history, according to the latest estimates by the International Energy Agency (see http://www.iea.org/index_info.asp?id=1959). However, A1FI is not yet considered by several relevant input variables such as LISFLOOD. Therefore, it was not possible to make use of this scenario for the exposure assessment.

[5]The relevant climate parameters frequently discussed in reports with respect to future climate change impacts relate to temperature and precipitation as well to wind speed (see IPCC, 2007b, pp. 872–879). Analyses focus mostly on changes in mean values as well as on extremes, which has been the basis for the choice of CCLM parameters as utilised within the exposure to climatic stimuli analysis to be carried out within the present research. Likewise, these fields are focussed on in a current report of the European Environmental Agency (EEA). Here, indicators are based on IPCC scenarios A1B and A2 and B2. Indicators in the field of atmosphere and climate include global and European temperature, European precipitation, temperature extremes in Europe and precipitation extremes in Europe as well as storms and storm surges and air pollution by ozone (see EEA, 2008, pp. 39–59).

The derived exposure indicators will be discussed in more detail within the subsequent sections. Generally, the change indicators always relate the climate conditions in the reference time period (1961–1990) to the climate conditions in the time period 2071–2100. The absolute or relative difference between these two periods constitutes the projected change for each climate parameter. These climatic variables reflect the wide range of climatic conditions, from temperature to hydrologic variables. Variables of pressure and heat fluxes have been disregarded due to the lack of direct relations with the sensitivity indicators. Data on storm events are subjected to large uncertainties on the European level. Mean wind speeds exhibit regional and large-scale biases, especially in Eastern Europe, at the west coast of Scandinavia, in France, parts of the Iberian Peninsula and parts of North Africa (compare with Hollweg *et al.*, 2008).

For hydrologic variables, relative changes have been considered to best account for the regional varying climatic conditions since, for example, small changes in summer precipitation can have much larger impacts in the Mediterranean region (with little absolute precipitation in summer), than a reduction of the same amount in Scandinavia, with considerably higher absolute precipitation levels.

The following eight indicators have been derived from the CCLM dataset for the purpose of this analysis:

- *Change in annual mean temperature*: Based on the CCLM parameter 'air temperature in 2 m above the surface' (T_2M_AV, yearly), average annual temperatures in degrees Celsius for the desired time frames have been calculated. This indicator serves to indicate regional variation of changes in temperature, as the main indicator for climate change.
- *Change in annual mean number of frost days*: Based on the CCLM parameter 'frost days' (FD, yearly), average annual number of frost days (days with minimum temperatures below $0\,^{\circ}$C) for the selected time frames have been calculated. This indicator serves to indicate changes in regional climate extremes with respect

to cold temperatures, which is, from a territorial perspective, particularly relevant for natural and agricultural systems.
- *Change in annual mean number of summer days*: Based on the CCLM parameter 'summer days' (SU, yearly), average annual number of summer days (days with maximum temperatures above $25\,^{\circ}$C) for the selected time frames have been calculated. This indicator serves to indicate changes in regional climate extremes with respect to summer temperatures. From a territorial perspective, this indicator is of relevance for the tourism sector as well as for the field of human wellbeing.
- *Relative change in annual mean precipitation in winter months*: Based on the CCLM parameter 'total precipitation' (PRECIP_TOT, monthly), average precipitation in $kg\,m^2$ for the chosen time frames has been summed up for the meteorological winter months (December, January and February). Seasonal averages have been calculated to account for the strong intra-annual variation of this variable. This indicator accounts for changes in winter precipitation. Taken together with precipitation in summer months, conclusions about water availability can be drawn.
- *Relative change in annual mean precipitation in summer months*: Based on the CCLM parameter 'total precipitation' (PRECIP_TOT, monthly), average precipitation in $kg\,m^{-2}$ for the selected time frames has been summed up for the meteorological summer months (June, July and August). This indicator represents regional exposure to changes in summer precipitation. Seasonal averages have been calculated to account for the strong intra-annual variation of this variable. Changes in summer precipitation are especially relevant for vegetation.
- *Change in annual mean number of days with heavy precipitation (R20MM, yearly)*: The average annual number of days with heavy precipitation (above $20\,kg\,m^{-2}$) for the selected time frames has been calculated. This indicator illustrates regional exposure to changes in heavy precipitation events and thus indicates hydrologic extremes. This variable has strong relevance for local

heavy rainfall events, especially when occurring over highly sealed surface areas.[6]

- *Relative change in annual mean evaporation*: Based on the CCLM parameter 'surface evaporation' (AEVAP_S, yearly), the average annual amount of water evaporating in a distinct area has been calculated. This indicator represents the changes in evaporation, and is from a territorial perspective of particular relevance for the natural systems, combining information on temperature and hydrologic conditions.

- *Change in annual mean number of days with snow cover*: Based on the CCLM parameter 'snow cover' (SNOW_COV), the average annual number of days with snow covering the surface has been calculated. This indicator serves to indicate changes in snow conditions, relevant from a territorial perspective, for example for the winter tourism sector.

In addition, two so-called 'triggered effects' (or secondary effects) were also included, constituting a culmination of several of the above variables:

- *Change of inundation through river flooding*: Some extreme weather events may be triggered by climatic stimuli related to precipitation, such as river flooding and mass movements (IPCC, 2007a; Prudhomme and Reynard, 2009). The impact of climate change on flooding is covered by JRC's LISFLOOD model. LISFLOOD is a GIS-based hydrological rainfall–runoff–routing model that is capable of simulating the hydrological processes that occur in a river catchment area (Van Der Knijff, Younis and De Roo, 2008). However, as is the case with any other pan-European model on highly complex and dynamic systems such as rivers, LISFLOOD's simulation outputs contain some degree of uncertainty. Besides the fact that in the case of ESPON Climate LISFLOOD data were only based on one climate model (CCLM), one may mention difficulties to account for effects of snowmelt, river regulation or dykes (for more

detailed discussions see Dankers and Feyen, 2009). Nevertheless, for the purpose of a pan-European, comparative assessment, the LISFLOOD simulations currently provide the best available dataset. The outputs of LISFLOOD used within this study are based on the A1B scenario and represent inundation heights on $100 \times 100\,$m grids along major European rivers. Using the outputs for the 1961–1990 and 2071–2100 time periods, the ESPON Climate project calculated the changes in inundation heights of a 100 year return flood event for each grid cell and subsequently for each NUTS 3 region.

- *Change of inundation through coastal storm surge based on projected sea-level rise*: Sea-level rise is not a climate change exposure indicator in the CCLM model, because it is, rather, a first-level effect triggered by changes in global temperatures and regionally also by land uplift and subsidence. Typically, vulnerability assessments focussing on sea-level rise concentrate on identifying areas located below critical elevations, which would consequently be affected by sea-level rise (compare with Nicholls, and Hoozemans, 2000; Nicholls *et al.*, 2007a). This requires use of elevation data from digital elevation models (DEMs) to identify low-lying land in coastal regions. However, sea-level rise during normal tides may not be the greatest challenge for these regions, because the most recent projections range only between 0.3 and 1.8 m (see Rahmstorf, 2007; Vermeer and Rahmstorf, 2009; Grinsted and Moore, 2009). However, during severe coastal storms such additions to storm surge heights may pose a great threat. It is not clear, though, exactly how coastal storms and sea-level rise interact. Furthermore, even though differences in sea-level rise exist between coastal regions in Europe (as measured by altimetry data since 1992), oceanologists have so far not been able to estimate how these differences would develop until, for example, the year 2100. Therefore, the ESPON Climate project decided to take a 'middle of the road' approach and base its coastal storm surge indicator on a uniform 1 m sea-level rise. This value lies in the middle range of the above cited projections (see

---

[6]In the following this variable will be referred to as 'heavy rainfall', which imposes a certain simplification and has to be kept in mind when considering the results related to this indicator.

also Nicholls, 2007b). In order to determine actual inundation heights the assumed sea-level rise of 1 m was added to the fine-grained regional DIVA projections of storm surge heights of a 100-year return event (compare with Vafeidis *et al.*, 2008). Using the digital elevation model Hydro1K it was calculated which areas would then be inundated by coastal flooding in comparison with storm surge flooding without sea-level rise.

4.4   Patterns of climatic changes across Europe

The exposure indicators presented in the preceding section have all been calculated based on the outputs of the respective parameters from the CCLM model runs or LISFLOOD and DIVA. The averaged CCLM projections for the time-slices 1961–1990 and 2071–2100. For the future projections, two climate model runs are available, for the reference period (1961–1990) three, respectively. In order to consider all available runs, the results from different runs have been averaged prior to further calculations of change indicators for each period of 30 years. The baseline change indicators presented in this chapter compare the future period 2071–2100 with the reference period 1961–1990 for the scenario A1B. The changes are calculated either as absolute changes, subtracting the averaged present value from the respective value for the simulated future period, or as relative changes in per cent, relating the absolute change value to the value for the reference period.

The outputs of climate models are usually provided at raster-cell level. Thus, in order to derive climate data for the European regions, the individual cell values have to be aggregated to the regional level (NUTS 3). To accomplish this task, different approaches may be taken. In order to ensure consistency throughout the whole ESPON space with its strong heterogeneity concerning the area of the NUTS 3 regions, the approach chosen by the project is based on an intersection of the administrative units with the CCLM cells. All of the results stemming from the CCLM model outputs presented in the following maps have been calculated based on this methodology.

4.4.1   Change in annual mean temperature

Annual mean temperatures are projected to increase between 2 and more than 4.1 °C in the ESPON territory (see Figure 4.2). The United Kingdom, Ireland, Denmark and parts of the Netherlands exhibit the lowest projected temperature changes of up to 3 °C. According to the model outputs, Western and Northern parts of France, Belgium, most parts of Germany, Poland, the Czech Republic, Slovakia and parts of Sweden and Norway and the Baltic states may be subject to temperature increases of between 3 and 3.5 °C. Southern and South-Eastern Europe as well as Northern Scandinavia and Finland are projected to experience the highest temperature changes with absolute changes of more than 3.5 °C. Among these, Spain, parts of Portugal and the Alpine region may even experience temperature changes of more than 4 °C.

4.4.2   Change in annual mean number of frost days

The averaged model outputs on number of frost days indicate roughly a South-West to North-East stretched pattern when considering the whole of Europe (see Figure 4.3). Coming from an already low level, Spain, most parts of France and Italy and Ireland exhibit comparatively slight decreases in the number of frost days. However, in particular, the alpine countries, most parts of Germany, Eastern Europe as well as the Baltic States, Scandinavia and Finland are projected to experience more significant decreases in the number of frost days with regional peaks of 60 days or more.

4.4.3   Change in annual mean number of summer days

The patterns on the projected changes of the annual mean number of summer days show almost the

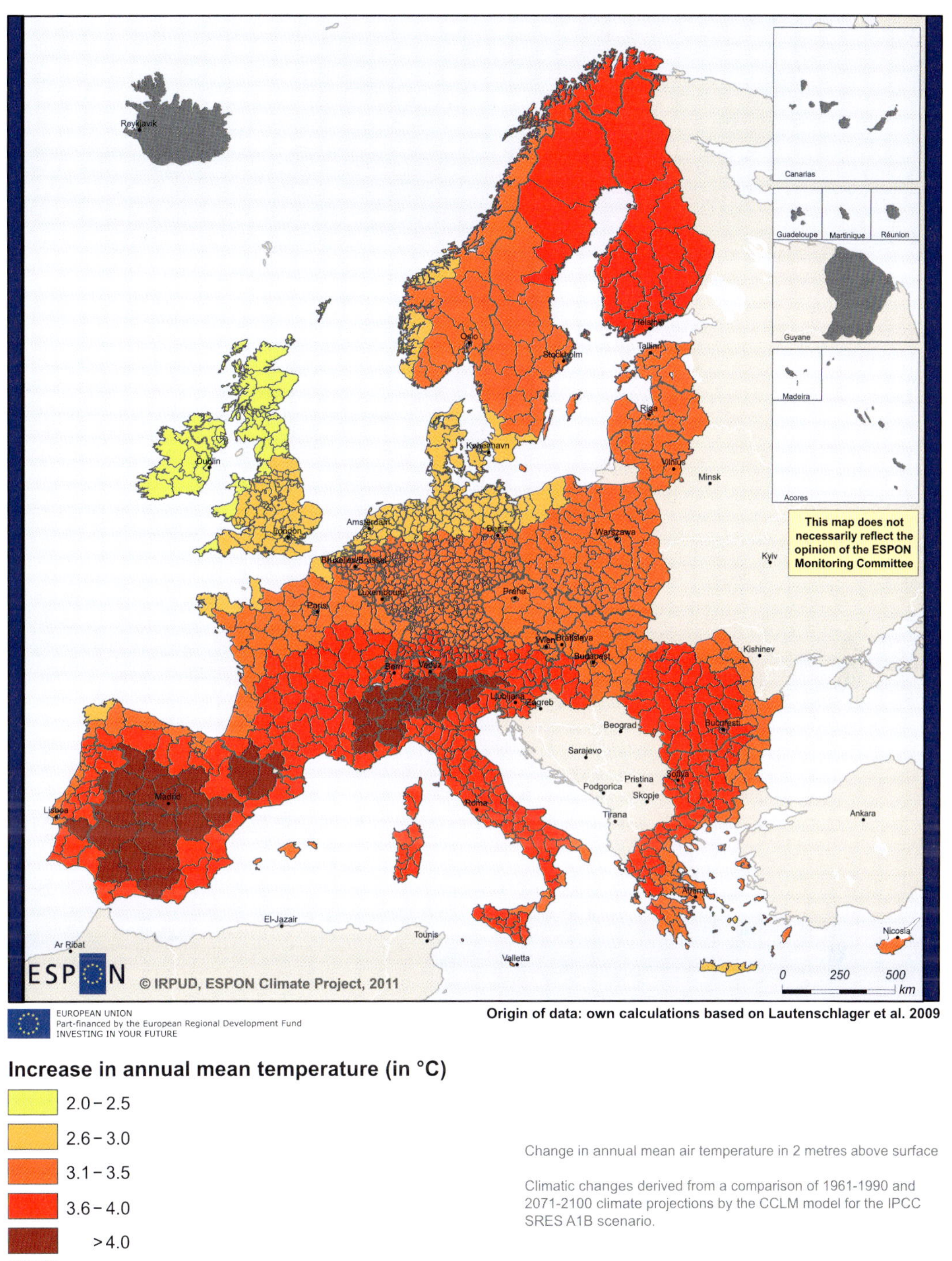

Increase in annual mean temperature (in °C)

- 2.0 – 2.5
- 2.6 – 3.0
- 3.1 – 3.5
- 3.6 – 4.0
- \>4.0
- no data

Change in annual mean air temperature in 2 metres above surface

Climatic changes derived from a comparison of 1961-1990 and 2071-2100 climate projections by the CCLM model for the IPCC SRES A1B scenario.

Figure 4.2 Change in annual mean temperature. © ESPON 2013, IRPUD, ESPON Climate Project, 2011.

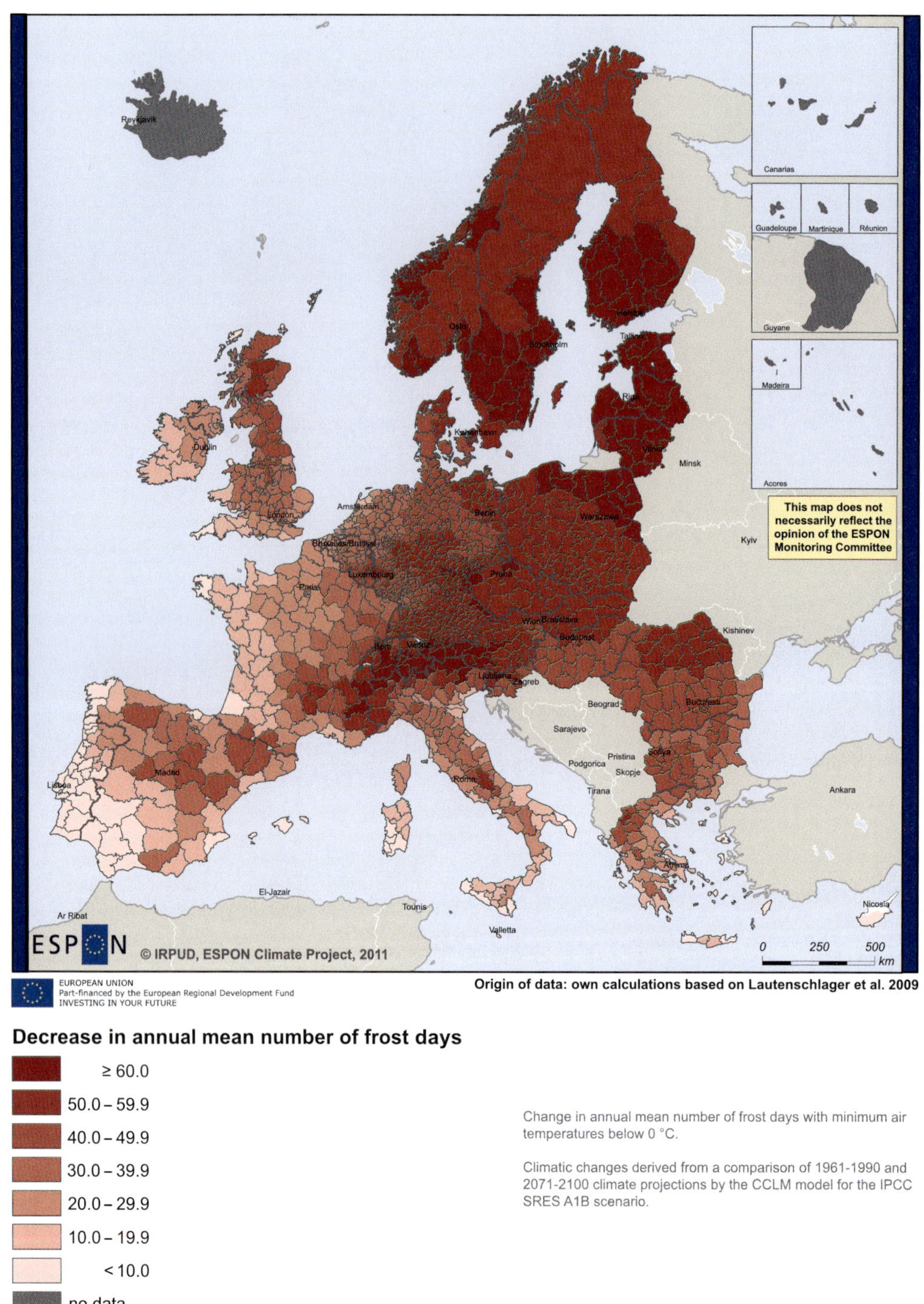

Decrease in annual mean number of frost days

- ≥ 60.0
- 50.0 – 59.9
- 40.0 – 49.9
- 30.0 – 39.9
- 20.0 – 29.9
- 10.0 – 19.9
- < 10.0
- no data

Change in annual mean number of frost days with minimum air temperatures below 0 °C.

Climatic changes derived from a comparison of 1961-1990 and 2071-2100 climate projections by the CCLM model for the IPCC SRES A1B scenario.

Figure 4.3   Change lin annual mean number of frost days. © ESPON 2013, IRPUD, ESPON Climate Project, 2011.

inverse picture compared with the changes in annual mean number of frost days (see Figure 4.4). For this indicator, increases of between less than 10 and more than 50 days per year on average are projected by the CCLM model. The comparatively slightest increases are projected for the North of Europe including Scandinavia, Finland, the Baltic States as well as parts of Denmark, the United Kingdom and Ireland, while predominantly France, Spain and Portugal may become subject to increases of more than 40 days per year on average.

4.4.4 Relative change in annual mean precipitation in winter months

For changes in winter precipitation, the CCLM projections exhibit a twofold pattern (see Figure 4.5). While in most parts of Northern and Central Europe winter precipitation is projected to increase by up to 20% or more, Southern Europe and particularly most parts of the Mediterranean may experience decreases in winter precipitation. Regions in Greece and Bulgaria as well as Cyprus are projected to experience the highest relative decreases.

4.4.5 Relative change in annual mean precipitation in summer months

The CCLM outputs on precipitation in summer month again reveal a twofold pattern considering the changes within the European territory (see Figure 4.6). While parts of Scandinavia and Finland as well as Northern United Kingdom are projected to experience increases of up to 40%, most of the ESPON space may experience decreases in summer precipitation of up to 40% or more. For parts of Scandinavia, the Baltic States, Poland, parts of the Czech Republic, Denmark, Ireland and parts of the United Kingdom, those decreases are projected to range up to 20%, while the rest of Europe, and here particularly France, Portugal, Spain, Italy and Greece are projected to experience the strongest relative decreases in annual summer precipitation.

4.4.6 Change in annual mean number of days with heavy rainfall

As for the previous precipitation-related indicators, the change in annual number of days with heavy rainfall also exhibits a twofold pattern when considering the whole of Europe. There is roughly a North–South divide, with a division at alpine latitudes becoming evident (see Figure 4.7). Most of the territory at lower latitudes is projected to experience average decreases in annual heavy rainfall of up to 5 days or more, whereas the territory north of this division line is projected to gain in the average number of days with heavy rainfall. For most of these regions, increases may amount up to 3 days but along the coastline of Norway as well as Western United Kingdom and Ireland and some parts of the Atlantic coast of France increases of between 4 and 13 days are projected by the CCLM model.

4.4.7 Relative change in annual mean evaporation

Projected European patterns on change in annual mean evaporation range from decrease of more than 15% to increases of up to 22% (see Figure 4.8). Most of the higher projected decreases can be found in Southern Europe, particularly in the Mediterranean and Romania. Strong increases, on the other hand, are predominantly projected for Scandinavia, Finland and the Baltic States as well as parts of Poland, but also the Alpine space and parts of the Czech Republic.

4.4.8 Change in annual mean number of days with snow cover

Snow cover is projected to decrease most significantly in Scandinavia, Finland, the Baltic States and the Alpine countries (see Figure 4.9). Furthermore, some parts of Eastern Europe are also projected to experience a comparatively strong decrease in the number of days with snow cover, whereas the rest

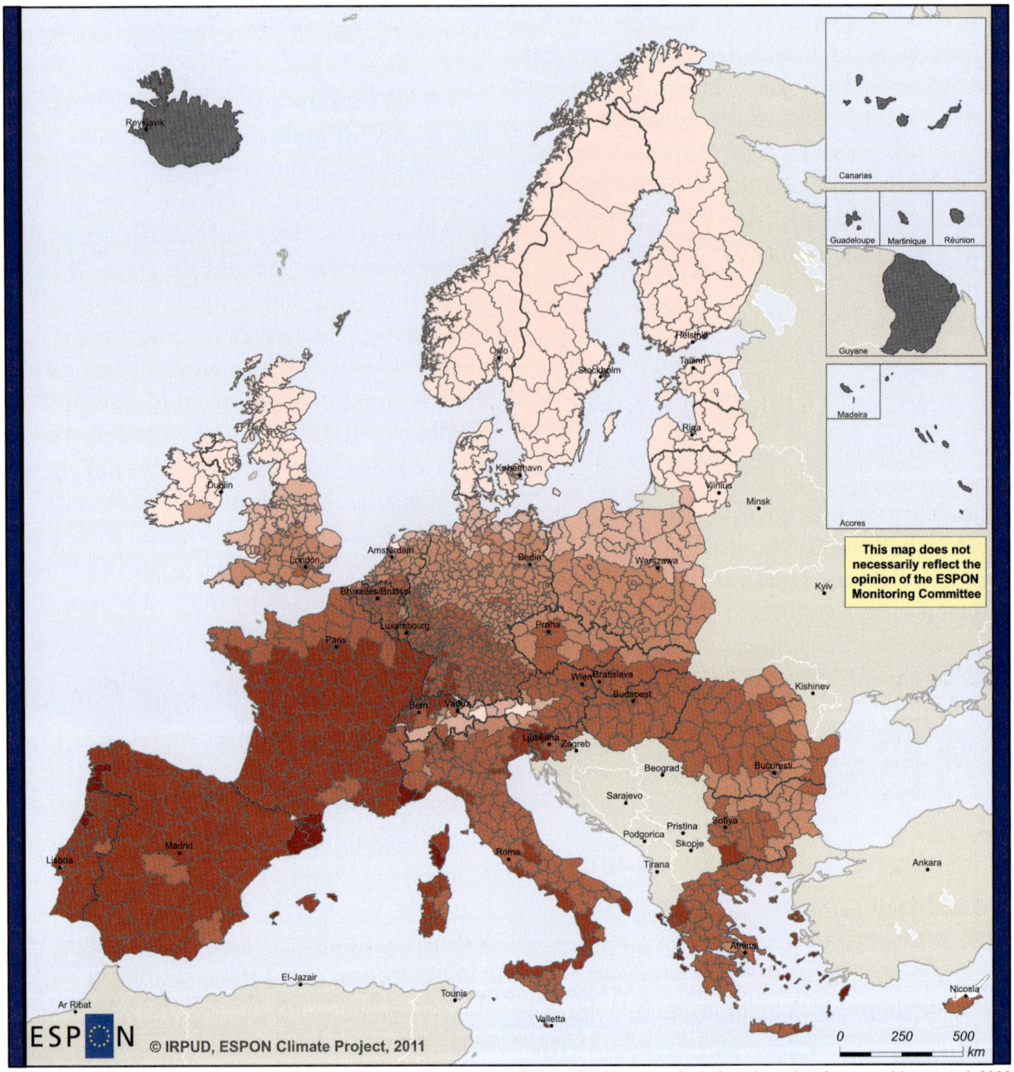

Figure 4.4   Change in annual mean number of summer days. © ESPON 2013, IRPUD, ESPON Climate Project, 2011.

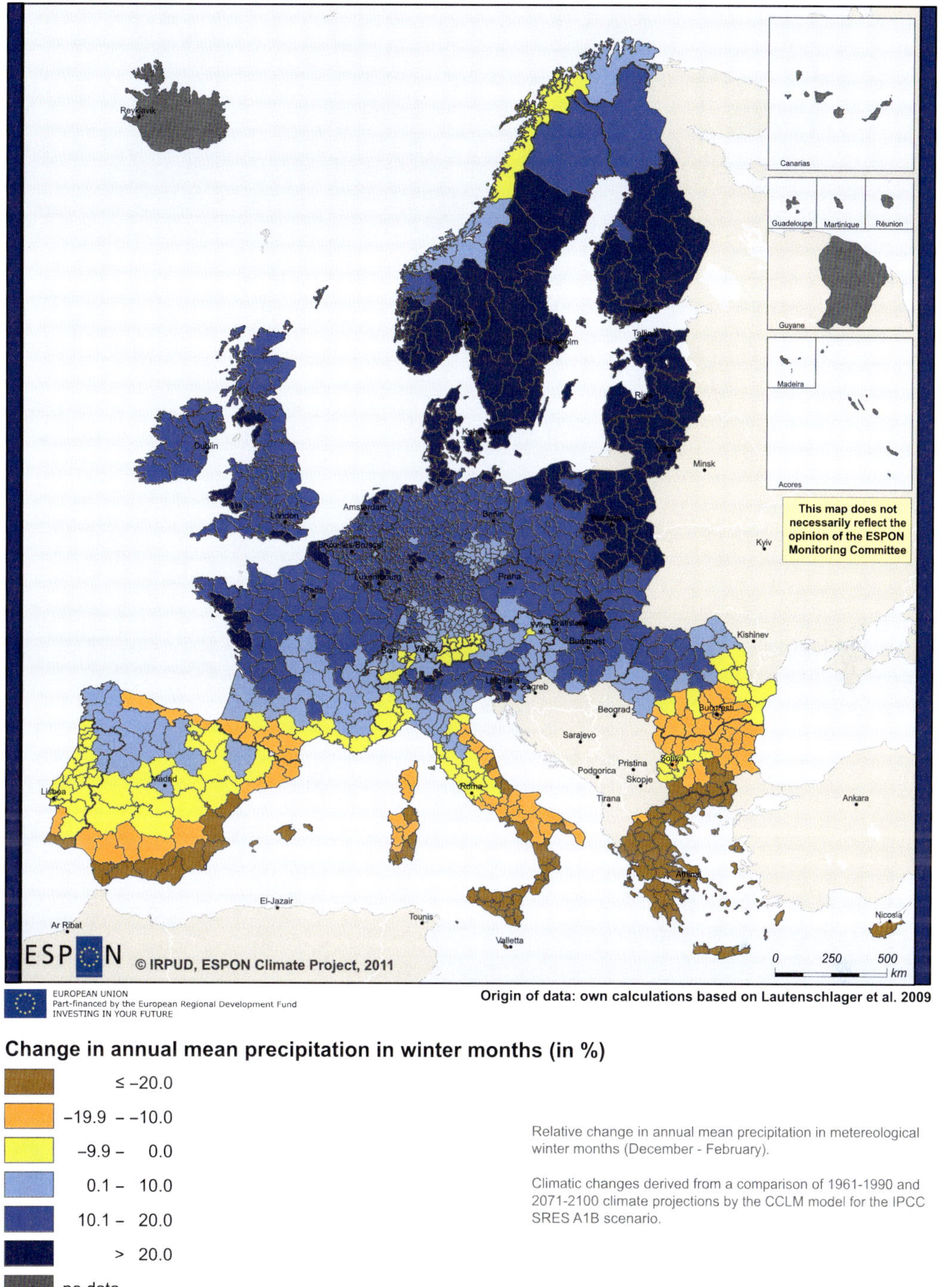

Change in annual mean precipitation in winter months (in %)

Figure 4.5   Relative change in annual mean precipitation in winter months. © ESPON 2013, IRPUD, ESPON Climate Project, 2011.

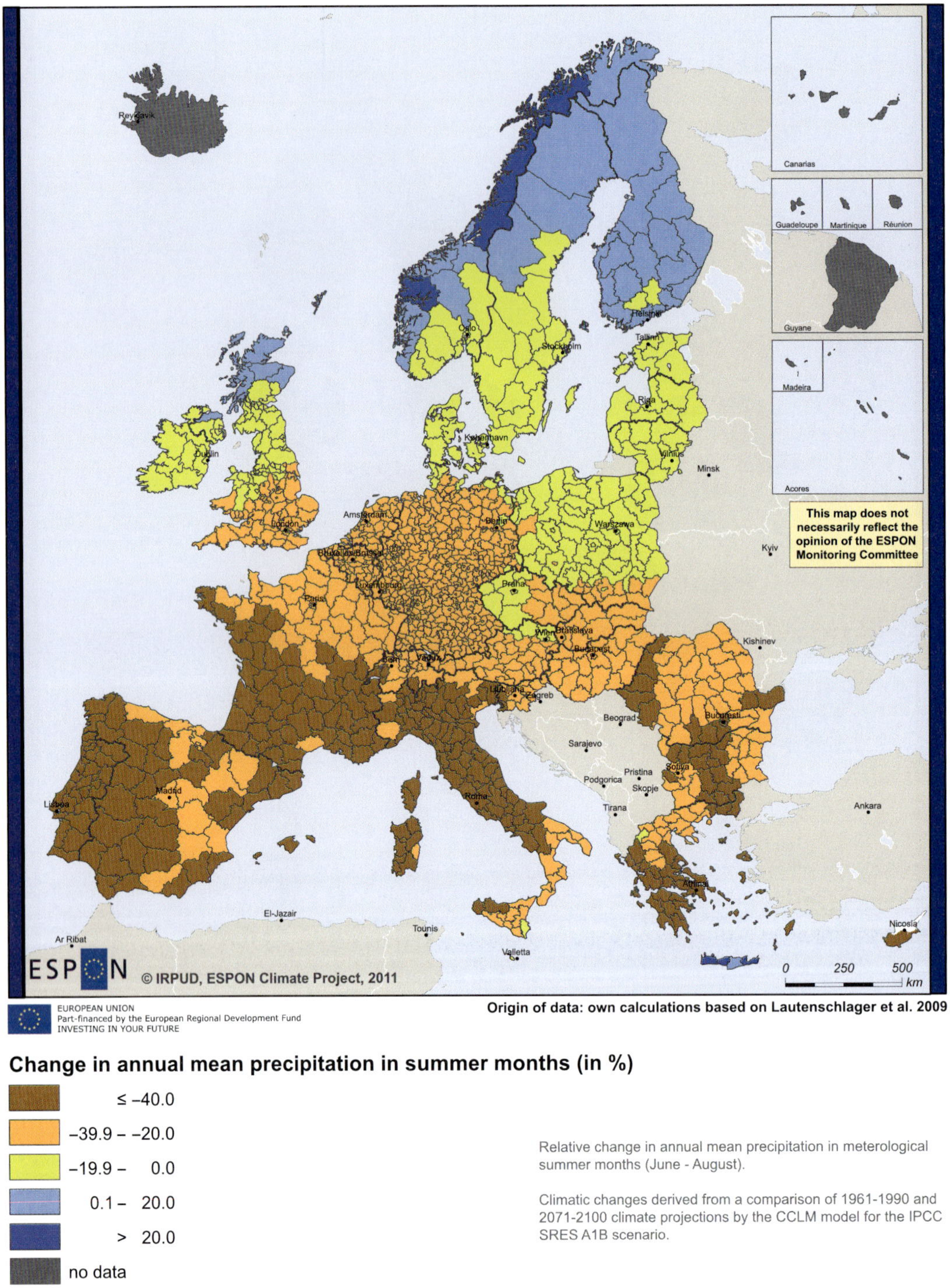

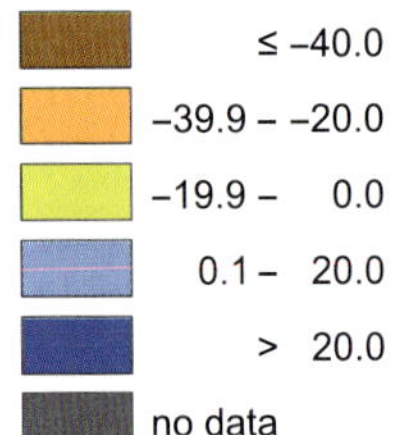

Change in annual mean precipitation in summer months (in %)

Figure 4.6 Relative change in annual mean precipitation in summer months. © ESPON 2013, IRPUD, ESPON Climate Project, 2011.

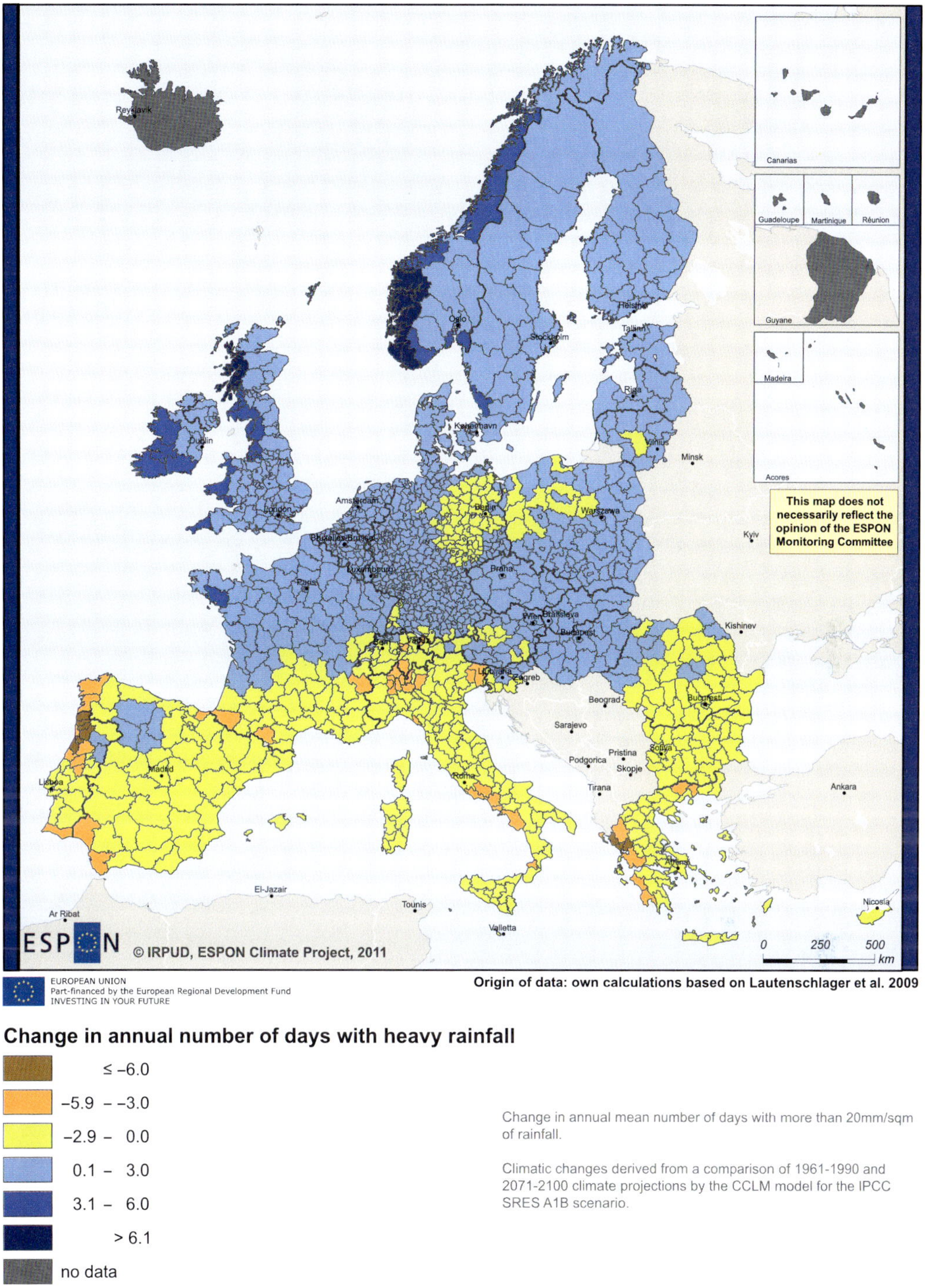

EUROPEAN UNION
Part-financed by the European Regional Development Fund
INVESTING IN YOUR FUTURE

Origin of data: own calculations based on Lautenschlager et al. 2009

Change in annual number of days with heavy rainfall

	≤ −6.0
	−5.9 − −3.0
	−2.9 − 0.0
	0.1 − 3.0
	3.1 − 6.0
	> 6.1
	no data

Change in annual mean number of days with more than 20mm/sqm of rainfall.

Climatic changes derived from a comparison of 1961-1990 and 2071-2100 climate projections by the CCLM model for the IPCC SRES A1B scenario.

Figure 4.7   Change in annual mean number of days with heavy rainfall. © ESPON 2013, IRPUD, ESPON Climate Project, 2011.

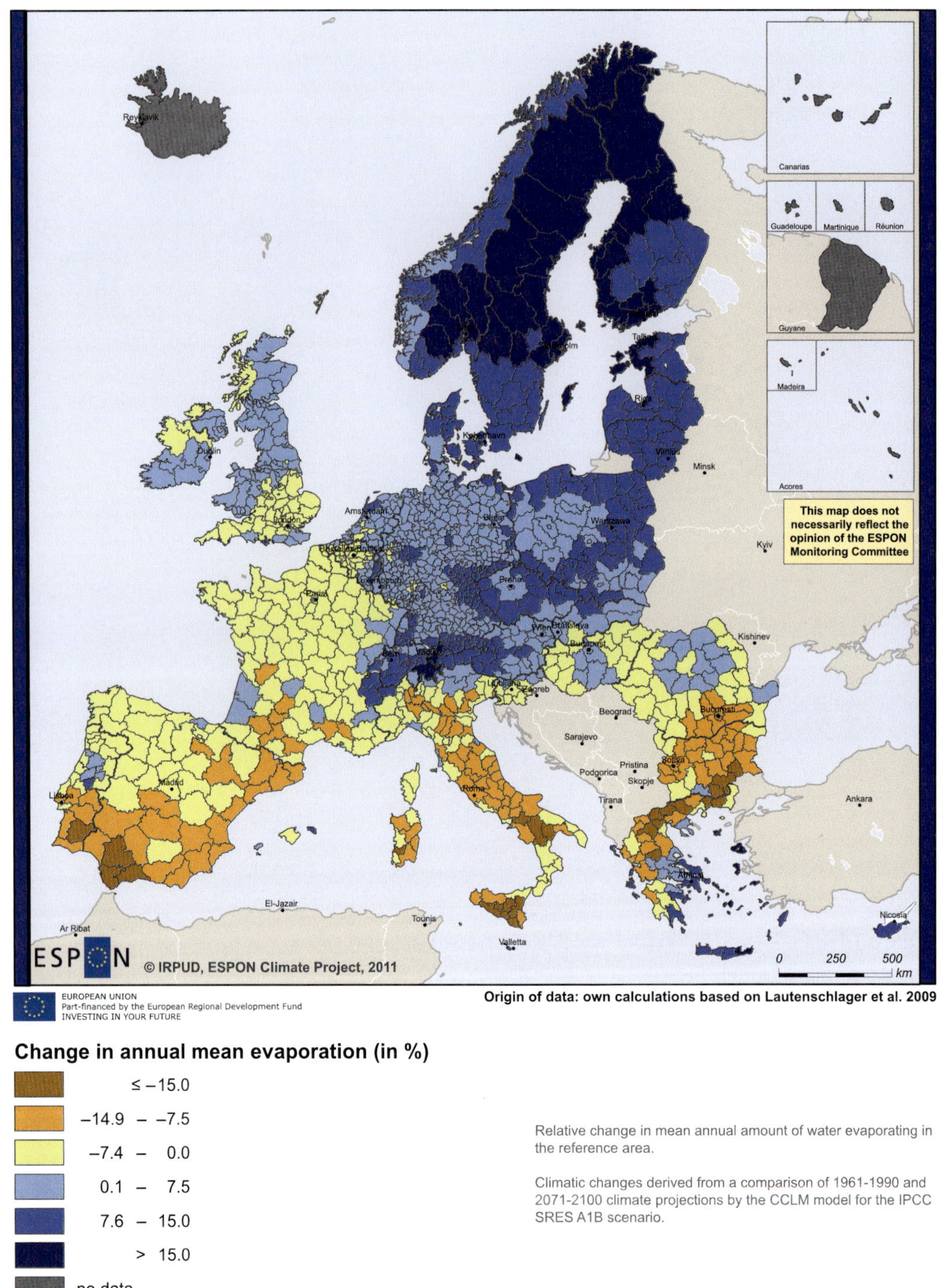

Figure 4.8   Relative change in annual mean evaporation. © ESPON 2013, IRPUD, ESPON Climate Project, 2011.

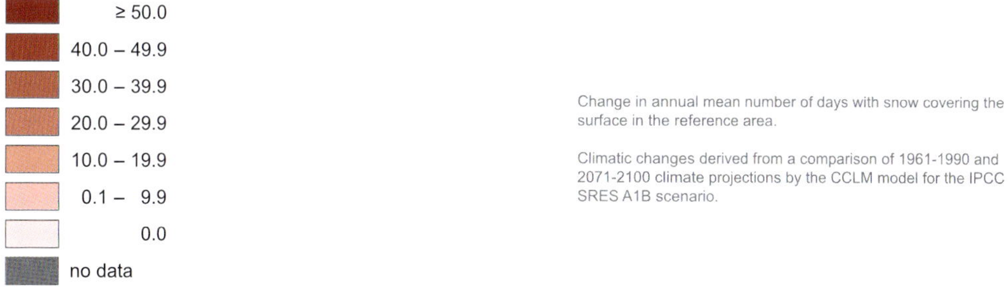

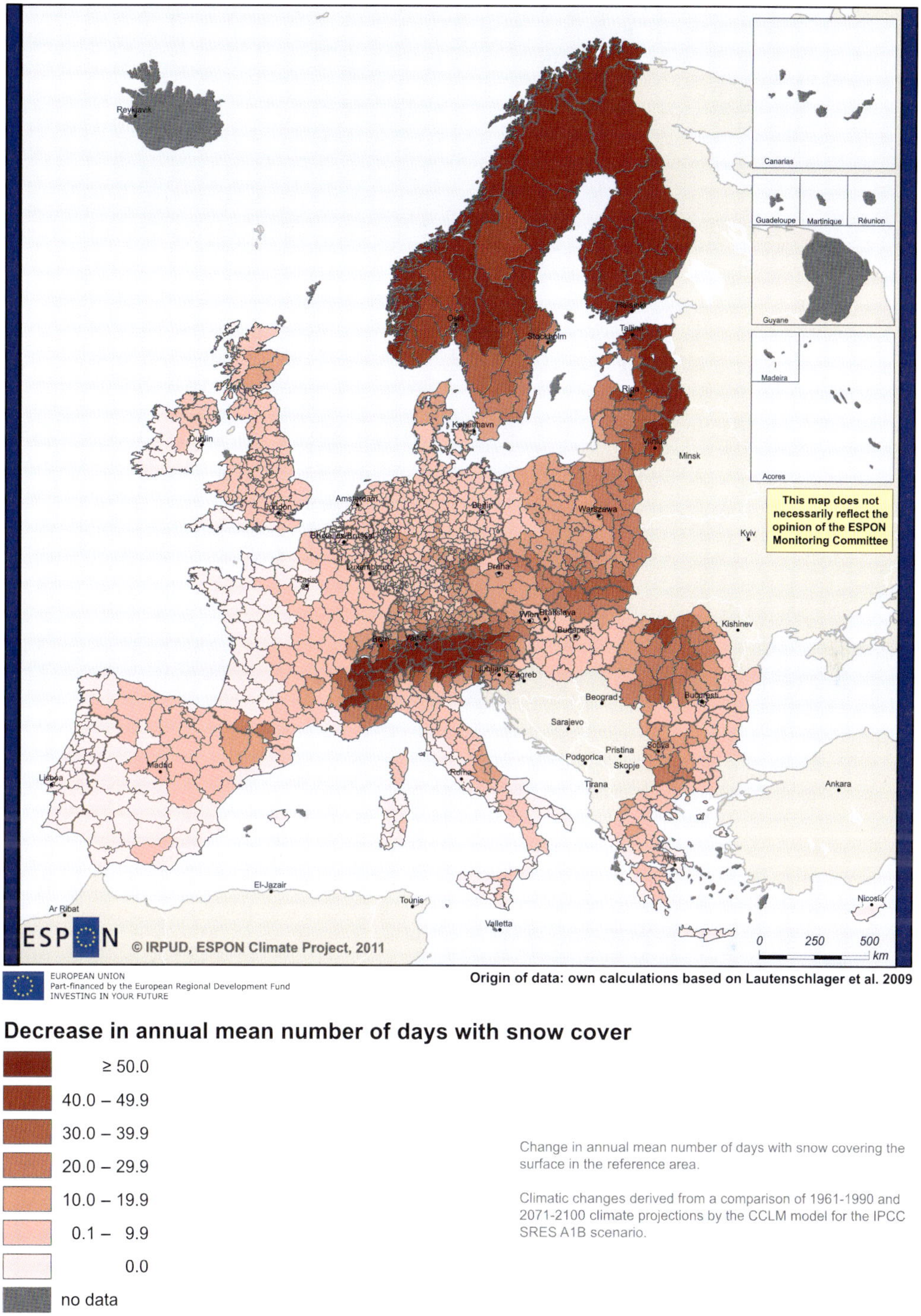

Figure 4.9   Change in annual mean number of days with snow cover. © ESPON 2013, IRPUD, ESPON Climate Project, 2011.

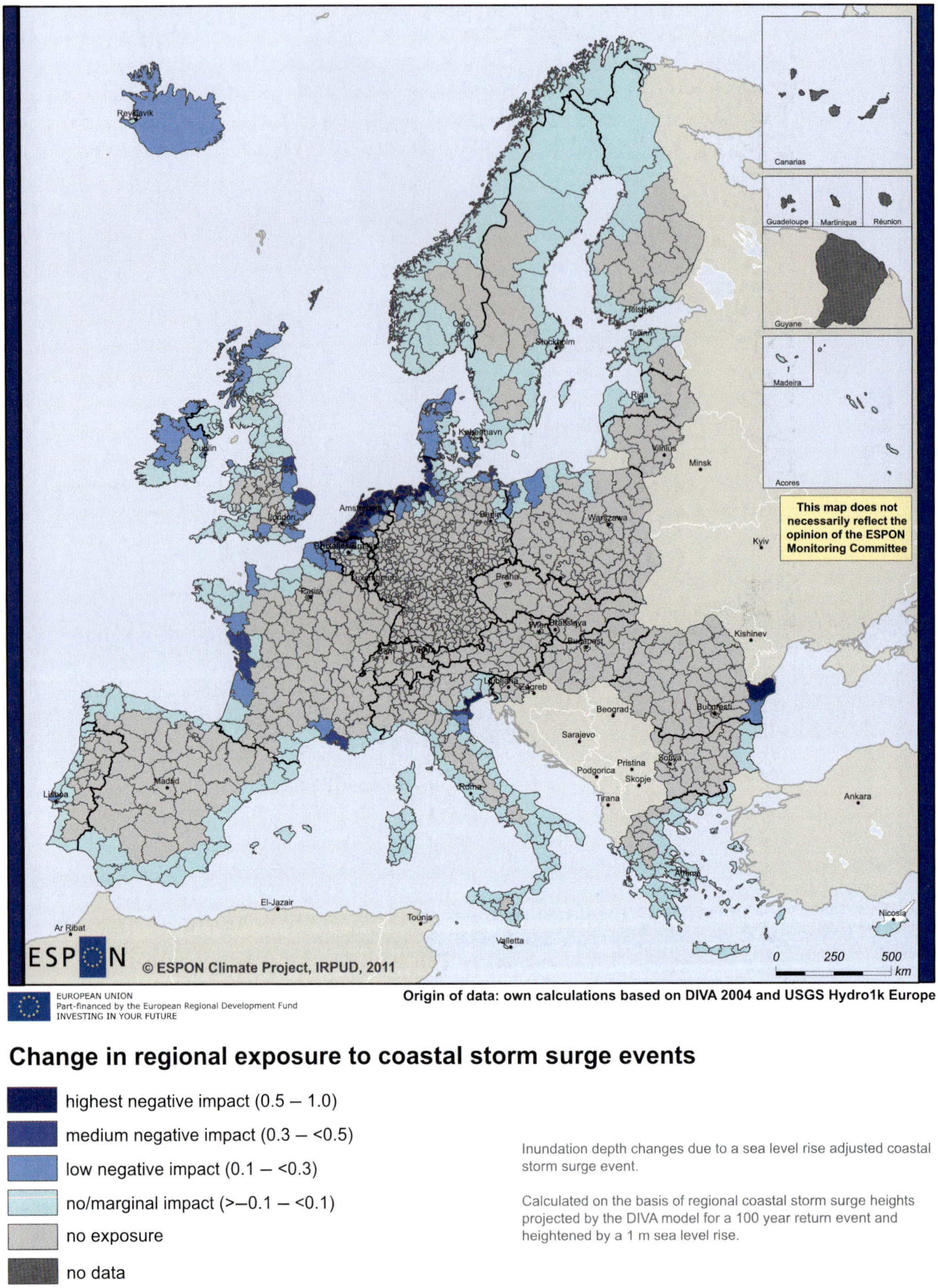

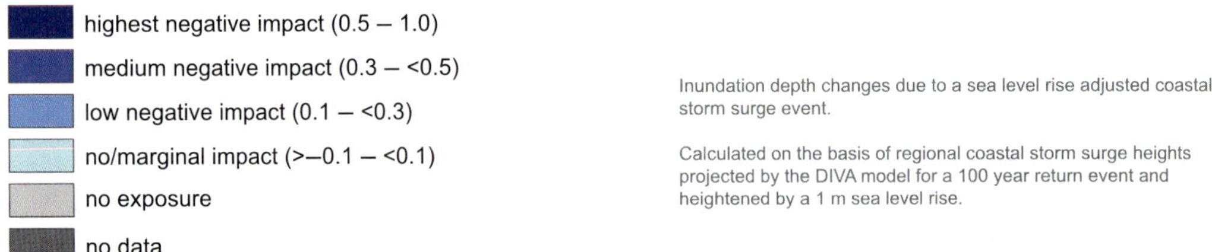

Change in regional exposure to coastal storm surge events

- highest negative impact (0.5 – 1.0)
- medium negative impact (0.3 – <0.5)
- low negative impact (0.1 – <0.3)
- no/marginal impact (>–0.1 – <0.1)
- no exposure
- no data

Inundation depth changes due to a sea level rise adjusted coastal storm surge event.

Calculated on the basis of regional coastal storm surge heights projected by the DIVA model for a 100 year return event and heightened by a 1 m sea level rise.

Figure 4.10   Change in regional exposure to coastal storm surge events. © ESPON 2013, IRPUD, ESPON Climate Project, 2011.

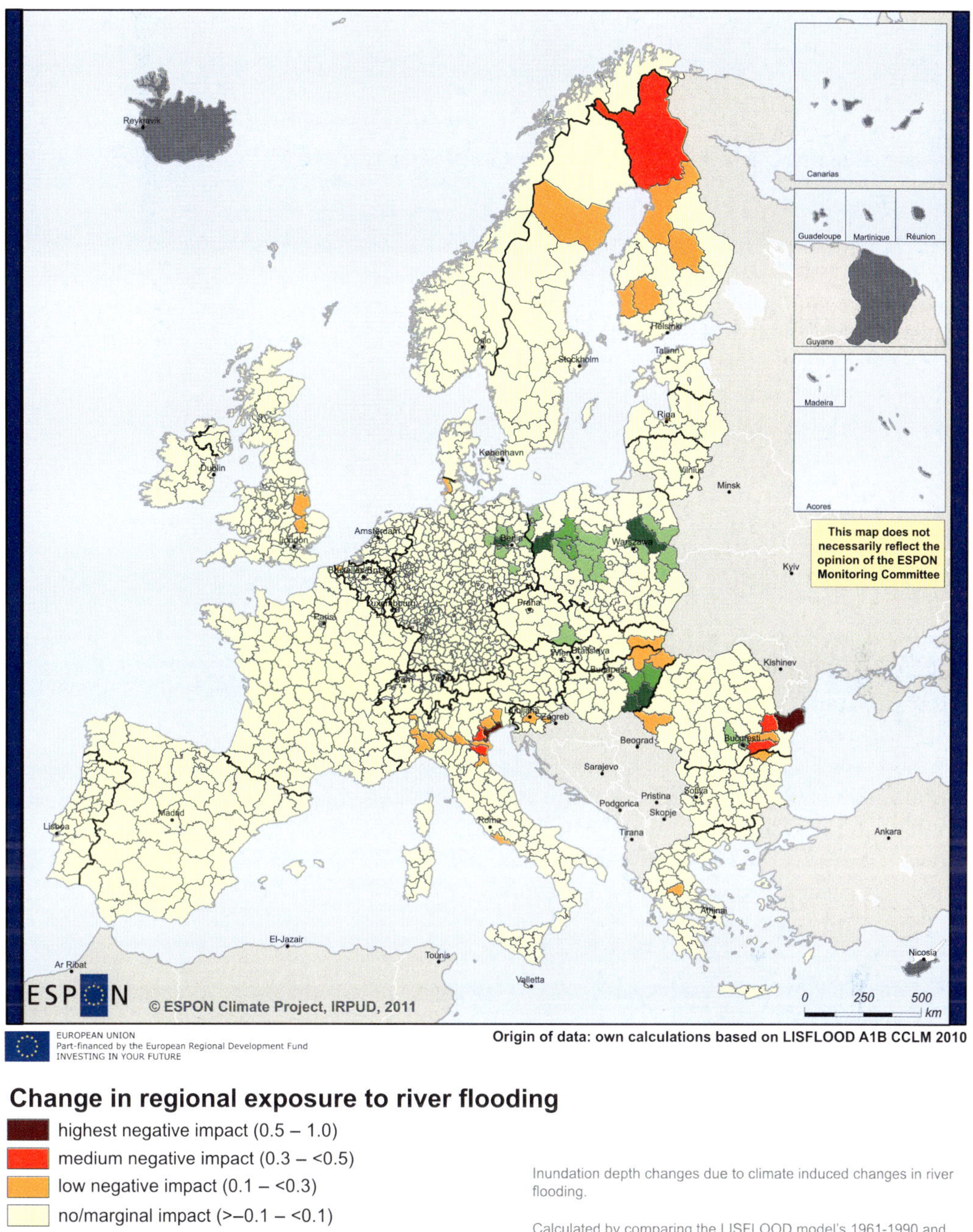

Change in regional exposure to river flooding

- highest negative impact (0.5 – 1.0)
- medium negative impact (0.3 – <0.5)
- low negative impact (0.1 – <0.3)
- no/marginal impact (>−0.1 – <0.1)
- low positive impact (−0.1 – >−0.3)
- medium positive impact (−0.3 – >−0.5)
- high positive impact (−0.5 – −0.86)
- no data*

Inundation depth changes due to climate induced changes in river flooding.

Calculated by comparing the LISFLOOD model's 1961-1990 and 2071-2100 projections for a 100 year return event based on climate projections by the CCLM model for the IPCC SRES A1B scenario.

Figure 4.11 Change in regional exposure to river flooding. © ESPON 2013, IRPUD, ESPON Climate Project, 2011.

of the European territory may rather experience decreases of up to 20 days.

4.4.9 Change in inundation through coastal storm surges

Sea-level rise adjusted coastal storm surge data display fairlyr marginal effects for most coastal regions. However, for some regions more severe changes may be expected (see Figure 4.10). This may primarily affect regions at the Dutch and German coastlines but also those in Denmark and France. Equally severe changes, however, may be expected for some regions in North-Eastern Italy and the Danube delta in Romania.

4.4.10 Change in inundation through river flooding

Most European regions are projected to be just marginally affected by river flooding while few regions exhibit more considerable changes (see Figure 4.11). These regions are predominantly located in Northern Scandinavia and Northern Italy as well as in Romania. In correspondence with the precipitation patterns, there are also some regions projected to experience decreases in exposure to river flooding, particularly in Eastern parts of Germany, in Poland and in Hungary.

4.5 Conclusion

The overall picture revealed by the exposure analysis depicts the strongest potential climatic changes for Southern and South-Eastern Europe. Here, temperature rise is likely to coincide with decrease in precipitation, which, particularly in the summer months, may worsen already critical situations. In particular, urban areas in these regions may be more severely struck by heat waves and hot weather conditions, which are likely to occur for longer periods and at higher frequencies. Increased precipitation patterns, particularly in the winter months,

along with more moderate temperature rises are projected for most of the European territory situated Northerly of the Alpine range, which itself may also be most severely affected by changes in snow cover, an effect which is also projected for Scandinavia. Sea-level rise induced coastal flooding may primarily threaten the coasts of the Netherlands and North-Western Germany but also some other low lying coastal sections across Europe. The patterns of changes in river flooding are likewise scattered. While some stronger increases are projected to occur in Scandinavia as well as in Northern Italy, some regions in Eastern Germany and particularly in Poland may even experience relief in terms of decreased flooding.

References

Dankers, R., and Feyen, L. (2009) Flood hazard in Europe in an ensemble of regional climate scenarios. *Journal of Geophysical Research*, **114**, D16108.

European Environment Agency (EEA) (2008) Impacts of Europe's changing climate – 2008 indicator-based assessment (EEA Report No 4/2008).

Füssel, H.M. and Klein, R.J.T. (2006) Climate change vulnerability assessments: An evolution of conceptual thinking. *Climatic Change*, **75**, 301–329.

Grinsted, A., Moore, J.C., and Jevrejeva, S. (2009) Reconstructing sea level from paleo and projected temperatures 200 to 2100 AD. *Climate Dynamics*, **34**, 461.

Hollweg, H.-D., Böhm, U., Fast, I., *et al.* (2008) Ensemble Simulations over Europe with the Regional Climate Model CLM forced with IPCC AR4 Global Scenarios, M&D Technical Report No. 3, Hamburg.

IPCC (2000) *Special Report on Emissions Scenarios*, Cambridge University Press, Cambridge.

IPCC (2007a) *Climate Change 2007: Synthesis report. Contribution of Working Groups I*, II and III to the Fourth Assessment Report of the Intergovernmental Panel on Climate Change. Cambridge University Press, Cambridge.

IPCC (2007b) *Climate Change 2007: The Physical Science Basis. Contribution of Working Group I to the Fourth Assessment*. Report of the Intergovernmental Panel on Climate Change. Cambridge University Press, Cambridge.

Lautenschlager, M., Keuler, K., Wunram, C., *et al.* (2009) Climate Simulation with CLM, Climate of the 20th Century (run no. 1, 2 and 3) and Scenarios A1B and B1 (run no.

1 and 2), *Data Stream 3: European region MPI-M/MaD,* World Data Center for Climate.

Nicholls, R.J. and Hoozemans, F.M.J. (2000) Global vulnerability analysis, in *Encyclopedia of Coastal Science* (ed. M. Schwartz), Kluwer Academic Publishers.

Nicholls, R.J. and de la Vega-Leinert, A.C. (eds) (2007a) Implications of sea-level rise for Europe's coasts. *Journal of Coastal Research*. Special issue.

Nicholls, R.J., Hanson, S., Herweijer, C., *et al.* (2007b) *Ranking of the world's cities most exposed to coastal flooding today and in the future*, OECD, Newark.

Prudhomme, C., and Reynard, N. (2009) Regionalised impacts of climate change on flood flows: rationale for definition of climate change scenarios and sensitivity framework. *Milestone report 2 – Project FD2020.* Joint Defra/EA Flood and Coastal Erosion Risk Management R&D Programme. http://nora.nerc.ac.uk/8637/1/FD2020_milestone2.pdf (accessed 5 March 2003).

Rahmstorf, S. (2007) A semi-empirical approach to projecting future sea-level rise. *Science*, **315**, 367–370.

USGS (2010), HYDRO1k Documentation http://eros.usgs.gov/#/Find_Data/Products_and_Data_Available/gtopo30/README (accessed 2 June 2010).

Wunram, C. (2007) Output parameter list for CLM regional climate model runs performed as community runs in the context of the BMBF funding priority. *'Research for Climate Protection and Protection from Climate Impacts' attended by Service Group Adaptation* (SGA). http://www.mad.zmaw.de/fileadmin/extern/SGA-Files/clm_parameter_weblist_0407.pdf (accessed 5 March 2013).

Van Der Knijff, J.M., Younis, J. and De Roo, A.P.J. (2008) LISFLOOD: a GIS-based distributed model for river basin scale water balance and flood simulation. *International Journal of Geographical Information Science*, **99999**:1, doi: 10.1080/13658810802549154.

van der Linden, P. and Mitchell, J.F.B. (eds) (2009) *ENSEMBLES: Climate Change and its Impacts: Summary of research and results from the ENSEMBLES project*, Exeter, UK.

Vafeidis, A.T., Nicholls, R.J., McFadden, L., *et al.* (2008) A new global coastal database for impact and vulnerability analysis to sea-level rise. *Journal of Coastal Research*, **24** (4), 917–924.

Vermeer, M. and Rahmstorf, S. (2009) Global sea level linked to global temperature. *Proceedings of the National Academy of Sciences of the United States of America*, **206**, 21527–21532.

Chapter 5

Physical, environmental, social and cultural impacts of climate change on Europe's regions

Johannes Lückenkötter, Christian Lindner, Florian Flex and Stefan Greiving

TU Dortmund University, Institute of Spatial Planning (IRPUD), August-Schmidt-Strasse 10, 44227 Dortmund, Germany

Abstract

This chapter deals with the sensitivities to and impacts of climate change – and how they vary across Europe. For this it needs to first be clarified, based on existing research, which particular elements of Europe's regions are potentially sensitive to climate-related stimuli. Due to the great physical, environmental, social and cultural variety of Europe's regions the sensitivities to climate change differ from one region to another so that the same climatic stimulus would already lead to different impacts. For example, the same increase in summer temperatures may affect the environment of two regions differently depending on the particular habitats and species currently present there. Regional differences in climate change impacts are further exacerbated (or moderated) by the fact that climate change is *not* occurring uniformly across Europe but, as shown in Chapter 4, is also projected to vary across Europe.

5.1 Introduction

ESPON Climate defined five dimensions of sensitivity and identified several indicators for each dimension. Climate change sensitivity may be defined as 'the degree to which a system is affected, either adversely or beneficially, by climate-related stimuli. The effect may be direct (e.g. a change in crop yield in response to a change in the mean, range, or variability of temperature) or indirect (e.g. damages caused by an increase in the frequency of coastal flooding due to sea-level rise)' (IPCC, 2001b, 993; IPCC, 2007b, 881). This chapter deals

European Climate Vulnerabilities and Adaptation: A Spatial Planning Perspective, First Edition.
Edited by Philipp Schmidt-Thomé and Stefan Greiving.
© 2013 John Wiley & Sons, Ltd. Published 2013 by John Wiley & Sons, Ltd.

with the physical, environmental, social and cultural dimensions. For each dimension several sensitivity indicators are discussed and specified, then related to corresponding exposure indicators and the resulting impact maps presented (see Chapter 2 for more methodological background). The economic dimension with its sensitivity and impact indicators is covered in Chapter 6.

5.2 Physical impacts of climate change

Physical impacts of climate change relate to all human artefacts that are important for territorial development and are potentially affected by climate change. This includes settlements (homes, public buildings, industrial facilities) and infrastructure (e.g. transport and energy infrastructure). These physical assets of a region are typically adapted to normal regional weather conditions and can thus withstand smaller climatic changes. However, buildings and infrastructure are especially sensitive to extreme weather events such as flash floods, large-scale river floods and coastal storm surges.

5.2.1 *Settlements prone to river flooding*

Human settlements are concentrations of dwellings but also of industrial and commercial buildings. A large proportion of a country's or region's population lives, and most social and economic activities take place, in settlements. Even though settlements have usually been adapted to their specific geophysical environment, they are nevertheless sensitive to extreme weather events, which can severely damage buildings, endanger the population and disrupt businesses.

One of these extreme weather events are river floods, which in Europe are mainly linked to prolonged or heavy precipitation and often also snowmelt, which culminate in high river flows. These hydrological parameters, in combination with temperature parameters, are probably affected by climate change. However, climatic changes in all regions of a river basin have to be taken into consideration, because the river flood occurring in one region is the result of climate changes in upstream regions. River floods often lead to catastrophic situations and severe damage because of the high concentrations of population and physical assets in river valleys.

While there is ample evidence that the number and frequency of river floods in Europe has significantly increased since 1960, there is no *general* trend with respect to climate change (Becker and Grunewald, 2003; Mudelsee *et al.*, 2003; Kundzewicz *et al.* 2005; Hisdal *et al.*, 2010; Wilson, Hisdal and Lawrence, 2010). Instead, regionally differentiated studies attribute changes, for example, in the frequency and seasonality of river flood events, to changes in snowmelt patterns in central Europe and the Nordic countries, as well as to changed precipitation patterns in the Mediterranean (Brázdil *et al.*, 2006; Cyberski *et al.*, 2006; Hisdal *et al.*, 2010; Wilson, Hisdal and Lawrence, 2010; Ramos and Reis, 2002). On the other hand, the occurrence and magnitude of river floods has also been shown to be significantly affected by human activity, for example, deforestation in river catchment areas, urbanisation in river valleys, loss of natural floodplain storage as well as river and flood management (Barnolas and Llasat, 2007). Consequently, research on river flood projections related to climate change has yielded complex results. The most comprehensive and sophisticated hydrological model on European river systems, LISFLOOD, projects an increase in river floods in western and eastern Europe, while warmer winters and shorter snow seasons reduce flood hazards in Central and North-Eastern Europe (Dankers and Feyen, 2008; 2009).

As outlined above, linking flood events to climate change requires careful consideration (and modelling) of hydrological processes in entire river basins. The most advanced modelling in this regard has been conducted by the EU's Joint Research Centre (JRC) within the framework of their LISFLOOD project. ESPON Climate compared LISFLOOD data on inundation heights of a 100 year return event for the time periods 1961–1990 and 2071–2100 and thus determined changes in inundation (see Chapter 4). The results were overlaid with CORINE Land Cover data (CORINE 2000/2006), which use the same 100×100 m grid cell raster. The changes in inundation height were multiplied with the respective area size of inundated settlement land and then added up for each NUTS 3 region. The aggregated values were then also related to the total settlement area of each NUTS 3 region.

5.2.2 Settlements prone to flash floods

Another extreme weather event to which settlements are very sensitive is flash floods, which can be defined as 'a local flood of great volume and short duration resulting from heavy rainfall in the immediate vicinity' (Merriam-Webster). Over natural watersheds they typically occur in cases of more than 200 mm of rain over less than 6 hours, while in built-up areas even precipitation of 50 mm within 1 hour can produce a local flash flood. What makes these flash floods especially damaging and dangerous are the short warning times and great water speeds.

Climate change affects the occurrence of flash floods through altering local precipitation patterns. The ESPON Climate project did not have detailed hourly precipitation projections at the required high resolution across Europe. A pragmatic assumption was therefore that there is a general linear relationship between 'days with heavy rainfall' (one of the project's exposure indicators) and the occurrence of more extreme and short-term heavy rainfall events. Besides the number of heavy rainfall days and the intensity of heavy rainfall there are other crucial factors, such as the topography of an area (e.g. steep slopes or narrow valleys), soil conditions and the coverage of the terrain (e.g. sealed surfaces or no vegetation cover) that facilitate or intensify flash floods.

Most studies on flash floods deal with specific flash flood events, the damage they caused and potential flood management measures. Large-scale or even pan-European studies and data on flash floods are rare and also more practice- and policy-oriented than data-driven and analytical (e.g. APFM 2007). However, a series of studies have been conducted within the framework of the Floodsite programme, which focused on flash flood occurrence in the Mediterranean region. The studies analysed historical data and modelled past and future flash flood events in the Mediterranean (Thielen *et al.*, 2000; Wobrock *et al.*, 2000). A more large-scale project that is currently on-going is compiling and analysing flash flood data from seven European hydrometereological regions (Gaume *et al.*, 2009). Initial findings of these studies show that flash floods are more severe in Mediterranean countries and regions than in more central European regions and that there is a strong seasonality for flash flood occurrence. However, flash floods are also a very regionally or even locally specific phenomenon with different climatic forcing mechanisms in each locale (Gaume *et al.*, 2009; see also Dankers and Feyen, 2008; Christensen and Christensen, 2003; Kundzewicz *et al.*, 2005). The URBAS project (2008) deals with such micro-scale parameters of flash floods in urban areas of Germany. Their methodology takes into account various climatic, topographical, infrastructural, land-use and building-related data that allow a fairly detailed analysis and forecasting of urban flash flood events.

Ideally one would determine flash flood sensitive settlements by performing a GIS (Geographic Information System) analysis that takes into account flow accumulation (water running down slopes and channelled into valleys) as well as water accumulation in sinks. However, after cross-checking results of such a detailed analysis, the ESPON project concluded that the currently available digital elevation models available across Europe do not have a sufficiently accurate vertical resolution for such a methodology. Consequently, flow accumulation would, for example, suddenly stop somewhere in a river valley, because the elevation model indicated a 1 m jump in elevation.

Therefore the flash flood risk was calculated as a general flash flood potential of a NUTS 3 region. Using the HYDRO1K digital elevation model the mean and standard deviation of slope steepness were calculated for each NUTS 3 region. Furthermore, using the EUROSOIL database the average hydrogeological class was calculated for each NUTS 3 region. The hydrogeological classification takes into account several relevant soil characteristics, such as the coarseness and water permeability of the topsoil and substratum material and the depth to an impermeable soil layer. In essence this indicator reflects how much and how easily the soil in a given location can absorb precipitation. As another component of flash flood sensitivity, the land cover was taken into account. Since the effects of heavy rainfall are generally mediated by natural vegetation, the share of area without forests, grass- or bushland was calculated, that is, the remaining areas included especially highly sealed areas such as settlements as well as agricultural areas (which are seasonally without any vegetation) (see also see also Krysanova *et al.*, 2000 on the SWIM model). The overall flash flood potential was calculated by adding the normalised slope steepness (weighted with 0.33), normalised standard deviation of slope steepness (0.33), the mean hydrogeological soil group (0.17) and non-natural land cover (0.17) (see also Lutz, 2009; Kwakand and Kondoh 2008 for similar methodologies). Then the absolute area and relative share of settlements areas in a NUTS 3 region was calculated and multiplied with the flash flood potential. Finally, these two sensitivity indicators were multiplied with the exposure indicator 'change in heavy rainfall days' as described in Chapter 4, and then normalised across Europe.

5.2.3 *Settlements prone to coastal storm surges*

In addition to river flooding and flash floods, which typically occur inland, sea-level rise and resulting coastal flooding have the potential to seriously affect coastal towns and villages. Coastal areas have always been the site of many major cities and urban agglomerations due to the importance of international maritime transport. Therefore, damage to buildings and settlements in coastal areas due to rising mean sea levels or more severe storm surges would possibly affect a large proportion of a country's urban centres.

Sea-level rise is being studied by many research groups world wide, whose findings were summarised and integrated into model projections by the IPCC (see IPCC, 2001a; 2007a). According to the IPCC, thermal expansion of sea waters is responsible for 70–75% of the sea-level rise projected by the various climate change scenarios. However, Rahmstorf (2007) has shown that observed sea-level rise from 1990 onwards is close to the upper limit of the projected global ranges. As regards Europe, various studies have validated that past sea-level changes range from (depending on the region) −0.3 mm to 2.8 mm per year during the 20th century (e.g. Guinehut and Larnical, 2008; Novotny and Groh, 2007; Church and White, 2006; Cazenave, 2006; Demirov and Pinardi, 2002). Sea-level rise projections for Europe indicate the greatest increases for the Baltic and Arctic coasts, which also partly experience a compensating process of land uplift due to post-glacial rebound effects, and Northern Mediterranean coasts (Johansson *et al.*, 2004; Meier *et al.*, 2006; Nicholls, 2004). The latest projections on sea-level rise on the basis of the A1B scenario indicate values for the year 2100 that are between 0.97 and 1.56 m above the 1990 annual mean sea level (Vermeer Rahmstorf, 2009; see also Nicholls *et al.*, 2011). Interestingly their projections for the other scenarios are also very close to this range, which was explained by the fact that air temperature increases projected by the various scenarios for the first half of the 21st century are very similar, and these air temperatures slowly translate into higher water temperatures in the second half of the century. Lastly, implications of sea-level rise for 13 European countries were discussed in separate articles and summarised by Nicholls and de la Vega-Leinert (2008), concluding among others that Mediterranean river deltas are expected to be 'hot spots' of sea-level rise impacts.

Based on the expert judgement of oceanographers specialised in climate change issues, the ESPON Climate project decided to use a standard value of 1 m sea-level rise for all European coasts. Based on this, the project assumed that every coastal settlement is – under normal weather conditions – prepared for a sea-level rise of 1 m. This assumption cannot be made, however, for extreme events like major coastal storms. Thus, using the DIVA model and the HYDRO1K digital elevation model, the area of land was determined that would be inundated after adding 1 m to the storm surge height projected by DIVA for a 100 year return event. Subsequently these areas were overlaid with the CORINE Land Cover data in order to determine the settlement areas located in these inundated areas. Finally, both the total size of these settlement areas and their ratio in relation to the total settlement area of each NUTS 3 region were calculated and then combined and normalised across Europe.

5.2.4 Combined potential impacts of climate change on settlements

Figure 5.1 shows the combined potential impacts due to river flooding, flash floods and costal storm surge flooding. It is evident that the sea-level rise adjusted coastal storm surge heights account for most of the high impacts in North-Western European regions bordering the Atlantic Ocean (sometimes exacerbated by fluvial and pluvial flooding). Projected increases in river flood heights are responsible for regional 'hot spots' in Italy, Slovenia and Hungary. However, large parts of Europe may not expect significant impacts on their infrastructure resulting from climate change. In fact, physical structures in some central and Southern European regions may even experience less climate-related impacts due to decreasing precipitation in these regions.

5.2.5 Transport and energy infrastructure prone to river flooding

Transport and energy infrastructures are of great importance for regional development. Roads, railways, airports, harbours and also power stations and refineries are the technical backbone of today's social and commercial life. Any disruption, damage or destruction of these infrastructures has severe economic and social consequences. Given the vast expanse of these infrastructure systems they are very sensitive to extreme weather events that have the potential to physically impact upon them.

As for settlements, river floods constitute such extreme weather events that may adversely affect transport infrastructures. Therefore, the above discussion on changing hydrological patterns in river basins and their linkage to climate change equally applies here.

Most studies concentrate on overall economic and human damages of flood events. A differentiated analysis of sensitivities and damages of settlements (concentrated areas) versus infrastructures (mostly linear) are usually not included. However, two recent FP7 research projects (WEATHER[1] and EWENT[2]) focussed specifically on the impacts of weather extremes on transport systems. The projects considered a wide range of impacts, such as accidents, time losses, safety impacts, damage and maintenance costs as well as indirect costs due to transport interruptions. According to the WEATHER project, rail transport is currently the most severely affected mode of transport in regard to extreme weather events (Bläsche *et al.*, 2010). The respective impacts on rail transport were also shown to differ markedly between European regions. The comparatively lower impacts on road transport are also more evenly distributed. The EWENT project assessed annual costs for the transport sector due to extreme weather events based on current and future climates (2041–2070). The dominating impacts in terms of costs were due to road accidents. However, both projects covered a very wide range of extreme weather events (i.e. they did not concentrate on river flooding), focussed on the transport sector only (i.e. they did not include energy infrastructure) and were also not conducted at a fine geographical resolution. Therefore ESPON Climate decided to utilise the LISFLOOD project, which provides the most up-to-date and comprehensive analysis of river flooding and climate change related flood scenarios in Europe.

For calculating the impact of river floods on transport and energy infrastructures the same indicator

---

[1] www.weather-project.eu
[2] www.ewent.vtt.fi

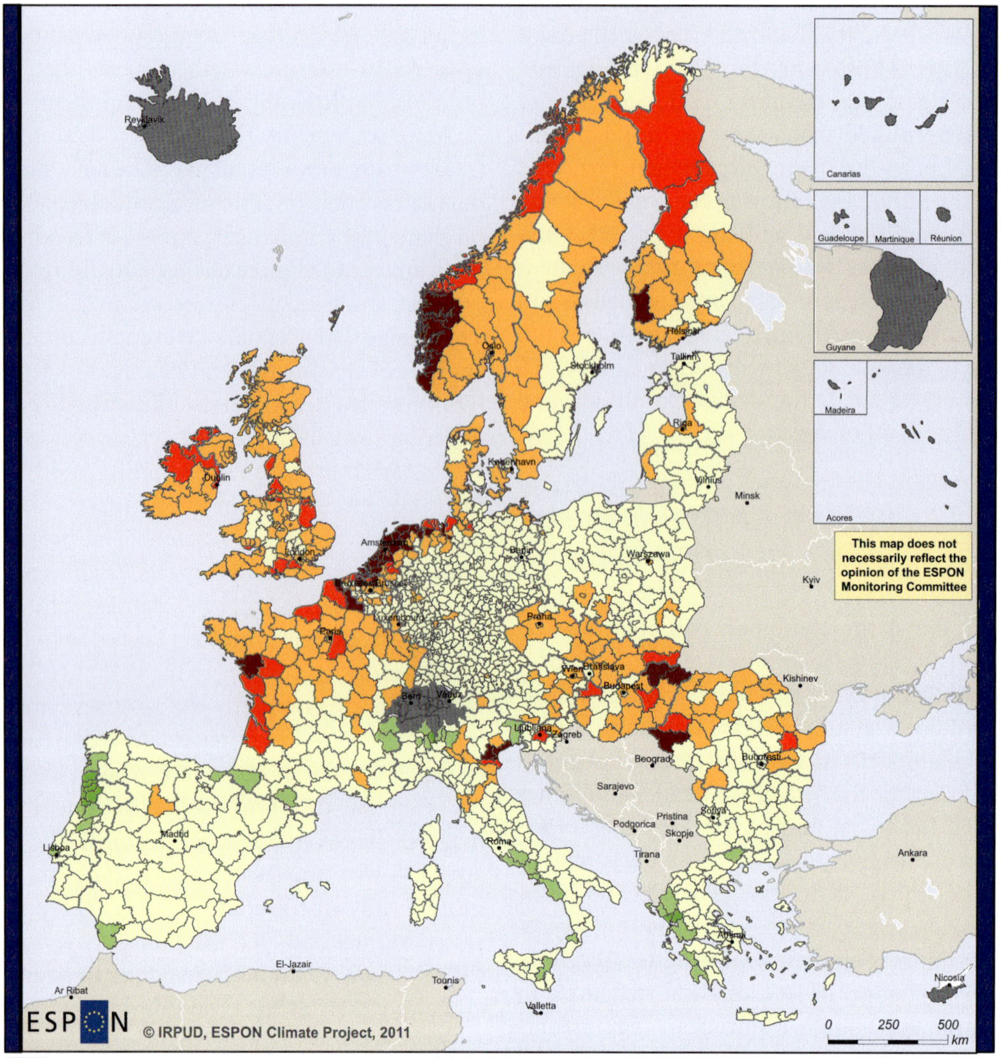

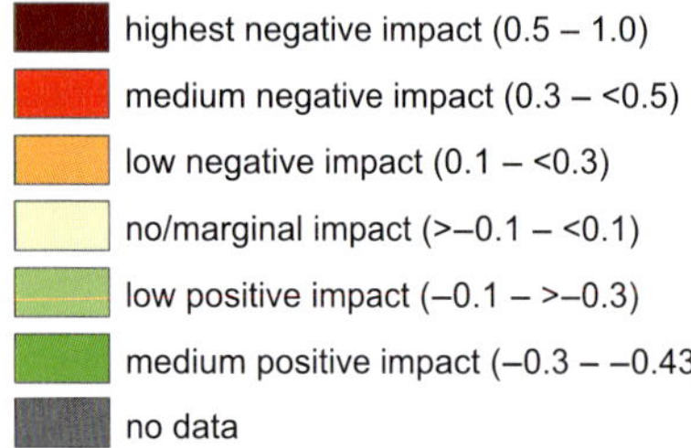

Potential impact of climate change on settlements

- highest negative impact (0.5 – 1.0)
- medium negative impact (0.3 – <0.5)
- low negative impact (0.1 – <0.3)
- no/marginal impact (>–0.1 – <0.1)
- low positive impact (–0.1 – >–0.3)
- medium positive impact (–0.3 – –0.43)
- no data

Combined potential impacts of changes in inundation depths of a 100 year river flood event and a sea level rise adjusted 100 year coastal storm surge event and changes in flash flood potential on settlements.

Fluvial inundation depth changes calculated by comparing 1961-1990 and 2071-2100 river flooding projections of the LISFLOOD model based on climate projections from the CCLM model for the IPCC SRES A1B scenario. Regional storm surge heights projected by the DIVA model adjusted with a 1 m sea level rise.

Figure 5.1   Potential impact of climate change on settlements. © ESPON 2013, IRPUD, ESPON Climate Project, 2011.

approach was adopted as described for settlements, namely using LISFLOOD results for a 100 year return flood event calculated for the A1B climate forcing scenario. On this basis and using geographical data on the various infrastructures, the area size or length of potentially affected portions of an infrastructure type was calculated, multiplied with the respective inundation height changes and finally related to the overall area or length of the respective infrastructure in a NUTS 3 region. Absolute and relative impacts were then combined – again for each infrastructure individually.

5.2.6   Transport and energy infrastructure prone to coastal storm surges

Coastal areas not only exhibit a high concentration of human settlements, but correspondingly also of transport and energy infrastructure systems. Thus streets, railways, airports, harbours, thermal power stations and refineries in coastal areas are likewise prone to the effects of sea-level rise. As explained previously, the disruption or destruction of these infrastructures would have great effects on the coastal population but also on industrial development, trade and tourism in coastal areas.

The most important studies regarding sea-level rise in Europe have been discussed in the section on settlements and equally apply to infrastructure systems.

To determine the impact of sea-level rise on transport and energy infrastructures in coastal areas the same procedures as for settlements were applied. The projected inundated areas and the corresponding changes in inundation heights were overlaid with a map of the respective infrastructure networks or facility locations. Absolute and relative impact indicators were calculated and then combined for each type of infrastructure separately.

5.2.7   Transport infrastructure prone to flash floods

As discussed above, transport infrastructures are vital for a society's social and economic functioning.

At the same time they are very sensitive to extreme weather events that have the potential to damage, disrupt or destroy them.

Flash floods, as already defined above, are one of these extreme weather events that have the capacity to seriously impact the facilities and the operation of infrastructure systems. Again, the short warning lead-time and great intensity are the most damaging and dangerous aspects of flash floods. As discussed above, climate change is affecting the occurrence and intensity of flash floods due to altering local precipitation patters, in particular heavy rainfall patterns.

Research on flash floods in Europe has already been discussed in the section on settlements prone to flash floods. Again, these studies concentrate more on flash floods as such and the overall economic and human damage, rather than differentiating sensitivities and damage of settlements versus particular types of infrastructure.

The same methodological procedures as discussed in the settlement section were applied for determining the regional impacts of flash floods on transport infrastructure. Because flash floods are usually more significant for linear infrastructures like roads and railways, potential impacts on airports and harbours were not calculated. Therefore, the regional flash flood potential of each NUTS 3 region was multiplied by the absolute kilometres of roads or railways, respectively, in a region and then the corresponding densities in relation to the total NUTS 3 area. Finally, these sensitivities were related to changes in the number of heavy rainfall days as the relevant ESPON Climate exposure indicator to arrive at the potential impacts of flash flood changes on roads and railways.

5.2.8   Potential impact of climate change on major roads and railways

The map in Figure 5.2 shows the combined potential impacts of changes in river flooding, coastal flooding and flash floods on major roads and railways. Apart from the highly impacted Italian regions on the Adriatic Sea, where coastal flooding and major river flooding from the river Po combine, most South-European regions' road and rail

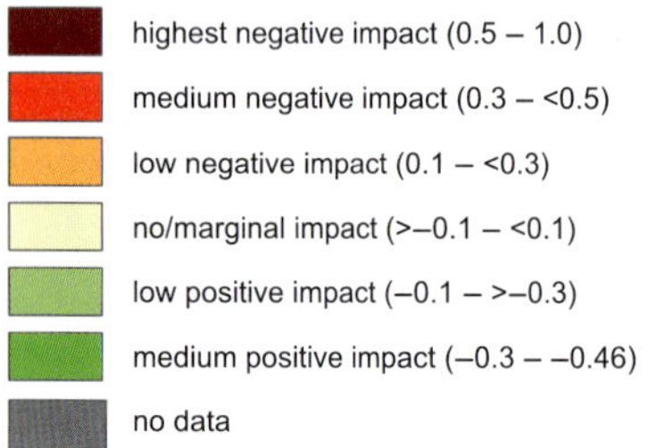

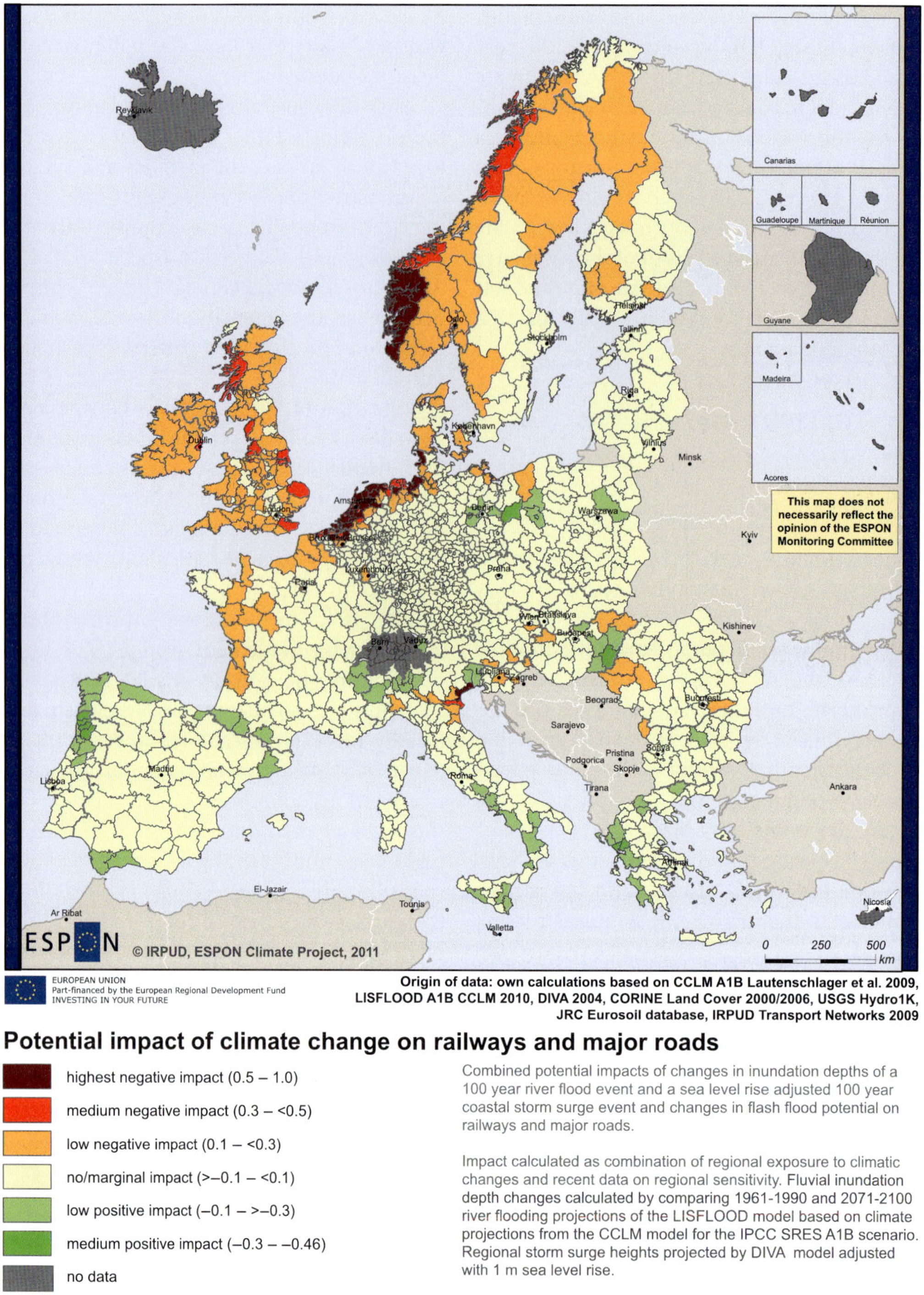

Potential impact of climate change on railways and major roads

Figure 5.2 Potential impacts of climate change on railways and major roads. © ESPON 2013, IRPUD, ESPON Climate Project, 2011.

infrastructure is projected to be only marginally affected, or would even be less subjected to flooding events than in the past. Medium to high negative impacts are mostly found in North-Western Europe along the coasts of the North Sea, where several types of flooding converge and transport infrastructures are highly concentrated. Roads and railways in Eastern and South-Eastern Europe are expected to be less affected by flooding than in the past.

5.2.9　Potential impact of climate change on airports and harbours

As one would expect, the sites of European airports and harbours seem to be generally located where they are less likely to be flooded. Almost across Europe the impacts of climate change on these point infrastructures are only marginal. This makes the few affected airports and harbours stand out even more in the map in Figure 5.3 (especially Venice, Lapland and the Dutch coastline).

5.2.10　Potential impact of climate change on thermal power stations and refineries

The map in Figure 5.4 shows the combined projected impacts of river flooding and coastal flooding on thermal power stations and refineries. Since these facilities are often located along rivers and coasts, there are many European regions with relatively high impacts. Thus power stations and refineries in North-Western Europe on the coastline of the Atlantic Ocean and the North Sea are projected to have a medium to high negative impact. A similarly negative impact is expected for facilities which are located along particular rivers (such as the Po and Tisza) for which the inundation heights for a 100 year flood event are projected to increase. In contrast, a positive impact is projected for facilities located in Barcelona and parts of Poland and the Czech Republic, where a decrease in inundations heights of river flood events is projected.

5.2.11　Combined physical impact of climate change

The map in Figure 5.5 depicts the combined impacts of climate change on settlements as well as transport and energy infrastructures. The 'hot spots' are almost all located on coasts and especially at river mouths (where coastal and river flooding impacts converge) and are mostly found in the North of Europe. In contrast, almost all regions projected to benefit from climate change in regard to settlements and infrastructures are located in Southern Europe due to decreasing river flood impacts.

5.3　Environmental impacts of climate change

Climate is an integral part of nature. Thus any changes to climate will directly or indirectly affect all parts of the natural environment. However, some environmental entities are more sensitive to climatic changes than others. The aim of this section is to identify these more sensitive elements and calculate the potential environmental impacts of climate change.

By definition the natural environment consists of all natural physical entities and biological life within the earth's biosphere. Relevant environmental impacts relate primarily to soils and species. With regard to species, one may differentiate between distributional and phenological changes.

Phenological changes comprise changes to periodic plant and animal life cycle events, for example, the date of the first blossoming of a flower species, the onset of leaf colouring and fall in certain tree species or the first appearance of migratory birds in an area. There is clear evidence of such phenological changes in Europe in recent decades (Parmesan and Yohe, 2003; Root *et al.*, 2003; Menzel *et al.*, 2006a). Many of these life cycle changes have been studied in detail and can be precisely measured (e.g. Menzel *et al.*, 2006a, 2006b; DEFRA, 2007;

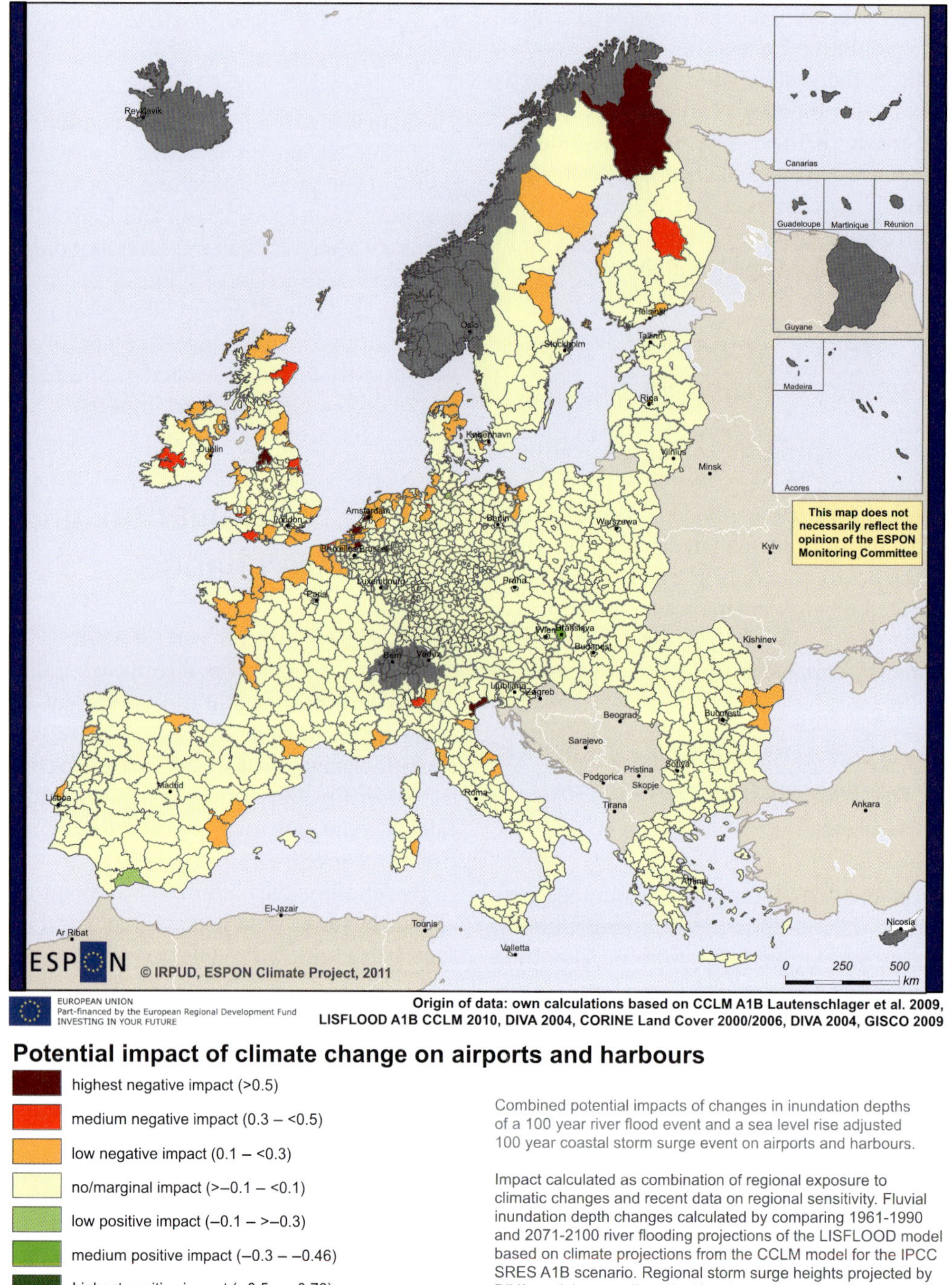

Potential impact of climate change on airports and harbours

Figure 5.3 Potential impact of climate change on airports and harbours. © ESPON 2013, IRPUD, ESPON Climate Project, 2011.

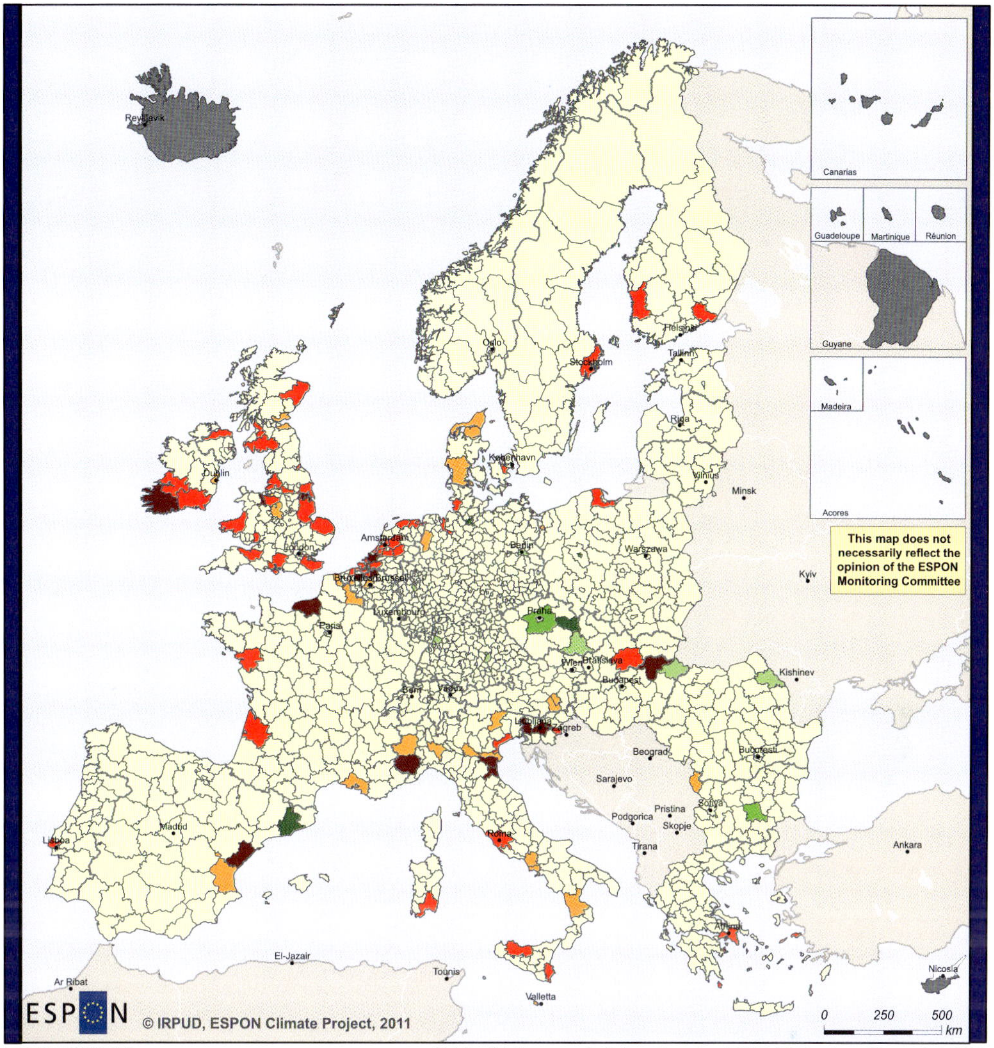

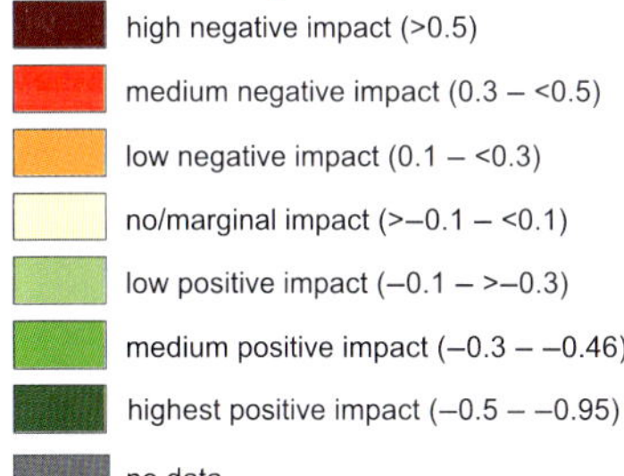

Potential impact of climate change on thermal power stations and refineries

Figure 5.4   Potential impact of climate change on thermal power stations and refineries. © ESPON 2013, IRPUD, ESPON Climate Project, 2011.

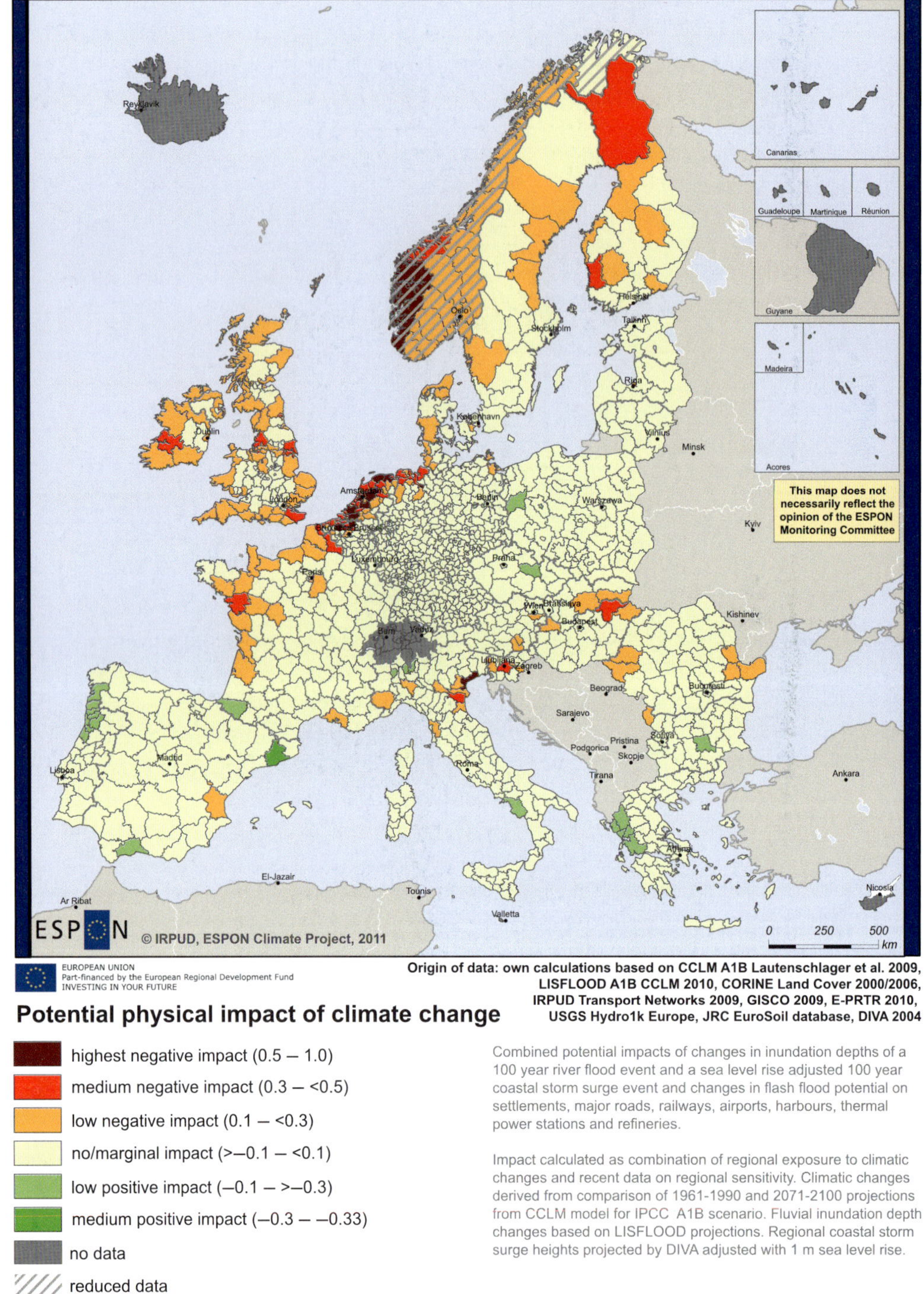

Figure 5.5   Potential physical impact of climate change. © ESPON 2013, IRPUD, ESPON Climate Project, 2011.

Hoye *et al.*, 2007) and most of them can even be reliably explained by climatic changes (van Vliet, 2008). However, reviewing the available scientific literature on the nexus between climate change to phenological change, a recent report by the European Environmental Agency (EEA, 2008) came to the conclusion: 'While advancing trends in seasonal events will continue as climate warming increases in the years and decades to come, it is uncertain how different species will respond when temperature thresholds are reached and whether linear relationships between temperature and growing season will be realised in the future' (p. 115). For this reason the scientific community has so far not made projections on future phenological changes, thus making it difficult to include life cycle event changes of species in the environmental assessment of the ESPON Climate project.

Distributional changes of plant and animal species are likewise highly related to climate change. Some species benefit from changing climatic parameters and are able to increase their populations and/or enlarge their habitats, while other species' habitats are shrinking and their populations are nearing extinction levels. Thus climate change has facilitated (in combination with other factors) completely new biodiversity patterns that will continue to change in the future. In particular the gradually warmer winters have led to and are projected to continue to extend the distribution areas of many species northwards and to higher altitudes (cf. Parmesan and Yohe, 2003; Walther *et al.*, 2005). Of interest to a comprehensive and pan-European assessment are, of course, not distributional changes of individual species but aggregate changes. With regard to plant distributions, such an aggregate analysis has been undertaken by Thuiller *et al.* (2005) and Bakkenes *et al.* (2002, 2006). They were able to model distributional changes for almost 1400 plant species across Europe until 2050 and 2080, respectively. Both models projected the greatest changes to occur in Mediterranean, Euro-Siberian and mountain regions and suggest up to 60% loss of plant species in some areas. However, both studies base their projections on climate models other than the

CCLM model used by the ESPON Climate project and, more importantly, on more extreme climate forcing scenarios.

Therefore, the indicators described below are mostly soil and ecosystem-based. Soils are made up of mineral and organic material, which serves as the natural medium for the growth of plants. Soils evolve over long time periods through complex interactions between the underlying rock formation, below surface micro-organisms, above surface plants and animals – and climatic factors such as moisture and temperature. Soils are therefore relatively stable environmental entities that are nevertheless climate sensitive, for example, to extreme weather events such as flash floods. Soils also form the basis for ecosystems, which may be defined as relatively stable systems characterised by particular functional relationships between plants, animals, micro-organisms and their physical environment in a particular area.

5.3.1   *Forests sensitive to forest fires*

Forests account for most of the earth's vegetation biomass. They contain a large variety of plant species and are the habitat of many micro-organisms and animals. Through the complex interaction of these forest species with the underlying soils and the local climate, forests play an important role in soil and water conservation. In addition to these natural functions, forests are also an economic asset, as a large proportion of European woodlands are used for forestry.

As regards climate change, it first needs to be noted that forests' biomass are the earth's major carbon pool. Thus forests (and changes to forests) have a significant effect on global CO_2 levels, which is one of the drivers of climate change. On the other hand, climate change affects forests in various and complex ways: in general, higher CO_2 levels have a 'fertilizer effect' on tree growth. Higher temperatures and thus longer growing seasons promote tree growth in some areas, but decreasing precipitation reduces tree growth in other areas. Changing local climate conditions may even enable or reduce the survival of certain tree species in particular locations – thus changing the geographical distribution of the various types of forests. Furthermore, other plant and animal species (in particular pests – but also pollinators) are likewise affected by climate change, leading to increasing or decreasing forest growth or damage. Much forest damage is typically caused by (winter) storms, whose patterns and intensity are expected to change due to climate change. Finally, changing temperature and moisture conditions affect forests' sensitivity to fires. Forest fires are most often human induced (and this human influence also differs between countries), but in any case they can more easily spread in warmer and drier conditions.

The climate change effects on forests outlined above are actually very complex and affect various types of trees and forests in different ways (see Kropp *et al.*, 2009 for a systematic overview). It is therefore very difficult to reliably model these effects – and even more so to geographically differentiate these effects across Europe. Historic studies have shown, however, that after centuries of forest exploitation growth rates of forest biomass have been recovering in Europe since the middle of the 20th century – in part due to better forest management (Spiecker *et al.*, 1996). Also, a North-East shift in the distribution of certain tree species has been observed (Bakkenes *et al.*, 2002; Harrison *et al.*, 2006). This shift is expected to continue under the further influence of climate change, as a complex modelling project undertaken by the European Commission's Joint Research Centre (JRC) indicates. Incorporating a wide range of tree, soil, geomorphological and climate data, the project simulated geographical distributional changes of habitats suitable for various types of forests (Casalegno *et al.*, 2011). Using the IPCC A1B climate scenario the simulation shows a general North-East shift in suitable forest habitat categories. Focusing on forest fire dangers in Europe (but using the IPCC A2 scenario) a modelling project by Flannigan *et al.* (2005) projects a significant increase of fire potentials, an enlargement of the fire-prone area and a lengthening of the fire season (see also Camia *et al.*, 2008). Further results can be derived from the LPJ Dynamic Global Vegetation Model, which is run on a global scale and also projects changes in functional forest types and carbon cycles.

Many of the possible climate change effects on forests are too complex or are so far not fully understood and have therefore not been modelled quantitatively. The two modelling projects referred to above are notable exceptions and have produced scenarios covering the entire European territory. However, the forest fire model is based on a different forcing scenario than the ESPON Climate project. The other modelling project only calculates changes of forest habitat suitability (i.e. where certain types of forest could or could not exist). Given the fact that most land in Europe is in one way or another under intensive human use, it seems unlikely that the overall coverage of forests will increase, that is, that forests will in the future cover all areas that are environmentally suitable habitats for forests. It is more reasonable to expect that any changes in the distribution of forest types will take place more or less within the boundaries of existing forests. A further problem not addressed in the above studies is the fact that anthropogenic factors often play an important role in the occurrence of forest fires. While new studies are beginning to model these factors by determining the proximity of forests to human infrastructures (Reineking *et al.*, 2010), a fully-fledged and tested model of forest fire occurrence that consistently incorporates anthropogenic factors has still not been developed yet.

Therefore, given the methodological shortcomings described above, the ESPON Climate project decided to refrain from any unwarranted and highly uncertain modelling and instead to take the past occurrence of forest fires as a starting point for determining the forest fire sensitivity. Using the detailed data of the ATSR World Fire Atlas (ESA, 2010) and the CORINE Land Cover data, the number of fires that occurred in forest areas of each NUTS 3 region was determined. Without trying to disentangle which exact natural or anthropogenic factors caused these fires, it was assumed that where fires occurred in the past there is also a certain likelihood that fires might occur in the future as well. The impact analysis then determined where climate conditions are changing in a way that would increase this likelihood. To calculate the forest fire sensitivity indicator, the number of forest fires and the ratio of these fires in relation to the total forest area of a NUTS 3 region were calculated and then combined. In order to determine potential impacts of climate change this sensitivity was then multiplied by changes in the number of summer days and in summer precipitation and then normalised across Europe.

5.3.2   Potential impact of climate change on forest fires

The map in Figure 5.6 shows the potential impacts of climate change on forest fire prone regions. It becomes evident that forest fires, although highly 'visible' in the media every summer, the problem is projected to increase only in Southern Europe in the future. In these areas, already dry conditions are exacerbated by increasing summer temperatures and declining precipitation. North-Western Spain, Portugal and some regions in southern Italy and Greece are projected to be most severely impacted by forest fires in 2071–2100.

5.3.3   Protected natural areas

Protected natural areas are clearly delineated geographical areas with legal protection status that aim to protect and conserve the most threatened plant and animal species and their habitats. With this general purpose, the NATURA 2000 network was set up by the European Union, creating a network of protected areas across Europe that conform to common selection and management criteria. Two types of areas are distinguished, namely: Special Protection Areas for Birds and Special Areas of Conservation (designated for other species and habitats). Currently there are a total of 27 661 protected areas covering about one sixth of the European Union's landmass that are part of the NATURA 2000 network (EEA, 2009).

These areas are important for protecting rare species and habitats, but they are not equally sensitive to climate change. Given the special characteristics and biological requirements of the endangered species and habitats, experts regard those habitats as most sensitive to the projected climate changes that rely on a certain amount of moisture (e.g. wetlands, humid grasslands). If precipitation and evaporation levels change, these habitats' uniquely adapted plants and animals would decline in numbers or even become locally extinct.

For assessing the sensitivity of protected areas and their habitats some studies have applied complex modelling approaches (e.g. Berry *et al.*, 2003; Normand *et al.*, 2007). Other studies opted for a simpler – but perhaps more robust – indicator approach, classifying the various habitat types on the basis of special habitat characteristics and especially temperature and moisture sensitive species (Kropp *et al.*, 2009; Holsten, 2007; Petermann *et al.*, 2007). This assumes that these habitats have to be conserved exactly as they are today and does not take into consideration possible natural adaptation mechanisms. So far only regional and national analyses have been conducted with the indicator-driven approach, but their methodologies make them in principle suitable for large-scale, pan-European studies using the NATURA 2000 statistical data. However, while national studies have developed and successfully implemented a methodology of classifying the climate change sensitivity of the habitat classes relevant for their country (e.g. Petermann *et al.*, 2007), such a classification does not yet exist for the over 230 European habitat classes of NATURA 2000.

Given the current state of research outlined above, the ESPON Climate project could not revert to a comprehensive classification of the individual NATURA 2000 habitat types with regard to climate change. One can argue, however, that NATURA 2000 protected habitats are protected just as they are today. Thus, even if some habitats are better able to adjust to climatic changes, this would nevertheless mean that these habitats have indeed changed. Thus, this would run counter to the conservation intended by NATURA 2000. Therefore, all NATURA 2000 areas were classified as generally sensitive to climate change. Hence the absolute size of NATURA 2000 areas and their ratio to the total NUTS 3 area

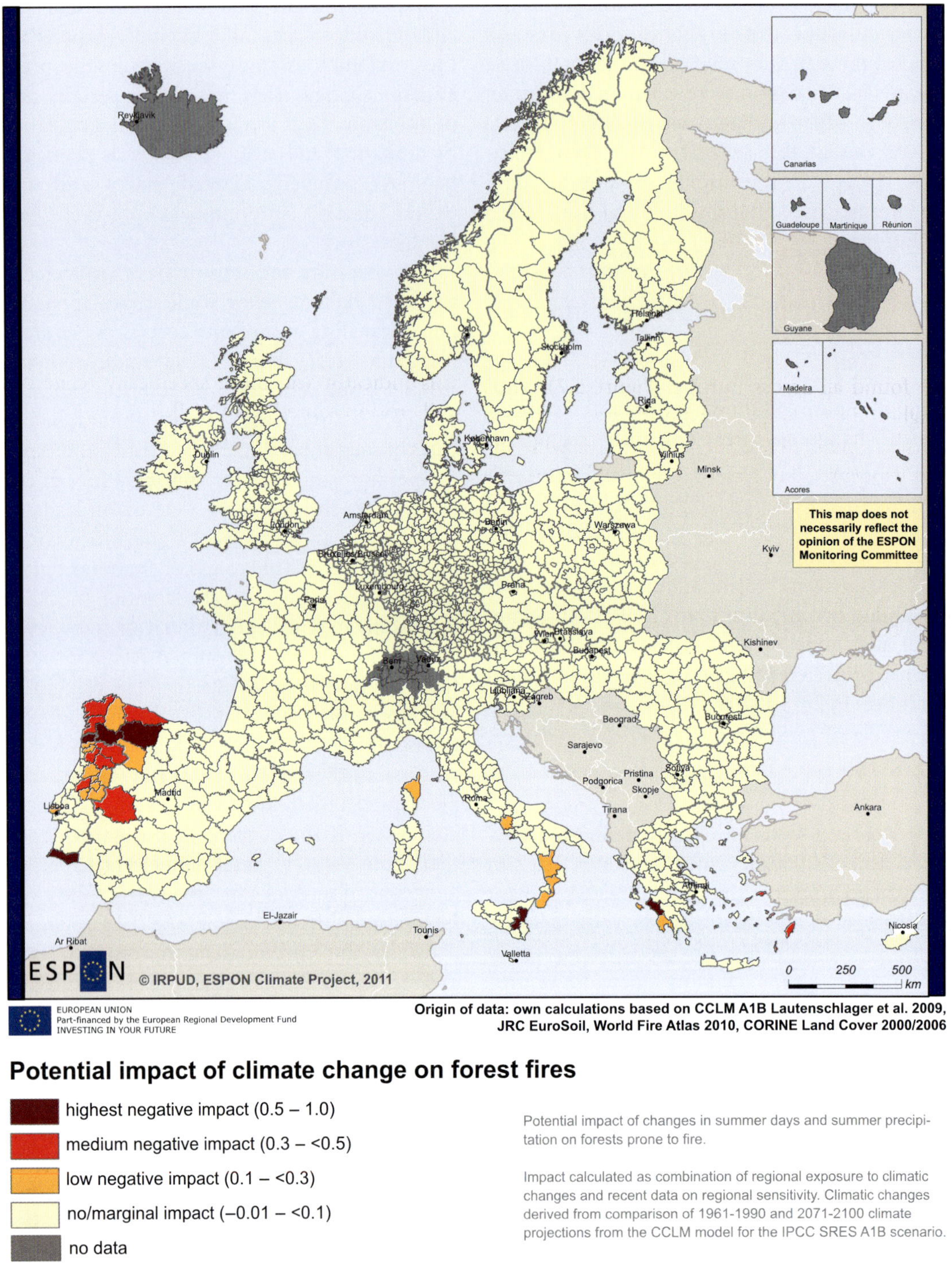

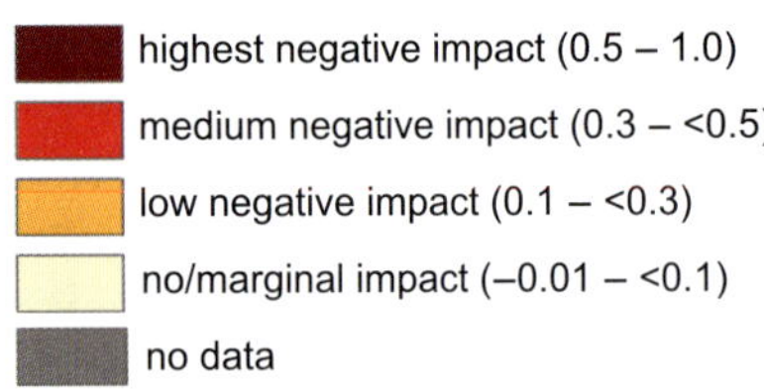

Origin of data: own calculations based on CCLM A1B Lautenschlager et al. 2009,
JRC EuroSoil, World Fire Atlas 2010, CORINE Land Cover 2000/2006

Potential impact of climate change on forest fires

- highest negative impact (0.5 – 1.0)
- medium negative impact (0.3 – <0.5)
- low negative impact (0.1 – <0.3)
- no/marginal impact (−0.01 – <0.1)
- no data

Potential impact of changes in summer days and summer precipitation on forests prone to fire.

Impact calculated as combination of regional exposure to climatic changes and recent data on regional sensitivity. Climatic changes derived from comparison of 1961-1990 and 2071-2100 climate projections from the CCLM model for the IPCC SRES A1B scenario.

Figure 5.6   Potential impact of climate change on forest fires. © ESPON 2013, IRPUD, ESPON Climate Project, 2011.

of each region was calculated and then combined. Finally, the combined sensitivity was related to all exposure indicators (as argued above) in order to determine the potential impacts of climate change on NATURA 2000 areas.

5.3.4 Potential impact of climate change on NATURA 2000 protected areas

Natural areas of high environmental value that are protected under the EU NATURA 2000 directive can be found all across Europe (Figure 5.7). Thus the highest impacts of climate change on these habitats are generally located in regions with the highest exposure changes. In Southern Europe this relates particularly to higher temperatures and less rain, in the North to less frost and snow cover days, which change living conditions for some species drastically. The high impacts in Sweden and Finland (Lapland) are in part also due to their very large protection areas.

5.3.5 Areas sensitive to soil erosion

Soil erosion may be defined as the wearing away of the land surface by natural or anthropogenic forces (Soil Science Glossary Terms Committee, 2008). Soil erosion is a natural, continuously occurring process, but its occurrence and intensity can dramatically increase if some of its driving forces change. At its worst, soil erosion can strip the topsoil from the underlying land and severely reduce the environmental (and thus also agricultural) function of an affected land area.

Several factors can be distinguished that determine the rate of soil erosion: Geological factors include primarily the soil and rock type (particularly its porosity and permeability) and the slope of the land. Biological factors include vegetation cover, the organisms living in and on the soil and the land use. For example, forested areas provide ground cover thus ameliorating the effect of rain, and the forest plants and other organisms also make forest soils more porous and permeable to rain water. Human land uses, such as agricultural use (cropland, grazing) reduce the vegetation ground cover permanently or seasonally and also decrease the water permeability due to soil compaction. Finally, climatic factors include the amount and intensity of precipitation, temperature, wind speed and storm frequency. All of these climatic factors are subject to processes of climate change, but wind speed and storm frequencies are more difficult to model and have thus have not been included in the exposure analysis of the ESPON Climate project. Therefore this indicator will more specifically relate only to soil erosion caused by rainfall.

At the European level (but often also at the national level) there are insufficient and non-comprehensive field data on soil erosion. Therefore most soil erosion assessments at the European scale have reverted to mathematical models of soil erosion (most notably the PESERA project). These models are used not only for predictive purposes but also for gauging the current state of soil erosion in Europe. According to the JRC, using model results from several research projects, an estimated 115 million hectares or 12% of the total EU land area was (in the year 2000) subject to rainfall-based soil erosion (EEA 2008). As to regional differentiation across Europe, the soil erosion focal point at the JRC concluded: 'The Mediterranean region is particularly prone to soil erosion, because it is subject to long dry periods followed by heavy bursts of erosive rain, falling on steep slopes with fragile soils. This contrasts with northwestern Europe where soil erosion is less because rain falling on mainly gentle slopes is evenly distributed throughout the year' (JRC, 2010a). There are also several projects attempting to model the effects of future climate change on soil erosion, for example, at the European level PESERA and MESALES. Generally erosion is projected to increase with increases in precipitation amount and intensity (heavy rainfall events) and further losses of vegetation cover. However, the models also 'show a non-linear spatial and temporal response of soil erosion to climate change' and are known to not

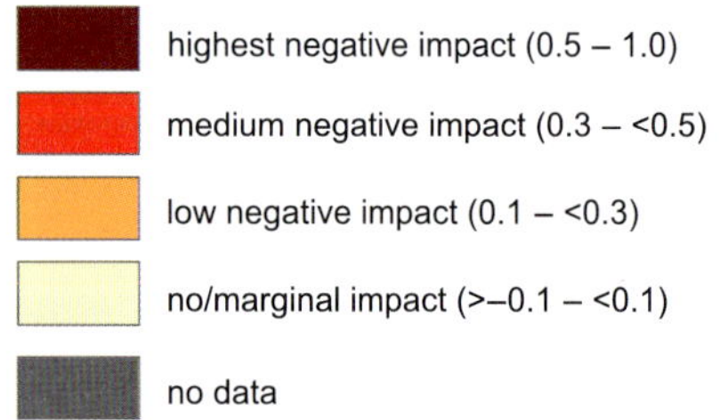

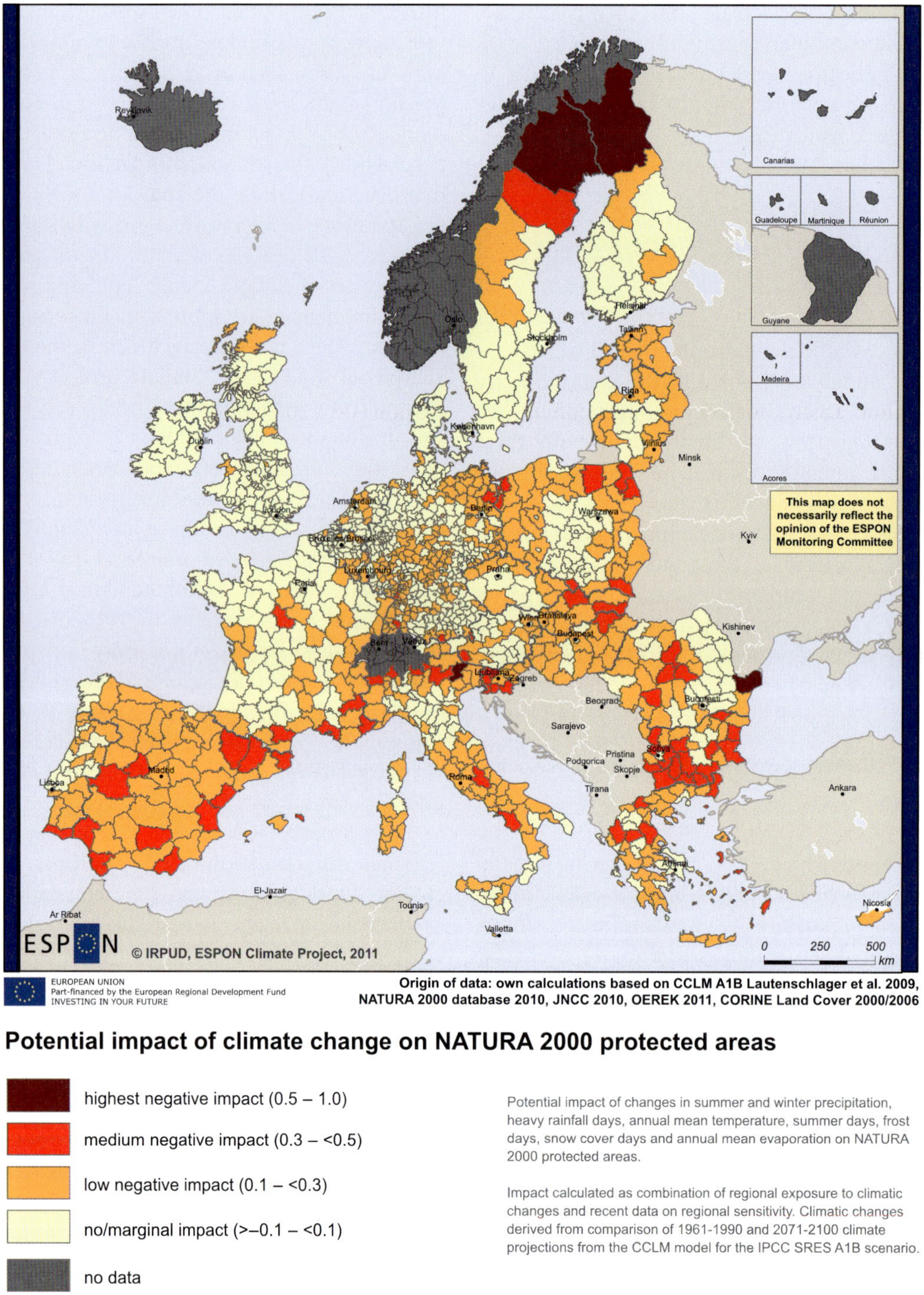

Potential impact of climate change on NATURA 2000 protected areas

Potential impact of changes in summer and winter precipitation, heavy rainfall days, annual mean temperature, summer days, frost days, snow cover days and annual mean evaporation on NATURA 2000 protected areas.

Impact calculated as combination of regional exposure to climatic changes and recent data on regional sensitivity. Climatic changes derived from comparison of 1961-1990 and 2071-2100 climate projections from the CCLM model for the IPCC SRES A1B scenario.

Figure 5.7   Potential impact of climate change on NATURA 2000 protected areas. © ESPON 2013, IRPUD, ESPON Climate Project, 2011.

adequately represent the increase of different types of storms (EEA, 2008, p. 131).

The soil erosion sensitivity indicator was calculated taking into account three main components: slope steepness, land cover and soil characteristics. The slope steepness component was calculated as a combination of mean and standard deviation of slope steepness in each NUTS 3 region using the HYDRO1K digital elevation model. The land cover component used the CORINE Land Cover database and concentrated on land *not* covered by forests and natural grass- and bushlands, because these vegetations are known to mediate heavy rainfall precipitation. Lastly, for the soil component the NUTS 3 average of the erodibility variable of the European Soil Database was calculated. This variable already combines several soil characteristics relevant for soil erosion. The three components were equally weighted when calculating the overall soil erosion sensitivity indicator of each NUTS 3 region. Finally, this sensitivity indicator was related to changes in the number of heavy rainfall days as the exposure indicator and the resulting impact indicator normalised across Europe.

5.3.6 *Potential impact of climate change on soil erosion*

The map in Figure 5.8 shows the results of the impact assessment with regard to soil erosion. Since precipitation levels are projected to be generally declining in the South of Europe, the soil erosion potential is projected to decline there as well. Regions expected to be adversely affected are located in Western France, the United Kingdom and Ireland and, most of all, in Norway. Here strong increases in precipitation, highly erodible soils and steep mountains converge to result in the highest soil erosion impacts in Europe.

5.3.7 *Soil organic carbon*

The major component of soil organic matter is soil organic carbon, which is derived from residual plant and animal material that is decomposed through complex biological and chemical processes. Organic carbon and organic matter more generally are a source of food for soil organisms (e.g. bacteria, fungi, microbes but also invertebrates such as worms, ants and termites) and thus supports soil biodiversity. Soil organic matter also contains various nutrients (e.g. nitrogen, phosphorus, sulfur) that contribute to soil fertility. Furthermore, organic matter absorbs six times its weight in water and thus constitutes an important water reservoir for plants. Finally, soil organic carbon improves the physical structure of soil, increasing water permeability and reducing compaction – both of which reduce the risk of soil erosion (JRC, 2009; EEA, 2008; see also JRC, 2010b).

Soil organic carbon is both affected by and affecting climate change. Organic carbon is essentially a net carbon sink and thus mitigates climate change: Plants build organic material using atmospheric CO_2, which then becomes encapsulated in the soil when the plants decay and are decomposed into soil organic carbon. In other words, the formation of soil organic carbon reduces CO_2 concentrations in the atmosphere and thus mitigates climate change. On the other hand climate change has a significant effect on soil organic carbon: basically organic matter decays more quickly at higher temperatures (leading to less organic carbon in warmer climates) and decays more slowly under more moist conditions (leading to more organic carbon accumulation in cooler climates). Both of these climatic variables are predicted to change significantly in the various climate change scenarios, leading to organic carbon gains or losses in different parts of Europe.

In the past the main driving force for the reduction of soil organic content has not been climate change but rather land conversion to cropland in combination with unsustainable agricultural land management practices (e.g. Sleutel *et al.*, 2003; Dersch and Boehm, 1997). Another factor has been the irrigation of peat land, which is of special importance because peat land is estimated to account for 60% of the entire carbon content in European soils (Byrne *et al.*, 2004; Lappalainen, 1996). Overall, research on European soil carbon content estimates that the organic content in European soils equals nearly 10% of the carbon accumulated in the atmosphere (EEA,

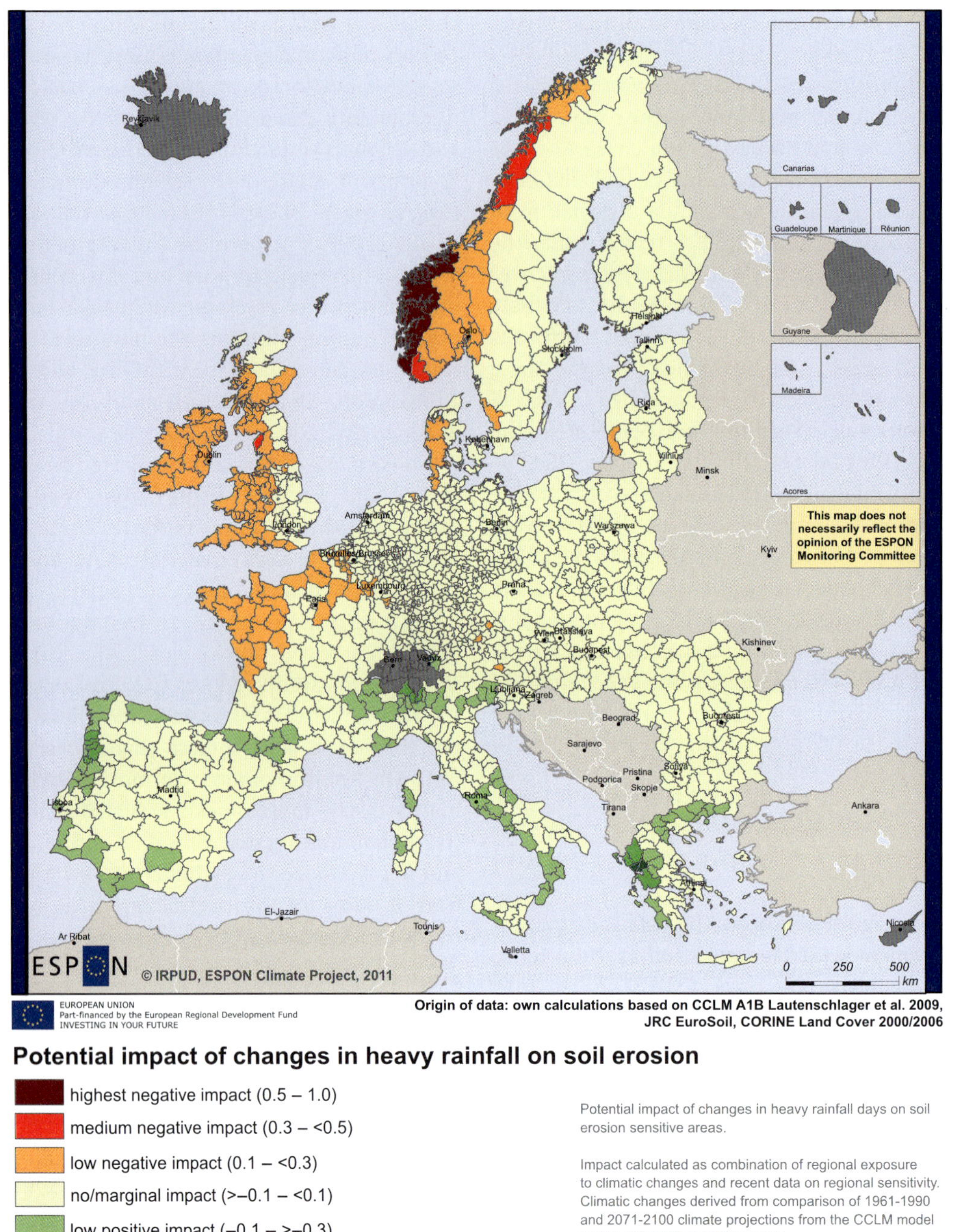

Potential impact of changes in heavy rainfall on soil erosion

- highest negative impact (0.5 – 1.0)
- medium negative impact (0.3 – <0.5)
- low negative impact (0.1 – <0.3)
- no/marginal impact (>−0.1 – <0.1)
- low positive impact (−0.1 – >−0.3)
- medium positive impact (−0.3 – >−0.5)
- highest positive impact (−0.5 – −0.54)
- no data

Potential impact of changes in heavy rainfall days on soil erosion sensitive areas.

Impact calculated as combination of regional exposure to climatic changes and recent data on regional sensitivity. Climatic changes derived from comparison of 1961-1990 and 2071-2100 climate projections from the CCLM model for the IPCC SRES A1B scenario.

Figure 5.8   Potential impact of changes in heavy rainfall on soil erosion. © ESPON 2013, IRPUD, ESPON Climate Project, 2011.

2008; Hiederer, 2009). As regards future changes, a study by Smith *et al.* (2005) projected for the year 2080 losses in soil organic carbon across Europe due to climate change, which could, however, in many areas be reversed through improved agricultural technology and practices.

There are no empirical pan-European data on soil organic content. However, the JRC soil focal point has combined comprehensive data from the European Soil Database with other databases on land cover, climate and topography and has thus been able to calculate data for the organic carbon content in the surface horizon of soils in Europe. On this basis, ESPON Climate calculated what the average soil organic carbon content was for each NUTS 3 region. This indicator was then related to two exposure indicators: higher temperatures leading to a reduction of soil organic content and more precipitation leading to an increase in soil organic content.

5.3.8 *Potential impact of climate change on soil organic carbon content*

The map in Figure 5.9 shows the projected impacts of climate change on soil organic carbon content. Since soil organic carbon consists of (decaying) organic matter, hotter and drier conditions accelerate its decomposition. Not surprisingly then, the soil's content of organic carbon is projected to decrease in the future in Southern and South-Eastern Europe. However, even in some lowlands in central Europe, warmer climatic conditions and the particular soil characteristics in these regions can create significant impacts on soil organic carbon. In the North, however, where the climate gets wetter, soil organic carbon will increase.

5.3.9 *Combined environmental impact of climate change*

Combining all environmental impacts determined and described above yields a map as shown in Figure 5.10. Accordingly, climate change is expected to have the highest negative environmental impacts

in the South and North of Europe – in particular in mountain regions. Important factors are the high slopes and specific soil characteristics that facilitate soil erosion there. In the Mediterranean the drier and hotter climate also increases the likelihood of forest fire occurrence. Soil organic carbon in river deltas and along coasts is also negatively impacted by climate change. The severe impacts in Northern Scandinavia are in part also due to their very large protected areas where *any* climatic change (in this case warmer and wetter climate) is considered as negatively affecting the specific ecosystems under environmental protection.

5.4 Social impacts of climate change

The term 'social' encompasses a wide spectrum of meanings, but in general refers to qualities of a human collective, for example, its composition or interactions. In a more narrow sense the term is also used in regard to socio-economic differences within a population and corresponding redistributive policies.

Within the scope of the ESPON Climate project, social sensitivity relates to human populations that may be adversely or positively affected by climate change. In particular, this includes climate-related sensitivities with regard to public health and personal mobility. Many of these sensitivities relate only to certain social groups, for example, senior citizens or spatially defined communities, such as urban populations.

With the above general definitions in mind, one might also expect to find certain socio-economic groups under this heading, such as poor households, but biologically they are equally sensitive to climate changes as other groups. However, poor households have a reduced economic capacity to cope with or adapt to climate changes, that is, they might not be able to afford better heating or air conditioning. Therefore, according to the ESPON Climate methodology such socio-economic population groups are covered in the adaptive capacity component of the assessment. Likewise,

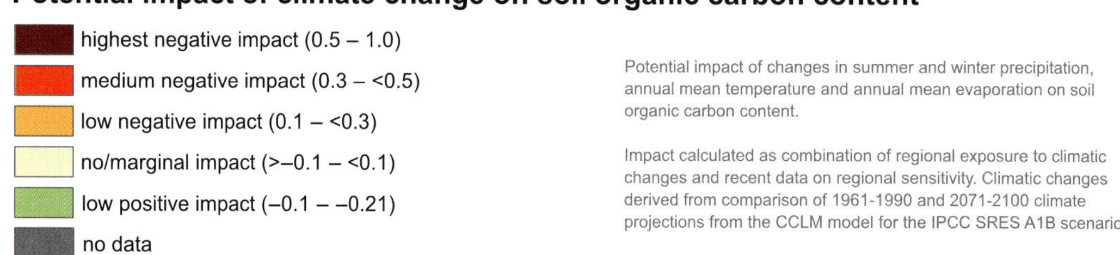

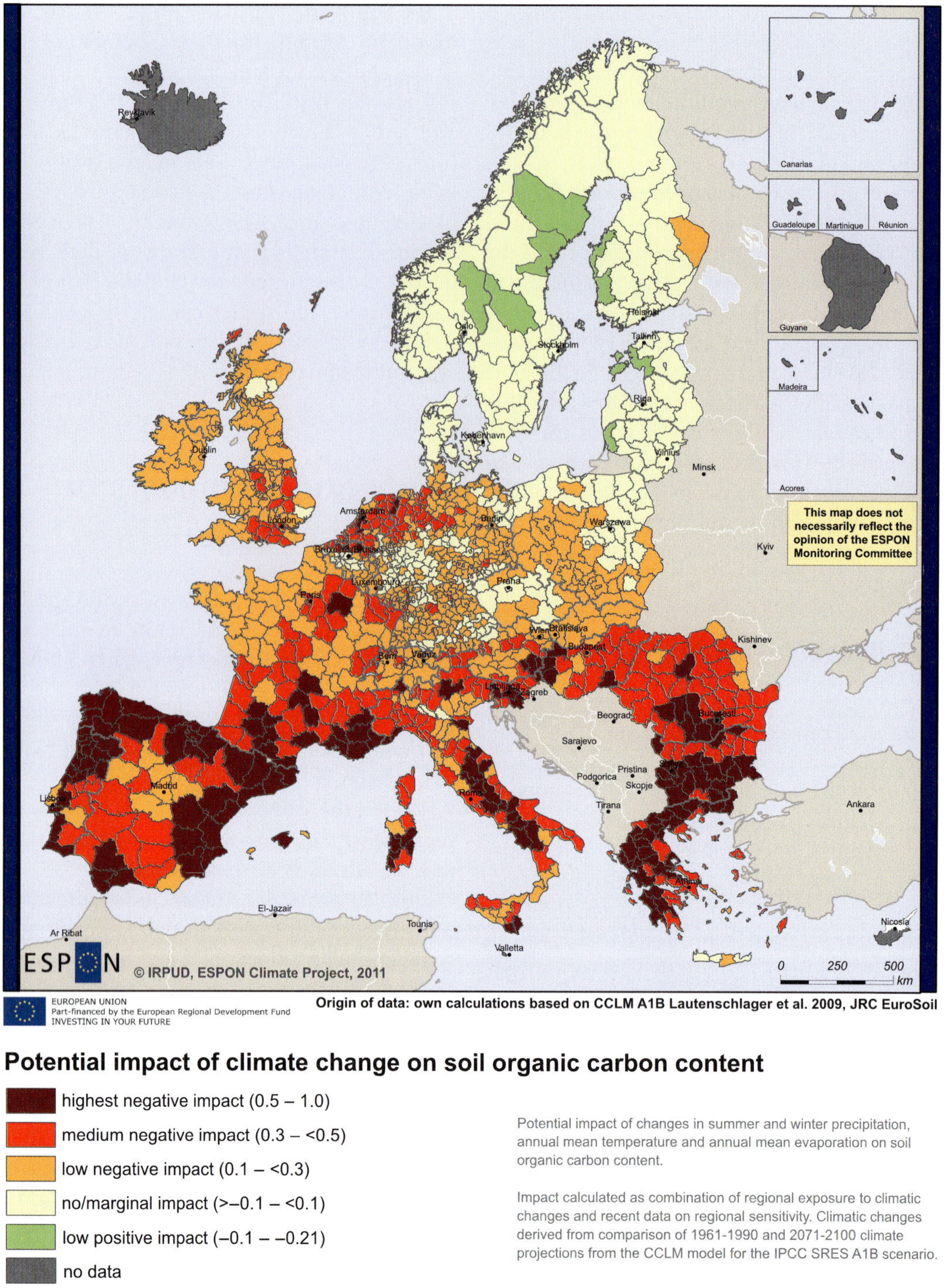

Potential impact of climate change on soil organic carbon content

- highest negative impact (0.5 – 1.0)
- medium negative impact (0.3 – <0.5)
- low negative impact (0.1 – <0.3)
- no/marginal impact (>−0.1 – <0.1)
- low positive impact (−0.1 – −0.21)
- no data

Potential impact of changes in summer and winter precipitation, annual mean temperature and annual mean evaporation on soil organic carbon content.

Impact calculated as combination of regional exposure to climatic changes and recent data on regional sensitivity. Climatic changes derived from comparison of 1961-1990 and 2071-2100 climate projections from the CCLM model for the IPCC SRES A1B scenario.

Figure 5.9 Potential impact of climate change on soil organic carbon content. © ESPON 2013, IRPUD, ESPON Climate Project, 2011.

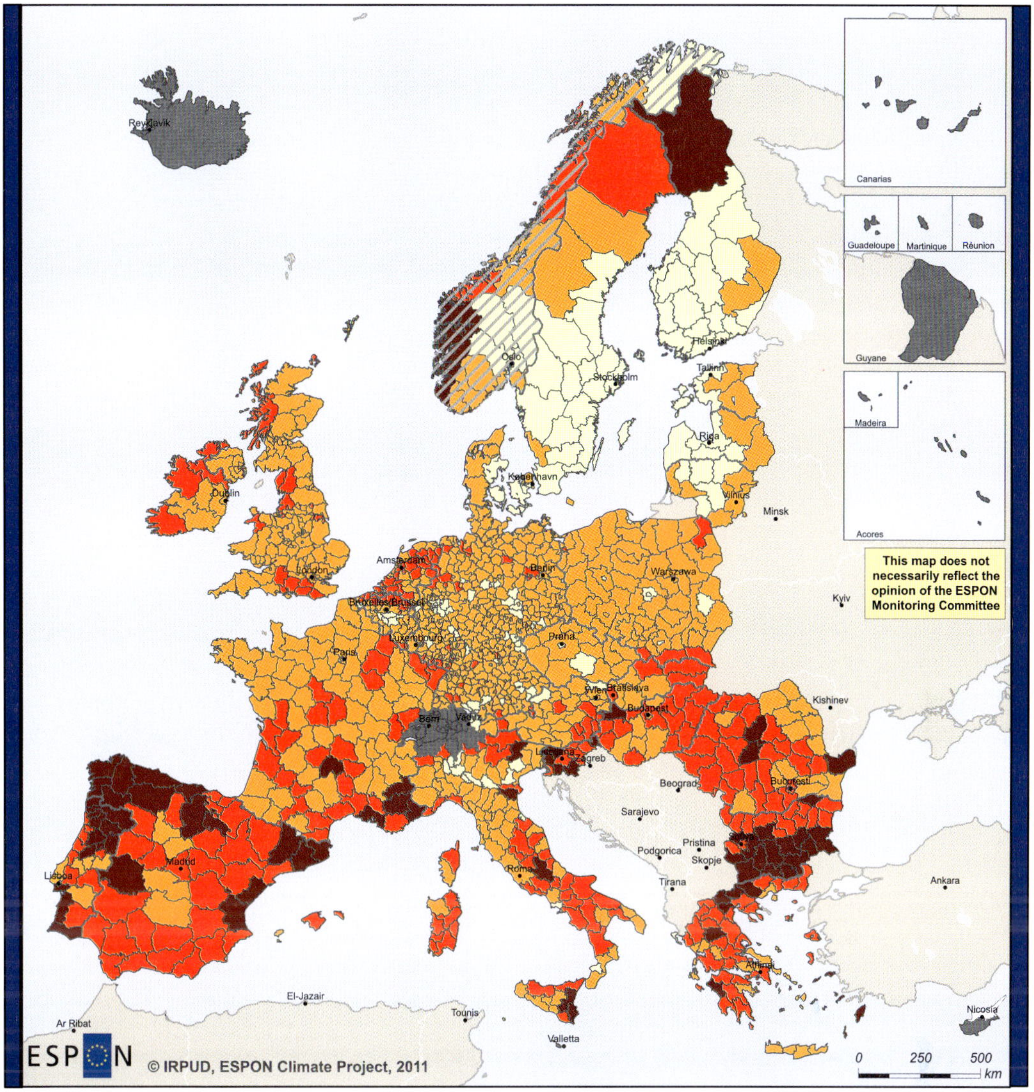

Potential environmental impact of climate change

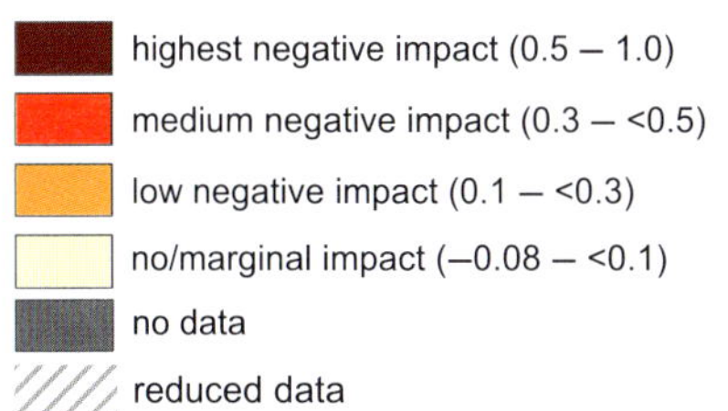

- highest negative impact (0.5 – 1.0)
- medium negative impact (0.3 – <0.5)
- low negative impact (0.1 – <0.3)
- no/marginal impact (−0.08 – <0.1)
- no data
- reduced data

Combined potential impacts of changes in summer and winter precipitation, heavy rainfall days, annual mean temperature, summer days, frost days, snow cover days and annual mean evaporation on soil erosion, soil organic carbon content, NATURA 2000 protected areas and forest fire sensitivity.

Impact calculated as combination of regional exposure to climatic changes and recent data on regional sensitivity. Climatic changes derived from comparison of 1961-1990 and 2071-2100 climate projections from the CCLM model for the IPCC SRES A1B scenario.

Figure 5.10   Potential environmental impact of climate change. © ESPON 2013, IRPUD, ESPON Climate Project, 2011.

particularly relevant social institutions such as hospitals are also dealt with in the adaptive capacity section of this report. Changes in the degree of accessibility of these and other institutions (schools, universities, municipal administrations, etc.) are reflected in the respective indicators on roads and railways.

For future projections of social sensitivities to climate change it would also be necessary to account for demographic changes. This includes regional population growth or loss as well as structural changes, for example, in the age composition of populations. For both types of demographic changes the ESPON Climate project was able to draw on population projections up to the year 2100 developed by ESPON's DEMIFER project, which deals precisely with such issues.

5.4.1   *Populations particularly sensitive to heat*

Humans are in general sensitive to high temperatures – especially when occurring over an extended period of time. Prolonged periods of high temperatures during the summer are a particular health issue for urban.

Of special importance in terms of heat sensitivity are senior citizens, who are usually defined as persons above the age of 65 years. This population group is expected to grow by an additional 58 million persons until the year 2050 (DG Regio, 2009) and is characterised, among others, by frequent and severe health problems and with age increasing mortality risk – even without any climate change.

Senior citizens living in urban areas are especially sensitive to heat. In rural areas hot temperatures are usually mediated by wind and vegetation, but in urban areas higher densities of buildings and infrastructure create a hotter and drier micro-climate. Owing to heat storage at night, there is also less cooling down of temperatures in urban areas. Thus, by default urban environments are more problematic for human health than rural environments.

Climate change will most likely exacerbate these urban conditions. For example, an increasing number of days with maximum temperatures above 25 °C are strenuous conditions for certain population groups (see senior citizen below). In densely built up areas with a high proportion of sealed surfaces this problem can become even more severe in so called heat islands.

There are many studies on the relationship between heat phenomena and human health. They show, for example, that urban populations of different cities have different temperature thresholds above which the daily mortality rate rises significantly, that is, residents of Athens have different heat sensitivities than residents of Stockholm (e.g. Baccini *et al.*, 2008, Kovats *et al.*, 2006). During extended or repeated heat waves the mortality rate increases even further, as studies on the 2003 heat waves in Europe proved (Robine *et al.*, 2007). Under climate change, with generally warmer temperatures the number of heat-related deaths is projected to rise by up to 20% until the year 2050 in Germany and more than fourfold by 2080 in Portugal (Koppe *et al.*, 2003 and Dessai, 2003, respectively). The PESETA study projected almost 86 000 net extra deaths per year in 2071–2100 in the EU-27 member states compared with the 1961–1990 EU-25 average – albeit under a severe climate change scenario (EEA, 2008).

In particular, the relationship between age and high temperatures is well researched. Older people are often not able to adequately regulate their body temperature during hot summer days. In addition, the human cardiovascular system, which is typically already weakened in elderly people, is under special stress during hot days. For these reasons senior citizens are more prone to heat-related illnesses and death (Koppe *et al.*, 2004; Havenith, 2005). Increasing numbers of senior citizens (see above) will therefore increase the proportion of population at risk (Confalonieri *et al.*, 2007).

The indicator for measuring the above described sensitivities is the absolute number and the share of urban residents older than 65 years living in high density urban areas. For identifying such urban heat islands, CORINE Land Cover data were used. The land cover type 'urban continuous' represents mostly inner-city areas with almost complete surface sealing. This analysis was conducted on a 100×100 m grid level. In order to rule out individual cells or small clusters of urban continuous cells, which might for example be located in small settlements where urban heat island effects clearly cannot develop, only those cells were retained in further calculations whose size was not smaller than the average of urban continuous clusters (about 90 ha) and that possessed a population density above the European average of urban settlements (3000 inhabitants per square kilometre). For this determination the CORINE Land Cover data and the disaggregated (also 100×100 m grid cells) population data developed by Gallego (Gallego, 2010 and unpublished 2011 update based on Gallego, 2009) were used. These data also allowed the absolute number of inhabitants in each urban heat island to be calculated. In a final step, applying Eurostat data on age composition, the absolute number of inhabitants above 65 years living in the identified urban heat islands and their share in relation to the total population of each NUTS 3 region was calculated, normalised and then combined. This sensitivity was then related to changes in the number of summer days (above 25 °C) in order to determine potential impacts.

5.4.2   *Potential impact of changes in summer heat on population*

As pointed out above, people older than 65 years living in urban heat islands are especially sensitive to increases in hot summer days. According to ESPON Climate's projections these impacts are concentrated in Southern Europe, where the local climate will generally be getting hotter by 2071–2100. In addition, most Mediterranean cities are much more compact than their Northern counterparts, thus accounting for more urban heat islands. The most severely impacted country, as can be seen in the map in Figure 5.11, is Spain and in particular its urban centres.

5.4.3   *Populations sensitive to sea-level rise adjusted costal storm surges*

Climate change is projected to lead to rising mean sea levels (see 'Settlements prone to coastal storm surges', Section 5.2.3, for more detail). This would – without further adaptive measures – result in low-lying coastal areas being permanently flooded or temporarily flooded during coastal storm surges. Of course current, new or improved dikes and other storm defences would mitigate or even prevent such flooding from occurring. However, according to ESPON Climate's overall methodology, such measures are covered in the adaptive capacity component of the project and will not be incorporated here.

Research on climate-related risks for coastal populations has been carried out for centuries in efforts to better determine risks and protect human lives. The outlook of drastic climate changes over the next decades has further intensified coastal research. The DINAS-COAST research consortium and the PESETA project have attempted to model the likely effects of sea-level rise and storm flooding – however, based on the A2 IPCC climate change scenario (not used by the ESPON Climate project). According to PESETA, by the year 2080 up to 1.3 million people on Europe's coasts would experience coastal flooding each year if no adaptation measures are implemented (EEA, 2008).

To calculate the respective impact indicator, the same methodology was applied as for settlements sensitive to coastal storm surges. The areas projected to be inundated by a 100 year return event of costal storm surge flooding and the corresponding changes in inundation heights were determined. Then, using the population disaggregation (Gallego, 2010 and unpublished 2011 update) the number of people living in these areas were determined

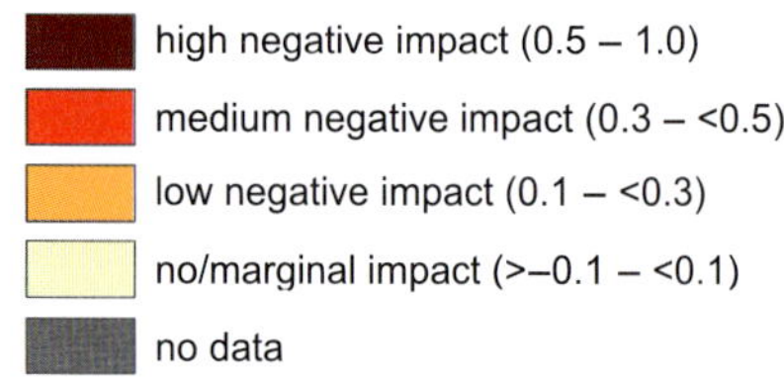

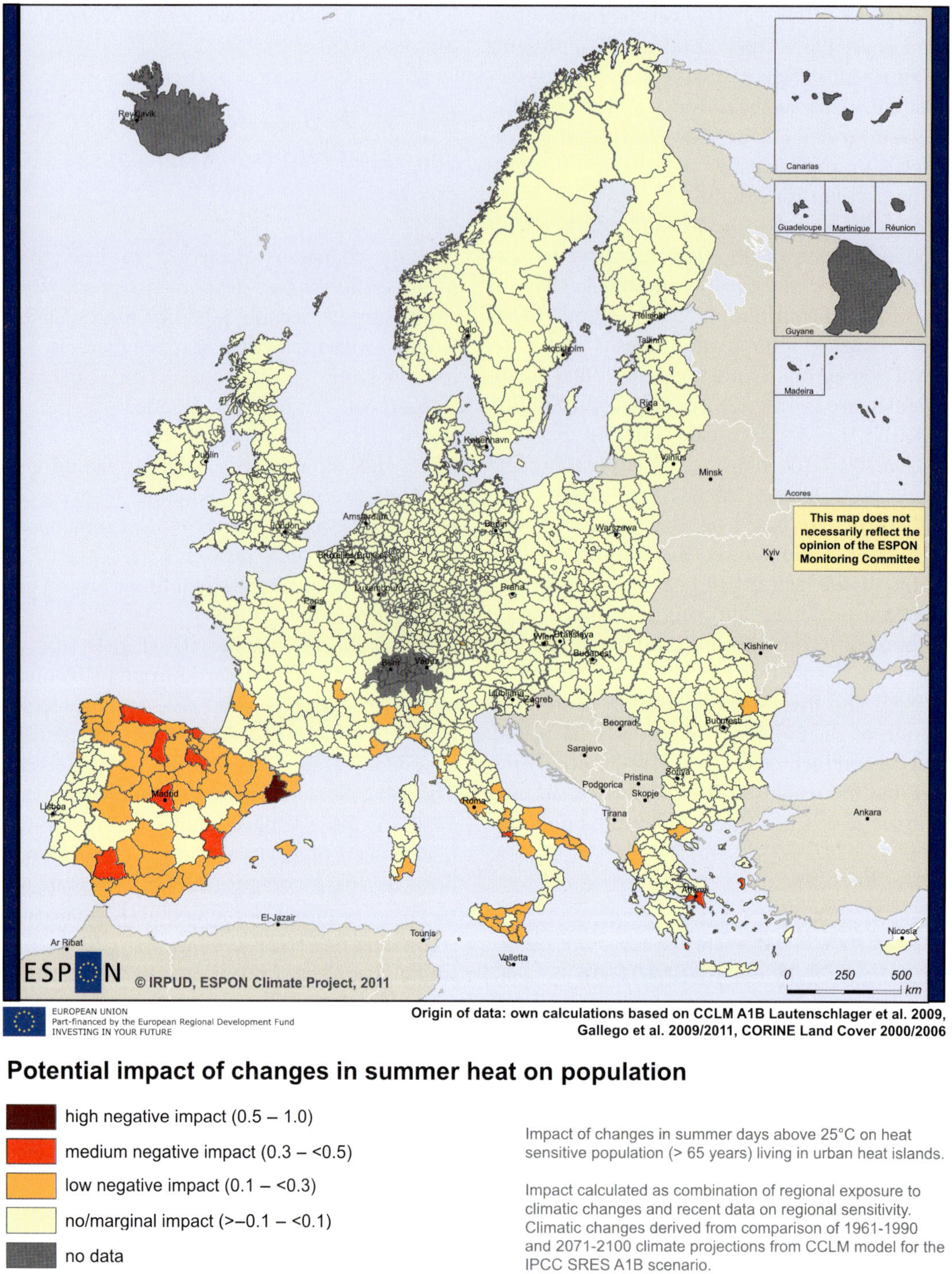

Potential impact of changes in summer heat on population

Figure 5.11   Potential impacts of changes in summer heat on population. © ESPON 2013, IRPUD, ESPON Climate Project, 2011.

and multiplied by the respective inundation height changes. Afterwards these values were also related to the total population of each NUTS 3 region. Finally, the absolute and relative indicators were combined and normalised across Europe.

5.4.4 *Potential impact of sea-level rise on population*

According to the map in Figure 5.12, most coastal regions of Europe are projected to be only marginally affected by sea level rise. However, considerable parts of the British coast and in particular the Belgium, Dutch and German coastal lowlands along the North Sea, with their high concentrations of population, are projected to be negatively affected.

5.4.5 *Populations sensitive to river flooding*

People living in river valleys are one of the most sensitive population groups in regard to climate change, because climatically induced hydrological changes taking place in entire river basins are accumulated and channeled through rivers. The most relevant studies on river flooding in Europe have already been discussed in the indicator description related to settlements prone to river flooding. The LISFLOOD and PESETA projects are the most advanced research projects to date that model river flooding changes in Europe up to the year 2100. They not only project the likelihood of occurrence of flood events but also model the exact geographic areas potentially affected by river water inundation. Lately LISFLOOD has been applied to climate data using the IPCC A1B forcing scenario. In particular the input of several climate models have been used, among them the CCLM model, which is also the basis of the ESPON Climate project.

The methodology for this indicator resembles the procedures used for the respective indicators for settlements and infrastructure. To determine the number of people living in flood prone river valleys, the spatially disaggregated population data

(Gallego, 2010 and unpublished 2011 update) based on the CORINE database was overlaid with the changes in inundation heights of a 100-year flood event projected by the LISFLOOD model. The resulting absolute indicator was then also related to the total population of each NUTS 3 region, before the relative and absolute impact indicators were combined and normalised across Europe.

5.4.6 *Potential impact of river floods on population*

Increased river flooding, as already seen with regard to settlements, will severely affect regions in Northern Italy, where climate changes in the Alps converge. Since these are also densely populated regions the negative impacts are projected to be very high. In contrast, some population centres in Spain and populous regions in central, South-Eastern and Eastern Europe are expected to benefit from less river flooding (Figure 5.13).

5.4.7 *Populations sensitive to flash floods*

Climate change affects the frequency of occurrence of heavy rainfall and the amount of water that precipitates in a short period of time.[3] The special dangers of flash floods for humans are the short warning period and great water speeds that are typical. It is therefore not uncommon that persons are washed off their feet and drown – even in relatively small, local flash flood events.

The most relevant studies regarding flash floods have been discussed in the respective settlement indicator section. It should be emphasized, however, that flash floods are a very locally specific phenomenon. It is thus an issue that was given a more thorough investigation in the case studies of the ESPON Climate project.

The methodology for this indicator again runs parallel to the equivalent indicator for settlements

---

[3]Climate change also affects another factor responsible for flash floods, namely the vegetation cover of flash flood prone areas.

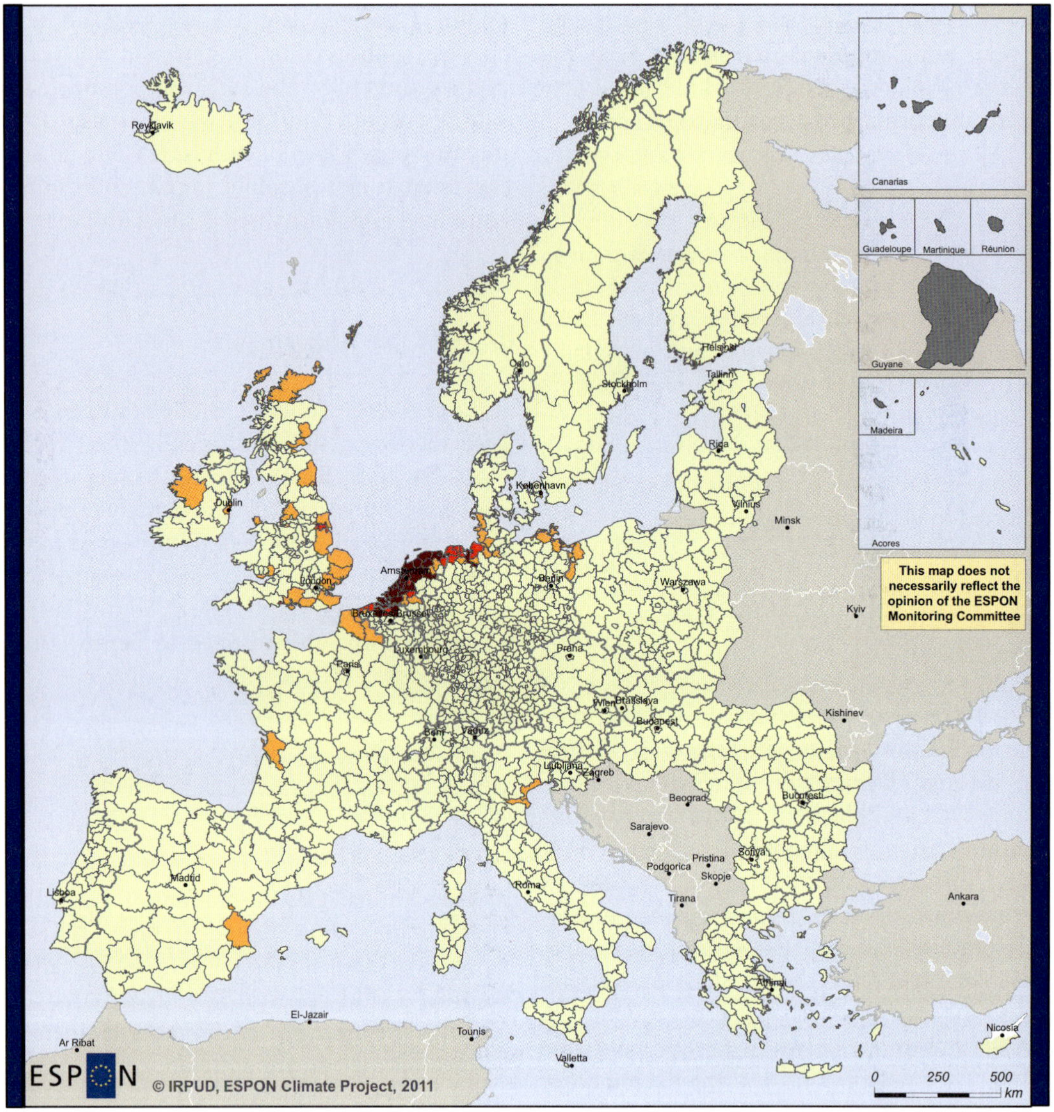

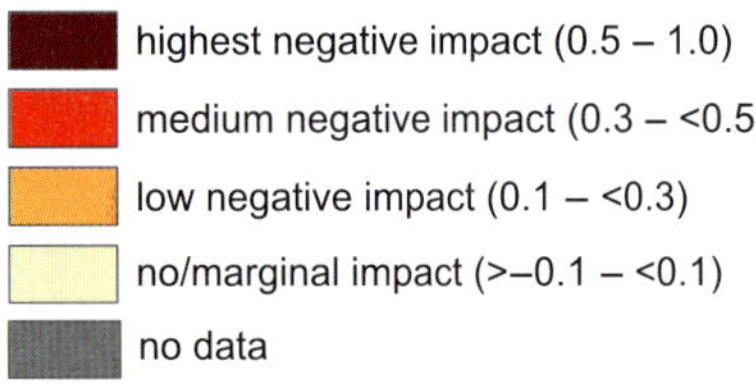

Potential impact of sea level rise on population

Figure 5.12   Potential impact of sea-level rise on population. © ESPON 2013, IRPUD, ESPON Climate Project, 2011.

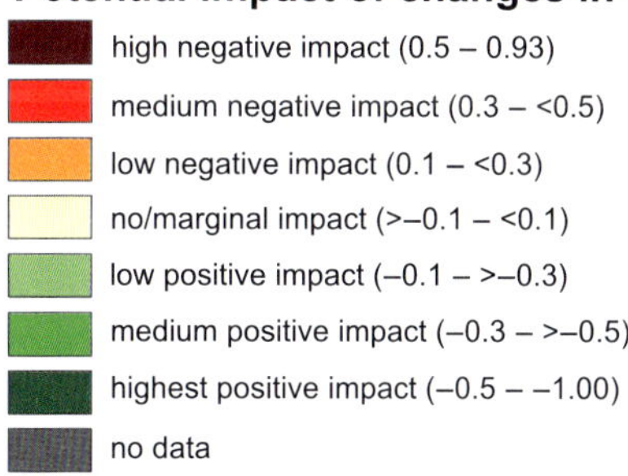

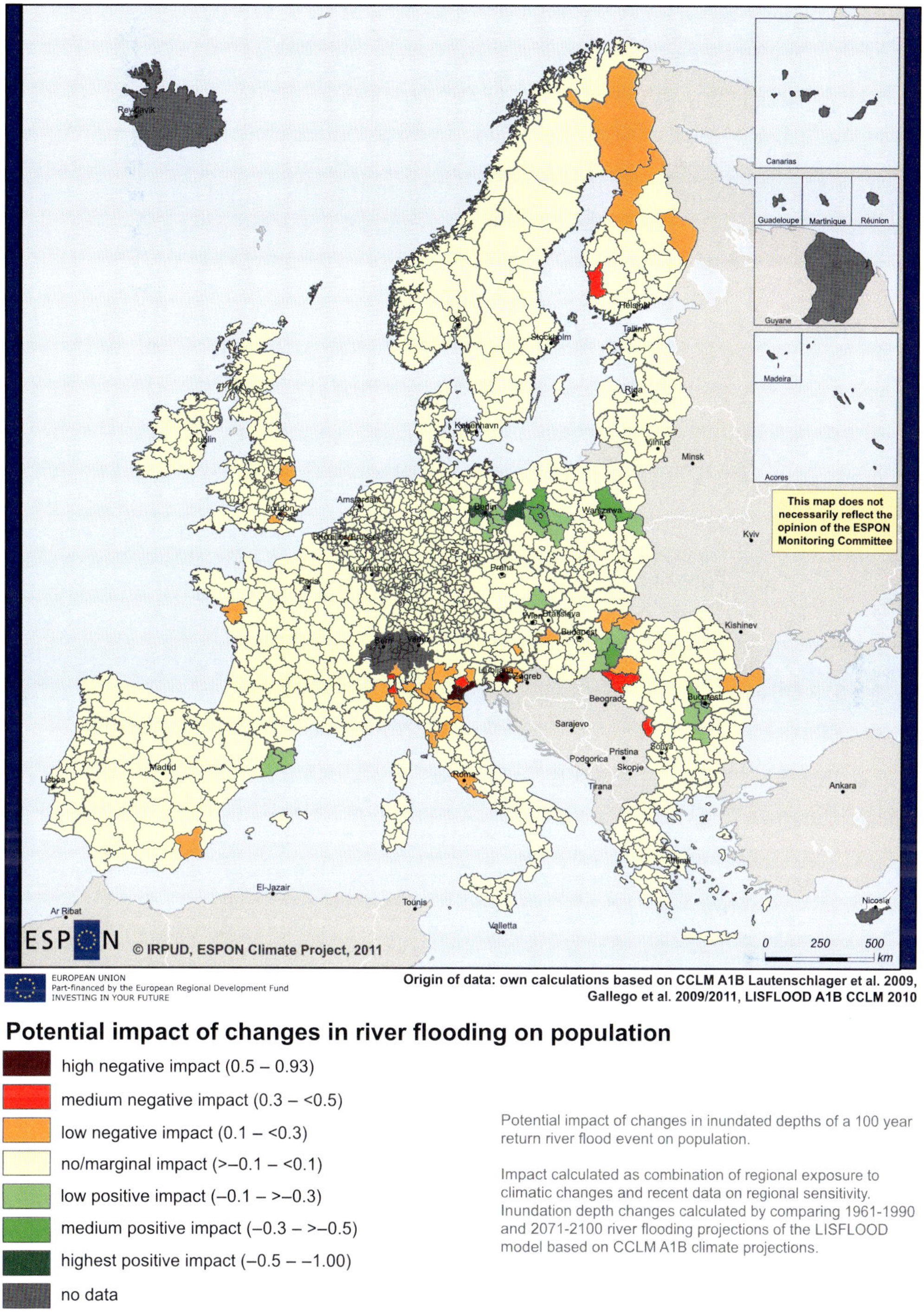

Potential impact of changes in river flooding on population

- high negative impact (0.5 – 0.93)
- medium negative impact (0.3 – <0.5)
- low negative impact (0.1 – <0.3)
- no/marginal impact (>–0.1 – <0.1)
- low positive impact (–0.1 – >–0.3)
- medium positive impact (–0.3 – >–0.5)
- highest positive impact (–0.5 – –1.00)
- no data

Potential impact of changes in inundated depths of a 100 year return river flood event on population.

Impact calculated as combination of regional exposure to climatic changes and recent data on regional sensitivity. Inundation depth changes calculated by comparing 1961-1990 and 2071-2100 river flooding projections of the LISFLOOD model based on CCLM A1B climate projections.

Figure 5.13　Potential impact of changes in river flooding on population. © ESPON 2013, IRPUD, ESPON Climate Project, 2011.

and infrastructures. First a flash flood potential was calculated for each NUTS 3 region on the basis of topographical, soil and land cover data. This was then related to the absolute number of inhabitants and the population density of each NUTS 3 region, before combining these absolute and relative sensitivity indicators with changes in the number of heavy rainfall days, to yield an overall impact indicator for potential flash flood changes.

5.4.8 Potential impact of flash floods on population

The impact patterns regarding flash flood and population reflect the generally decreasing number of days with heavy rainfall in Southern Europe and its major population concentrations. In Northern Europe, populations will be especially affected in Ireland and the Western United Kingdom and parts of Western France (Figure 5.14). The highest negative impacts of flash floods are projected for Norway with its steep slopes and especially flash-flood prone soils.

5.4.9 Combined social impact of climate change

In combination, the impacts of changes in river flooding, sea-level rise adjusted coastal flooding and flash floods on population yield an already familiar picture (Figure 5.15). Coastal regions in North-Western Europe in general and coastal cities in particular are projected to be highly impacted. Inhabitants in most inland regions, except in Spain, would only be marginally affected, some even enjoying positive impacts due to generally drier climates resulting in decreasing river flood risks.

5.5 Cultural impact of climate change

The terms culture and cultural heritage refer to a wide range of tangible artefacts and intangible attributes. Tangible artefacts include – among others – buildings, other built structures (e.g.

bridges of historic value), monuments, works of art and books, but also special landscapes that have been shaped by human use over centuries and have thus acquired certain cultural or historical qualities. Intangible aspects of culture encompass music, folklore, language, literature but also shared attitudes, values and practices of a group, organisation or community.

In principle, all of these cultural assets and intangibles can be sensitive to climate change. For example, monuments, churches and castles are sensitive to all types of flooding, but also to precipitation and temperature changes. The same applies even more to landscapes and open archaeological sites. Similarly, one can investigate the sensitivity and geographical locations or clusters of public and private institutions 'delivering' culture, for example. art workshops, galleries, museums, theatres, operas, musical halls and all the people working in the production, maintenance and dissemination of various forms of art and culture. Thus, one may also be able to assess the climate change sensitivity of the 'cultural economy'.

However, for conceptual as well as practical reasons only a small subset of the above aspects could be included in the project's sensitivity analysis.

Firstly, on a conceptual level all intangible cultural aspects of a region were defined as part of that region's adaptive capacity: norms and values are not primarily *affected by* climate change, but are important cultural assets that influence how well or quickly a society can *adjust to* climate change. Thus these aspects are taken up in the adaptive capacity module (see Chapters 4 and 7).

Secondly, on a more practical level, for many of the above cultural assets there are no or only insufficient databases for the type of analysis carried out by the ESPON Climate project. One has to keep in mind that several assessments are performed on a very fine-grained geographical level (e.g. 100×100 m raster cells for flooding events) that require exact geographical locations of institutions and places of work (and respective employment data). Thus, it is not sufficient to know that 'the town of Firenze is a cultural cluster', but one needs exact positions of the respective public and private cultural institutions or firms. Furthermore, in order to later link certain climate change variables to particular cultural assets

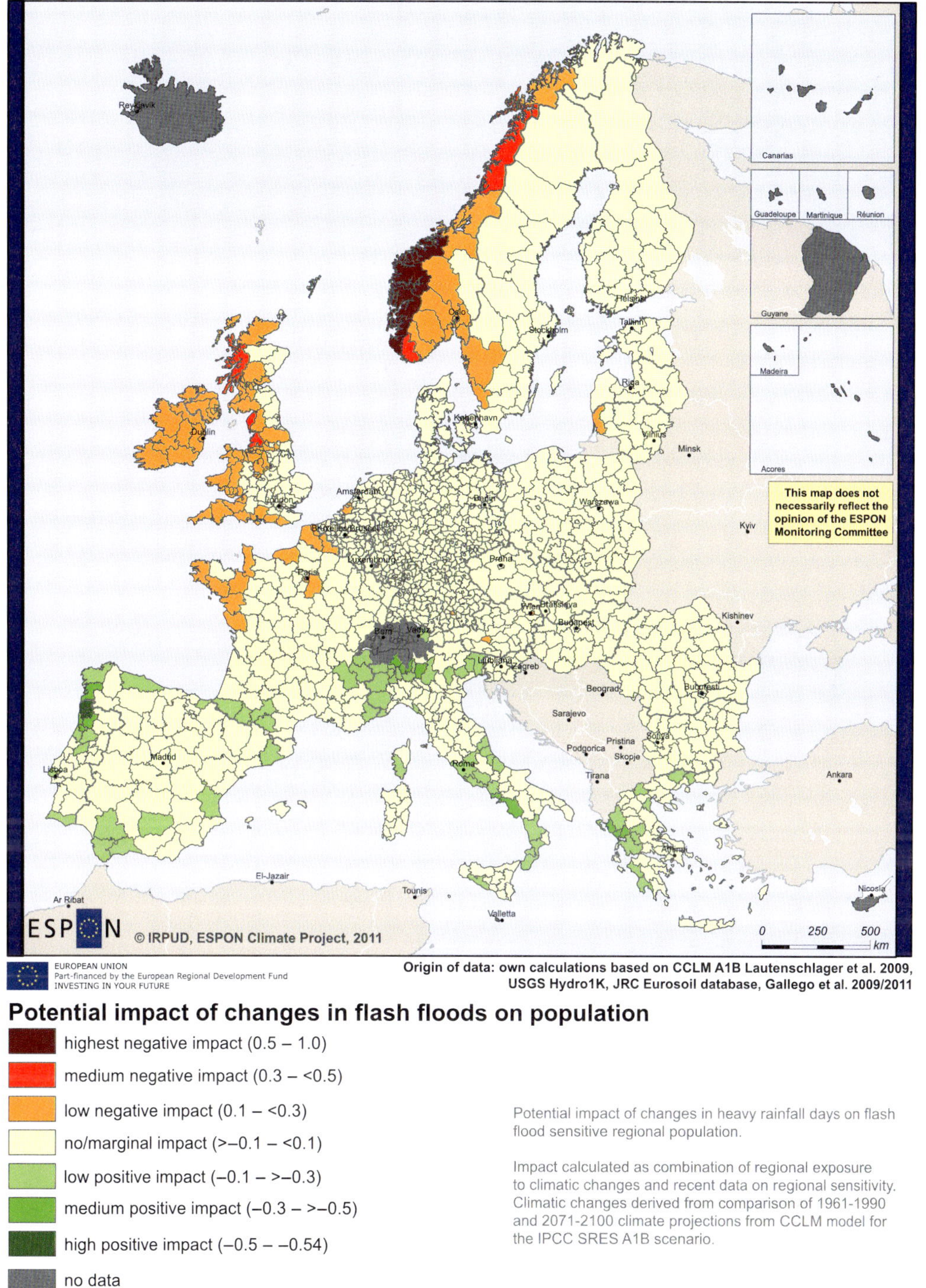

EUROPEAN UNION
Part-financed by the European Regional Development Fund
INVESTING IN YOUR FUTURE

Origin of data: own calculations based on CCLM A1B Lautenschlager et al. 2009,
USGS Hydro1K, JRC Eurosoil database, Gallego et al. 2009/2011

Potential impact of changes in flash floods on population

- highest negative impact (0.5 – 1.0)
- medium negative impact (0.3 – <0.5)
- low negative impact (0.1 – <0.3)
- no/marginal impact (>–0.1 – <0.1)
- low positive impact (–0.1 – >–0.3)
- medium positive impact (–0.3 – >–0.5)
- high positive impact (–0.5 – –0.54)
- no data

Potential impact of changes in heavy rainfall days on flash
flood sensitive regional population.

Impact calculated as combination of regional exposure
to climatic changes and recent data on regional sensitivity.
Climatic changes derived from comparison of 1961-1990
and 2071-2100 climate projections from CCLM model for
the IPCC SRES A1B scenario.

Figure 5.14   Potential impact of changes in flash floods on population. © ESPON 2013, IRPUD, ESPON Climate Project, 2011.

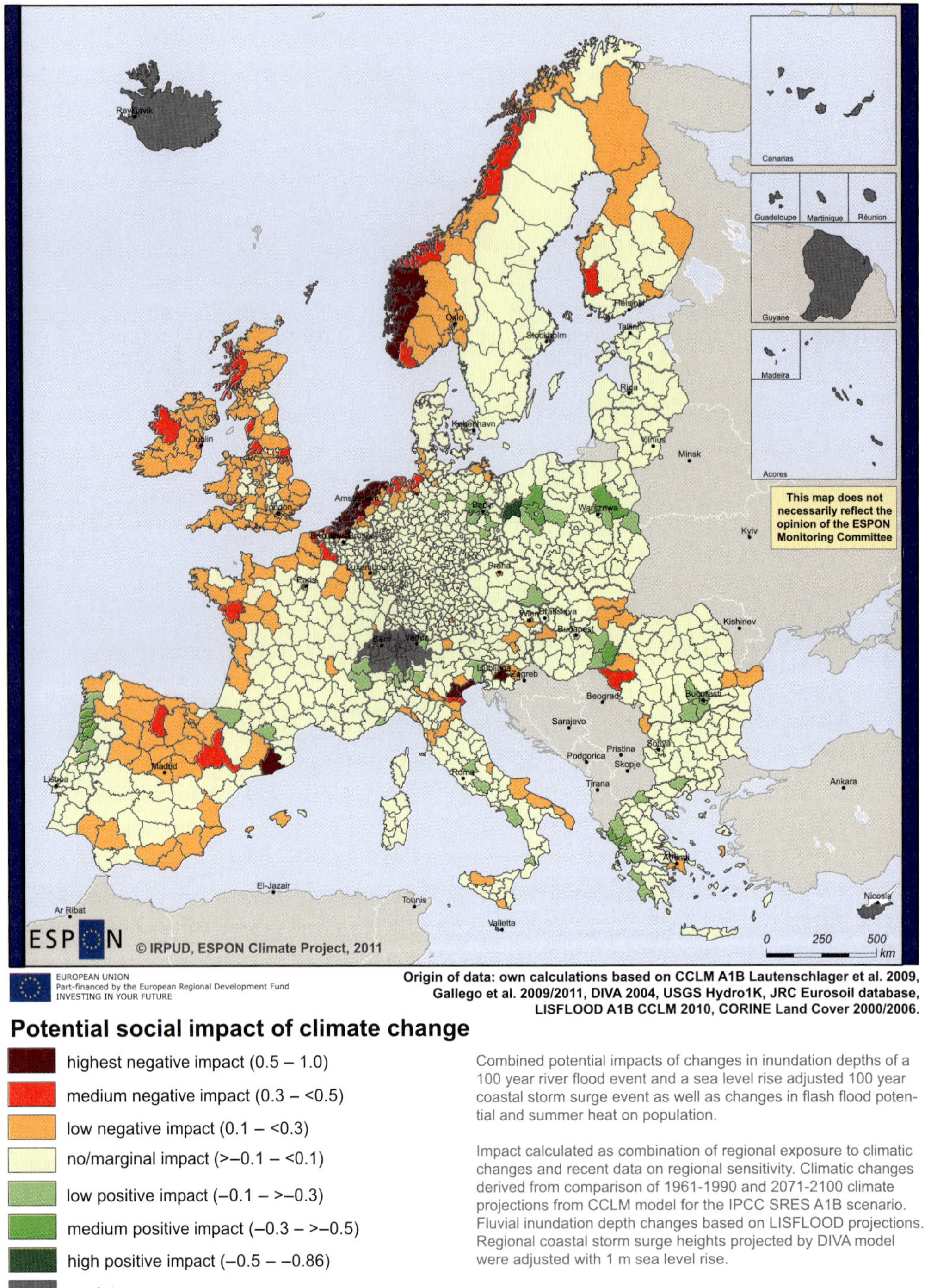

Potential social impact of climate change

Combined potential impacts of changes in inundation depths of a 100 year river flood event and a sea level rise adjusted 100 year coastal storm surge event as well as changes in flash flood potential and summer heat on population.

Impact calculated as combination of regional exposure to climatic changes and recent data on regional sensitivity. Climatic changes derived from comparison of 1961-1990 and 2071-2100 climate projections from CCLM model for the IPCC SRES A1B scenario. Fluvial inundation depth changes based on LISFLOOD projections. Regional coastal storm surge heights projected by DIVA model were adjusted with 1 m sea level rise.

Figure 5.15   Potential social impact of climate change. © ESPON 2013, IRPUD, ESPON Climate Project, 2011.

and thus determine the cultural impact, one also needs information on the quality of the various cultural assets. Such qualities would be, for example for a historic building, the age, architectural structure, outer building materials, underground sealing and underground materials. Without such basic data as well as coherent definitions and datasets on a pan-European level, one cannot determine sensitivities to climatic changes of such cultural assets.

Thirdly, at a more rudimentary level, for many cultural assets referred to above there is not even complete and coherent European or even national *listings*. Plus, even where such national listings exist, they are often not compatible with each other and require crude 'rule of thumb' adjustments, as the ESPON Cultural Heritage project found out for national listings of historic monuments, which reflected more the differences in national preservation legislation and bureaucracy than the actual existence and distribution of historic monuments. However, attempts to collect data from various sources and create integrated directories quickly run into conceptual and semantic problems (for example, there are many different types and corresponding names for 'castles' – not just in English but in every other European language).

In the end only two types of cultural assets that are sensitive to two types of climatic stimuli remained, which could be adequately assessed by the ESPON Climate project. A detailed database including geographical positions of all World Heritage Sites has been compiled and maintained by the UNECO's World Heritage Commission. This list is based on a coherent and systematic assessment that is applied to all submitted nominations worldwide. In addition, a list of all museums in Europe was compiled by the ESPON Climate project team on the basis of national telephone directories and Google Earth for exact geographical positioning of postal addresses. Only institutions that bear the term 'museum' (in the respective national or regional language) in their name were included. Both museums and World Heritage sites were only analysed with regard to river flooding and coastal flooding, because the available data on each site/building did not allow sensitivities to be analysed with regard to other climate change

aspects, for example. temperature or precipitation impacts on building materials, etc.

5.5.1 *UNESCO Cultural World Heritage Sites sensitive to river flooding*

Cultural World Heritage sites are monuments, buildings or (parts of) cities that are listed by UNESCO as being of global cultural importance. The sites have to either (a) represent a masterpiece of human creativity, (b) give testimony to a past or present cultural tradition or civilization, (c) exemplify a type of building, architectural ensemble or settlement that illustrates an important stage in human history or (d) be directly associated with traditions, ideas, artistic and literary works of global importance. Within Europe (including Turkey and European parts of Russia) there are currently 354 sites that have been officially listed as Cultural World Heritage Sites. (Note that the official nomination criteria for these sites have undergone several changes and are generally now more condensed: compare for example WHC (1992) and (WHC 2012)). Disruption, damage or destruction of these sites has to be considered a significant loss to human culture and history.

Cultural World Heritage Sites are especially sensitive to climate change. For example, historic buildings are often built of organic building materials that are sensitive to temperature and moisture changes. Furthermore, walls and floors are often directly grounded on the underlying soil and react to soil moisture changes. In both cases cosmetic and even structural damage can occur to the sites. More intense or frequent river or coastal flooding brought about by climate change can damage archaeological sites and historic buildings through water erosion. Also, building materials of monuments and historic buildings are often not designed to withstand prolonged immersion in water. After a flood event there can also be damage caused by micro-organisms such as mould.

In general Sabbioni *et al.* (2006), summarising the results of the Noah's Ark project, evaluated the impacts of climate change on historic building materials and the bio-deterioration of built heritage. Accordingly, sensitivity to climate change induced temperature and moisture changes depends to a large degree on the specific building materials and the soil types upon which the historic buildings were built. With regard to World Heritage Sites, there are very few publications on the effects of climate change, both globally and in Europe. Studies on individual sites or sites in a particular country exist as well as UNESCO reactions, for example, to severe river floods in Europe (UNESCO, 2002). The most comprehensive report on climate change effects and World Heritage Sites was published by the World Heritage Committee itself (WHC, 2006). However, the report only provided a type of conceptual overview that served as the basis for recommendations on how the WHC and member states should react to expected climate change effects. The empirical basis of the report consisted of a simple survey among the states that signed the World Heritage Convention. According to the survey, 125 World Heritage sites were mentioned as threatened by climate change. Of these threatened sites, 42 are Cultural World Heritage Sites such as archaeological ruins, churches, temples and historic city centres, and so on. The most relevant climatic threats identified for these sites were (in order of importance): (i) hurricanes, storms and lightening, (ii) sea-level rise, (iii) wind and water driven erosion, (iv) flooding, (v) rainfall increase, (vi) drought, (vii) desertification and (viii) temperature rise (WHC, 2006 p. 33). Unfortunately, the results were not disaggregated by continent or country. More detailed and pan-European data can finally be expected from an EU research project called 'Climate for Culture' that investigates the impacts of climate change on World Heritage Sites in Europe and North Africa.

Temperature and moisture sensitivity of World Heritage sites is difficult to determine without detailed data on the specific building materials and soil conditions. However, it can be more easily assessed which sites are sensitive to climate change due to their location in areas likely to be affected by river flooding. The specific calculation method runs parallel to the respective indicator for sensitive settlements. Geographical data for each World Heritage Site were obtained from the WHC and buffered according to the type of heritage site (building, historic city centre, historic park, landscape). These areas were then overlaid with the map containing the changes in inundation heights due to changes in river flooding and the respective values multiplied by each other. This impact indicator was then related to the total area of World Heritage Sites in each NUTS 3 region in order to also have a relative indicator. The absolute and relative impact indicators were finally combined and normalised across Europe.

5.5.2 *UNESCO Cultural World Heritage Sites sensitive to coastal flooding*

As discussed above, the only relevant literature on the topic of World Heritage Sites and their relation to climate change is only very general and typically analyses individual sites or sites located in one specific country.

To measure the sensitivity of World Heritage Sites in coastal areas due to sea-level rise, the same procedures as for settlements were applied. The projected inundation changes (based on DIVA model results of storm surge heights of a 100 year return event, which were then adjusted by 1 m sea-level rise) were overlaid with a map of all (buffered) World Heritage Sites. Potentially affected WHC site areas were aggregated to the NUTS 3 level and also related to all WHC site areas in each region. Finally, the absolute and relative indicators were combined to an overall impact indicator, which was normalised across Europe.

5.5.3 *Potential impact of climate change on Cultural World Heritage Sites*

The map in Figure 5.16 shows the combined potential impacts of river flooding and sea-level rise adjusted coastal flooding on World Heritage Sites.

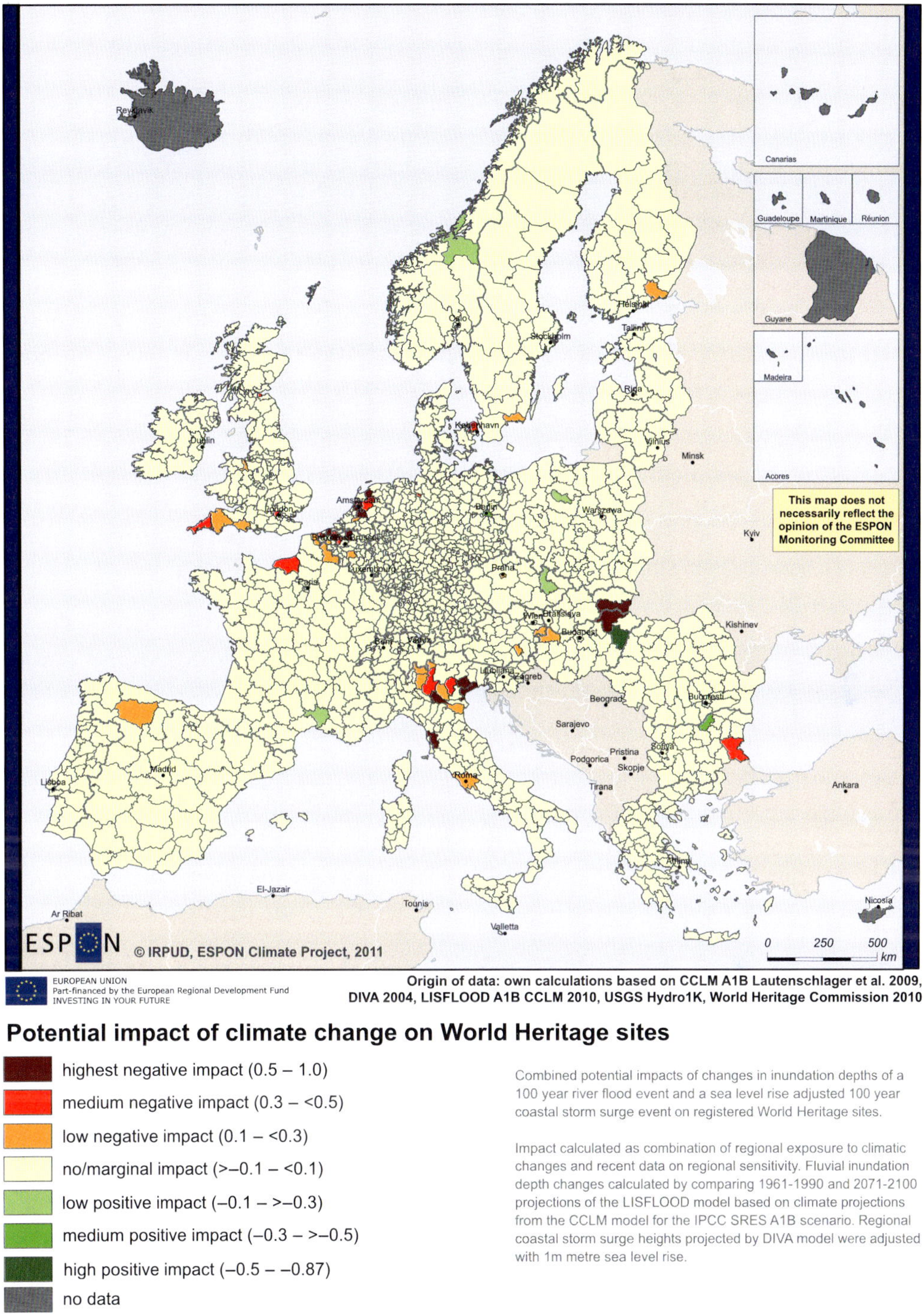

Potential impact of climate change on World Heritage sites

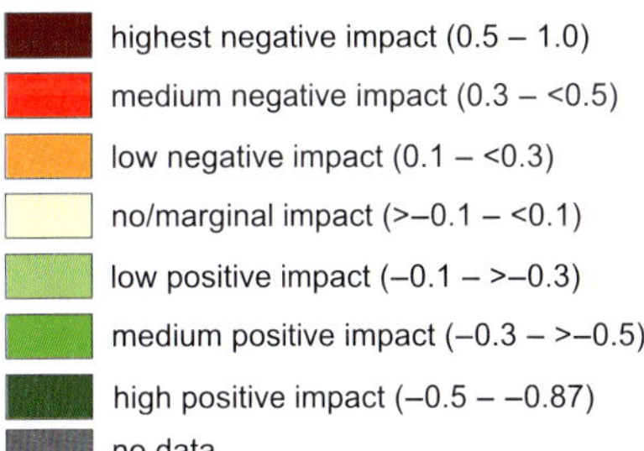

Combined potential impacts of changes in inundation depths of a 100 year river flood event and a sea level rise adjusted 100 year coastal storm surge event on registered World Heritage sites.

Impact calculated as combination of regional exposure to climatic changes and recent data on regional sensitivity. Fluvial inundation depth changes calculated by comparing 1961-1990 and 2071-2100 projections of the LISFLOOD model based on climate projections from the CCLM model for the IPCC SRES A1B scenario. Regional coastal storm surge heights projected by DIVA model were adjusted with 1m metre sea level rise.

Figure 5.16 Potential impact of climate change on World Heritage Sites. © ESPON 2013, IRPUD, ESPON Climate Project, 2011.

Clearly, historic sites registered by the WHC are not found in every European region. Hence it is not surprising to only find scattered impacts across Europe. However, the 'hot spots' in some way coincide with the patterns found concerning settlements and population, because most major urban centres have long historical roots and thus possess historic sites of great value. This applies especially to many Italian regions, but also parts of the Netherlands and border regions between Slovakia and Hungary.

5.5.4 Museums sensitive to river flooding

Museums are cultural institutions that accommodate collections of artefacts of artistic, historical or scientific importance. The mandate of museums is to make their artefacts available for public display. Any damage or destruction of museum buildings inhibits public access and poses a potential threat to its unique contents.

Thus the sensitivity of museums to climate change relates primarily to the physical structure of the buildings in which they are accommodated. One can assume that the level of moisture and temperature changes projected until the year 2100 does not pose any major threats to these buildings. However, as has often happened in the past, large-scale river flooding and coastal storm floods have the potential to cause serious damage to museums, thus incapacitating their operation and endangering their contents.

Research and publications on the topic of past or future weather related impacts on museums are so far rare. Most scientific literature on the topic consists of special reports written in the aftermath of a catastrophic flood event that also affected cultural institutions. For example, there are reports on the great river flood in Florence (1966) in which millions of art works were damaged or destroyed, or on the Elbe river flood of 2002, which affected art collections of state museums. However, pan-European studies on such climate related events and damage, and even less on future projections, do not seem to exist. Nevertheless, the ESPON 2006 project on cultural heritage provides a first basis for further analysis, as it includes an overview, data and geographical analysis of the distribution of museums, galleries, libraries and theatres – both separately and combined. Unsurprisingly the study shows that these cultural institutions are generally concentrated in cities and that there are significant differences between countries: some countries have a tradition of a dense network of museums, while others may have much less museums. It can thus be expected that there are clearly distinguishable patterns of museums' sensitivity to climate change.

To determine the sensitivity of museums to river flooding a similar methodology was used as for settlements and infrastructures. Based on the LISFLOOD model, areas were identified with inundation changes when comparing past and future model results. Afterwards it was determined which museums are located in the relevant $100 \times 100\,\mathrm{m}$ raster cells. For the necessary geographical database of the museums' locations, ESPON Climate initially made use of OpenStreetMap data. The indicator combined the absolute number of sensitive museums and their ratio to the total number of museums in each NUTS 3 region. While this allowed a first and partly satisfactory analysis, it was determined that it would be necessary to have a more comprehensive database of European museums. Unfortunately this did not exist. The project therefore generated its own database on the basis of postal addresses of museums listed in the national yellow pages (or their equivalents). Using the internet-based Google Earth software and 'harvesting' the yellow page data on a $50 \times 50\,\mathrm{km}$ raster grid it was possible to create a pan-European GIS database with exact geographic locations of 20 000 museums in Europe. For the purposes of further analysis, each data point was buffered with 25 m. Afterwards these areas were overlaid with the map containing the changes in inundation heights due to changes in river flooding. The absolute sensitivity indicator was calculated by multiplying the affected museum areas with the respective changes in inundation heights. The resulting value was then also related to the total area of museums in each NUTS 3 region. Combining

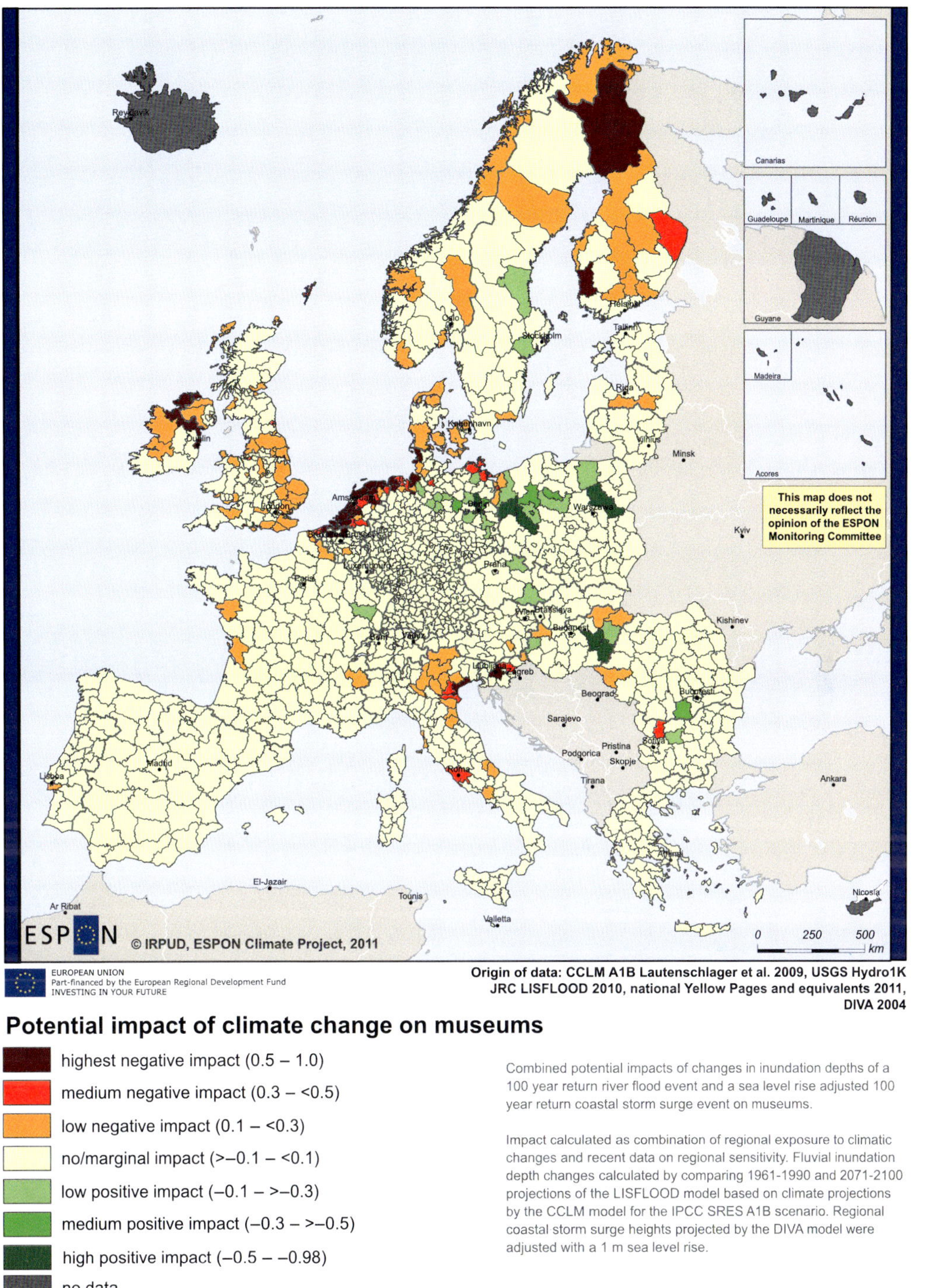

Potential impact of climate change on museums

Figure 5.17   Potential impact of climate change on museums. © ESPON 2013, IRPUD, ESPON Climate Project, 2011.

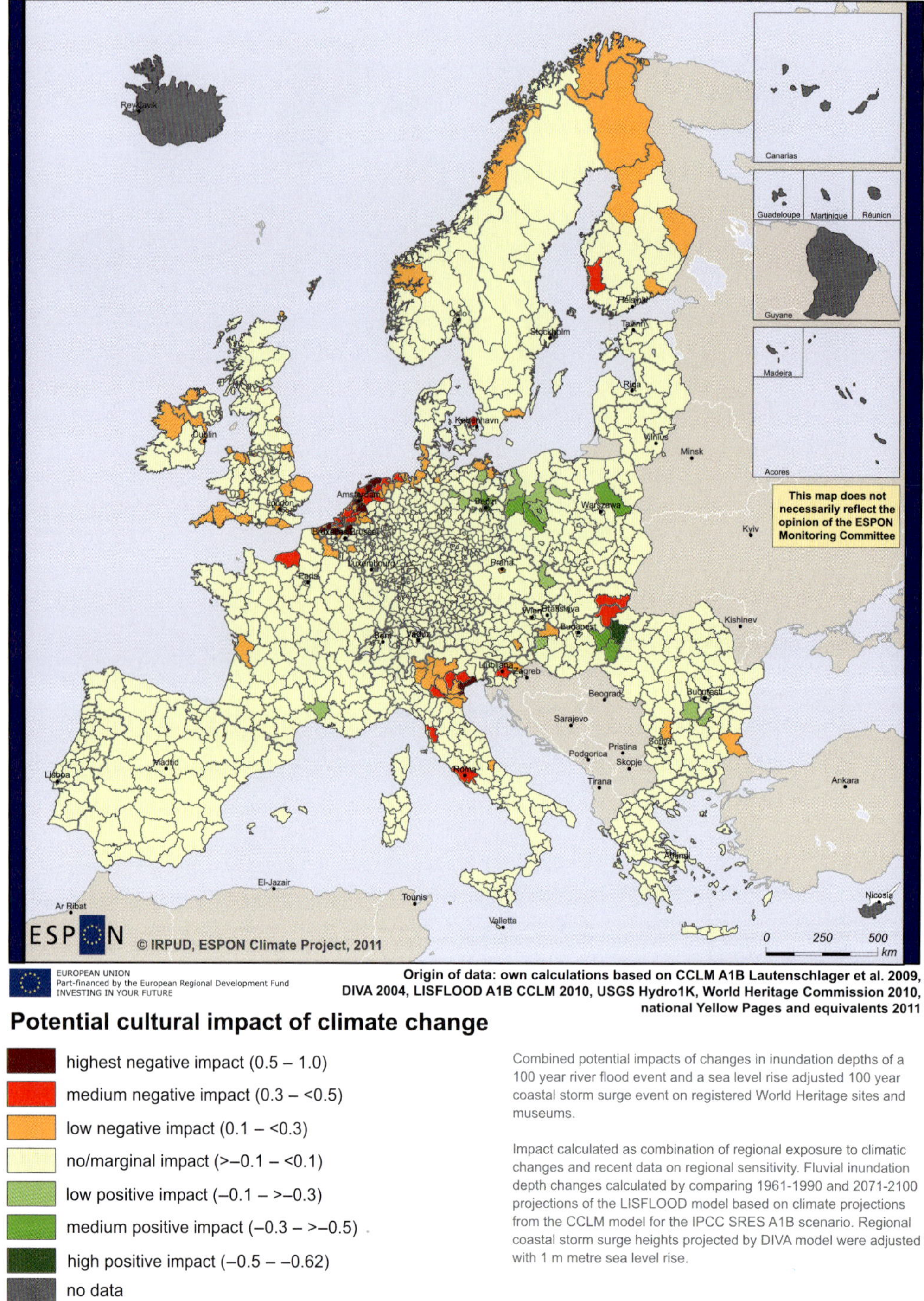

Origin of data: own calculations based on CCLM A1B Lautenschlager et al. 2009, DIVA 2004, LISFLOOD A1B CCLM 2010, USGS Hydro1K, World Heritage Commission 2010, national Yellow Pages and equivalents 2011

Potential cultural impact of climate change

- highest negative impact (0.5 – 1.0)
- medium negative impact (0.3 – <0.5)
- low negative impact (0.1 – <0.3)
- no/marginal impact (>–0.1 – <0.1)
- low positive impact (–0.1 – >–0.3)
- medium positive impact (–0.3 – >–0.5)
- high positive impact (–0.5 – –0.62)
- no data

Combined potential impacts of changes in inundation depths of a 100 year river flood event and a sea level rise adjusted 100 year coastal storm surge event on registered World Heritage sites and museums.

Impact calculated as combination of regional exposure to climatic changes and recent data on regional sensitivity. Fluvial inundation depth changes calculated by comparing 1961-1990 and 2071-2100 projections of the LISFLOOD model based on climate projections from the CCLM model for the IPCC SRES A1B scenario. Regional coastal storm surge heights projected by DIVA model were adjusted with 1 m metre sea level rise.

Figure 5.18   Potential cultural impact of climate change. © ESPON 2013, IRPUD, ESPON Climate Project, 2011.

these absolute and relative impact indicators yielded an overall impact indicator which was normalised across Europe.

5.5.5 Museums sensitive to coastal flooding

As indicated above, literature on the sensitivity of museums to extreme weather events is extremely rare and usually focused on (i) specific museums and/or (ii) specific catastrophes. ESPON Climate's analysis seems to be the first effort at the pan-European level to assess European museums' sensitivities to costal storm surges.

For determining the impact of sea-level rise adjusted coastal storm surges on museums in coastal areas the same procedures as for World Heritage Sites were applied. The inundated areas according to DIVA and adjusted by a 1 m sea-level rise were overlaid with the geographical information on museums' locations. Then the affected museum area in these inundated areas was summed up and subsequently also calculated as a ratio in relation to the total museum area of each NUTS 3 region. Finally, the absolute and relative impact indicators were combined and normalised across Europe.

5.5.6 Potential impact of climate change on museums

The map in Figure 5.17 shows the combined potential impacts of changes in river flooding and coastal flooding on museums in Europe. There are museums in all regions of Europe, so that the impact patterns are more comprehensive and differentiated than for the more sparse World Heritage Sites. The high impact regions – often due to a convergence of river flooding and coastal flooding – are located in Italy, the Belgian, Dutch and German coastal lowlands and Scandinavian and British coastal regions. In contrast, museums in some central and Eastern European regions are projected to benefit from decreasing river flooding.

5.5.7 Combined cultural impact of climate change

The map in Figure 5.18 shows the combined potential impact of changes in river flooding and coastal flooding on museums and cultural World Heritage Sites. The patterns naturally resemble those of the previous two maps and show high impact areas especially in river valley and coastal areas of Italy and along the Belgian and Dutch coast. Other coastal areas in Northern Europe and the mountainous border region between Slovakia and Hungary are also negatively affected. On the other hand, cultural assets in some Central, Eastern and South-Eastern European regions are projected to benefit from decreasing river flooding.

The following Chapter 6 discusses in detail the economic sensitivities and impacts of climate change on Europe's regions. The overall impact, combining all five impact dimensions, is presented in Chapter 9, which also relates impact to adaptive capacity in order to determine the overall vulnerability of Europe's regions to climate change.

References

APFM (Associated Programme on Flood Management) (2007) Guidance on Flash Flood Management: Recent Experiences from Central and Eastern Europe.

Baccini M., Biggeri, A., Accetta, G., *et al.* (2008) Effects of apparent temperature on summer mortality in 15 European cities: results of the PHEWE project. *Epidemiology*, **19** (5).

Bakkenes, M., Alkemade, J.R.M., Ihle, F., *et al.* (2002) Assessing effects of forecasted climate change on the diversity and distribution of European higher plants for 2050. *Global Change Biology*, **8** (4), 390–407.

Bakkenes, M., Eickhout, B. and Alkemade, R. (2006) Impacts of different climate stabilisation scenarios on plants species in Europe. *Global Environmental Change*, **16**, 19–28.

Barnolas, M. and Llasat, M.C. (2007) A flood geodatabase and its climatological implications: the case of Catalonia for the last century. *Natural Hazards and Earth System Sciences*, **7**, 271–281.

Becker, A. and Grunewald, U. (2003) Disaster management – flood risk in Central Europe. *Science*, **300** (5622), 1099–1099.

Berry, P.M., Dawson, T.P., Harrison, P.A., *et al.* (2003) The sensitivity and vulnerability of terrestrial habitats and species in Britain and Ireland to climate change. *Journal for Nature Conservation*, **11**, 15–23.

Bläsche, J., Kreuz, M., Mühlhausen, T., *et al.* Nokkala, M., Athanasatos, S., Michaelides, S., Papadakis, M. and Ludvigsen, J. (2010) Consequences of extreme weather, EWENT Consortium (http://ewent.vtt.fi/Deliverables/D3/EWENT_D34_v12_20120209.pdf) (accessed 15 November 2012).

Brázdil, R., Kotyza, O. and Dobrovolný, P. (2006) July 1432 and August 2002 – two millennial floods in Bohemia? *Hydrological Sciences Journal*, **51** (5), 848–863.

Byrne, K.A., Chojnicki, B., Christensen, T.R., *et al.* (2004) *EU peatlands; Current carbon stocks and trace gas fluxes.* Carbo-Europe report 4.

Camia, A., Amatulli G. and San-Miguel-Ayanz, J. (2008) Past and future trends of forest fire danger in Europe. EUR Technical Report.

Casalegno S., Amatulli G., Bastrup-Birk A., *et al.* (2011) Modelling and mapping the suitability of European forest formations at 1 km resolution. *European Journal of Forest Research*, **130** (6), 971–981.

Cazenave, A. (2006) How fast are the ice sheets melting? *Science*, **314**, 1250–1252.

Christensen, O.B. and Christensen, J.H. (2003) Severe summertime flooding in Europe. *Nature*, **421**, 805–806.

Church, J.A. and White, N.J. (2006) A 20th century acceleration in global sea-level rise. *Geophysical Research Letters*, **33**, L01602.

Confalonieri , U., Menne, B., Akhtar, R., *et al.* (2007) Human health. in (eds M.L. Parry, O.F. Canziani, J.P. Palutikof, P.J. van der Linden and C.E. Hanson). *Climate Change 2007*: *Impacts, Adaptation and Vulnerability*. Contribution of Working Group II to the Fourth Assessment Report of the Intergovernmental Panel on Climate Change, Cambridge University Press, Cambridge, pp. 391–431.

Cyberski, J., Grześ, M., Gutry-Korycka, M., *et al.* (2006) History of floods on the River Vistula. *Hydrological Sciences Journal*, **51** (5), 799–817.

Dankers, R. and Hiederer, R. (2008) Extreme Temperatures and Precipitation in Europe: Analysis of a High-Resolution Climate Change Scenario, Office for Official Publications of the European Communities, Luxembourg.

Dankers, R. and Feyen, L. (2008) Climate change impact on flood hazard in Europe: An assessment based on high resolution climate simulations. *Journal of Geophysical Research*, **113**, D19105.

Dankers, R. and Feyen, L. (2009) Flood hazard in Europe in an ensemble of regional climate scenarios. *Journal of Geophysical Research*, **114**, D16108.

DEFRA (2007) Conserving Biodiversity in a Changing Climate: Guidance on Building Capacity to Adapt, DEFRA, UK.

Demirov E. and Pinardi N. (2002) Simulation of the Mediterranean Sea circulation from 1979 to 1993: Part I. The inter-annual variability. *Journal of Marine Systems*, **33–34**, 23–50.

Dersch, G. and Boehm, K. (1997) *Bodenschutz in Österreich*, (eds W.E.H. Blum, E. Klaghofer, A. Loechl and P. Ruckenbauer), Bundesamt und Forschungszentrum für Landwirtschaft, Austria, pp. 411–432.

Dessai, S. (2003) Heat stress and mortality in Lisbon Part II. An assessment of the potential impacts of climate change. *International Journal of Biometeorology*, **48** (1), 37–44.

DG Regio (2009) Regions 2020 – The Climate Challenge for European Regions. Background Document to the Regions 2020 Report, Brussels.

EEA (European Environment Agency) (2008) Impacts of Europe's Changing Climate – 2008 Indicator-based Assessment. EEA Report 4/2008, Copenhagen.

EEA (2009) Greenhouse Gas Emission Trends and Projections in Europe 2009. Tracking Progress Towards Kyoto Targets, European Environment Agency, Copenhagen.

ESA (European Space Agency) (2010) ATSR World Fire Atlas http://due.esrin.esa.int/wfa/ (accessed 15 November 2012).

ESPON (2006) The Role and Spatial Effects of Cultural Heritage and Identity. Final Report, Luxembourg.

Flannigan, M.D., Amiro, B.D., Logan, K.A., *et al.* (2005) Forest fires and climate change in the 21st century. Mitigation and adaptation strategies for global change. *Mitigation and Adaptation Strategies for Global Change*, **11**, 847–859.

Gallego J. (2009) A Downscaled Population Density Map of the EU from Commune Data and Land Cover Information. JRC-Ispra. Accessible at http://epp.eurostat.ec.europa.eu/portal/page/portal/research_methodology/documents/S14P3_JAVIER_GALLEGO_DOWNSCALED_POPULATION_DENSITY.pdf (accessed 15 November 2012).

Gallego, F.J. (2010) A population density grid of the European Union. *Population and Environment*, **31** (6): 460–473.

Gaume, E., Bain, V., Bernardaza, P., *et al.* (2009) A compilation of data on European flash floods. *Journal of Hydrology*, **367**, 70–79.

Guinehut, S. and Larnicol, G. (2008) Produced for EEA by Collecte Localisation Satellites (CLS). http://www.cls.fr/ (accessed 15 November 2012).

Harrison, P.A. and Butterfield, R.E. (2000) Modelling climate change impacts on wheat, potato and grapewine in Europe, in (eds T.E. Downing, P.A. Harrison, R.E. Butterfield and K.G. Lonsdale), *Climate Change, Climate Variability and Agriculture in Europe: An Integrated Assessment, Research*

Report 21, Environmental Change Unit, University of Oxford, pp. 367–390.

Harrison, P.A., Berry, P.M., Butt, N. and New, M. (2006) Modelling climate change impacts on species' distributions at the European scale: implications for conservation policy. *Environmental Science & Policy*, **9**, 116–128.

Havenith, G. (2005) Temperature regulation, heat balance and climatic stress, in (eds W. Kirhc, B. Menne and R. Bertollini), *Extreme Weather Events and Public Health Responses*, World Health Organization, *pp.* **69–80**.

Hiederer, R. (2009) Distribution of Organic Carbon in Soil Profile Data. JRC Technical and Scientific Reports. Office for Official Publications of the European Communities, Luxembourg.

Hisdal, H., Holmqvist, E., Kuusisto, E., *et al.* (2010) *Has Streamflow Changed in the Nordic Countries?* Norges Vassdrags- og Energidirektorat, Oslo.

Holsten, A. (2007) Ökologische Vulnerabilität von Schutzgebieten gegenüber Klimawandel – exemplarisch untersucht für Brandenburg, Institut für Geowissenschaften, University of Tübingen, Tübingen.

Hoye, T., Post, E., Meltofte, H., Schmidt, *et al.* (2007) Rapid advancement of spring in the High Arctic. *Current Biology*, **17**, R449–R451.

IPCC (Intergovernmental Panel on Climate Change) (2001a) Climate Change 2001. The Scientific Basis. Intergovernmental Panel on Climate Change, Cambridge University Press, Cambridge.

IPCC (Intergovernmental Panel on Climate Change) (2001b) Climate Change 2001: Impacts, Adaptation and Vulnerability, Cambridge University Press, Cambridge.

IPCC (Intergovernmental Panel on Climate Change) (2007a) Climate Change 2007: The Physical Science Basis. Contribution of Working Group I to the Fourth Assessment. Report of the Intergovernmental Panel on Climate Change, Cambridge University Press, Cambridge.

IPCC (Intergovernmental Panel on Climate Change) (2007b) Climate Change 2007-Impacts, Adaptation and Vulnerability. Contribution of Working Group II to the Forth Assessment Report of the Intergovernmental Panel on Climate Change, Cambridge University Press, Cambridge.

IPCC (Intergovernmental Panel on Climate Change) (2008) Technical paper on climate change and water. Accessible at www.ipcc.ch/meetings/session28/doc13.pdf (accessed 15 November 2012).

Johansson, M.M., Kahma, K.K. and Bowman H. (2004) Scenarios for sea level on the Finnish coast. *Boreal Environment Research*, **9**, 153–166.

JRC (Joint Research Centre) (2009) Organic matter decline. Sustainable Agriculture and Soil Conservation Fact sheet No. 3. http://eusoils.jrc.ec.europa.eu/projects/SOCO /FactSheets/ENFactSheet-03.pdf (accessed 15 November 2012).

JRC (Joint Research Centre) (2010a) Soil Erosion. http://eurosoils.jrc.ec.europa.eu/library/ themes/erosion (accessed on 5 July 2010).

JRC (Joint Research Centre) (2010b) Soil Organic Carbon Content. http://eusoils.jrc.ec.europa.eu/ESDB_Archive /octop/octop_download.html (accessed on 6 May 2010).

Koppe, C., Kovats, R.S., Jendritzky, G. and Menne, B. (2004) Heat Waves: Risks and Responses. Health and Global Environmental Change, World Health Organization.

Koppe, C., Jendritzky, G., Pfaff, G. (2003) Die Auswirkungen der Hitzewelle 2003 auf die Gesundheit, (ed. DWD), Klimastatusbericht, pp. 152–162

Kovats, S., Jendritzky, G., *et al.* (2006) Heat Waves and Human Health, in (eds B. Menne and K.L. Ebi), *Climate Change and Adaptation Strategies for Human Health*, WHO Regional Office for Europe, Steinkopff Verlag, Darmstadt, pp. 63–90.

Kropp, J., Holsten, A., Lissner, T., *et al.* (2009) Klimawandel in Nordrhein-Westfalen – Regionale Abschätzung der Anfälligkeit ausgewählter Sektoren. Abschlussbericht des Potsdam-Instituts für Klimafolgenforschung (PIK) für das Ministerium für Umwelt und Naturschutz, Landwirtschaft und Verbraucherschutz Nordrhein-Westfalen (MUNLV).

Krysanova, V. and Wechsung, F. (2000) *Soil and Water Integrated Model (SWIM) – User Manual*, Potsdam, PIK Report 69.

Kundzewicz, Z.W., Ulbrich, U., Brücher, T., *et al.* (2005) Summer floods in Central Europe – Climate change track? *Natural Hazards*, **36**, 165–189.

Kwak, Y. and Kondoh, A. (2008) A study on the extraction of multi-factor influencing floods from RS image and GIS data – A case study in Nackdong Basin, South Korea. *The International Archives of the Photogrammetry, Remote Sensing and Spatial Information Sciences*, **XXXVII**. Part B8.

Lappalainen, E. (1996) Global Peat Resources (International Peat Society, Jyskä), Finland.

Lutz, W. (2009) What can demographers contribute to understanding the link between population and climate change. *Popnet Population Network Newsletter* (International Institute for Applied Systems Analysis) **41**, 1–2.

Meier, H.E.M., Broman, B., Kallio, H. and Kjellström, E. (2006) Projections of future surface winds, sea levels, and wind waves in the late 21st Century and their application for impact studies of flood prone areas in the Baltic Sea region, in (ed. P. Schmidt-Thomé), *Sea Level Change Affecting the Spatial Development of the Baltic Sea Region*, Geological Survey of Finland, *Special Paper*, **41**, 23–43.

Menzel, A., Sparks, T.H., Estrella, N., *et al.* (2006a) European phenological response to climate change matches the warming pattern. *Global Change Biology*, **12** (10), 1969–1976.

Menzel, A., Sparks, T.H., Estrella, N. and Roay, D.B. (2006b) Altered geographic and temporal variability in phenology in response to climate change. *Global Ecology and Biogeography*, **15** (5), 498–504.

Merriam-Webster, http://www.merriam-webster.com /dictionary/flash%20flood, accessed on 15-11-2012.

Mudelsee, M., Börngen, M., Tetzlaff, G. and Grünewald, U. (2003) No upward trends in the occurrence of extreme floods in Central Europe. *Nature*, **425**, 166–169.

Nicholls, R.J. and de la Vega-Leinert, A.C. (eds) (2008) Implications of sea-level rise for Europe's coasts. *Journal of Coastal Research*, **24**, 285–287.

Nicholls, R.J. (2004) Coastal flooding and wetland loss in the 21st century: changes under the SRES climate and socio-economic scenarios. *Global Environmental Change*, **14**, 69–86.

Nicholls, R.J., Marinova, N, Lowe, J.A., *et al.* (2011) Sea-level rise and its possible impacts given a 'beyond 4 °C world' in the twenty-first century. *Proceedings of the Royal Society of London. Series A: Mathematical and Physical Sciences*, **369**, 161–181.

Normand, S., Svenning, J.-C. and Skov, F. (2007) National and European perspectives on climate change sensitivity of the habitats directive characteristic plant species. *Journal for Nature Conservation*, (**15**), 41–53.

Novotny, K. and Groh, A. (2007) Untersuchung von Pegelreihen zur Bestimmung der Änderung des mittleren Meeresspiegels an den europäischen Küsten; Technical University of Dresden. Internal report prepared for the German Federal Environment Agency (UBA).

Parmesan, C. and Yohe, G. (2003) A globally coherent fingerprint of climate change impacts across natural systems. *Nature*, **421**, 37–42.

Petermann, J., Balzer, S., Ellwanger, G., *et al.* (2007) Klimawandel – Herausforderung für das europaweite Schutzgebietssystem Natura 2000, in (eds S. Balzer, M. Dieterich and B. Beinlich), *Natura 2000 und Klimaänderungen*, Bundesamt für Naturschutz, Naturschutz und Biologische Vielfalt, Bonn – Bad Godesberg, vol. **46**.

Rahmstorf, S. (2007) A semi-empirical approach to projecting future sea-level rise. *Science*, **315**, 367–370.

Ramos, C. and Reis, E. (2002) Floods in Southern Portugal: their physical and human causes, impacts and human response. *Mitigation and Adaptation Strategies for Global Change*, **7**, 267–284.

Reineking, B., Weibel, P., Conedera, M. and Bugmann, H. (2010) Environmental determinants of lightning- vs. human-induced forest fire ignitions differ in a temperate mountain region of Switzerland. *International Journal of Wildland Fire*, **19**, 541–557.

Robine, J.M., Cheung, S.L., Le Roy, S. *et al.* (2007) Report on excess mortality in Europe during summer 2003. EU Community Action Programme for Public Health, Grant Agreement 2005114. 2003 Heat wave project.

Root, T.L., Price, J.T., Hall, K.R., *et al.* (2003) Fingerprints of global warming on wild animals and plants. *Nature*, **421**, 57–60.

Sabbioni, C., Cassar, M., Brimblecombe, P., *et al.* (2006) Global climate change impact on built heritage and cultural landscapes, in (eds R. Fort, M. Alvarez de Buergo, M. Gomez-Heras, C. Vazquez-Calvo (eds.), *Heritage, Weathering and Conservation*. Taylor-Francis, London, pp. 395–401.

Sleutel, S., De Neve, S. and Hofman, G. (2003) Estimates of carbon stock changes in Belgian cropland, *Soil Use & Management*, **19**, 166–171.

Smith J., Smith P., Wattenbach, M., *et al.* (2005) Projected changes in mineral soil carbon of European croplands and grasslands, 1990–2080. *Global Change Biology*, **11** (12), 2141.

Spiecker, H., Mielikainen, K., Kohl, M. and Skovsgaard, J.P. (1996) *Growth Trends in European Forests: Studies from 12 Countries*, Springer-Verlag GmbH & Co., Berlin and Heidelberg, p. 372.

Soil Science Glossary Terms Committee (2008) *Glossary of Soil Type Terms 2008*. Soil Science Society of America. https://www.soils.org/publications/soils-glossary (accessed 15 November 2012).

Thielen, J., Boudevillain, B. and Andrieu, H. (2000) A short-term rainfall prediction model for urban areas – An evaluation using meso-scale meteorological modelling. *Journal of Hydrology*, **239** (1–4), 97–114.

Thuiller, W., Lavorel, S., Araujo, M., Sykes, M. and Prentice, C. (2005) Climate change threats to plant biodiversity in Europe. *Proceedings of the National Academy of Sciences of the United States of America*, **102** (23), 8245–8250.

UNESCO (2002) Floods in Europe: Damages to libraries and archives. http://portal.unesco.org/ci/en/ev.php-URL _ID=3603&URL_DO=DO_TOPIC&URL_SECTION =201.html (accessed 14 November 2012).

URBAS Projekt (2008) *Vorhersage und Management von Sturzfluten in urbanen Gebieten*. Schlussbericht, Aachen.

van Vliet, A. (2008) Monitoring, analysing, forecasting and communicating phenological changes. PhD thesis. University of Wageningen, The Netherlands.

Vermeer, M., and Rahmstorf, S. (2009) Global sea level linked to global temperature. *Proceedings of the National*

Academy of Sciences of the United States of America, **206**, 21527–21532.

Walther, G-R., Beissner, S. and Burga, C. A. (2005) Trends in the uphill shift of alpine plants. *Journal of Vegetation Science*, **16**, 541–548.

WHC (World Heritage Committee) (1992) Operational Guidelines for the Implementation of the World Heritage Convention.

WHC (World Heritage Committee) (2006) *Reconciling Nature and Culture in a Global Context? Lessons from the World Heritage* List, Cairns.

WHC (World Heritage Committee) (2012) Operational Guidelines for the Implementation of the World Heritage Convention.

Wilson, D., Hisdal, H., Lawrence, D. (2010) Has streamflow changed in the Nordic countries? – Recent trends and comparisons to hydrological projections. *Journal of Hydrology*, **394** (3–4), 334–346.

Wobrock, W., Flossmann, A., Thielen, J. and Farley, R. (2000) Modelling of severe precipitation events in North-Eastern Italy, in *Conference Proceedings of 13th Conference on Clouds and Precipitation*, Reno 2000, Ag. 14–18, Vol. **II**, 1077–1080.

Chapter 6

Economic impacts of climate change on Europe's regions

Johannes Lückenkötter[1], Ove Langeland[2], Bjørg Langset[2], Emmanouil Tranos[3], Simin Davoudi[4] and Christian Lindner[1]

[1]*TU Dortmund University, Institute of Spatial Planning (IRPUD), August-Schmidt-Strasse 10, 44227 Dortmund, Germany*

[2]*Norwegian Institute for Urban and Regional Research (NIBR), Gaustadalléen 21, 0349 Oslo, Norway*

[3]*VU University Amsterdam, Faculty of Economics and Business Administration, De Boelelaan 1105, 1081 HV Amsterdam, The Netherlands*

[4]*Newcastle University, School of Architecture Planning and Landscape, Claremont Towe, Newcastle upon Tyne, NE1 7RU, United Kingdom*

Abstract

Climate change can potentially impact on a wide range of economic sectors and activities. This could be changes in profitability in agriculture or forestry, changes in tourism supply and demand patterns, loss of production due to flooding or the costs of re-building infrastructure after extreme weather events. Economic effects can, to a large extent, be thought of as second-order effects of other dimensions of sensitivity. A region that is physically sensitive to climate change will also be economically sensitive to it. For example, if a region's transport infrastructure is sensitive to flooding, this will have effects on its economy as well. In other words, a region's physical and social sensitivity to extreme weather events is usually highly correlated with its economic sensitivity to climate change.

ESPON Climate's analysis is focused on those economic sectors that are directly affected by climate change, namely the agriculture, forestry, tourism and energy sectors. For each of these sectors the economic impacts were determined by linking particular indicators of climate change to the sector's importance in a regional economy. After a literature review the relevant indicators are defined, the results of a pan-European assessment of the potential impacts of climate change on the different sectors and regions are presented and discussed.

European Climate Vulnerabilities and Adaptation: A Spatial Planning Perspective, First Edition.
Edited by Philipp Schmidt-Thomé and Stefan Greiving.
© 2013 John Wiley & Sons, Ltd. Published 2013 by John Wiley & Sons, Ltd.

6.1   Introduction

Climate change can potentially impact on a wide range of economic sectors and activities. This could be changes in profitability in agriculture or forestry, changes in tourism supply and demand patterns, loss of production due to flooding or the costs of rebuilding infrastructure after extreme weather events. The primary sector, for instance, will be directly affected by changes in temperature and precipitation patterns. Manufacturing industries might be affected indirectly through supply and demand chains. Infrastructure services, as a third sector, will primarily be affected as a result of extreme weather events, such as flooding, but may also be affected by gradual long-term changes in temperature and precipitation.

These examples indicate that the economic sensitivity of a region will be largely dependent on a region's physical, environmental, social and cultural characteristics. Economic effects can to a large extent be thought of as second-order effects of other dimensions of sensitivity. A region that is physically sensitive to climate change will also be economically sensitive to it. For example, if a region's transport infrastructure is sensitive to flooding, it will have effects on its economy as well. Damage to transport infrastructure will result in repair costs as well as costs due to weakened regional economic development, because reduced regional accessibility, at least in the short term, will have a negative impact on the profitability of firms operating in that region. Therefore, a region's physical and social sensitivity to extreme weather events will be highly correlated with its economic sensitivity.

Indirect economic effects of climate change, that is, those related to complex economic supply chains, are difficult to quantify – even more so across Europe with its diverse regional economies (Hallegatte *et al.*, 2008a; 2008b). ESPON Climate's analysis has therefore focused on those economic sectors that are directly affected by climate change, namely the agriculture, forestry, tourism and energy sectors. For each of these sectors the economic impacts were determined by linking particular indicators of climate change to the sector's importance in a regional economy. This regional importance of a sector can be measured by its relative share of employment or of the gross value-added (GVA) in a region. Both of these indicators shed light on different aspects of a regional economy and have therefore been considered in combination. For example, a region with a large share of its GVA and/or employment coming from the agriculture sector can be seen as dependent on (or specialised in) agriculture. If the regional climate changes affect agriculture (e.g. through changes in growing seasons or changed water availability) in a negative or positive way, that region can be labelled as agriculturally (and by implication economically) sensitive to climate change.

This chapter focuses only on the directly affected economic sectors identified above. For each sector the existing literature is reviewed regarding the sector's economic relevance in the European economy and the ways in which it is potentially impacted by climate change. After defining the relevant indicators, the results of a pan-European assessment of the potential impacts of climate change on the respective sector are presented and discussed.

6.2   Agriculture and forestry

In quantitative terms agriculture and forestry only account for small fractions of the European economy. In the EU (27 countries) only about 1.7% of total GVA accrued from agriculture in 2010, and its share has decreased by 0.6% between 2000 and 2010. However, the primary sector's share of GVA differs considerably between countries: it accounts for a larger part of the GVA in the South and East of Europe, for example 3.4% in Greece, 4.0% in Poland and 7.0% in Romania, but only 0.9% in Germany and 0.6% in the United Kingdom (all 2011 data from DG Agri, 2012). The same applies to forestry, which only employs 1.4 million workers in the EU, measured in full-time equivalents, and whose economic importance, likewise, varies substantially between European countries and regions (Blombäck, Poschen and Lövgren, 2003). However, beyond these two economic indicators, agriculture and forestry clearly have a high relevance for rural areas: 'With over 56% of the population in the 27 Member States of the European Union (EU) living in rural areas, which cover 91% of the territory […] farming and forestry remain crucial for land use and the management of natural resources in the EU's rural areas, and as a platform for economic diversification in rural communities' (DG Agri, 2011).

Climate change is likely to have strong implications for agriculture and forestry in Europe in both positive and negative ways (CEC, 2009; Alcamo *et al.*, 2007; Shaver *et al.*, 2000; Blennow and Sallnäs, 2002; Askeev *et al.*, 2005; Kellomäki and Leinonen, 2005; Maracchi, Sirotenko and Bibdi, 2005). In general, plants respond to both increased temperatures and increased CO_2. Higher CO_2 levels leads to higher productivity for all plant growth but can reduce the nutritional value of plants due to an increased C/N ratio. Higher temperatures and longer growing seasons will benefit plant growth in some regions, whereas increased temperature combined with decreased precipitation will limit growth in other regions. Climate variability might also be of concern, as both trees and food crops are especially sensitive to climatic factors in some specific stages of growth. Forests may be a special case, because of the long growth time of trees. Trees planted today will grow under future climate conditions for several decades, which may vary substantially. High temperatures and drought will increase the risk of forest fires and this may lead to substantial damage in the Mediterranean forests. Increased frequency and intensity of storms might also harm the forestry sector.

6.2.1 *Existing studies on climate change effects*

In terms of geographical impacts, there seems to be a consensus that agriculture and forestry in Europe are especially sensitive to climate change in northern and southern regions (Maracchi, Sirotenko and Bibdi, 2005). However, the impacts are quite different: 'In northern areas climate change may produce positive effects on agriculture through introduction of new crop species and varieties, higher crop production and expansion of suitable areas for crop cultivation' due to longer growth periods and increased temperatures (Olesen, 2008, p. 1). The southern regions, however, will experience limited benefits and large disadvantages. Some of the negative effects of increasingly limited water resources may be compensated by higher efficiency of water use. In general though 'lower harvestable yields, higher yield variability and reduction in suitable areas of traditional crops are expected for these areas' (Maracchi, Sirotenko and Bibdi, 2005, p. 117). The IPCC Fourth Assessment Report (AR4) reaches a similar conclusion, that agriculture in Southern Europe will be especially and negatively affected by climate change (Alcamo *et al.*, 2007, p. 543). In contrast, agricultural systems in Western Europe are assumed to be less sensitive to climate change, and model-based projections indicate better opportunities with regard to yield increases and wider agricultural crops for Northern Europe (EEA, 2008).

In summary, the uneven distribution of the economic importance of agriculture and forestry goes hand in hand with the uneven distribution of climatic impacts on these sectors. The Southern and Eastern parts of Europe are relatively dependent on the primary sector and – according to most climate scenarios – are precisely the ones that will be most affected by climate change. Therefore, the negative effects of climate change on the primary sector in Southern Europe, combined with the relative greater importance of the sector, is expected to lead to larger income losses in these regions than in the rest of Europe.

More detailed analyses explore the complex and diverging effects of climate change on livestock and different types of crops in different regions (e.g. Maracchi, Sirotenko and Bibdi, 2005). Several methodologies have been developed to study the response of agriculture to climate change (see Iglesias *et al.*, 2009 for an overview). Of these, the process-based crop models are the most comprehensive, causally tested and empirically validated approach. It basically consists of controlled experiments, where crops are grown in laboratories or fields under controlled climatic conditions. How crop growth is affected by different environmental factors – such as different soil quality, climate, topographic differences and management, that is, sawing date/growth period (e.g. Porter and Seminov, 2005; Ferrera *et al.*, 2010) can thus be tested. Thus, the process-based models link different climate exposures directly to agricultural production and yields. The following paragraphs summarise the main findings of process-based studies for different types of agricultural products.

With regard to cereals, Maracchi, Sirotenko and Bibdi (2005), referring to agro-climatic studies (Harrison and Butterfield, 2000; Nonhebel, 1996), show that increased temperatures will only cause a small reduction in yield in the case of wheat. On the other hand, more CO_2 will lead to a big yield increase. The net effect of both temperature and CO_2 increase within a modest climate change scenario is a large yield increase. Type of soil and topographic environment may also affect yields (Popova and Kercheva, 2005; Ferrara *et al.*, 2010). In combination with these factors an influential study projected climate change to result in significant increases in yields in Southern Europe, particularly in Northern Spain, Southern France, Italy and Greece, and Fenno-Scandinavia (cf. Maracchi, Sirotenko and Bibdi, 2005). In the rest of Europe, yields may show small increases except for some small areas where they are projected to decrease, such as in Southern Portugal, Southern Spain and the Ukraine. For other types of crops, such as maize, another study simulated future climate scenarios for selected sites in different agricultural zones in Europe and suggested increases in yield for northern areas (up to Denmark; the Boral region was not included in this particular study) and decreases in southern areas (Wolf and van Diepen, 1995). According to Maracchi, Sirotenko and Bibdi (2005, p. 125) '[t]his is due to a small effect of increased CO_2 concentration on growth [...] and a negative effect of temperature on the duration of growing seasons'.

Vegetables are also sensitive to changes in temperature and CO_2. For example for onions temperature increases reduce the period of crop growth and accordingly the yield, while growth and yield for carrots improves by an increase in temperature. However, the effects of a changing climate on different types of vegetables vary a lot according to the type of yield and the response of phenological development to temperature change (Farrar, 1996; Komor *et al.*, 1996; Fuquay, 1989; Wolf, 2000).

A comprehensive study (PESETA) on the potential effects of climate change on agricultural crop production in Europe, derived crop production functions from process-based calibrated and validated models for Europe (Iglesias *et al.*, 2009). The study assessed six major crops grown in different agri-economic zones in Europe and also included information on irrigation, technology and agricultural management. Unfortunately, the PESETA study used different climate change scenarios than the ESPON Climate Project and hence its results could not be directly used in the latter. Nevertheless, PESETA's projections show large decreases in yields in Southern Europe for the period 2071–2100. In Nordic countries, increasing yields are expected mainly due to a longer growing season and higher minimum temperatures in the winter. This supports the IPCC Fourth Assessment Report (AR4), which states that agriculture in Southern Europe will have to deal with rising demand for water for irrigation. Furthermore, increases in crop-related nitrate leaching would impose additional restrictions on agriculture in these areas (Alcamo *et al.*, 2007, p. 543), an effect that is not fully accounted for in the PESETA model.

6.2.2 Indicator methodology

As discussed above, the impact of climate change on plant and animal growth is complex. Ideally one would use sensitivity indicators that reflect the regions' different dependency on different types of crops. However, this would require not only fine-grained weather data, but also detailed agricultural management and output data for all regions. In addition one would need a complex model that would *inter alia* include different thresholds for each type of crop or forest. In the absence of such a differentiated model, the ESPON Climate project assessed the impact of climate change on agriculture in total – and the same for forestry – by focussing on the soil characteristics of Europe's NUTS 3 regions as a proxy variable for the sensitivity to drought. The chosen indicator from the European Soil Data Centre (ESDC) reflects on the available water capacity of soils. For each NUTS 3 region the average value of this soil indicator was calculated for all land used for agriculture and then for forestry (on the basis of the CORINE Land Cover database).

In addition to this environmental indicator, two economic variables were used: (i) the percentage of employment in the primary sector (NACE A-B) and (ii) the share of the primary sector GVA at the NUTS 3 level. Both of these variables could potentially be used separately for the dependency of regional economies on the primary sector. However, they are expected to perform differently across regions due to the diverse structure of regional economies: while the share of employment is expected to highlight regions with a more labour-intensive primary sector, the share of GVA is expected to highlight regions with high output and high productivity in the primary sector. Therefore, both indicators were combined to represent a more holistic concept of economic dependency of a region on the primary sector. Thus this indicator captures both employment-related social effects as well as output-related economic effects. Subsequently the combined agricultural sensitivity indicator was related to changes in annual evaporation, which represents the most relevant exposure indicator of the ESPON Climate indicator set (see Chapter 4). This finally constitutes the potential impacts of climate change on the agriculture and forestry sectors.

6.2.3 *Potential impact of climate change*

Figure 6.1 shows the potential impacts of climate change on agriculture and forestry normalised across Europe. The impacts broadly reflect the regional variations of precipitation and temperature changes are agriculture and forestry in many Southern European areas are projected to experience a hotter and especially drier climate in the future. Southern Portugal and Spain, Southern Italy and Northern Greece seem to be particularly affected by temperature increases, also because of the relatively great importance of these sectors in the respective local economies. In contrast, much of Central,

Eastern and Northern Europe are projected to become warmer and wetter. Economic gains are most pronounced in Scandinavian regions that have a relatively large forestry sector. Overall, the geographical patterns of potential impacts of climate change on agriculture and forestry are to a high degree in line with and confirm the results of previous studies.

6.3   Tourism

According to the UNWTO (United Nations World Tourism Organization), tourism may be defined as '[a]ctivities of persons travelling to and staying in places outside their usual environment for not more than one consecutive year for leisure, business and other purposes not related to the exercise of an activity remunerated from within the place visited' (ESPON, 2006, p. 10). In Europe tourism shows a strong seasonality, with a peak in the summer season (July–September) and generally lower levels of activity in the winter season (October–March). While international tourism is usually in the spotlight, domestic tourism, that is, travel by residents in their own country, makes up almost three-quarters of tourism consumption (OECD, 2010).

The tourism sector accounts for approximately 9.7 million jobs or 5.2% of the total workforce in Europe. In total, the European tourism industry generates more than 5% of EU GDP, and this figure has been steadily rising (ECORYS SCS Group, 2009). This makes tourism the third largest service sector in the EU after the trade and distribution and the construction sectors. However, tourism is strongly interlinked with other sectors such as distribution, construction, transport and the cultural sector. Considering these links, tourism's contribution to GDP is probably much bigger than shown in the formal tourism statistics. It has therefore been estimated to generate more than 10% of Europe's GDP and 12% of all jobs (CEC, 2010).

In some European countries the tourism sector is even more important. According to official statistics, tourism in Spain and Portugal contribute to more than 10% of GDP and also account for a large share of employment. Tourism makes up almost 13% of total employment in Spain, 10% in

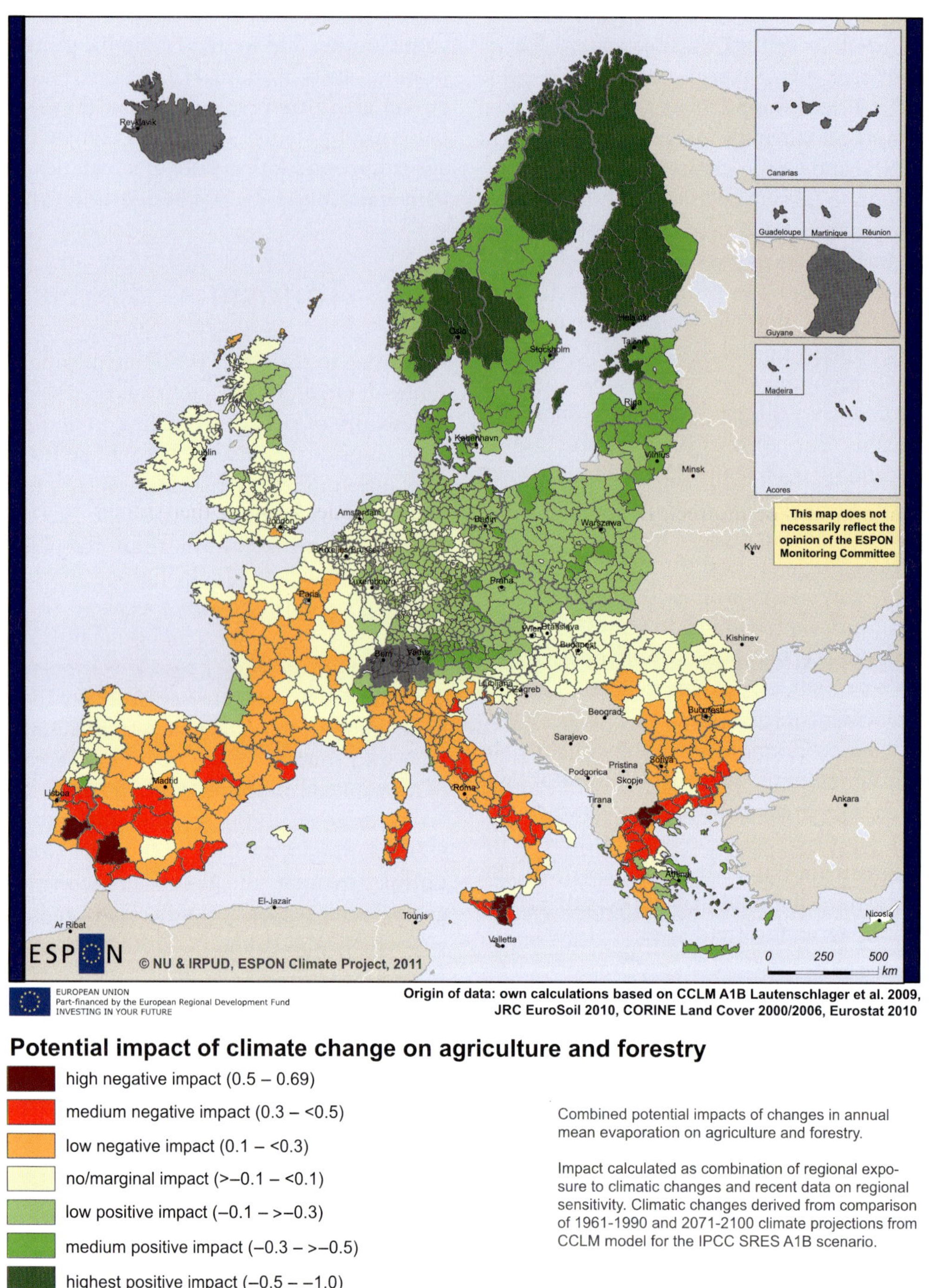

Potential impact of climate change on agriculture and forestry

- high negative impact (0.5 – 0.69)
- medium negative impact (0.3 – <0.5)
- low negative impact (0.1 – <0.3)
- no/marginal impact (>−0.1 – <0.1)
- low positive impact (−0.1 – >−0.3)
- medium positive impact (−0.3 – >−0.5)
- highest positive impact (−0.5 – −1.0)
- no data

Combined potential impacts of changes in annual mean evaporation on agriculture and forestry.

Impact calculated as combination of regional exposure to climatic changes and recent data on regional sensitivity. Climatic changes derived from comparison of 1961-1990 and 2071-2100 climate projections from CCLM model for the IPCC SRES A1B scenario.

Figure 6.1 Potential impact of climate change on agriculture and forestry. © ESPON 2013, IRPUD, ESPON Climate Project, 2011.

Italy, more than 9% in Hungary and 8% in Portugal (OECD, 2010). Although these figures clearly illustrate the differences in importance of tourism between different countries, the variations would be vastly larger if the figures were regionalised. In popular tourist regions in Spain, Portugal, Greece, Italy and France both the share of GDP and employment is far above the national average.

For most types of tourist activities a region's attractiveness for potential tourists depends heavily on the local weather and climate. Thus future changes in climate can be expected to have a strong effect on the tourism sector. Climate change may make some destinations more attractive to visitors by making what were previously less hospitable climates more attractive. However, the impact of climate change on tourism may also be indirect. Destinations may become less attractive due to climate induced 'loss of biodiversity, impacts on the natural and built environment and on tourism-related infrastructure' (OECD, 2010, p. 55).

6.3.1 Existing studies on climate change effects

An important foundation for many recent studies was the development of the 'Tourism Comfort Index' (TCI). This index is used as a composite measure for systematically assessing the climatic elements that are most relevant to the quality of the tourism experience for the 'average' summer tourist. The original TCI (Mieczkowski, 1985) was based on previous research related to climate classifications for tourism and recreation, and on theoretical considerations regarding human comfort. Meteorological data limitations reduced the number of climate variables that were integrated into the TCI to seven (monthly means for maximum daily temperature, mean daily temperature, minimum daily relative humidity, mean daily relative humidity, total precipitation, total hours of sunshine and average wind speed). These seven climate variables were combined into five sub-indices that now comprise the TCI. A standardised rating system, ranging from 5 (optimal) to -3 (extremely unfavourable), was devised to provide a common basis for the measurement of each of the sub-indices. Although devised on the basis of available literature, the rating systems of the five sub-indices and their relative weightings within the TCI are subjective, and the single most important variable in the index is temperature. Changes in the TCI are used as exposure variables in several new studies that investigate climate change effects on tourism (Amelung and Moreno, 2009; Perch-Nielsen, Amelung and Knutti, 2010).

In recent years a growing and diverse body of literature has been focusing explicitly on the relation between tourism and climate change. Some studies focus on particular tourist destinations and investigate how the regional tourism sector is exposed to and changed by climatic changes (e.g. Scott, Wall and McBoile, 2004). Other studies build statistical models of behaviour focusing on the tourism demand of certain types or groups of tourists as a function of weather and climate (Maddison, 2001; Lise and Tol, 2002; Bigano *et al.*, 2006), which can be thought of as sensitivity analyses (Perch-Nielsen, Amelung and Knutti, 2010). Changes in tourism demand have also been addressed in simulation models, where projected changes in tourism flows depend on how climate change affects the attractiveness of a place relative to its competitors (Hamilton, Maddison and Tol, 2005). A climate change vulnerability framework has even been developed for the tourism sector (Perch-Nielsen, Amelung and Knutti, 2010), using beach tourism as an example, because sunbathing and swimming are more linked to specific weather conditions than other tourism activities such as sightseeing.

One of the fundamental findings of these studies is that tourism is very sensitive to climatic changes both in tourism source and destination regions. It is also sensitive to climate seasonality, and according to Alcamo *et al.* (2007, p. 556) referring to Viner (2006) it is 'the seasonal contrasts that drive the demand for summer vacations in Europe'. Hamilton *et al.* (2005) point out that a climate change scenario of $1\,^{\circ}\mathrm{C}$ increase in mean temperature would likely lead to a gradual move of tourist destinations further north and to more mountainous areas. This would influence the preferences of sun and beach lovers living in Western and Northern Europe. As a result, the relative coolness of high-lying and mountainous areas of France, Italy and Spain could gain increased popularity (Ceron and Dubois, 2000).

In the Mediterranean climate, change may result in seasonal changes and extended tourist seasons. Tourism may decrease in the summer due to increased temperature during this period of the year, whereas tourism in spring and perhaps autumn may increase (Amelung and Viner, 2006). Alcamo *et al.* (2007, pp. 556–557) state, referring to Amelung and Viner (2006), that '[o]ccupancy rates associated with a longer tourism season in the Mediterranean will spread demand evenly and thus alleviate the pressure on summer water supply and energy demand'. Thus climate change may reduce demand peaks in the summer and increase occupancy rates in the 'shoulder seasons' and in the end even be beneficial for the Mediterranean tourism industry (EEA, 2008).

In central Europe, climate change will affect traditional winter tourism regions. In the winter significant reductions in natural snow cover is expected to shorten the season and thereby significantly impact on the ski industry. Skiing will have to start later and finish earlier in the season (Elsasser and Burki, 2002). According to Hantel, Ehrendorfer and Haslinger (2000) an increase of mean temperatures of $1\,^{\circ}$C in the most sensitive elevations in the Austrian Alps will reduce skiing days in winter time by four weeks, and by six weeks in spring. An estimate from Beniston, Keller and Goyette (2003) indicates that a temperature increase of $2\,^{\circ}$C and no change in precipitation would cut down the seasonal snow cover at a Swiss Alpine site by 50 days per year.

However, the economic effects of climate change on tourism depend very much on the question of whether holiday seasons in Europe remain fixed or would possibly shift. For example, a more flexible timing of holidays among a large proportion of the population would alter projected impacts significantly.

In the Nordic Countries, summer tourism is likely to benefit from improved conditions. Increased temperatures are expected to make this region more attractive to international tourists during summer.

The effect on winter tourism is more uncertain as it depends on the number of days with snow at the winter sports centres. Also, since in Scandinavia 'one of the projected effects of climate change is higher temperatures and less snow, this may have a negative effect on winter tourism in some areas' (Aaheim *et al.*, 2009, p. 30). At the country level, or even lower at the NUTS 3 level, the positive effect on summer tourism is expected to outweigh the potential negative effects on winter tourism so that the total effects are considered to be positive or neutral.

For the Atlantic north region (United Kingdom and Ireland) it is probable that climate change will lead to a small decrease in international arrivals of tourists. However, the negative effect of less international tourists is assumed to be outweighed by an increase in domestic tourists (Hamilton, Maddison and Tol, 2005). The total effect might therefore be slightly positive.

Looking at climate change impacts on international versus domestic tourism, Hamilton, Maddison and Tol (2005) expect domestic tourism to increase and international tourism to decrease. Thus they project an increase in the tourism industry in the Baltic Sea Region. On the other hand, Southern Europe and the Iberian Peninsula are the European regions with the highest share of international tourists. In Malta and Cyprus the shares of international arrivals already reach levels of 90 and 79%, respectively. Summer temperatures are already high in Southern Europe, thus climate change with even higher temperatures might make these regions less attractive to international tourism, potentially leading to decreases in international arrivals. Domestic tourism is expected to increase but is not expected to outweigh the decrease in international tourism.

In contrast, an increase in international tourism is probable in the Alpine region (Hamilton, Maddison and Tol, 2005). This is because of the assumption that localities at higher altitudes will become more attractive for summer tourism (Hamilton and Tol, 2007). On the other hand, the effect on winter tourism in Austria and Switzerland is more uncertain as the ski industry in central Europe is likely to be afflicted by further reductions in natural snow cover.

This will primarily be a problem at the beginning and the end of the skiing season.

6.3.2 *Indicator methodology*

Unfortunately, the economic data needed for a comprehensive international assessment of tourism in Europe is very limited – at the NUTS 3 level. Hence it is very difficult to estimate the sensitivity of this sector or activities to climatic changes (see Greiving *et al.*, 2011).

However, data are available at the NUTS 3 level for number of beds in hotels and similar accommodations. This indicator can be used as a proxy indicator for estimating the significance of the tourism sector in a regional economy. Number of beds should be measured both in absolute and relative numbers (in relation to regional population). To relate the sector to the climate change model we assume that the tourism sector in a region is related to either summer or winter tourism, and not to both.

6.3.2.1 Winter tourism

In order to identify which regions are winter tourism regions, the DG Regional Policy's typology of mountain regions was used. The ESPON Climate project included all four types of regions of the typology.[1] Subsequently those NUTS 3 regions were identified, which, according to the CCLM, include areas that have at least 100 days of snow cover, as this is considered a crucial threshold for profitable winter sports tourism (Elsasser and Bürki, 2002). Number of beds in these regions reflect the tourism intensity with respect to winter sports, and thus represent the winter tourism sensitivity. This sensitivity indicator was then related to the indicator 'changes in snow cover days' as the relevant climate exposure indicator.

6.3.2.2 Summer tourism

For all non-winter tourism regions the indicator 'number of beds in hotels and similar

accommodations' was used as a sensitivity indicator for summer tourism – both as an absolute and a relative indicator. This rests on the assumption that in principle all regions in Europe can be, and are, destinations for summer tourism, notwithstanding major concentrations of summer tourism in the warmer, Mediterranean climate zones. Also, coasts with sandy beaches are prime summer tourism destinations. However, not all tourists are attracted to coastal areas, there is also significant summer tourism in mountain regions – and in regions with otherwise attractive landscape and/or historic cities. Nevertheless, a major factor for the suitability and attractiveness of a region as a summer tourism destination clearly is 'pleasant summer weather'. As described above, the Tourism Comfort Index has defined and quantitatively modelled what weather conditions are preferred by summer tourists. Therefore it makes sense not to just take one climatic variable of the CCLM model (e.g. mean temperatures) and use its changes as the relevant exposure variable for summer tourism. Instead the necessary climatic variables underlying the TCI were selected and the future TCI score of each NUTS 3 region during the summer season was modelled based on the A1B CCLM projections for 2071–2100. Then the TCI changes were calculated for each region and used as the composite exposure variable for summer tourism, which was then related to the sensitivity indicator of the number of beds available for tourists in each region.

Some may argue that the focus on the summer season is not justified. Accordingly, an increase of temperatures may not only lead to hotter climate in the summer but also extend the tourism season before and after the summer period. Thus negative effects during the summer months would be offset by positive effects before and after the summer. However, the absolute quantity of accommodations, the highest occupancy rates, the highest prices and thus the main revenue generation in a tourism region is determined by the tourism peak in the summer months when most tourists want and can (e.g. due to school holidays) spend their main vacation away from home.

---

[1] The four types of mountain regions are: predominantly mountainous regions that are (a) remote or (b) under urban influence; moderately mountains regions that are (c) remote or (d) under urban influence.

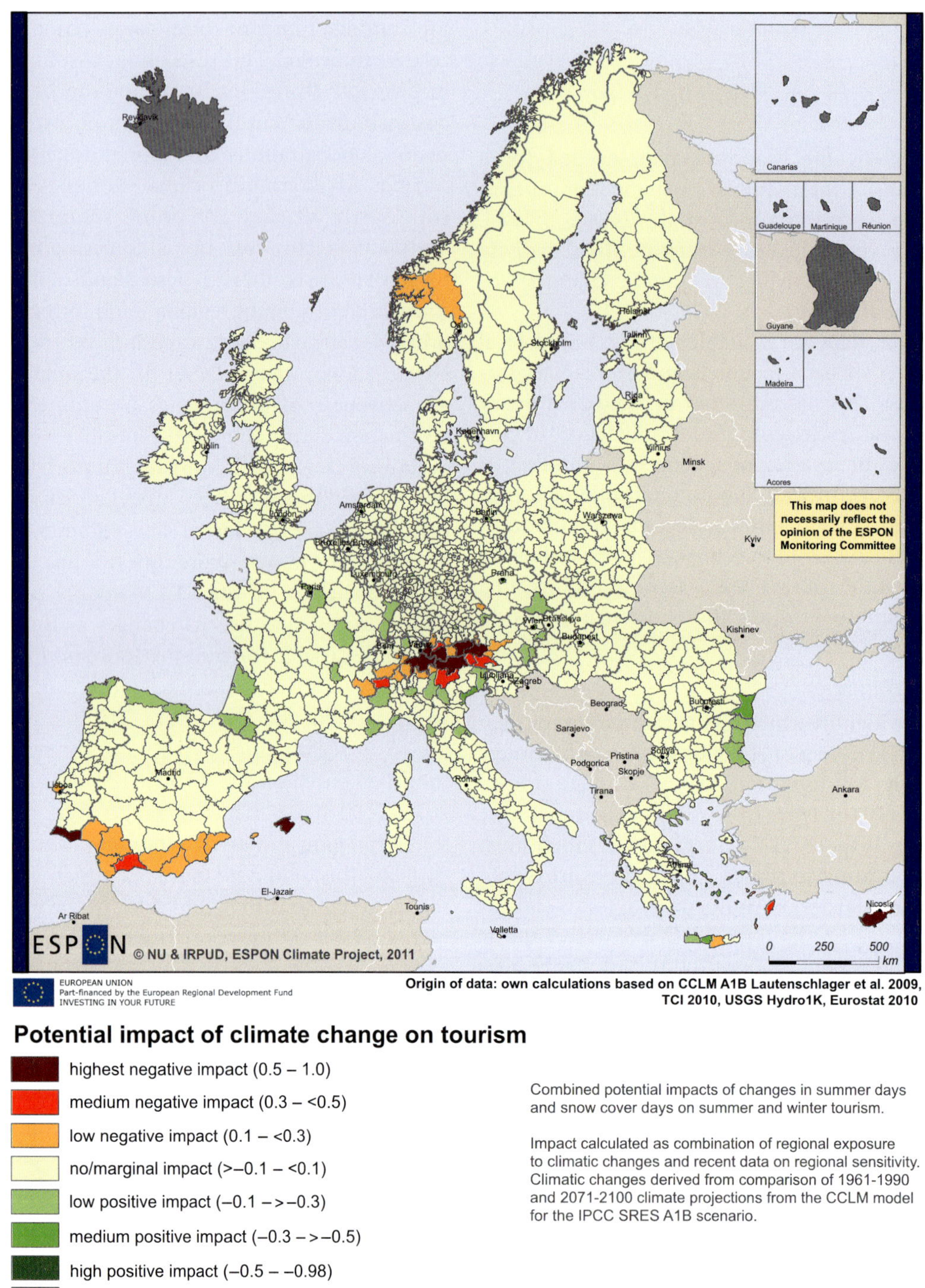

Figure 6.2   Potential impact of climate change on tourism. © ESPON 2013, IRPUD, ESPON Climate Project, 2011.

6.3.3   Potential impact of climate change on tourism

Figure 6.2 presents the results of the relationship between the relevant climatic changes for summer and winter tourism (changes in snow cover days and changes in the TCI) and the number of beds in hotels and similar accommodations. Thus the map shows the potential impacts of climate change on the tourism sector normalised across Europe.

Potential negative impacts are possibly concentrated in Cyprus, Mallorca, Southern Spain and Portugal because of the declining TCI scores and subsequent negative effects on summer tourism. On the other hand, the 'hot spots' in the Alps are strongly impacted because of the declining snow cover days, with adverse effects on winter tourism. Otherwise, the tourism sectors in some other European regions (particularly along the coasts and the middle range mountain areas) might benefit slightly from a warmer climate.

6.4   Energy

While in 2009 only 2.4% of EU-27 GVA was due to the energy, gas and water supply sector (Eurostat, 2010c), the importance of this sector is much greater in the overall economy given its contribution to other economic sectors. However, the overall trend shows that the energy intensity of the EU's economy (gross inland energy consumption divided by GDP) is steadily decreasing. The decoupling of economic growth from a rise in energy consumption is an important aspect of sustainable development.

In general, primary energy production decreased by 4.7% during the period 2008–2009, following a downward trend in the previous decade. The decrease relates to both natural gas (−10.1%) and coal (−9.2%). In contrast, renewable energy recorded a substantial increase of +8.3% and accounted for 18.4% of total EU-27 primary energy production. Nuclear energy continues to be the main energy source accounting for 28% of total EU-27 energy production. With regard to the net energy imports, a decrease of 5.7% took place during 2008 and 2009. The EU-27 total energy dependence rate (EDR) slightly decreased from 54.8% in 2008 to 54.7% in 2009 (all figures are from Eurostat, 2010b).

This mix of energy production is sensitive to climatic changes in two key ways. Firstly, hydropower is the main renewable energy source today and it is highly dependent on water availability and quantities. Climate change with changing precipitation patterns may have regional effects on the hydro-electric production potential (Lehner, Czisch and Vassolo, 2005). In addition the generation of electric power in thermal power stations (in particular coal-fired and nuclear facilities) relies on large volumes of water for cooling (e.g. Förster and Lilliestam, 2010). The use of cooling water may be restricted if certain thresholds for water temperature are exceeded during heat waves or drought periods. This would force plant operators to reduce a facility's capacity or even temporarily close it down.

Energy demand is likewise dependent on climatic conditions. For example, outside temperatures often lead to additional or reduced cooling or heating. This applies particularly to households, but also to companies in the service and industry sectors. Because of an increase in mean average temperature in Europe, models indicate fewer days with heating but an increase in days with cooling. However, in the short to medium term, changes in energy and economic costs taken as a whole are estimated to be modest. This is due to aggregated effects of reduced demand in winter heating and increased demand for summer cooling (EEA, 2008).

6.4.1   Existing studies on climate change effects

Since the electricity sector's demand and supply side have physical underpinnings (infrastructure facilities, residential, commercial and industrial buildings), the energy sector will be affected by the physical impacts of climate change (Eskeland and Mideksa, 2009). The regional effects will vary depending on the specific climatic changes, infrastructure, socioeconomic conditions and energy use profiles of each region (Amato *et al.*, 2005).

Consequently the regional distribution of the positive and negative effects of climate change on the energy sector is likely to be uneven across Europe. Increased electricity demand due to more cooling in summer will occur in Southern Europe, while reduced heating energy demand due to more moderate winters will occur in Northern Europe (EEA, 2008). This translates into a likely net benefit to Northern Europe and a net loss for Southern Europe. Drawing on estimates made by Giannakopoulos *et al.* (2005), Alcamo *et al.* (2007, p. 556) suggest that: 'Around the Mediterranean, two to three fewer weeks a year will require heating but an additional two to three (along the coast) to five weeks (inland areas) will need cooling by 2050'. Peak electricity demand is likely to shift in some locations from winter to summer.

On the supply side, there are concerns regarding the potential climate change effects on the production of thermoelectric power. Decreasing precipitation in many regions, changing energy consumption patterns and competing water demand from other sectors increase the pressure on power generators to reduce water use. Feeley *et al.* (2008) analysed how projected energy demand patterns affect the freshwater withdrawal and consumption rates for various cooling systems. Changing precipitation patterns and higher temperatures might also have impacts on the cooling capacity for power plants. Förster and Lillenstam (2009) model how thermal power plants with once-through cooling could be affected by changing river temperatures and steam flows. They report that even if climate change may not have severe effects on power production in the autumn, winter and spring, power generation could be severely constrained in the summer months due to a changing climate.

The demand side (cooling and heating) is closely connected to outside air temperatures. In general, energy demand relates to outside air temperatures in a U-shaped function. Therefore, energy demand for heating and cooling varies across space due to temperature differences. For example, in rural areas hot temperatures are usually mediated by wind and vegetation, but in urban areas temperatures

can be even higher due to the high proportion of sealed surface (Santamouris, Demosthenes and Asimakopoulos, 2001). This is particularly the case when so called 'heat islands' develop in densely built-up areas, where air temperatures (also at night) are considerably higher than in surrounding areas. The heat-related phenomena affects both the urban population and the economy through, for example, extra costs for residential and commercial cooling systems.

In a study on the effect of climate change on energy demand Bigano *et al.* (2008) found that an increase in temperature affects the demand in households and in production sectors differently. They report that households demand for cooling and heating responds to temperature changes, and that the magnitude of the response in energy demand to changes in temperature depends on the existing temperature level in the region. For service and industry sectors they report that the effects are generally small and statistically insignificant.

6.4.2   *Indicator methodology*

The first group of energy related indicators focuses on the demand side of the energy sector and is specified as follows.

6.4.2.1   Changes in demand for cooling, overall population

One approach for quantitatively assessing the energy demand in relation to outdoor temperatures is through the concept of heating degree days (HDD) and cooling degree days (CDD) (see Isaac and van Vuuren, 2009; Hekkenberg, Moll and Schoot Uiterkamp, 2009). HDD and CDD describe the departure of daily temperature from some threshold representing the human comfort zone. The default threshold is often defined as 18 °C – while in reality this value of course varies from country to country (for example temperatures already perceived as cold by Southern Europeans would feel comfortable to Scandinavians). In the absence of detailed HDD and CDD data, ESPON Climate used the number of summer days (number of days with temperatures of more than 25 °C) as a proxy indicator for changes in cooling demand. The starting point for the sensitivity indicator is the population in each NUTS 3 region and the average number of summer days in the period 1961–1990. The sensitivity indicator was calculated as the product of multiplying the normalised values of these two variables. For determining climate change impacts, this composite sensitivity indicator was then related to changes in the average number of summer days between the time periods 1961–1990 and 2071–2100.

6.4.2.2　Changes in demand for cooling – heat island effect

The same methodology was applied for this indicator, too, with the only exception being that instead of using the overall population, only the population living in urban heat islands (see definition and identification method under 'social impacts' in Chapter 5) was determined for each NUTS 3 region. Again, the corresponding exposure variable was changes in the number of summer days between the time periods 1961–1990 and 2071–2100.

6.4.2.3　Changes in demand for heating, overall population

The focus here was not on demand for cooling but rather on demand for heating. Thus, instead of using the number of summer days, the number of frost days as an average during the period 1961–1990 was used. Apart from this, the methodology and the data used were the same as for the other sensitivity indicators for the overall population. Naturally, the relevant exposure indicator needed for calculating the climate change impacts is changes in the average number of frost days between the time periods 1961–1990 and 2071–2100.

6.4.2.4　Changes in demand for cooling, service sector

Apart from residential users, the service sector is the other main occupier of urban buildings and responsible for a great proportion of the overall demand

for cooling. The sensitivity and exposure indicators utilised for determining the climate change impacts are calculated using the same methodology and data used for the corresponding indicator for overall population, but instead of using population data the average GVA and data on employment in the service sector were used.

6.4.2.5　Changes in demand for in heating, service sector

The justification and methodology of this indicator is basically the same as for heating demand of the overall population of a region. The only difference is that instead of inhabitants of a region, the average GVA and employment in the service sector was used.

6.4.2.6　Changes in conditions for energy supply

This indicator captures the sensitivity of energy production *vis-á-vis* the projected climatic changes as well as the resulting potential impacts based on a set of assumptions. As a first step, the existence of thermal power stations is used as the main sensitivity indicator. The location of the power plants is important for determining the sensitivity of the energy production because energy production is closely related to physical resources located in a region. In particular, water availability is a crucial factor for a thermal power station as water is necessary for cooling purposes. Lack of sufficient cooling water can lead to downscaling or temporarily closing the energy production of a power plant.

For determining climate induced changes of cooling water availability, the most appropriate spatial unit of analysis is the river basin, as power plants located in the same basin would likely face similar water availability problems (Förster and Lillenstam, 2009). Based on this argument, two CCLM climatic exposure indicators were calculated for all European river basins, namely 'decrease in summer precipitation' and 'increase in summer days'. An underlying assumption is that the river temperatures increase when summer days (days with temperatures over 25 °C in summer) increase. Thus, both exposure variables reflect on the availability of cooling water resources during the summer: The higher

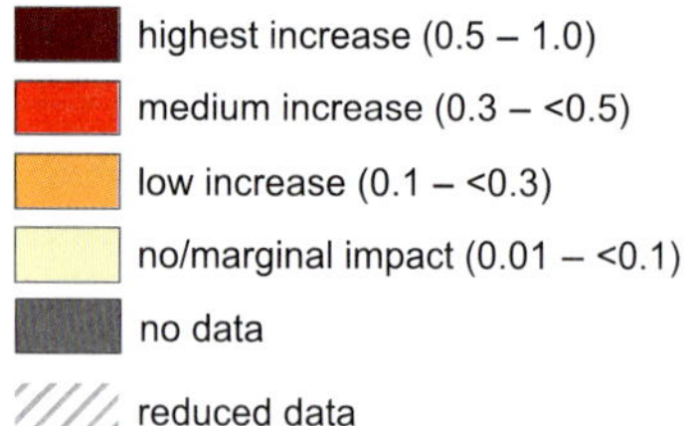

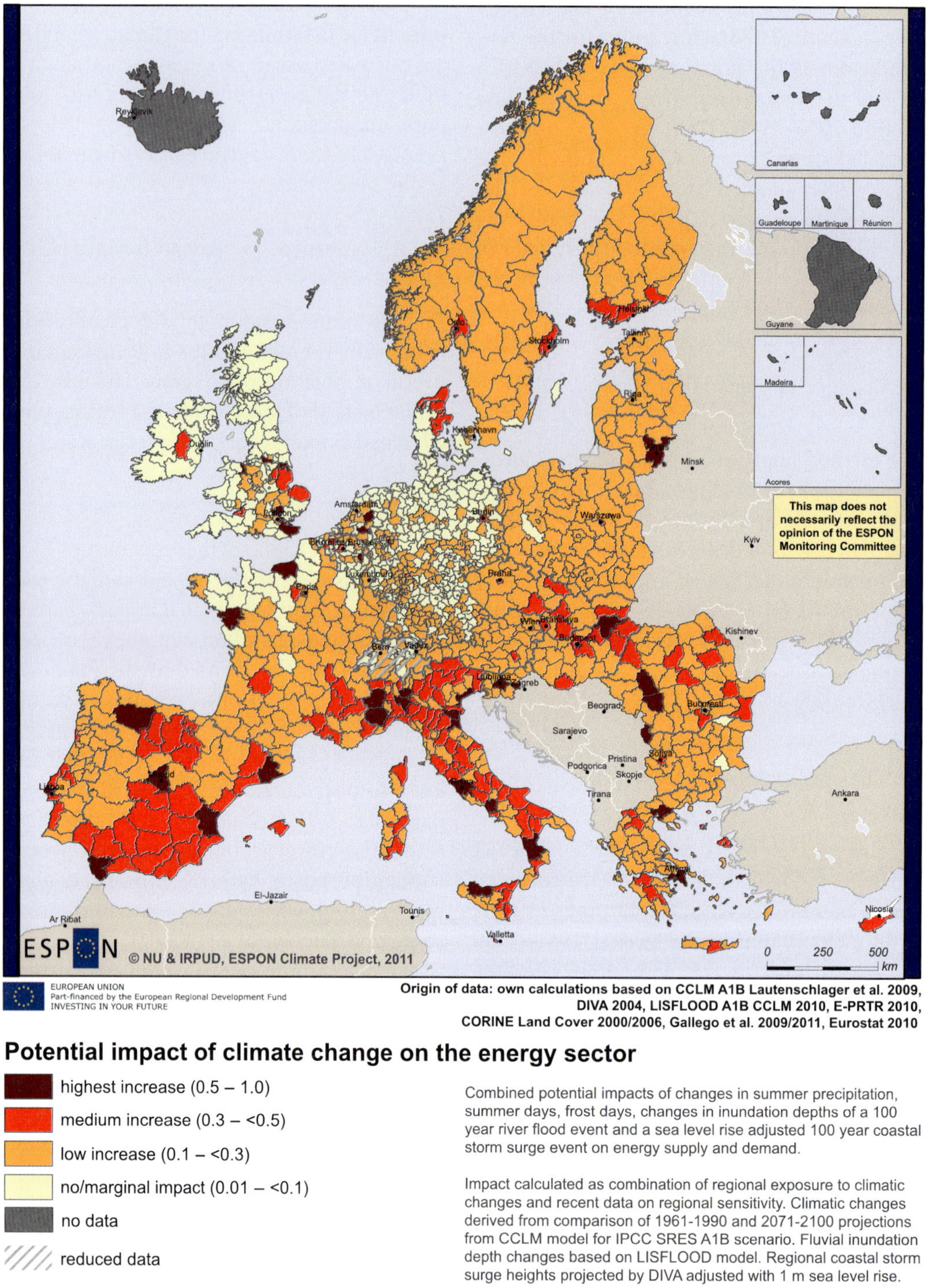

Potential impact of climate change on the energy sector

Figure 6.3 Potential impact of climate change on the energy sector. © ESPON 2013, IRPUD, ESPON Climate Project, 2011.

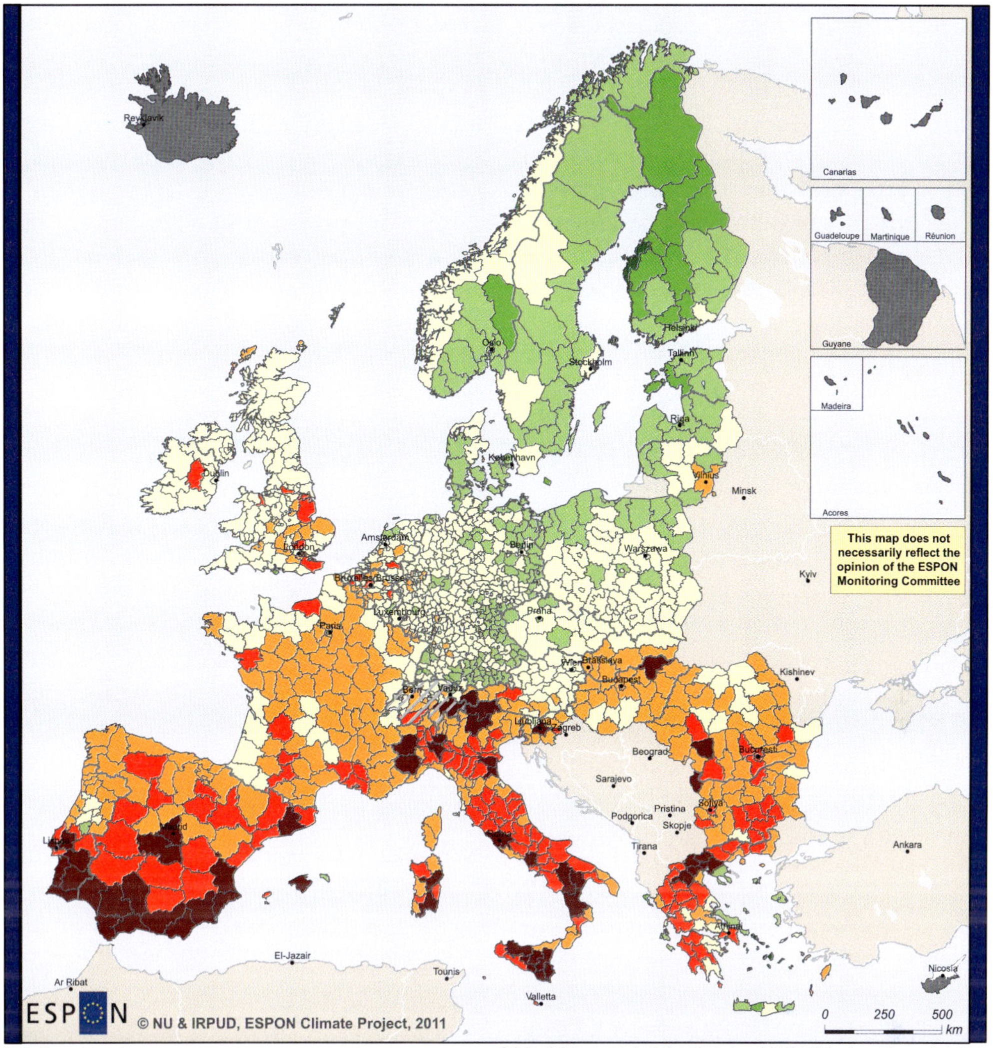

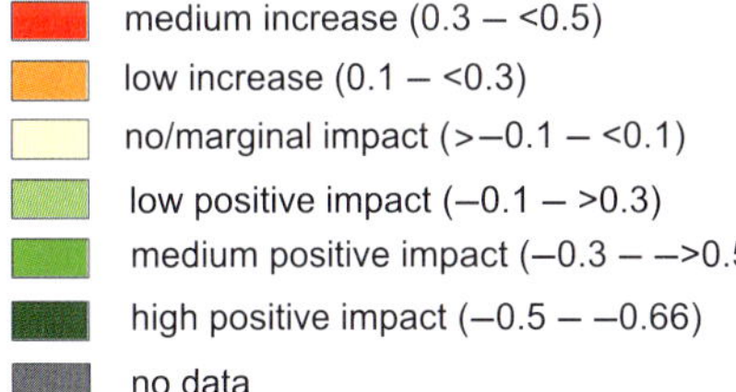

Potential economic impact of climate change

Combined potential impacts of changes in annual mean evaporation,
summer days, snow cover days, frost days, changes in inundation
heights of a 100 year river flood event and a sea level rise adjusted
100 year coastal storm surge event on agriculture, forestry, summer
and winter tourism, energy supply and demand.

Impact calculated as combination of regional exposure to climatic
changes and recent data on regional sensitivity. Climatic changes
derived from comparison of 1961-1990 and 2071-2100 climate projections from CCLM model for the IPCC SRES A1B scenario. Fluvial
flooding changes based on LISFLOOD model. Regional coastal storm
surge heights projected by DIVA were adjusted with 1 m sea level rise.

Figure 6.4   Potential economic impact of climate change. © ESPON 2013, IRPUD, ESPON Climate Project, 2011.

the increase in summer days and the decrease in summer precipitation, the higher would be the negative impact for energy production. After calculating the above at the river basin level, the data were disaggregated to the NUTS 3 level. Then the exposure was calculated as the average of the normalised values of the two exposure indicators[2] and afterwards related to the existence or non-existence of thermal power plants in each region. The result represents the potential impact of climate change on the supply side of the energy sector.

In addition, power plants also face the risk of coastal and river flooding. In order to address this issue, the portion of power plants per NUTS 3 regions that would be affected by a 100 year river and coastal flooding event was calculated.

6.4.3 *Potential impact of climate change on the energy sector*

Figure 6.3 presents the combined potential impacts of climate change on the supply and demand side of the energy sector. According to this assessment, Europe's energy sector is projected to be heavily affected by climate change. While heating demand will generally decline, the warmer temperatures across Europe and especially in the South will drive up energy demands for cooling. Urban agglomerations in Mediterranean and South-Eastern European countries are expected to be most severely impacted. But declining water levels in major rivers would also impact on power stations, which require water for cooling their combustion facilities. The affected regions in this regard are located in Southern, and South-Eastern as well as Western European countries.

6.5 Combined economic impact of climate change

Overall the economic impacts of climate change show a clear south–north gradient (Figure 6.4):

most regions in economically strong countries may expect only a low to marginal negative impact on their economy (e.g. large parts of the United Kingdom and France) or even a positive impact, which applies to many regions in Germany, Poland, the Czech Republic, the Baltic countries and almost all of Scandinavia. However, large parts of Southern Europe are projected to be negatively affected because they are dependent on (summer) tourism, but also on agriculture and forestry. These sectors are expected to significantly suffer from the projected increases in temperature and decreases in precipitation. Moreover, energy demands come into play through the increased need for cooling. The Alps as a premier tourist destination are also identified as a 'hotspot', which mainly results from the projected decreases in snow cover days. The negative economic impacts in South-Eastern Europe are mainly due to the impacts on agriculture and forestry, which still play a relatively important role in the economy of this region. In contrast, the environmental conditions for agriculture and forestry in North-Eastern and Northern Europe can be expected to improve.

The economic impact results and the impacts presented in the previous chapter were later combined to determine the overall impact of climate change on Europe's regions. This overall impact will be presented and discussed in Chapter 9, where it serves as a stepping stone for determining regional vulnerabilities to climate change.

References

Aaheim, A., Amundsen, H., Dokken, T., Ericson, T. and Wei, T. (2009) A macroeconomic assessment of impacts and adaptation to climate change in Europe, CICERO Report 2009:06, Oslo, Norway.

Alcamo, J., Moreno, J.M., Nováky, B., *et al.* (2007) Europe, in (eds M.L. Parry, O.F. Canziani, J.P. Palutikof, P.J. van der Linden and C.E. Hanson), *Climate Change 2007*: *Impacts*, Adaptation and Vulnerability. Contribution of working group II to the Forth Assessment report of the Intergovernmental Panel on Climate Change, Cambridge University Press, Cambridge, pp. 541–580.

Amato, A., Ruth, M., Kirshen, P. and Horwitz, J. (2005) Regional energy demand responses to climate change:

---

[2]Note that in order to address the negative impact of precipitation (decrease in precipitation leads to increase in exposure) the normalised values were multiplied by −1.

methodology and application to the commonwealth of Massachusetts, *Climatic Change*, **71**, 175–201.

Amelung, B. and Moreno, A. (2009) Impacts of Climate Change in Tourism and Europe. PESETA-Tourism Study, European Commission: Joint Research Centre, Institute for Prospective Technological Studies, Seville.

Amelung, B. and Viner, D. (2006) Mediterranean tourism: Exploring the future with the tourism climatic index. *Journal of Sustainable Tourism*, **14**, 349–366.

Askeev, O.V., Tischin, D., Sparks, T.H. and Askeev, I.V. (2005) The effect of climate on the phenology, acorn crop and radial increment of pedunculate oak (*Quercus robur*) in the middle Volga region, Tatarstan, Russia. *International Journal of Biometeorology*, **49**, 262–266.

Beniston, M., Keller, F. and Goyette, S. (2003) Snow pack in the Swiss Alps under changing climatic conditions: an empirical approach for climate impact studies. *Theoretical and Applied Climatology*, **74**, 19–31.

Bigano, A., Hamilton, J.M., Maddison, D. and Tol, R.S.J. (2006) Predicting tourism flows under climate change. An editorial comment on Gössing and Hall (2006). *Climatic Change*, **79**, 175–180.

Bigano, A., Hamilton, J.M. and Tol, R.S.J. (2008) Climate change and tourism in the Mediterranean, *Working paper* FNU-157.

Blennow, K. and Sallnäs, O. (2002) Risk perception among non-industrial private forest owners. *Scandinavian Journal of Forest Research*, **17**, 472–479.

Blombäck P., Poschen, P. and Lövgren, M. (2003) Employment trends and prospects in the European forest sector. A study prepared for the European Forest Sector Outlook Study, Geneva Timber and Forest Discussion Papers, 29, UN.

CEC (2009) Adapting to climate change: Towards a European framework for action. Impact Assessment, Brussels.

CEC (2010) Europe, the world's No. 1 tourist destination – a new political framework for tourism in Europe, Brussels.

Ceron, J.-P. and G. Dubois (2000) Tourisme et changement dimantique, in *Impacts Potentiels du Changement Climatique en France au XXIéme Siècle*. 2nd edn, Premier Ministre, Ministére due l'aménagement du territoire et de l'environment, 1998, pp. 104–111.

DG Agri (European Commission, Directorate General for Agriculture and Rural Development) (2011) Rural Development Policy 2007–2013. http://ec.europa.eu/agriculture/rurdev/index_en.htm (accessed 15 November 2012).

DG Agri (European Commission, Directorate General for Agriculture and Rural Development) (2012) Rural Development in the European Union – Statistical and Economic Information, Report 2012 : 99, Brussels.

ECORYS SCS Group (2009) Study on the competitiveness of the EU tourism industry. Rotterdam.

EEA (2008) Impacts of Europe's changing climate – 2008 indicator-based assessment, Report no. 4/2008, Copenhagen.

Elsasser, H. and Bürki, R. (2002) Climate change as a threat to tourism in the Alps. *Climate Research*, **20**, 253–257.

Eskeland, G.S. and Mideksa, T.K. (2009) Climate change and residential electricity demand in Europe. Working paper, CÌCERO, version June 29 2009.

ESPON (2006) Territory matters for competitiveness and cohesion. ESPON, Luxembourg.

EUROSTAT (2010a) Statistical aspects of the energy economy in 2009. Luxembourg.

EUROSTAT (2010b) Statistical data. Luxembourg.

Farrar, J.F. (1996) Sinks, integral parts of a whole plant. *Journal of Experimental Botany*, **47**, 1273–1280.

Feeley T.J., Skone, T.J., Stiegel Jr, G.J., *et al.* (2008) Water: a critical resource in the thermoelectric power industry. *Energy*, **33**, 1–11.

Ferrara, R.M, Trevisiol, P., Acutis, M., *et al.* (2010) Topographic impacts on wheat yields under climate change: two contrasted case studies in Europe, *Theoretical and Applied Climatology*, **99**, 53–65.

Förster, H. and Lillenstam, J. (2010) Modeling thermoelectric power generation in view of climate change, *Regional Environmental Change*, **10** (4), 327–338.

Fuquay, J.W. (1989) Heat stress as it affects animal production. *Journal of Animal Science*, **52**, 164–174.

Giannakopoulos, C., Bindi, M., Loriondo, M., *et al.* (2005) *Climate Change Impacts in the Mediterranean Resulting from a 2 °C Global Temperature Rise*. WWF Report, Gland.

Greiving, S., Flex, F., Lindner, C., *et al.* (2011) ESPON Climate: Climate Change and Territorial Effects on Regions and Local Economies, Scientific Report, Luxembourg.

Hallegatte, S., Henriet, F. and Corfee-Morlot, J. (2008a) 'The economics of climate change impacts and policy benefits at city scale: A conceptual framework', OECD Environment Working Papers, No. 4, OECD Publishing, doi: 10.1787/230232725661.

Hallegatte, S., Patmore, N., Mestre, O., *et al.* (2008b) Assessing climate change impacts, sea level rise and storm surge risk in port cities: a case study on Copenhagen. Environment Working Papers No. 3, OECD, Paris.

Hamilton, J.M., Maddison, D.J. and Tol, R.S.J. (2005) Climate change and international tourism: A simulation study, *Global Environmental Change*, 253–266.

Hamilton, J.M. and Tol, R.S.J. (2007) The impact of climate change on tourism in Germany, the UK and Ireland: a simulation study, *Regional Environmental Change*, **7**, 161–172.

Hantel, M., Ehrendorfer, M. and Haslinger, A. (2000) Climate sensitivity of snow cover duration in Austria. *International Journal on Climatology*, **20**, 615–640

Harrison, P.A. and Butterfield, R.E. (2000) Modelling climate change impacts on wheat, potato and Grapewine in Europe, in (eds T.E. Downing, P.A. Harrison, R.E. Butterfield and K.G. Lonsdale), *Climate Change, Climate Variability and Agriculture in Europe: An Integrated Assessment*, Research Report 21, Environmental Change Unit, University of Oxford, pp. 367–390.

Hekkenberg, M., Moll, H.C. and Schoot Uiterkamp, A.J.M. (2009) Dynamic temperature dependence patterns in future energy demand models in the context of climate change, *Energy*, **34**, 1797–1806.

Iglesias, A., Garrote, L., Quiroga, S. and Moneo, M. (2009) *Impacts of climate change in agriculture in Europe*. PESETA-Agriculture study, JRC Scientific and Technical Reports.

Isaac, M., and van Vuuren D.P. (2009) Modeling global residential sector energy demand for heating and air condition in the context of climate change. *Energy Policy*, 507–521.

Kellomäki, S. and S. Leinonen (eds) (2005) *Management of European Forests under Changing Climatic Conditions*. Final Report of the Project Silvistrat. University of Joensuu, Research Notes 163, Joensuu, Finland, 427 pp.

Komor, E., Orlich, G., Weig, A. and Kockenberger, W. (1996) Phloem loading – not metaphysical, only complex: towards a unified model of phloem loading, *Journal of Experimental Botany*, **47**, 1155–1164.

Lehner, B., Czisch, G. and Vassolo, S. (2005) The impact of global change on the hydropower potential of Europe: a model-based analysis. *Energy Policy*, **33**, 839–855.

Lise, W. and Tol, R.S.J. (2002) Impacts of climate on tourist demand (2002). *Climatic Change*, **55**, 429–449

Maddison, D. (2001) In search of warmer climates? The impact of climate change on flows of British tourists. *Climate Change*, **49**, 193–208.

Maracchi, G., Sirotenko, O. and Bindi, M. (2005) Impacts of present and future climate variability on agriculture and forestry in the temperate regions: Europe. *Climatic Change*, **70**, 117–135.

Mieczkowski, Z. (1985) The tourism climatic index: a method of evaluating world climates for tourism. *The Canadian Geographer*, **29** (3), 220–33.

Nonhebel, S. (1996) Effects of temperature rise and increase in CO_2 concentration on simulated wheat yields in Europe. *Climatic Change*, **34**, 73–90.

OECD (2010) OECD Tourism Trends and Policies 2010, Paris.

OECD and EEA (2010) OECD/EEA database on instruments used for environmental policy and natural resources management. Available at http://www2.oecd.org/ecoinst/queries/ (accessed 12 December 2010).

Olesen, J. (2008) *Climate Change as a Driver for European Agriculture*, Danish Institute of Agricultural Sciences, Tjele.

Perch-Nielsen, S.L, Amelung, B. and Knutti, R. (2010) Future climate resources for tourism in Europe based on the daily Tourism Climatic Index. *Climate Change*, **103**, 363–381.

Popova, Z. and Kercheva, M. (2005) CRES model application for increasing preparedness to climate variability in the agriculture planning – risk analyses, *Physics and Chemistry of the Earth*, **30**, 117–124.

Porter, J.R., and Semenov, M.A. (2005) Crop responses to climatic variation, *Philosophical Transactions of the Royal Society*, **360**, 2021–2035.

Santamouris, M., Demosthenes, D. and Asimakopoulos, N. (2001) *Energy and Climate in the Urban Built Environment*, James and James, London.

Scott, D., Wall, G. and McBoile, G. (2004) Climate change and the distribution of climatic resources for tourism in North America. *Climate Research*, **27**, 105–117.

Shaver, G.R., Canadell, J., Chapin III, F.S., *et al.* (2000) Global warming and terrestrial ecosystems: a conceptual framework for analysis. *Bioscience*, **50**, 871–882.

Viner, D. (2006) Tourism and its interactions with climate change. *Journal of Sustainable Tourism*, **14**, 317–322.

Wolf, J. (2000) Modelling climate change impacts at the site scale on potato, in (eds T.E. Downing, P.A. Harrison, R.E. Butterfield and K.G. Lonsdale), *Climate Change, Climate Variability and Agriculture in Europe: an Integrated Assessment*, Research report 21, Environmental.

Wolf, J. and van Diepen, C.A. (1995) Effects of climate change on grain maize yield potential in the European community, *Climatic Change*, **29**, 299–331.

Chapter 7

Assessing adaptive capacity to climate change in European regions

Sirkku Juhola[1,2], Lasse Peltonen[1,3] and Petteri Niemi[1]

[1]*Aalto University, Department of Real Estate, Planning and Geoinformatics, P.O. Box 12200, 00076 Aalto, Espoo, Finland*
[2]*University of Helsinki, Department of Environmental Sciences, P.O. Box 65, 00014 Helsinki, Finland*
[3]*Finnish Environment Institute, P.O. Box 140, 00251 Helsinki, Finland*

Abstract

The ability of a society to adapt to the challenges of climate change is dependent on the adaptive capacity of that society. Adaptive capacity, according to the widely accepted definitions in the literature, consists of several determinants that reflect the political, economic, technological and infrastructural capacity and willingness to adapt to the impacts of climate change. Assessing adaptive capacity is important not only in terms of its role in a vulnerability assessment but also to establish a Europe-wide baseline in terms of adaptive capacity. This chapter contributes to the Europe-wide vulnerability assessment complementing the analysis of exposure and sensitivity assessment that is carried out in the ESPON Climate project. In order to do so, the theoretical basis of adaptive capacity and the complexities associated with assessing it are discussed. Subsequently, indicators are developed to assess generic adaptive capacity of European regions. Discussion and conclusions outline general measures to enhance adaptive capacity and challenges associated with it.

7.1   Introduction

The inertia of the earth's climate system means that there is a need to adapt, irrespective of the mitigation measures undertaken to reduce the emissions of greenhouse gases (IPCC, 2007b). Adaptation is seen as a response strategy to climate change, involving the adjustments to reduce vulnerability of communities, regions or activities to the impacts of climate change. Adaptation refers to the processes, practices or structures to moderate or offset potential damage or to take advantage of opportunities associated with the changing climate (Smit and Pilifosova, 2001). The Intergovernmental Panel on Climate Change (IPCC) defines adaptation as 'adjustment in natural or human systems in response to actual or expected climatic stimuli or their effects, which moderates harm or exploits beneficial opportunities' (IPCC, 2007a).

Adaptation to climate change has been characterised as a process that can take place in different ways. For example, adaptation can occur as anticipatory before climate change events take place. Reactive responses to the changes occur as a result of observed climate change impacts. Alternatively, adaptation can take place through planned adaptation actions and measures that are undertaken by different actors, that is, private actors or public interests. Private actors are generally considered to undertake autonomous adaptations without interventions from the government (Leary, 1999). Hence, autonomous adaptation has been termed as private and planned adaptation as public adaptation in the Third Assessment Report of the IPCC (Smit and Pilifosova, 2001).

A majority of the European countries have now begun the process of drafting national adaptation strategies and there is also a trend towards regional and local strategies of adaptation. Adaptive capacity as a concept has been used in order to explore and understand how adaptation processes take place and what types of resources and processes enable adaptation to take place and what processes and structures hinder it. Adaptive capacity, therefore, consists of determinants that underlie the ability to adapt. There is thus an increasing need to understand adaptive capacity of different regions in order to understand whether they can adapt to climate change. The ability of a region to adapt is dependent on the adaptive capacity of that region and this can vary significantly across different regions.

In this chapter, the assessment of generic adaptive capacity is linked to the Europe-wide vulnerability assessment of the NUTS3 regions.[1] Adaptation is a crucial component of impact and vulnerability assessments, as a system's vulnerability is not only based on the exposure of the system to changes but also the ability of the system to adapt to changes experienced or projected. Thus, combining exposure, sensitivity and adaptive capacity, it is possible to examine the vulnerability of the European regions. The second section of this chapter discusses the literature related to adaptive capacity and its determinants. The chapter also elaborates on the development of indicators for assessing adaptive capacity, and presents the methodology adopted in this study. The results of the assessment are presented in a variety of maps that show how adaptive capacity is distributed across the regions. The discussion and the conclusion present general recommendations for the enhancement of adaptive capacity.

7.2   Adaptation and adaptive capacity

Adaptation of a society is dependent on the adaptive capacity of that particular society, irrespective of whether adaptation is autonomous or planned. The existence of adaptive capacity has been shown to be a necessary prerequisite for the design and implementation of adaption strategies that effectively reduce the likelihood of adverse effects from climate change (Brooks, Adger and Kelly, 2005). Thus, adaptive capacity also enables a society to take advantages of the opportunities that are created through changes in the climate.

Most commonly, adaptive capacity is defined as the ability or potential of a system to respond successfully to climate variability and change, and includes adjustments in both behaviour as well as in resources and technologies (IPCC, 2007a). Defined in this manner, adaptive capacity has a distinct context or a site-specific flavour. Thus, a system's adaptive

---

[1]The NUTS classification (Nomenclature of Territorial Units for Statistics) is a hierarchical system for dividing up the economic territory of the EU. In most ESPON projects, NUTS 3 scale is used as it gives an overview of European regions below the national level.

capacity is first and foremost determined by a locally determined set of resources and conditions that constrain or facilitate the ability of the system to successfully adapt to the changes in climate (Adger, Arnell and Tompkins, 2005; Smit and Wandel, 2006).

The IPCC's Fourth Assessment report also outlines adaptive capacity to have two dimensions, a generic one and a specific one (IPCC, 2007a). Generic adaptive capacity refers to the general ability and capacity of a system to respond to climate change, reflecting its socio-economic status, whereas specific capacity relates to a specific climate change impact, such as a drought or a flood that poses a threat to the system. This is closely related to the idea that adaptive capacity refers not only to the ability of a system to plan for hazards and opportunities in advance (anticipatory adaptation) but also its ability to respond to or cope with the effects (reactive adaptation) (Smit and Pilifosova, 2001).

7.2.1 *Determinants of adaptive capacity*

As adaptive capacity refers to the characteristics contributing to adjustments in natural or human systems in response to actual or expected environmental change and external stress, much research effort has been expended on understanding what the characteristics of a system are that affect its ability or propensity to act. Researchers have put forward and stressed the economic aspects (Williamson, Hesseln and Johnston, 2012) or institutional aspects (Gupta

et al., 2010) of adaptive capacity. It has been argued that adaptive capacity, first and foremost, is context-specific and varies from country to country and region to region and within social groups and individuals, as well as over time (Smit and Wandel, 2006). Furthermore, adaptive capacity varies according to its value, but also in terms of its nature in that the scales of adaptive capacity are not independent or separate (Smit and Wandel, 2006). This means, for example, that capacities of regions are tied to the capacity of countries in terms of enabling or constraining environments for adaptation (Smit and Wandel, 2006).

Irrespective of the complex nature of the concept, there have been several further studies that have tried to identify the determinants of adaptive capacity. There have been assessments of adaptive capacity at the national level (Moss, Brenkert and Malone, 2001; Yohe and Tol, 2002; Adger, Arnell and Tompkins, 2005; Haddad, 2005) at the local level (Posey, 2009), across all levels of governance (Westerhoff, Keskitalo and Juhola, 2011), or for a specific sector (Williamson, Hesseln and Johnston, 2012). Although it is acknowledged that adaptive capacity is a dynamic concept, it is possible to identify a set of determinants that affect a region's ability to adapt (Smit and Pilifosova, 2001), see Table 7.1. These six determinants have become the most widely used definition since the publication of the IPCC's Third Assessment Report.

In most studies, determinants of adaptive capacity have been linked to levels of national development, political stability and economic wellbeing,

Table 7.1 Determinants of adaptive capacity.

Economic resources	Economic assets, capital resources, financial means and wealth
Technology	Technological resources enable adaptation options
Information and skills	Skilled, informed and trained personnel enhances adaptive capacity and access to information is likely to lead to timely and appropriate adaptation
Infrastructure	Greater variety of infrastructure enhances adaptive capacity
Institutions	Existing and well-functioning institutions enable adaptation and help to reduce the impacts of climate-related risks
Equity	Equitable distribution of resources contributes to adaptive capacity

Source: Adapted from Smit and Pilifosova, 2001.

and human and social capital and proxy indicators are used for human and civic resources (IPCC, 2007a). For example, a study by Yohe and Tol (2002) attempts to operationalise a working definition of adaptive capacity at the national level, utilising the IPCC TAR list of determinants. The study applies them to national level data and discusses the capacities in terms of two examples of flooding.

In addition to the national level studies, there have been some local case studies that have analysed the adaptive capacity of a particular region or a community. These studies argue for the need to assess adaptive capacity at the regional or local level because the decisions to adapt are made at that level. The lessons from these are important but they always represent context-specific cases from which extrapolation of findings can be difficult. The lessons that have been drawn from the local level stress the importance of relationships within community members through social networks as the ability to participate in decision making (Tompkins, Adger and Brown, 2002). Engle and Lemos (2010) analysed the adaptive capacity of river basin management in Brazil and based their study on nine broad categories of determinants, highlighting the role of institutions in adaptive capacity.

However, there are at least three issues that have been raised in the literature that need to be recognised in terms of defining adaptive capacity and choosing the determinants to include in the study that need to be made explicit. All these issues relate to the choice of determinants and indicators, and the role of researchers in it.

Firstly, it is important to note the role that the choice of indicators plays in determining the outcome of the study. Brooks, Adger and Kelly (2005) state that the choice of indicators in vulnerability and adaptive capacity indicators is often based on assumptions about the factors and processes that lead to vulnerability, informed literature reviews and intuitive understandings of the human–environment relationship, thus to some extent is based on a subjective choice of the authors. For example, a study by Alberini, Chiabai and Muehlenbachs (2005) used an expert judgement to assess adaptive capacity at the national level by selecting determinants based on a review of the literature and on consultations with public health and climate change researchers.

Secondly, little research has been undertaken to explore the linkages between determinants. Yohe and Tol (2002) recognise that little is known about the linkages between the different capacities, and their relative importance to each other, further complicating the choice of which determinants to include. Similar insights have been put forward by Haddad (2005), who analyses the adaptive capacity of nations to fulfil their given national aspirations. National aspirations are defined as a set of defensible principles by which choices can be made between policies that focus on climate change adaptation.

Thirdly, although interesting in terms of the theoretical development of adaptive capacity, it is not surprising that different capacities are needed for different aspirations in terms of adaptation, thus leading to changing rankings. It is necessary to acknowledge that the specific national level determinants of adaptive capacity remain a contested issue (IPCC, 2007a). Although Haddad's study demonstrates the difficulties of using adaptive capacity to rank or compare countries in terms of their adaptive capacity, it does not automatically mean that indicators for adaptive capacity are rendered obsolete. In fact, it rather highlights the complexity and context dependency of adaptive capacity.

7.3   Assessing adaptive capacity

Given the complexities surrounding adaptive capacity, it is not surprising that 'the literature lacks consensus on the usefulness of indicators of generic adaptive capacity and the robustness of the results' (IPCC, 2007a, p. 728). There are naturally also difficulties in developing indicators for measuring or assessing social phenomena (Hinkel, 2011). According to Hinkel, vulnerability, as well as adaptive capacity, are both examples of social phenomena of which a measurement is almost impossible, given that the choice of determinants can vary.

For instance, an assessment of five vulnerability studies demonstrates that the 20 countries ranked 'most vulnerable' show little consistency across studies (Eriksen and Kelly, 2007). Haddad (2005) also points out this problem, whereby an exhaustive ranking of countries in terms of adaptive capacity is dependent on the objectives of their adaption policies. Haddad demonstrates the fact that a nation's ability to adapt is altered when their aspirations in terms of adaptation are changed, leading to different outcomes in the ranking of countries in terms of their adaptive capacity.

In terms of the methodological issues, Malone and Engle (2011) provide a good overview of the critique that can be directed towards indicator-based assessments of vulnerability, and these are naturally relevant to adaptive capacity assessments too. It is important to keep in mind that there are several assumptions that are made when choosing the indicators whilst some indicators may be left out altogether (Malone and Engle, 2011). One of the difficulties in developing and designing an indicator for adaptive capacity below the national level is the availability of data. This has also been recognised by Yohe and Tol (2002), who acknowledge that adaptive capacity is essentially a local characteristic but admit that availability of data does not allow analysis below the national level.

7.3.1 Methodology of ESPON adaptive capacity indicators

In this chapter, the objective is to develop a combined adaptive capacity index for the regions within the boundaries of the ESPON project, based on a selection of available indicators that assess the generic adaptive capacity of each region. A regional focus at the NUTS 3 level, according to the emerging literature, is favourable as more information of the regional level is necessary. This claim is based on the notion that many of the decisions related to adaptation are likely to be taken at the sub-national level.

This study considers adaptive capacity, along the lines of previous research of the Advanced Terrestrial Ecosystem Analysis and Modelling (ATEAM) (Schröter *et al.*, 2004), to consist of three parts: awareness, ability and action, which are further comprised of determinants of adaptive capacity as defined by the IPCC and others, see Figure 7.1. Similar to Schröter *et al.*, knowledge and awareness as a determinant of adaptive capacity play an important role in the present study in terms of identifying vulnerabilities in relation to climate change and enable the identification of adaptation measures. In order to move from awareness to action, ability is necessary, which consists of technology and infrastructure within a given society. Finally, the ability to achieve action is supported by economic resources and institutions that enable a society to carry out the adaptation measures that have been defined. Equity, the sixth IPCC determinant is not considered in this study. The IPCC considers that equity in relation to gender, socio-economic status and political institutions and other aspects is crucial (IPCC, 2001). These are considered within the economic and institutional determinant with separate indicators. For more details on how the framework for assessment of adaptive capacity is constructed, see Figure 7.1.

In this project, the focus is on generic determinants of adaptive capacity that can be assessed across the regions in Europe. It is accepted that some determinants are generic in that they enable adaptation across the localities and countries irrespective of their location and climate impacts. For example, factors such as education, income and health are considered to be contributing, in general, to a higher adaptive capacity of any given society. Whilst other capacities are more specific to particular climate change impacts, such as heavy precipitation and drought (IPCC, 2007a), solutions of which require specialised technical knowledge or technological capacity, which may not be reflected in this study. Although the focus here is on generic capacity, the coping capacity to sudden impacts of climate change, such as extreme weather events which require coping measures, will be assessed with specific indicators, that is, number of hospital beds per capita per region.

Three indicators each assess all five determinants. Educational commitment, computer skills and attitudes towards climate change reflect the

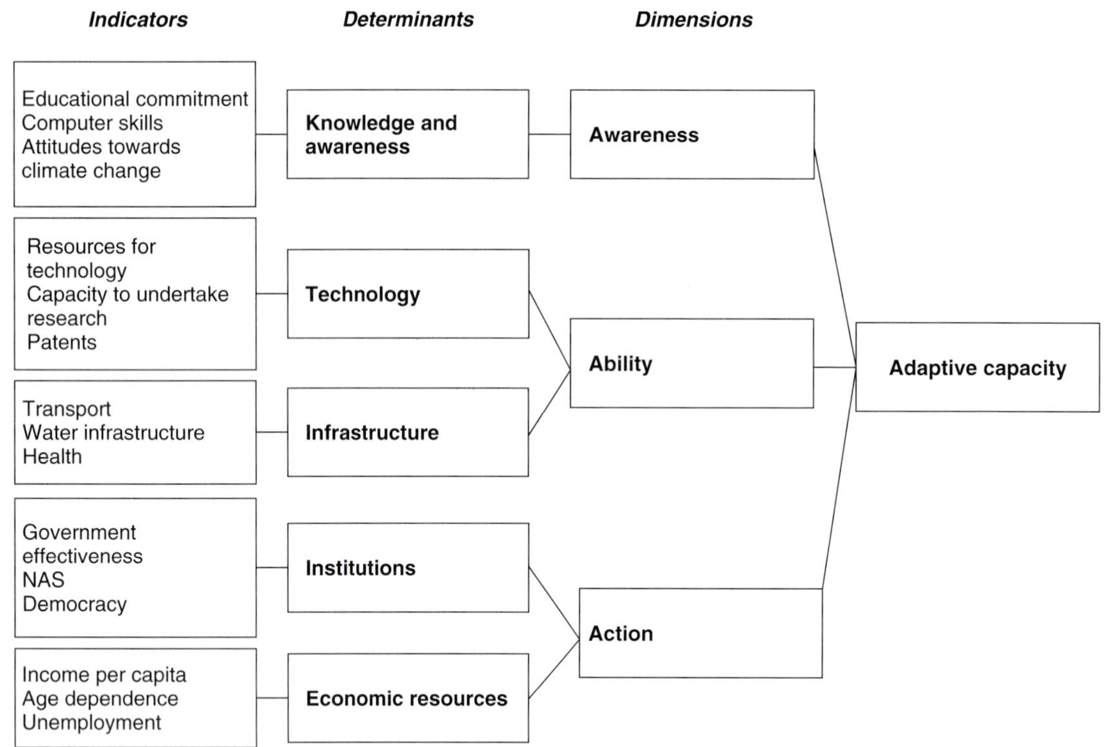

Figure 7.1   Dimensions, determinants and indicators of adaptive capacity. Modified after Schröter *et al.* (2004).

knowledge and awareness determinant. Resources for technology, capacity to undertake research and the number of patents reflect the technology determinant. The existence of a road network, water infrastructure and hospital beds represent the infrastructure determinant. Government effectiveness, the existence of a national level adaptation strategy and a gender weighted democratisation index are used to assess the institutional determinant. Finally, economic resources are assessed by using income per capita, age dependence and the level of long-term unemployment. Owing to the constraints of space to elaborate on each indicator, the reader is directed to the original project final report for further discussion on the choice of indicators (Greiving *et al.*, 2011).

The index is formed as a combination of adaptive capacity and impact indices. In order to capture the variety that is thought to form adaptive capacity in societies, the five determinants adopted from the IPCC and Schröter *et al.* (2004) must also have a broad focus, looking into many different sectors of activities. Thus, the selected indicators (Figure 7.1) will also form a very heterogenic set of variables. Consequently, actual data availability presents a major challenge for implementing this type of index on European regions. Data available on the NUTS 3 level for the entire ESPON space are scarce and some generalisation has to be performed. Descriptions of the indicator data and regional levels used can be found in the final report of ESPON Climate. The data have been acquired as close to year 2006 as possible.

In order to create an index from the heterogenic set of variables, the indicators were first normalised relative to the largest values. The determinants were then calculated as averages of the corresponding indicators and the resulting values were also normalised. Finally, the adaptive capacity index was calculated as the weighted average of the five determinants. The weights for the determinants were drawn from a Delphi survey conducted in ESPON

Table 7.2 Adaptive capacity determinant weights from Delphi survey.

Determinant	Knowledge and awareness	Technology	Infrastructure	Institutions	Economic resources	Total
Weight	23	23	16	17	21	100

Climate and are presented in Table 7.2. The Delphi was carried out with the members of the monitoring committee of the ESPON programme in order to have a Europe-wide perspective on how the different capacities rate in relation to each other. The three adaptive capacity dimensions (Figure 7.1) were also calculated as weighted averages of respective determinants using the same weights as for the overall adaptive capacity index.

It is acknowledged here that the indicators chosen and applied in this report do not fully reflect the nature of each determinant of adaptive capacity, and that the efforts to enhance adaptive capacity solely based on factors that are described by the indicators in this report may not lead to improvements in the determinants of adaptive capacity (Table 7.3).

7.4   Mapping the adaptive capacity of European regions

This section presents the results of the assessment. In addition to the map that combines all indicators, maps are presented for each of the three dimensions of adaptive capacity, awareness, ability and action (Figures 7.2–7.5).

7.4.1   *Adaptive capacities of European regions*

The map of adaptive capacity shows the adaptive capacity that European regions possess across the continent. Overall, there is variation in adaptive capacity between countries and within countries. At the European level, there are several trends that can be seen from the map. When analysing the map, there are several methodological issues that need to

be kept in mind. First, absence of regional level data has led to the fact that for some indicators, national level data has been used, which reduces regional variation since the data for regions in countries is the same. Second, for some indicators, particularly those related to institutions are by nature national, that is, government effectiveness and the existence and quality or progress in relation to a national adaptation strategy.

First, in analysing the maps, a difference in adaptive capacity can be distinguished between Northern Europe and Southern Europe. Overall, the Nordic countries have higher capacities than most of the Southern European countries. Norway and Iceland have no regional variation in terms of adaptive capacity of individual regions. Most of Western and Central Europe have a relatively high capacity when one considers the European average. In comparison, Eastern European countries, on the whole, have lower capacities than Western or Northern European countries. Overall, the countries around the Mediterranean appear to have lower capacities than the countries around the Baltic Sea region.

Similar trends can also be identified at the country and regional level throughout Europe. Firstly, it can be noted that in all countries, capital city regions, overall, have higher capacity than most regions within that country. This is also true even in cases where the country itself as a whole has a lower capacity. This is due to the fact that the capital city regions tend to have concentrations of wealth and other resources. The Baltic regions, however, are a curious exception in this regard with adaptive capacity being uniform throughout the three countries with no regional variation at all. Similarly, Iceland and Norway have no regional variation either.

The regional variation within countries also shows how existing regional patterns are reflected in the way that adaptive capacity is spread across

Table 7.3 Indicators for adaptive capacity.

Determinant	Proxy	Relevance	Methodology
Knowledge and awareness			
	Educational commitment	Skilled and trained personnel and population aids in assessing and implementation of measures required to adapt	Education expenditure per capita within a region (NUTS 0 data)
	Computer skills	ICT based approaches can play an important role in coping and adapting to climate change	Percentage of people who have never used a computer (NUTS 2 data)
	Attitudes towards climate change	Attitudes influence whether adaptation actions will be undertaken	Eurobarometer questions surveyed in 2008 and 2009 (NUTS 1,2,3 data)
Technology			
	Resources of technology	Ability to develop new technologies for adaptation is an important element of adaptive capacity	Percentage of GDP in R&D investment (NUTS 2 data)
	Capacity to undertake research	Capacity to undertake research is important for technological capacity	Number of scientists and engineers in R&D per million labour force (NUTS 2 data)
	Patents	Utilisation of research findings for further purposes to develop products and services	Number of patent applications per million inhabitants (NUTS 2 data)
Infrastructure			
	Transport	The existence of a road network can enhance coping and adaptive capacity	Kilometres of road per square kilometre (NUTS 2 data)
	Water exploitation	Existence and well-functioning water infrastructure is important for adaptive capacity	Water exploitation index by European Environment Agency (NUTS 0 data)
	Hospital beds and doctors	Important for coping capacity in case extreme weather events increase	Number of hospital beds per 100 000 inhabitants (NUTS 2 data)
Institutions			
	Government effectiveness	The efficiency of government is likely to impact positively the adaptive capacity of a region	Government effectiveness index that is available from the World Bank Group (NUTS 0 data)
	Democracy	Adaptive capacity will be greater if resources and power in governing resources for adaptation are equitably allocated	Gender weighted democratisation index (Vanhanen, 2010). Data for this is available from FSD (The Finnish Social Science Data Archive; NUTS 0 data)
	National adaptation strategies	The existence of a national adaptation strategy (NAS) is likely to increase adaptive capacity of a region	Data from IVM used to build an indicator (concerns, recommendations, measures) (NUTS 0 data)
Economic resources			
	Income per capita	Economic assets can be used to fund and support adaptation measures and strategies	GDP per capita (€ PPP) of a NUTS 3 region (NUTS 2 data)
	Long-term unemployment	Long-term unemployment can reduce adaptive capacity by leading to lower standards of living as well as to increases in vulnerable groups	Percentage of population who are long-term unemployed (NUTS 3 data)
	Age dependency ratio	The extent to which a part of a region's population is dependent on other members	Age dependency ratio available from Eurostat (NUTS 3 data)

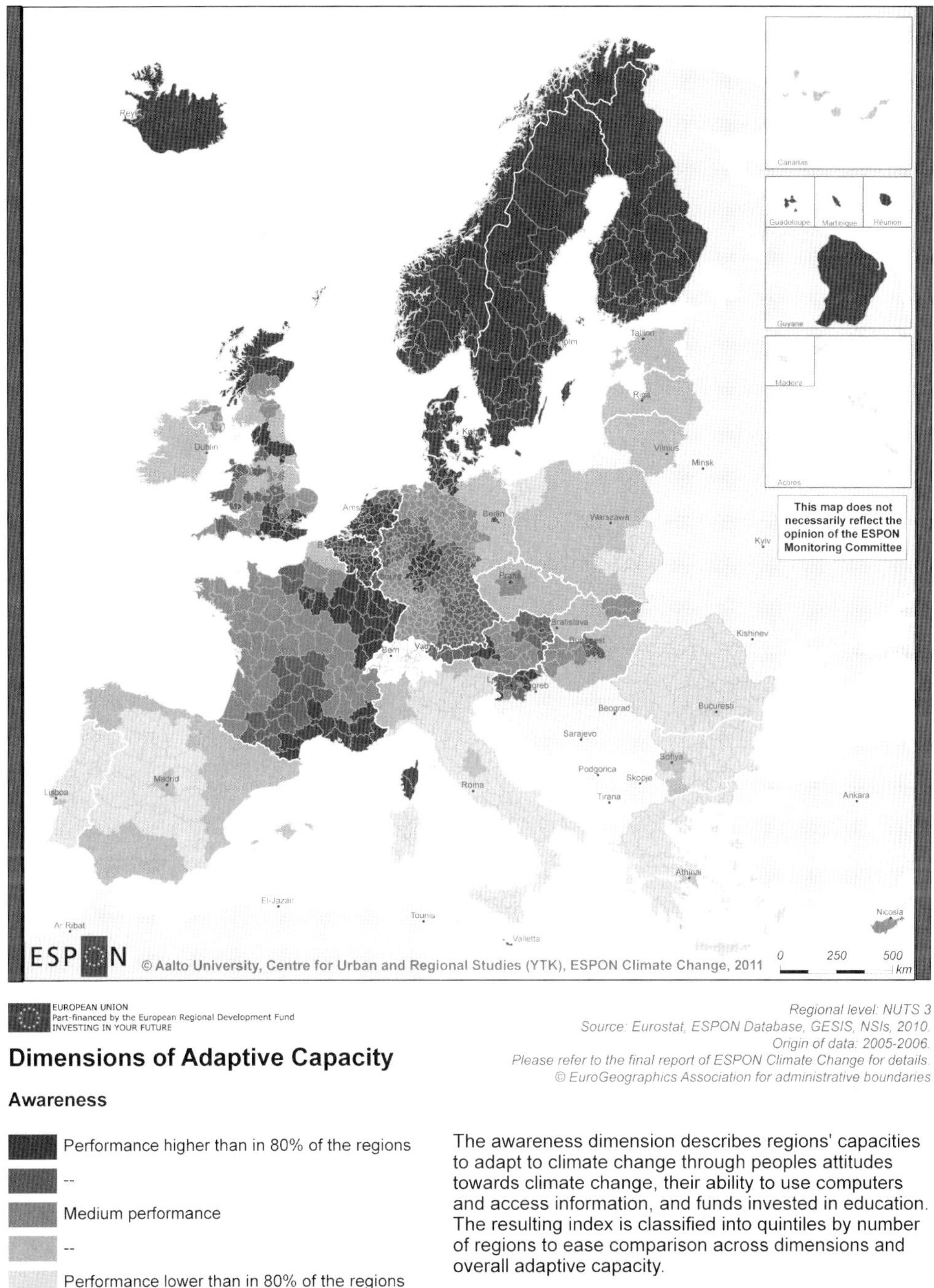

Dimensions of Adaptive Capacity

Awareness

- Performance higher than in 80% of the regions
- --
- Medium performance
- --
- Performance lower than in 80% of the regions
- No data

The awareness dimension describes regions' capacities to adapt to climate change through peoples attitudes towards climate change, their ability to use computers and access information, and funds invested in education. The resulting index is classified into quintiles by number of regions to ease comparison across dimensions and overall adaptive capacity.

Please refer to the Final report of "ESPON Climate Change" for details on the methodology

Figure 7.2   Awareness dimension. © ESPON 2013, IRPUD, ESPON Climate Project, 2011.

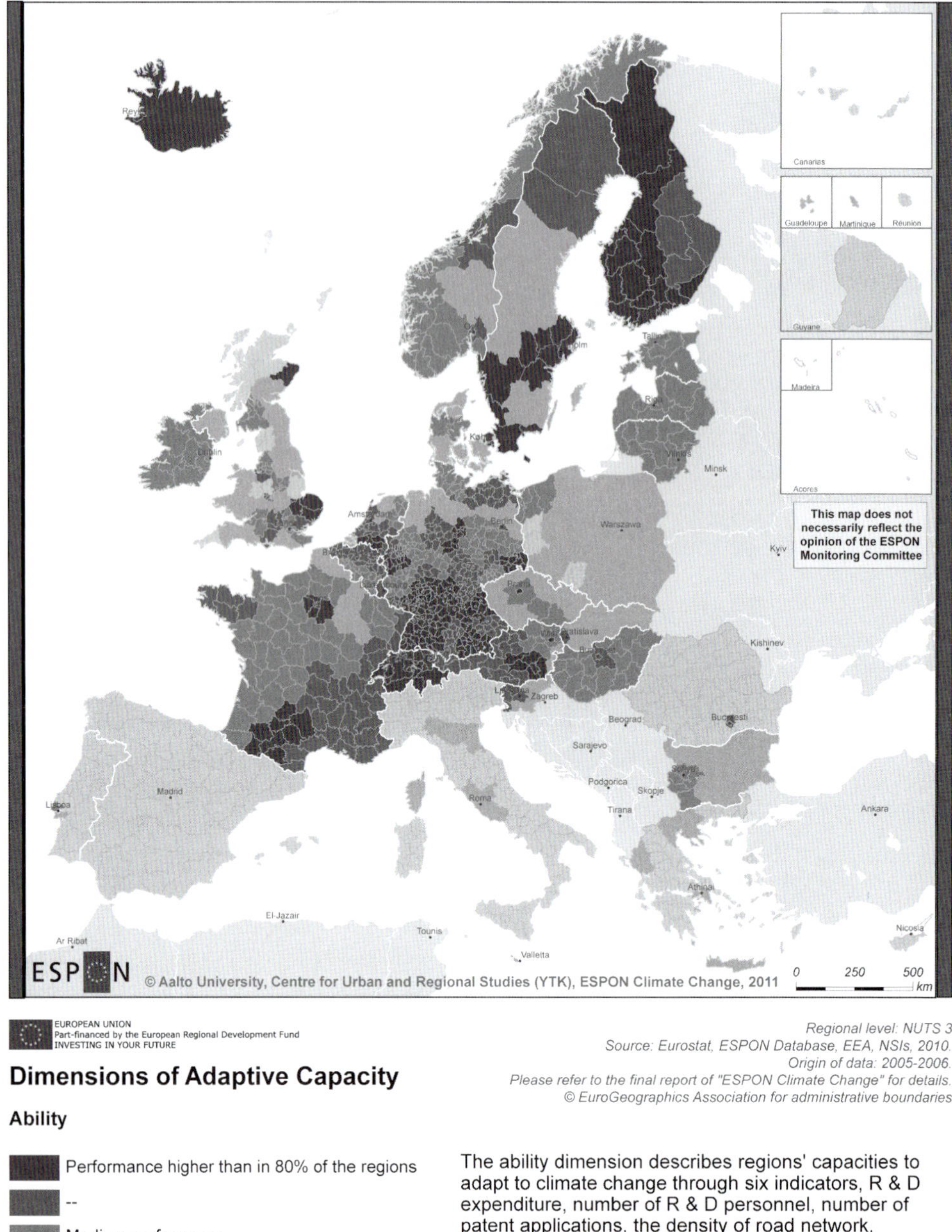

EUROPEAN UNION
Part-financed by the European Regional Development Fund
INVESTING IN YOUR FUTURE

Regional level: NUTS 3
Source: Eurostat, ESPON Database, EEA, NSIs, 2010.
Origin of data: 2005-2006.
Please refer to the final report of "ESPON Climate Change" for details.
© EuroGeographics Association for administrative boundaries

Dimensions of Adaptive Capacity

Ability

- Performance higher than in 80% of the regions
- --
- Medium performance
- --
- Performance lower than in 80% of the regions
- No data

The ability dimension describes regions' capacities to adapt to climate change through six indicators, R & D expenditure, number of R & D personnel, number of patent applications, the density of road network, sustainability of water use and health care capacity. The resulting index is classified into quintiles by number of regions to ease comparison across dimensions and overall adaptive capacity.

Please refer to the Final report of "ESPON Climate Change" for details on the methodology.

Figure 7.3 Ability dimension. © ESPON 2013, IRPUD, ESPON Climate Project, 2011.

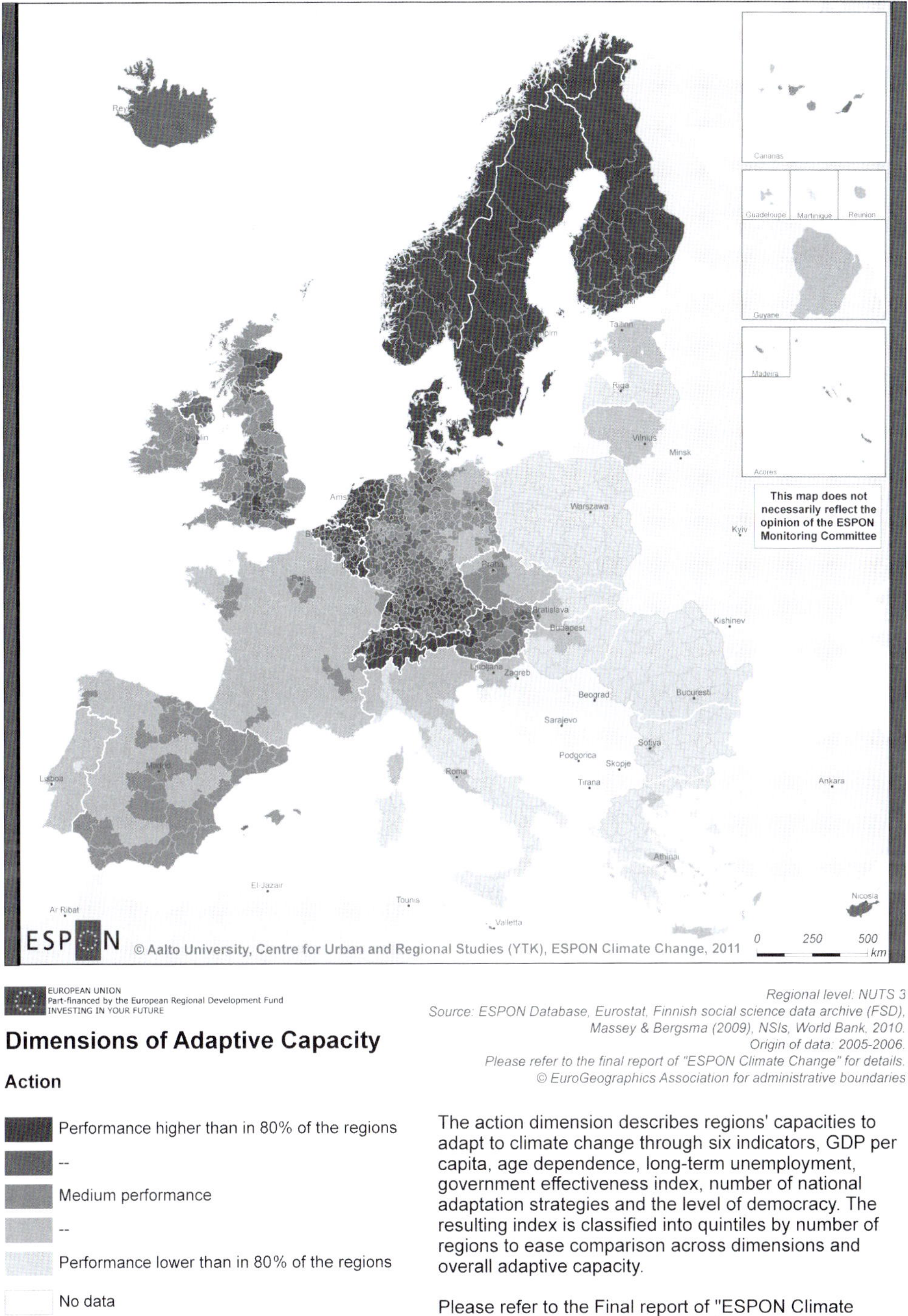

Dimensions of Adaptive Capacity

Action

- Performance higher than in 80% of the regions
- --
- Medium performance
- --
- Performance lower than in 80% of the regions
- No data

The action dimension describes regions' capacities to adapt to climate change through six indicators, GDP per capita, age dependence, long-term unemployment, government effectiveness index, number of national adaptation strategies and the level of democracy. The resulting index is classified into quintiles by number of regions to ease comparison across dimensions and overall adaptive capacity.

Please refer to the Final report of "ESPON Climate Change" for details on the methodology.

Figure 7.4   Action dimension. © ESPON 2013, IRPUD, ESPON Climate Project, 2011.

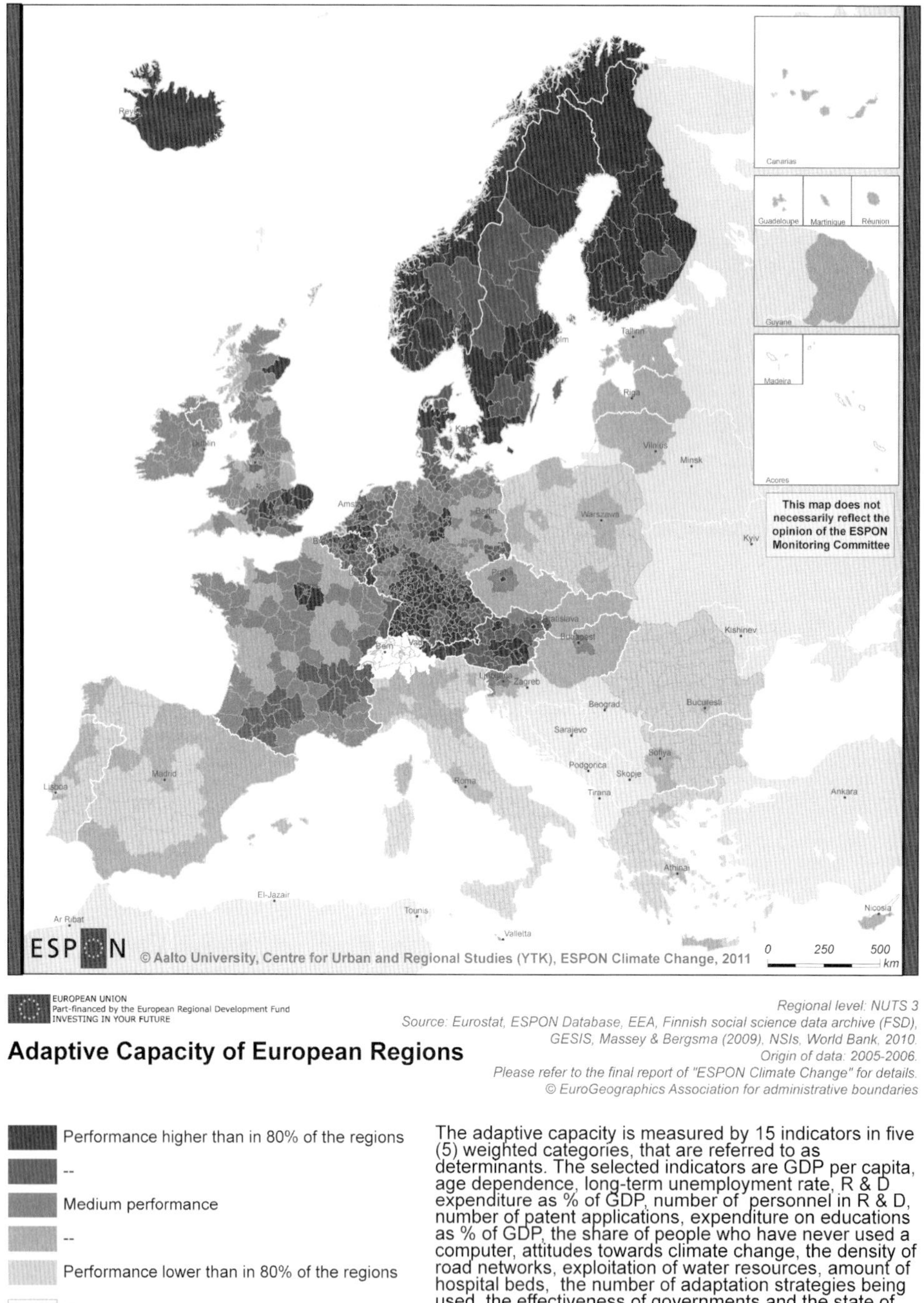

Adaptive Capacity of European Regions

The adaptive capacity is measured by 15 indicators in five (5) weighted categories, that are referred to as determinants. The selected indicators are GDP per capita, age dependence, long-term unemployment rate, R & D expenditure as % of GDP, number of personnel in R & D, number of patent applications, expenditure on educations as % of GDP, the share of people who have never used a computer, attitudes towards climate change, the density of road networks, exploitation of water resources, amount of hospital beds, the number of adaptation strategies being used, the effectiveness of governments and the state of democracy.

Please refer to the Final report of "ESPON Climate Change" for details on the methodology.

Figure 7.5   Aggregate adaptive capacity dimension. © ESPON 2013, IRPUD, ESPON Climate Project, 2011.

the countries. North–South or East–West divisions can be seen in the maps by reflecting the overall development patterns of a particular country. Those regions, which are less developed, can also be seen to have less adaptive capacity.

It is also possible to analyse the adaptive capacity of European regions in terms of the dimensions of adaptive capacity, hence focusing on awareness, ability and action. In terms of Northern Europe, where aggregated adaptive capacity is generally high, differences between regions can be seen in all three different dimensions. For example, Sweden scores high on awareness and action but has a lower ability to adapt. Finland on the other hand has a high ability but scores lower on awareness and action. Similar trends can be observed in Western and Central Europe. Ability and action are high but awareness is lower in comparison with the other two dimensions.

For Eastern Europe, all three dimensions are lower than in other parts of Europe and significant differences exist between the three different dimensions. However, indicators used for measuring action are consistently low across the regions within Eastern Europe. The Mediterranean region overall has a lower capacity than the more northern regions in Europe. The Iberian Peninsula scores low on awareness and ability but scores higher in terms of the action dimension. Mediterranean regions in France have a high ability but score lower on awareness and action. Similar trends can also be observed within regions in Italy.

Capital cities also emerge as having high adaptive capacity from the aggregated map. Interestingly, there is also variation in terms of the dimensions of adaptive capacity in the capital city regions in Europe. When comparing the three dimensions in each capital city region, it appears that ability is the highest across the board. This is compared with the action dimension, which is also relatively high and appears to be higher than awareness.

It is also useful to compare the results of this analysis with results from other research efforts that have mapped adaptive capacity on a European scale. The ATEAM produced adaptive capacity maps and published them in their final report (Schröter *et al.*, 2004). Overall, the results of this ESPON study and the ATEAM study show similar trends. This is partially because the construction of the indicators is similar with this ESPON study utilising similar indicators as the ATEAM. Both maps show that the northern parts of Europe have higher capacities than Southern Europe. The ATEAM maps did not calculate adaptive capacity of Eastern European countries whereas this study does. The ATEAM project also projected changes of adaptive capacity into the future, which was not done in this project. The ATEAM does not explain their methodology, so it is difficult to comment on how this was done.

7.5 Enhancement of adaptive capacity in Europe

Enhancement of adaptive capacity is important for a society in order to improve its ability to adapt to climate change since adaptive capacity lays the foundation for actions on adaptation. Enhancing adaptive capacity has been discussed in the literature to varying degrees in recent years, and it has been discussed in relation to the development of societies in general. The IPCC's Third and the Fourth Assessment Report discuss adaptation mainly with regards to developing countries, and, as a result, enhancement of adaptive capacity is mainly discussed in relation to, and considered to be compatible with, the goals of sustainable development. It is admitted that the processes needed for enhancing adaptive capacity are similar to those of sustainable development processes, including social, economic and environmentally sustainable growth. On the other hand, underdevelopment can hamper and result in a lower adaptive capacity for a society.

The Third Assessment Report (TAR) outlines several issues that contribute to the building of adaptive capacity that draw mostly on developing country context but are also relevant in the developed country context (IPCC, 2001). These include reduction of poverty within the society, as well as the moderation of structural inequities. Lowering of inequities in the distribution of wealth and resources among groups, as well as intergenerationally, is considered important, as well as improving access to resources in general. Overall, improved institutional capacity and efficiency also contribute to the enhancement of adaptive capacity. Improvements in technology and infrastructure naturally play a part in the enhancement in adaptive capacity, but case studies have also shown the value of local experience together with active participation to match the local resources and needs. As vulnerability to climate change will be observed across multiple levels in society, adaptive capacity is also distributed across levels of decision-making in a society. Therefore, the enhancement of adaptive capacity can also take place across several scales within a society (IPCC, 2001).

The Third Assessment Report also outlines key findings related to adaptation and adaptive capacity on different continents (IPCC, 2001). For Europe, the report recognises that in general, the adaptive capacity and potential for adaptation is relatively high in comparison with other continents. This is mainly caused by strong economic conditions and a stable population with a possibility to migrate. In addition, well-developed political, institutional and technological systems can support adaptation to the impacts of climate change. The TAR, however, does point out that equity issues are also of concern in Europe, since more marginal and less wealthy areas are less likely to adapt. Furthermore, without appropriate policies, responses to climate change can potentially lead to greater inequities.

Adaptation policy can of course be used to enhance adaptive capacity, and this is also addressed in some national level adaptation strategies (Keskitalo, 2010). A recent review of developments in relation to adaptation at the regional level in Europe concludes that adaptation is only just beginning to emerge as an important issue. Although action on adaptation is mostly focused on strategy building at the moment, there is a need to improve implementation issues, both in the selection of policy instruments and addressing the issue of costs (Ribeiro *et al.*, 2009). There are naturally tools available, facilitated by the EU, that can be used to support the regional adaptive capacity. These activities, such as knowledge development, monitoring of regional strategies and its implementation and awareness-raising, can be supported

from existing EU mechanisms. These include, for example, regional development, economic and social cohesion funds, such as the European Regional Development Fund (ERDF), The European Social Fund (ESF), LIFE + and INTERREG funding mechanisms (Ribeiro *et al.*, 2009).

7.5.1   *Policy recommendations*

As already mentioned, adaptive capacity and the enhancement of it as a whole, as well as its determinants enables or improves the ability of societies to adapt. The majority of the studies that deal with enhancing adaptive capacity focus either on societal development in general or on a particular determinant. Thus, far too little attention has been paid to the relations between the different determinants and how these interact with each other. For example, it is still uncertain whether it is necessary to have a particular capacity to develop and improve other kinds of capacity, or whether it is necessary to improve the capacity of all determinants in order to achieve enhancement of adaptive capacity as a whole. This naturally raises the question of how the dynamics between the determinants, if at all, affect the efforts to enhance adaptive capacity. Thus, it is acknowledged here that it is nearly impossible to discern the way in which the different determinants of adaptive capacity interrelate.

In this chapter, adaptive capacity is considered to consist of three components, awareness, ability and action. Measures to enhance adaptive capacity here, naturally, relate to the development of awareness, ability or action in a broader manner than just by focusing on the aspects that are measured here by indicators of adaptive capacity as discussed in the methodology. Thus, improvements in one aspect of one determinant do not necessarily mean that one determinant has a direct influence on the overall adaptive capacity. Thus far, as discussed above, examples of cases where adaptive capacity has been enhanced are still rare in the relevant literature, and much of the literature focuses on developing country cases, although some examples from municipal and regional levels in the developed countries are beginning to emerge.

In terms of raising the awareness of climate change, its impacts and the possibilities for adaptation, successful cases and early movers on adaptation have generally highlighted the changes in awareness and thinking and leadership in terms of responding to the challenge of climate change (Saavedra and Budd, 2009). Other examples have shown the role of climate information, and the provision of data to be crucial for building adaptive capacity (Twomlow *et al.*, 2008). In terms of the EU, the clearing house mechanism for adaptation knowledge to be established as a result of the White paper on adaptation is a good example of steps towards increasing the knowledge of climate change. In addition, at the national level, there are several examples of national portals, for example, Denmark and Finland, where climate change and adaptation related information is being made available for stakeholders at the sub-national level.

Ability is composed of technological solutions for adaptation, as well as existing infrastructure and new developments for solving problems posed by climate change impacts. The role that technology can play in enhancing adaptive capacity has been highlighted in some cases. For example, the use of irrigation related technologies can play a major role in agricultural areas (Pittman, Wittrock, Kulshreshtha and Wheaton, 2011). Similarly, new technologies in tilling the land have been shown to increase the adaptive capacity of resources users (Hagmann and Chuma, 2002). However, the authors also point out that issues related to knowledge and dissemination are crucially related to the uptake of new technology, highlighting that the existence of technology does not itself mean that a society possesses a higher adaptive capacity. Infrastructure is important for adaptation, taking the form of public goods that societies provide for their citizens. These public goods, such as land use, coastal defence systems and early warning systems of extreme events, are important for enhancing adaptive capacity (Tompkins *et al.*, 2010). Infrastructure in Europe in general is quite well developed but much of this has been designed before climate change had become a concern, and new developments need to take the projected climate change impacts into account.

Action in this report is composed of two determinants, institutions and economic resources that enable a society to act on adaptation. In comparison with the other determinants, more has been written about the role of institutions, including social capital and decision-making at different scales of social organisation (Spiess, 2008; Eakin, Lerner and Murtinho, 2010; Gupta *et al.*, 2010; Tompkins *et al.*, 2010). Collaborative learning and sharing of experiences and practices is considered to be important and can enhance adaptive capacity (Pelling and High, 2005; Marshall, 2010), and according to the literature, this is particularly important in the developing country context (Eakin, Lerner and Murtinho, 2010). Institutions perhaps play a larger role in enhancing adaptive capacity in Europe, where the other determinants are considered to be quite high already. Gupta *et al.* (2010) even outline the adaptive capacity of institutions themselves to comprise of 22 characteristics that can be used to analyse institution's adaptive capacity. For example, within the forest sector, socio-economic factors that affect the adaptive capacity of the sector relate to the management traditions and decision-making structures (Lindner *et al.*, 2010), and measures to enhance capacity need to address these management systems. There has not been much written in terms of enhancing the economic resources for adaptation. Some examples in terms of business do exist with the focus on different sectors (Berkhout, Hertin and Arnell, 2004).

In terms of the assessment conducted here, the adaptive capacity maps can give an initial indication in terms of what dimension scores lowest in the indicators. For Northern Europe, both awareness and ability seem to be lower than action, thus indicating that in terms of policy recommendations, measures that focus on knowledge and awareness as well technology and infrastructure can have a positive effect on adaptive capacity. In the United Kingdom, awareness appears to be lowest of all three dimensions, although quite significant variation between regions can be distinguished. Here policy recommendations to enhance adaptive capacity need to focus on all three dimensions. In terms of enhancing adaptive capacity in Central Europe, ability appears to be the dimension with the highest score, and therefore measures can be targeted towards enhancing the awareness and action dimensions of adaptive capacity. Regions around the Mediterranean appear to have lowest scores in terms of awareness and to some extent ability. However, given that on a European scale, the regions score lower on average, policy measures should target each of the three dimensions. Similarly, regions within Eastern Europe have lower capacities all around, but specific policy measures can be targeted on action, since this appears to have the lowest capacity of the three dimensions.

7.6 Conclusion

This chapter presents a generic adaptive capacity assessment of European regions as part of a vulnerability assessment. Adaptive capacity, defined as the ability of a system to adapt and take advantage of climate change, combined with exposure and sensitivity, form an understanding of the vulnerability of regions. In order to understand society's adaptive capacity, this report reviews literature on adaptive capacity and constructed an index to assess it. As a result, this report presents maps of adaptive capacity of European regions on a NUTS 3 level. The maps show how adaptive capacity varies between and within European countries. Northern, Western and Central Europe have higher adaptive capacities than Southern Europe. Eastern Europe on the whole has a lower capacity than other parts of Europe. Within countries, adaptive capacity is higher in the capital city regions in comparison with other regions in the countries. Only a few countries have uniform adaptive capacities across all regions.

Adaptive capacity underlies society's action and determines the response to the changing climate and enables action on climate change, but as already discussed, having high adaptive or mitigative capacity does not necessarily mean that action on climate change takes place. It is important to note that whilst adaptive capacity is integral in the process of adapting to climate change, it has been argued that it is not enough on its own for adaptation to take place (Smith, Vogel and Cromwell III, 2009). Recently, there have been efforts to produce a list of general outlines required for planned adaptation (Füssel, 2007), and the role of institutions is also advocated by some (Gupta *et al.*, 2010). Furthermore, it should be noted that high capacity at the national level is not necessarily reflected as high capacity at the lower levels of governance (O'Brien, Eriksen, Sygna and Næss, 2006), which is of particular relevance in the future.

References

Adger, W.N., Arnell, N.W. and Tompkins, E.L. (2005) Successful adaptation to climate change across scales. *Global Environmental Change Part A*, **15** (2), pp.77–86.

Alberini, A., Chiabai, A. and Muehlenbachs, L. (2005) *Using Expert Judgment to Assess Adaptive Capacity to Climate Change: Evidence from a Conjoint Choice Survey*. Fondazione Eni Enrico Mattei, Milano.

Berkhout, F., Hertin, J. and Arnell, N. (2004) *Business and Climate Change: Measuring and Enhancing Adaptive Capacity*. Tyndall Centre for Climate Change Research, Norwich.

Brooks, N., Adger, W.N. and Kelly, M.P. (2005) The determinants of vulnerability and adaptive capacity at the national level and the implications for adaptation. *Global Environmental Change Part A*, **15** (2), 151–163.

Eakin, H., Lerner, A.M. and Murtinho, F. (2010) Adaptive capacity in evolving peri-urban spaces: Responses to flood risk in the Upper Lerma River Valley, Mexico. *Global Environmental Change*, **20** (1), 14–22.

Engle, N.L. and Lemos, M.C. (2010) Unpacking governance: Building adaptive capacity to climate change of river basins in Brazil. *Global Environmental Change*, **20** (1), 4–13.

Eriksen, S.H. and Kelly, P. (2007) Developing credible vulnerability indicators for climate adaptation policy assessment. *Mitigation and Adaptation Strategies for Global Change*, **12** (4), 495–524.

Füssel, H.-M. (2007) Adaptation planning for climate change: concepts, assessment approaches, and key lessons. *Sustainability Science*, **2** (2) 265–275.

Greiving, S., Flex, F., Lindner, C., *et al.* (2011) Climate change and territorial effects on regions and local economies. Final Report, Applied research 2013/1/4, ESPON & IRPUD TU Dortmund, Dortmund.

Gupta, J., Termeer, C., Klostermann, J., *et al.* (2010) The adaptive capacity wheel: A method to assess the inherent characteristics of institutions to enable the adaptive capacity of society. *Environmental Science & Policy*, **13**, 459–471.

Haddad, B.M. (2005) Ranking the adaptive capacity of nations to climate change when socio-political goals are explicit. *Global Environmental Change Part A*, **15** (2), 165–176.

Hagmann, J. and Chuma, E. (2002) Enhancing the adaptive capacity of the resource users in natural resource management. *Agricultural Systems*, **73** (1), 23–39.

Hinkel, J. (2011) "Indicators of vulnerability and adaptive capacity": Towards a clarification of the science–policy interface. *Global Environmental Change*, **21** (1) 198–208.

IPCC (2001) Climate Change 2001: Impacts, Adaptation and Vulnerability, Cambridge University Press, Cambridge.

IPCC (2007a) Assessment of Adaptation Practices, Options, Constraints and Capacity, Cambridge University Press, Cambridge.

IPCC (2007b) Climate Change 2007: The Physical Science Basis. Contribution of Working Group I to the Fourth Assessment. Report of the Intergovernmental Panel on Climate Change, Cambridge University Press, Cambridge.

Keskitalo, E.C.H. (ed.) (2010) *The Development of Adaptation Policy and Practice in Europe: Multi-level Governance of Climate Change*. Springer Verlag, Dordrecht.

Leary, N.A. (1999) A framework for benefit-cost analysis of adaptation to climate change and climate variability. *Mitigation and Adaptation Strategies for Global Change*, **4** (3-4), 307–318.

Lindner, M., Maroschek, M., Netherer, S., *et al.* (2010) Climate change impacts, adaptive capacity, and vulnerability of European forest ecosystems. *Forest Ecology and Management*, **259** (4), 698–709.

Malone, E.L. and Engle, N.L. (2011) Evaluating regional vulnerability to climate change: purposes and methods. *Wiley Interdisciplinary Reviews: Climate Change*, **2** (3), 462–474.

Marshall, N.A. (2010) Understanding social resilience to climate variability in primary enterprises and industries. *Global Environmental Change*, **20** (1), 36–43.

Moss, R.H., Brenkert, A. and Malone, E.L. (2001) Vulnerability to Climate Change: A Quantitative Approach. Richland, Pacific Northwest National Library, Washington.

O'Brien, K., Eriksen, S., Sygna, L. and Næss, L.O. (2006) Questioning complacency: Climate change impacts, vulnerability, and adaptation in Norway. *AMBIO: A Journal of the Human Environment*, **35** (2), 50–56.

Pelling, M. and High, C. (2005) Understanding adaptation: what can social capital offer assessments of adaptive capacity? *Global Environmental Change*, **15** (4), 308–319.

Pittman, J., Wittrock, V., Kulshreshtha, S. and Wheaton, E. (2011) Vulnerability to climate change in rural Saskatchewan: Case study of the Rural Municipality of Rudy No. 284. *Journal of Rural Studies*, **27** (1), 83–94.

Posey, J. (2009) The determinants of vulnerability and adaptive capacity at the municipal level: Evidence from floodplain management programs in the United States. *Global Environmental Change*, **19** (4), 482–493.

Ribeiro, M., Losenno, C., Dworak, T., *et al.* (2009) Design of guidelines for the elaboration of Regional Climate Change Adaptation Strategies. Ecologic Institute, Vienna.

Saavedra, C. and Budd, W.W. (2009) Climate change and environmental planning: Working to build community resilience and adaptive capacity in Washington State, USA. *Habitat International*, **33** (3), 246–252.

Schröter, D., Acosta-Michlik, L., Arnell, A.W., *et al.* (2004) ATEAM Final report. Section 5 and 6 and Annex 1 to 6. Detailed report, related to overall project duration. Reporting period: 01.01.2001–30.06.2004. Potsdam.

Smit, B. and Pilifosova, O. (2001) Adaptation to climate change in the context of sustainable development and equity, in IPCC, 2001. *Climate Change 2001: Impacts, adaptation, and vulnerability- Contribution of the Working Group II to the Third Assessment report of the Intergovernmental Panel on Climate Change*. Cambridge University Press, Cambridge.

Smit, B. and Wandel, J. (2006) Adaptation, adaptive capacity and vulnerability. *Global Environmental Change*, **16** (3), 282–292.

Smith, J.B., Vogel, J.M. and Cromwell III, J.E. (2009) An architecture for government action on adaptation to climate change. An editorial comment. *Climatic Change*, **95** (1-2), 53–61.

Spiess, A. (2008) Developing adaptive capacity for responding to environmental change in the Arab Gulf States: Uncertainties to linking ecosystem conservation, sustainable development and society in authoritarian rentier economies. *Global and Planetary Change*, **64** (3-4), 244–252.

Tompkins, E.L., Adger, W.N. and Brown, K. (2002) Institutional networks for inclusive coastal management in Trinidad and Tobago. *Environmental Planning A*, **34**, 1095–1111.

Tompkins, E.L., Adger, W.N., Boyd, E., *et al.* (2010) Observed adaptation to climate change: UK evidence of transition to a well-adapting society. *Global Environmental Change*, **20** (4), 627–635.

Twomlow, S., Mugabe, F.T., Mwale, M., *et al.* (2008) Building adaptive capacity to cope with increasing vulnerability due to climatic change in Africa – A new approach. *Physics and Chemistry of the Earth, Parts A/B/C*, **33** (8-13), 780–787.

Vanhanen, M. (2010) *FSD1289 Measures of Democracy 1810-2010*. Finnish Social Science Data Archive, Tampere.

Westerhoff, L., Keskitalo, E.C.H. and Juhola, S. (2011) Capacities across scales: enabling local to national adaptation policy in four European countries. *Climate Policy*, **11** (4), 1071–1085.

Williamson, T., Hesseln, H. and Johnston, M. (2012) Adaptive capacity deficits and adaptive capacity of economic systems in climate change vulnerability assessment. *Forest Policy and Economics*, **15**, 160–166.

Yohe, G. and Tol, R.S.J. (2002) Indicators for social and economic coping capacity – moving toward a working definition of adaptive capacity. *Global Environmental Change*, **12** (1), 25–40.

Chapter 8

Exploring mitigative and response capacities to climate change in European regions

Sirkku Juhola[1,2], Lasse Peltonen[1,3], Petteri Niemi[1] and Jarmo Vehmas[4]

[1]*Aalto University, Department of Real Estate, Planning and Geoinformatics, P.O. Box 12200, 00076 Aalto, Espoo Finland*
[2]*University of Helsinki, Department of Environmental Sciences, P.O. Box 65, 00014 Helsinki, Finland*
[3]*Finnish Environment Institute, P.O. Box 140, 00251 Helsinki, Finland*
[4]*Finland Futures Research Centre, University of Turku, 20014 Turku, Finland*

Abstract

Mitigation, alongside adaptation, is an important societal response to climate change. Although mitigation does not explicitly play a role in a vulnerability assessment, the ability of regions to mitigate greenhouse gas emissions is an important, yet under-researched, issue. The determinants that underlie the ability to mitigate climate change, it is argued in the literature, are to some extent similar to those required to adapt to the impacts of climate change. This chapter reviews the literature and builds the concept of mitigative capacity further, and develops a set of indicators to assess the ability to reduce greenhouse gas emissions at the regional level. Furthermore, this chapter explores the idea of response capacity, by combining the idea of adaptive and mitigative capacity at the regional level. The concept of response capacity in terms of climate change is relatively new and this chapter presents the first illustrations of what response capacity could potentially look like on a European regional scale.

8.1   Introduction

Thus far this volume has focused on the vulnerability of European regions in relation to the expected climate change impacts. However, mitigation of greenhouse gas (GHG) emissions is also important, and considered to be a crucial societal response to global climate change (IPCC, 2007a). The determinants that underlie the capacity to mitigate climate change, that is, to reduce the emission of GHGs, are to a certain extent the same as the ones needed to adapt to the impacts of climate change. Indeed, it is possible to further examine the concept of response capacity (Tompkins and Adger, 2005), and this chapter explores this concept in more detail. Response capacity refers to idea of a system's ability to respond to the dual challenge of mitigation and adaptation. The concept itself has been explored relatively little in the literature, and this chapter contributes to the discussion in terms of combining adaptive and mitigative capacity indicators and illustrating the response capacity of European regions.

Hence, the aim of this chapter is twofold. First, a set of indicators to measure mitigative capacity of European regions are develpoed. In order to do this, this chapter firstly reviews the literature on mitigative capacity and the determinants that underlie it. Second, the chapter discusses the regional GHG emissions, as the ability to reduce emissions is naturally dependent on the level of emissions. Third, indicators for measuring mitigative capacity that mirror the adaptive capacity indicators that have been developed for the pan-European vulnerability analysis are discussed. Fourth, the concept of response capacity is introduced. Response capacity, in this case, refers to the ability of regions to respond to the climate challenge by combining aspects from both mitigative and adaptive capacity. Finally, this report discusses response capacity in terms of European regions.

8.2   Regional capacities to mitigate climate change

Climate change mitigation in general refers to all societal attempts to mitigate the emissions of GHGs that contribute to climate change. In practice, mitigation activities strongly focus on decreasing net GHG emissions into the atmosphere, stressing the preventive nature of climate policy. Owing to this, current climate policy also includes adaptation

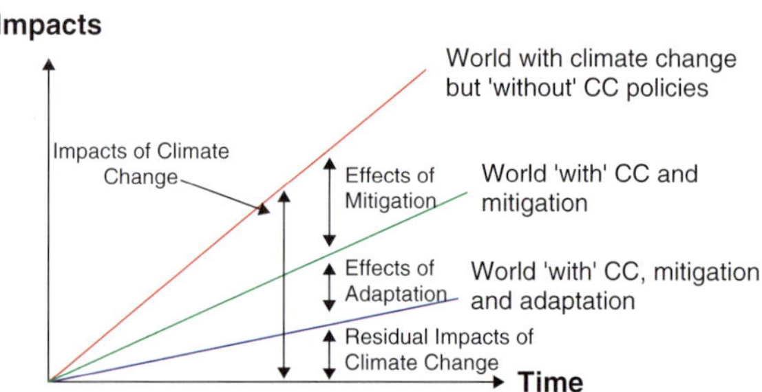

Figure 8.1   The impacts of climate change. *Sources*: Kirkinen *et al.* (2005), modified from Rothman, Amelung and Polomé (2003).

to climate change, which focuses the inevitability of climate change and its undesirable effects (see Figure 8.1). Impacts and the time taken for societies to respond either through mitigation or adaptation, or both, are linked in terms of how severe the impacts are likely to be in the future.

In this chapter, the focus is on developing a set of indicators from a territorial and/or regional perspective. Region here, as in other chapters in this volume, is considered to be the NUTS 3 region in Europe. It is important to keep in mind that climate change mitigative capacity, just like adaptive capacity, lacks a commonly agreed definition and is thus open for various types of interpretations. In this kind of situation, selected indicators and their data availability may more or less define or even determine the issue (cf. Rosenström, 2009, pp. 9 and 10). These theoretical concerns, as well as methodological issues, are similar to those when assessing adaptive capacity as discussed in this volume. These concerns are not reviewed here but the reader is advised to read the chapter on adaptive capacity (Chapter 7), as well as many recent publications that summarise these issues (Engle, 2011; Hinkel, 2011; Malone and Engle, 2011).

Mitigative capacity was initially discussed by the Third Assessment Report of the IPCC (2001) and by Yohe (2001), both employing definitions that are rather generic. Winkler *et al.* (2007) further clarified the definition of mitigative capacity, stating that '(W)e define mitigative capacity simply as a country's ability to reduce anthropogenic GHG emissions or enhance natural sinks' (p. 694). Winkler *et al.* (2007) also point out that mitigative capacity is the ability to reduce GHG emissions in either absolute or relative terms.

Given this definition, mitigative capacity is not intended to simply explain the degree to which countries do in fact mitigate GHG emissions, but it is about *how much countries could mitigate* – their theoretical possibilities to reduce emissions and enhance natural GHG sinks. The latter refers to mitigative *potential*, which could be used as a synonym for mitigative capacity. In this chapter, mitigative capacity is defined as the capacity to include elements from awareness, ability and action, mirroring the definition by Winkler *et al.* (2007) as well as the definition of adaptive capacity discussed in Chapter 7 of this book. This includes also *action*, which refers to what countries actually do or have done in order to mitigate GHG emissions, or to enhance natural sinks that are available to them.

8.3 Regional GHG emissions

In order to examine regional mitigative capacity, it is necessary to take into account the regional emissions. GHGs that will be dealt with here in the context of mitigative capacity, include those identified in the UNFCCC Kyoto Protocol (UNFCCC, 1998). These GHGs are carbon dioxide (CO_2), methane (CH_4), nitrous oxide (N_2O), hydrofluorocarbons (HFCs), perfluorocarbons (PFCs) and sulfur hexafluoride (SF_6), all measured in tonnes of CO_2 equivalents.

Regional GHG emissions inevitably have an influence on the mitigative capacity of that region (Figure 8.2). It can be even argued that the larger the regional GHG emissions, the larger the driver for improving the regional mitigative capacity. On the other hand, climate change is a global phenomenon, and GHG emissions contributing to this problem are not dependent on their geographical origin. Thus, regional GHG emissions are an important element from the mitigative capacity perspective. It is

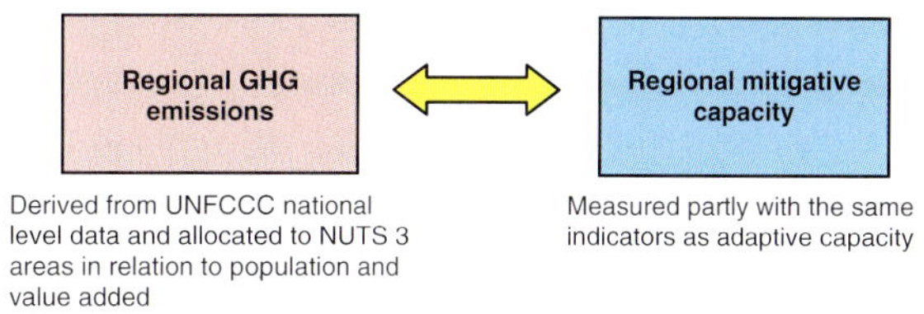

Figure 8.2   Relationship between regional GHG emissions and mitigative capacity in the region.

important to calculate them and include them in the assessment here. Moreover, it is interesting to compare them with the regional mitigative capacity and evaluate their relationship. This comparison gives the opportunity to examine whether areas with high emissions also have high capacity, or areas with high emissions have low capacity to mitigate.

8.3.1   *Indicator methodology*

The data for this study of European regions are derived from the UNFCCC. The UNFCCC Secretariat collects national level GHG data through the national reports that countries submit (see UNFCCC, 2010). The available data from the Secretariat covers each GHG emission category and includes time series from 1990 onwards at the national level (up to 2007 at the moment). Given that emissions data are only available at the national level, it presents a challenge to this study where the focus is on the NUTS 3 level. Generally, GHG emissions data at regional/local level is usually not available, the national level data need to be allocated to regions in relation to a variable that reflects the size of the region, such as land area, amount of population and regional GDP or value added.

Allocating all national GHG emissions to regions in relation to one variable provides a very rough estimate on regional GHGs. In industrialised countries, such as EU Member States, CO_2 from energy use represents 80–90% of all GHG emissions. Applying regional value added (provided by EUROSTAT, 2009) to energy consumption of industry, services, agriculture and non-energy use (provided by IEA, 2009), and population (provided by EUROSTAT, 2010) to energy consumption of transport and the residential sector (provided by IEA, 2009) allocates sectoral energy consumption to regions, and assuming that the same allocation holds with GHG emissions, we get a slightly better estimate on regional GHG emissions. The results of this calculation will be added to the map on mitigative capacity, see Figure 8.5 and also Figure 8.6.

8.4   Determinants of mitigative capacity

The literature on mitigative capacity and its determinants is not as wide as the literature on adaptive capacity and its determinants because less attention has been paid to mitigative capacity in the literature on the whole. Yohe (2001) was the first to attempt to define determinants of mitigative capacity, and his work has been referred to by the IPCC in the Third Assessment Report (2001). Yohe's list of determinants of mitigative capacity has been influenced by the work on adaptive capacity. The following generic determinants are presented, which influence the ability of a system to reduce emissions:

1. The range of viable technological options for reducing emissions.
2. The range of viable policy instruments with which it might affect the adoption of these options.
3. The structure of critical institutions and the derivative allocation of the decision-making authority.
4. The availability and distribution or resources required to underwrite their adoption and the associated, broadly defined opportunity cost of devoting those resources to mitigation.
5. The stock of human capital, including education and personal security.
6. The stock of social capital including the definition of property rights, the country's access to risk spreading processes and the ability of decision-makers to manage information, the processes by which these decision-makers determine which information is credible.

7. The credibility of the decision-makers themselves.

Yohe (2001) states that this is essentially the same list of determinants that he established for adaptive capacity in the same article, but their application to the emissions side of the climate issue is not the same. This suggests that the indicators of mitigative capacity cannot be the same as indicators of adaptive capacity, but he does not develop a list of indicators further.

However, Winkler *et al.* (2007) have further developed Yohe's determinants and also defined a set of indicators in relation to the factors that influence mitigative capacity. The authors use the term factor instead of determinant, perhaps to reflect the difficulty of commanding all aspects of climate change mitigation and squeezing them into a compact list of determinants. Thus, Winkler *et al.* (2007) provide a list of determinants of mitigative capacity and related indicators of which some are quantitative, and some qualitative, see Table 8.1 for more details.

In addition to Yohe (2001) and Winkler *et al.* (2007), Richerzhagen and Scholz (2008) have dealt with capacities for climate change mitigation in their analysis that focused on China. The authors identify structural determinants of climate capacities that are similar to the determinants identified by the previous studies. The structural determinants that are identified include economic–technological factors, such as the structural features of economy and society, as well as the availability of climate-relevant economic, financial and technological resources. Second, the authors consider political–institutional factors to be important and these are, for example, different kinds of governance arrangements, administrative structures and procedures as well as climate-related policies and laws that influence participation and the coordination of policies and activities of public administration and other relevant stakeholders. Finally, Richerzhagen and Scholz (2008) consider that cognitive-informational issues, that is, the existence of climate-related information and the degree of public concern about climate problems, are crucial factors in the ability to reduce GHG emissions.

Table 8.1 Determinants of mitigative capacity and related indicators.

Determinant		Indicators
Economic factors	Ability to pay Abatement costs	GDP_{ppp} per capita Average abatement cost of reducing GHG emissions 20% from business as usual until the year 2030 (data from the POLES project)
	Opportunity costs	The best foregone alternative use of the money. Change in production possibility frontier (PPF), which represents the trade-off between expenditure on mitigation and on other purposes. No quantitative indicators presented
Institutional factors	Regulatory effectiveness and market rules	Ability of the Government to formulate and implement sound policies and measures such as CDM projects; effective rules for markets such as carbon tax or tradable permit system; capacity to monitor and enforce regulations; effectiveness of the court system in enforcing contracts. No quantitative indicators presented
	Education and skills base	Adult literacy rate; enrolment ratio (% of population of school age enrolled in education)
	Public attitudes and awareness	Influence, effectiveness and agenda of media (in relation to climate change); GHG emission targets at municipal level: reliance on science of climate change; reliance on climate-oriented NGOs
Technological factors		Number of researchers per million inhabitants; electricity consumption per capita; number of telephones per 1000 people; Internet users per 1000 people

Source: Winkler *et al.* (2007).

This chapter follows the idea of mirroring adaptive and mitigative capacity by using the same dimension and determinant categories. As in adaptive capacity, studies on mitigative capacity have also broadly identified and focused on factors that enable mitigation either by increasing awareness, hence supporting the ability to and encouraging action. However, as suggested by Yohe (2001) and particularly by Winkler *et al.* (2007), the same indicators are not used to measure both capacities but some different indicators are developed to particularly reflect factors that support mitigative actions.

Again, as in the context of adaptive capacity, this chapter adopts the definition by Schröter *et al.* (2004) and divides mitigative capacity into three determinants – awareness, ability and action. The actual determinants include knowledge and awareness representing the awareness dimension; infrastructure and technology representing the ability dimension and economic resources and institutions representing the action dimension (see Figure 8.3). For details of the indicators chosen and the proxies used, see Table 8.2.

In terms of the first determinant, knowledge and awareness are considered to be important. Recognition of climate change and the necessity to mitigate, gathering knowledge of available options, and the ability to assess and implement the policies and measures are crucial for mitigative capacity (IPCC, 2001). Skilled, informed and trained personnel enhances mitigative capacity and access to information is likely to lead to development of mitigation options that are timely and appropriate, whilst lack of trained and unskilled personnel can lower a nation's mitigative capacity (Yohe, 2001; Winkler et al., 2007). In the European context, awareness also plays an important role as it can be argued awareness and the willingness to act on climate change rather than just being educated are more important in the European context.

The second dimension that consists of the technology and infrastructure determinants assesses the availability of technology and of new technologies. Technological resources enable different mitigation options, and consequently lack of access and development of technology can lead to lower mitigative

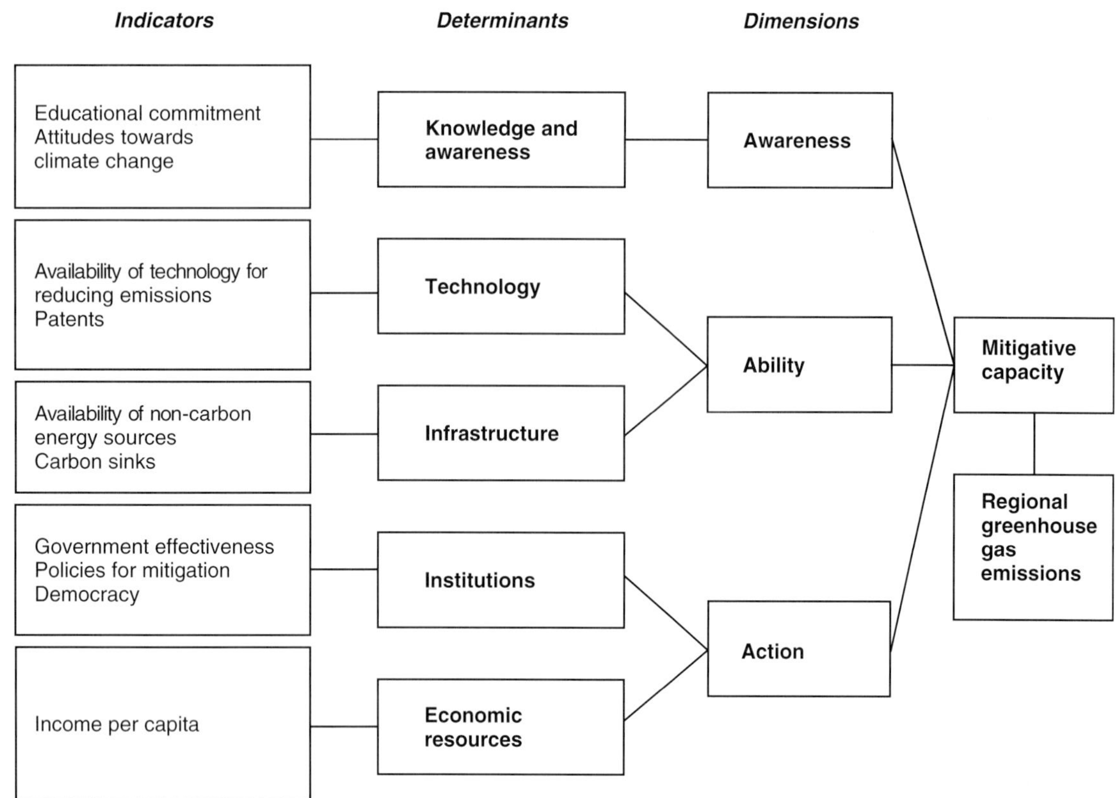

Figure 8.3 Dimensions, determinants and indicators of mitigative capacity. Adapted from Schröter *et al.* (2004).

capacity, since many of the strategies identified in response to climate change involve technology (IPCC, 2001). Both the public and private sector undertake the development of technologies, and innovation is considered an important factor in this. However, it is necessary to keep in mind the difference between general technological capacity, and specific technological response to which a specific GHG component can be developed for, or specific type of carbon sink. However, given the difficulties in accessing data, the focus here is on general aspects of technology without a specific focus on particular GHG or sink type.

Infrastructure is important for mitigative capacity as it can facilitate and enable the move to the use of non-fossil fuel energy resources. Natural resources, particularly related to the utilisation of renewable energy, available for a region are part of the energy infrastructure that underlies a region's ability to mitigate GHGs. Similarly, land use is important from the perspective of climate change mitigation capacity. For example, changes in land use cover impact the emissions of a region because forests are an important carbon sink in the climate system.

The final and third dimension, action, consists of institutional capacity as well as economic resources. Institutions, defined as social rules and norms, are considered to play an important part developing responses to climate change, and also are a determinant of mitigative capacity. Well-developed institutions enable societies to plan and execute policies related to the reduction of GHGs. In addition, the efficiency of government is an important factor overall in decision-making, and is likely to impact positively the mitigative capacity of a region. If decision-making is effective and carried out in

Table 8.2   Indicators for mitigative capacity.

Determinant	Proxy	Relevance	Methodology
Knowledge and awareness			
	Educational commitment	Skilled and trained personnel and population aids in assessing and implementation of measures required to mitigate emissions	Education expenditure per capita within a region
	Attitudes towards climate change	Attitudes influence whether mitigation actions will be undertaken	Eurobarometer questions surveyed in 2008 and 2009
Technology			
	Resources of technology	Ability to develop new technologies for adaptation is an important element of mitigative capacity	Percentage of GDP in R&D investment
	Patents	Utilisation of research findings for further purposes to develop products and services	Number of patent applications per million inhabitants
Infrastructure			
	Carbon sinks in the region	Carbon sinks help mitigation by absorbing CO_2 from the atmosphere	National estimates of carbon sinks are available at the UNFCCC database
	Non-carbon energy resources available in the region	Use of renewable energy sources aids in decreasing of CO_2 emissions	Data on 'PV output' and 'wind energy potential' from the ESPON ReRisk project
Institutions			
	Government effectiveness	The efficiency of government is likely to impact positively the mitigative capacity of a region	Government effectiveness index that is available from the World Bank Group at NUTS 0 level
	Democracy	Mitigative capacity will be greater if resources and power in governing resources for mitigation are equitably allocated	Gender weighted democratisation index (Vanhanen, 2010). Data for this are available from FSD (The Finnish Social Science Data Archive)
	Policies and measures in use for climate change mitigation in the region	Existing policies and measures aimed at climate change mitigation reflect a political willingness to mitigate climate change	The number of policies and measures in use in the region from EEA database
Economic resources			
	Income per capita	Economic assets can be used to fund and support mitigation measures and strategies	GDP per capita (€ PPP) of a NUTS 3 region

that manner, it is also likely that decisions related to mitigation will be taken.

Furthermore, existing policies and measures aimed at climate change mitigation reflect a political willingness to mitigate climate change. Climate change related policies are designed, including the use of legally based instruments, at the national level. In addition, some regions may have their own GHG emission targets, economic instruments, voluntary agreements and other instruments. Policies and measures are often in use at national level, so the national data availability is applicable at a regional level in most of the cases. Thus, there is no specific need to provide a regional indicator on policies and measures due to the fact that policies and measures are mostly implemented at a national level, following a common EU policy.

Economic assets are an important part of mitigative capacity as they enable the use of capital resources and financial means to be used in the development of techniques for low carbon technologies. They play an important role in mitigative capacity as they can be used to fund and support mitigation measures and strategies, and also in the development of technologies to reduce GHG emissions. They can also further increase mitigative capacity by investment in other capacities, such as information dissemination and education, amongst others.

8.4.1 *Methodology of indicators*

In terms of operationalising a set of indicators for mitigative capacity at the regional level, an obvious difficulty is the availability and the challenge of gathering regional data. To some extent mathematical allocation of national data to regions may be reasonable, but obviously gathering original data at the regional level requires access to original data collected for national statistics. This is often not available, which strongly limits the possibilities to provide relevant and reliable information on regional climate change mitigation capabilities, and should be kept in mind when interpreting the results.

The general methodology of estimating regional mitigative capacity in this chapter is identical to that of adaptive capacity described in Chapter 7. While 15 indicators are used to calculate adaptive capacity ten are used for mitigative capacity. Owing to data availability issues, eight of those ten indicators are the same as those used in the adaptive capacity set. Regions are ranked on all individual indicators and classified into quintiles. The assigned quintile values (positions) are then used for further calculations. Determinants of mitigative capacity are calculated as un-weighted averages of indicators specified in Figure 8.3. The determinants are then normalised relative to the largest values. The mitigative capacity index is then calculated as the weighted average of the determinants, using the same Delphi weights as in the adaptive capacity.

8.5 Territorial potentials for mitigation of climate change

The territorial potential for mitigation can be illustrated with two different maps. Firstly, it is possible to illustrate, similarly to adaptive capacity, the mitigative capacity of European regions. This map includes both the GHG emissions of regions as well as the indicators that are used to assess mitigative capacity. Secondly, it is possible to explore and compare the regional GHG emissions in relation to the mitigative capacity. With this method, it is possible to identify regions that have high emissions but also high capacity to mitigate and vice versa.

The mitigative capacity in Europe is illustrated in Figure 8.4. In terms of the whole of Europe, the Nordic countries appear to have the highest mitigative capacity. Also France and parts of Italy have high capacities. The lowest capacity in mitigation can be observed in Portugal and the former Eastern European countries. Thus, a clear north–south and east–west divide can be observed on the European scale in terms of mitigative capacity.

Many of the capital city regions again appear to have high capacities, as with the results of the adaptive capacity assessment (Chapter 7). To what

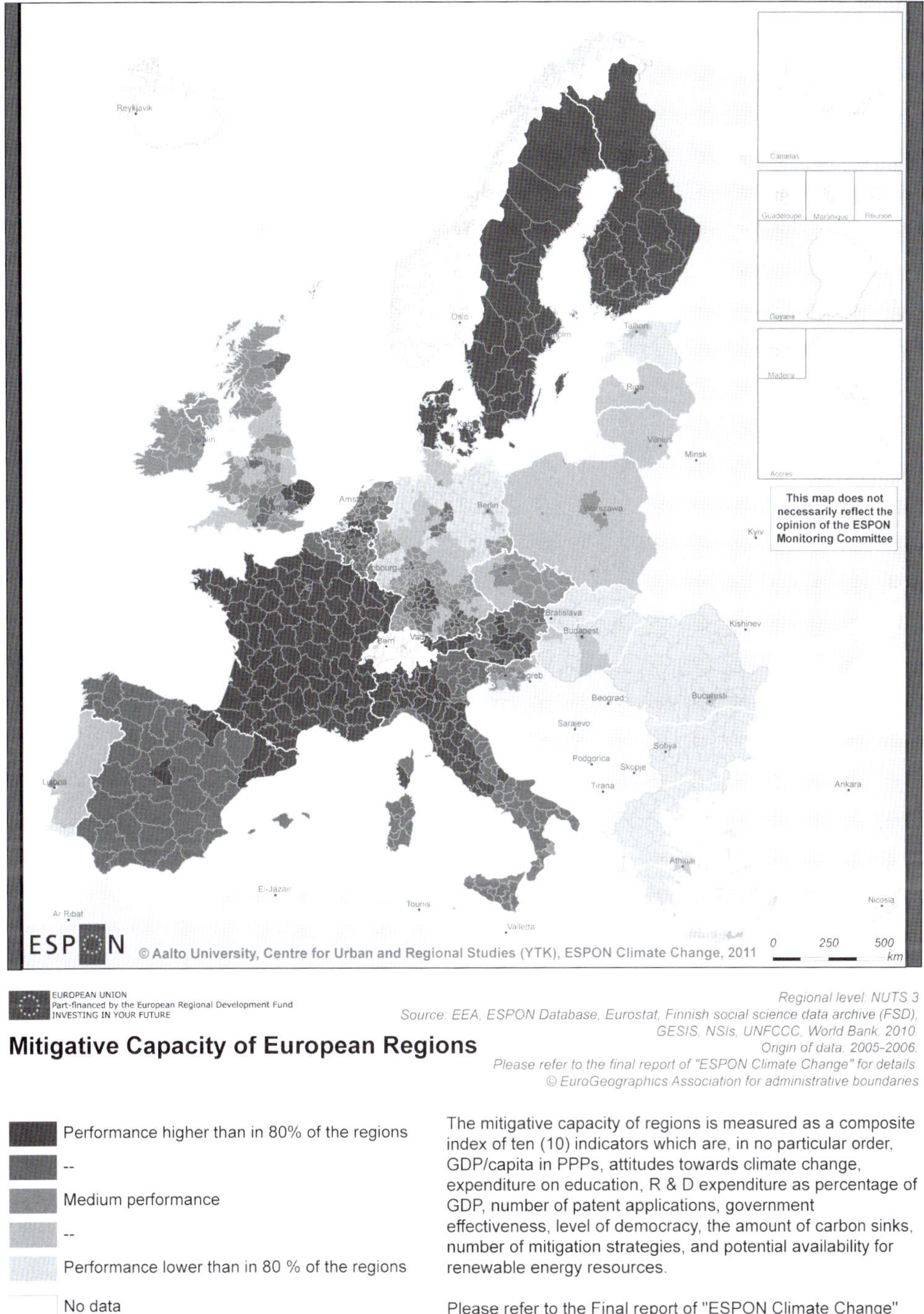

Figure 8.4   Mitigative capacity of European regions. © ESPON 2013, IRPUD, ESPON Climate Project, 2011.

extent this is a result of indicators, such as GDP or investment in research and technology, or the existence or availability of renewable energy resources is of course debatable. As always with illustrations of data, care should be taken when interpreting the results. Because many of the indicators used are from the national level, there is no significant variation within many countries. However, regional variation in mitigative capacity can be observed in Germany, the United Kingdom and Italy in particular. The Nordic countries, according to this assessment, have no regional variation at all.

As mentioned above, it is also possible to examine regions by comparing their GHG emissions and their mitigative capacity, see Figure 8.5. In terms of the results, regions can be classified into four different types, depending on whether they have high or low levels of GHG emissions and whether they simultaneously have high or low mitigative capacity. A high value is considered to be average value and these are calculated from the case study area. Similarly, low GHG emissions mean that the rate of emissions is lower than the average of the case study areas.

It appears that regions that have low emissions and high mitigative capacity are mostly located in Northern parts of Europe, and parts of France and the Iberian Peninsula. Regions that have high emissions and high mitigative capacity can be found in Western Europe as well as in parts of Scandinavia. Regions that have low emissions and low mitigative capacity can mostly be found in Eastern Europe as well as in Scotland and Portugal. Regions that have high emissions and low mitigative capacity are of course the most crucial in terms of reduction of GHG emissions. These regions can be found in Eastern Europe, and in the United Kingdom Isles and Ireland. Also, some regions in Southern Italy fall into this category.

As this is very much exploratory work and given the constraints and limitations related to data and the methodology, it is premature to draw any firm conclusions based on the findings. In terms of enhancing mitigative and response capacity, similar conclusions can be drawn when discussing the enhancement of adaptive capacity. Measures that target the ability to respond to the climate change challenge can be complementary to both goals as the same determinants can underlie the action on both mitigation and adaptation. However, it is as yet unknown whether some capacities are necessary to create or support the emergence of other capacities. Overall, measures that support the awareness, ability and action on mitigation, can contribute to the mitigative capacity as a whole, and the response to climate change in general.

8.6 Regional response capacity to deal with climate change

Mitigation of GHG emissions and adaptation to climate change impacts have generally been dealt with in separate policy domains, not only on the global level but also by national governments. However, at the regional and local level, adaptation and mitigation are often dealt with together in a regional or a local climate change strategy (Ribeiro *et al.*, 2009). These strategies, in most cases, have begun as strategies for mitigation but have then incorporated adaptation into the strategy process. Given the previous analysis of mitigative and adaptive capacity (Chapter 7), this raises the broader question of what is the capacity of the European regions to deal with climate change.

The relationship between mitigation and adaptation has increasingly become of interest to researchers. The Fourth Assessment Report of the IPCC states that the capacities to adapt and mitigate are driven by similar sets of factors (IPCC, 2007b). These factors, according to the IPCC, represent a generalised response capacity that can be mobilised both for adaptation and mitigation. A society's response capacity is closely related to the development path chosen, and it is argued that pursuing sustainable development can be a way of promoting both mitigation and adaptation. Furthermore, it is pointed out that response capacity, whilst relating to climate policy, is also closely tied

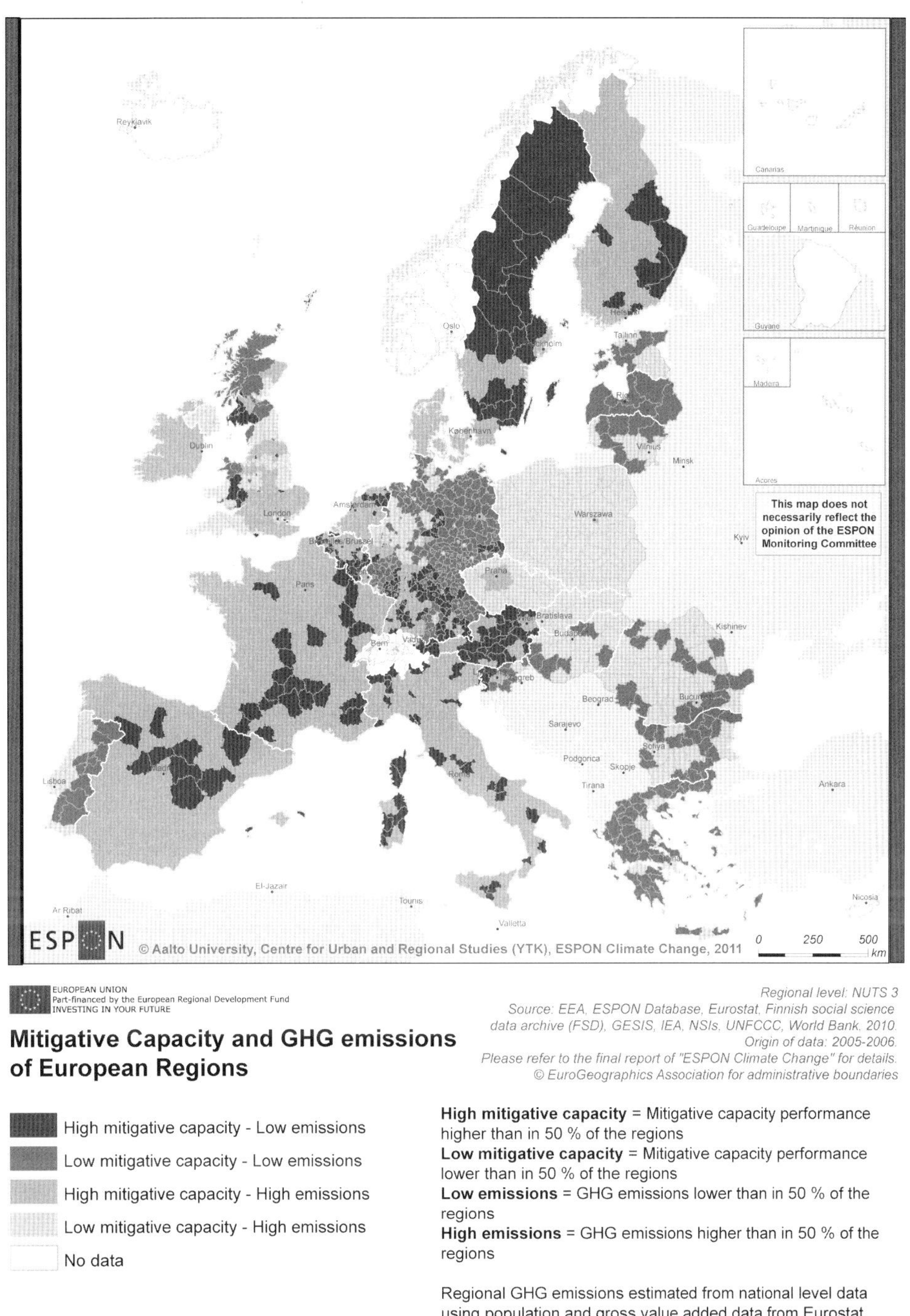

EUROPEAN UNION
Part-financed by the European Regional Development Fund
INVESTING IN YOUR FUTURE

Regional level: NUTS 3
Source: EEA, ESPON Database, Eurostat, Finnish social science
data archive (FSD), GESIS, IEA, NSIs, UNFCCC, World Bank, 2010.
Origin of data: 2005-2006.
Please refer to the final report of "ESPON Climate Change" for details.
© EuroGeographics Association for administrative boundaries

Mitigative Capacity and GHG emissions of European Regions

- High mitigative capacity - Low emissions
- Low mitigative capacity - Low emissions
- High mitigative capacity - High emissions
- Low mitigative capacity - High emissions
- No data

High mitigative capacity = Mitigative capacity performance higher than in 50 % of the regions
Low mitigative capacity = Mitigative capacity performance lower than in 50 % of the regions
Low emissions = GHG emissions lower than in 50 % of the regions
High emissions = GHG emissions higher than in 50 % of the regions

Regional GHG emissions estimated from national level data using population and gross value added data from Eurostat. Please refer to the final report of "ESPON Climate Change" for details.

Figure 8.5   Mitigative capacity and GHG emissions of European regions. © ESPON 2013, IRPUD, ESPON Climate Project, 2011.

to the underlying socio-economic and technological development paths of a given society.

Tompkins and Adger (2005) have further explored the notion of response capacity in order to highlight the unnecessary dichotomy between mitigation and adaptation, and to aid the formulation and implementation of climate policy. Tompkins and Adger consider that creating a false dichotomy between adaptation and mitigation slows down the response to the climate challenge. Rather, it is more useful to focus on the two together as part of the management of risk and resources in a society. Response, according to the authors, is defined as any actions that are taken by any region, nation, community or an individual, to tackle or manage environmental change, either before the change occurs or before the change has taken place (Tompkins and Adger, 2005). In defining response capacity, the authors avoid an explicit reference to climate policy in order to emphasise the fact that there are also many other drivers of decision-making and that climate issues should not be analysed in isolation from wider developments in societies.

The authors identify two factors that drive response capacity, mainly the availability and penetration of new technology and willingness and capacity of a society to change or adopt this new technology (see Figure 8.6). Thus, response capacity 'describes the ability to manage both the cause of environmental change and the consequences of that change' (Tompkins and Adger, 2005, p. 564). These two factors create a response space within which responses to climate change takes place, combining the availability of new technology and willingness to change existing practices.

Tompkins and Adger do not explicitly address or explore the relationship between adaptive and mitigative capacity to response capacity in their paper. This chapter acknowledges that this relationship is most likely complex and that it may well be too simplistic to assume that combining adaptive and mitigative capacity in a simple manner equals response capacity on a conceptual level. However, response capacity denotes the capacity to deal with the causes of environmental change, that is, mitigation, and the consequences of that change, that is

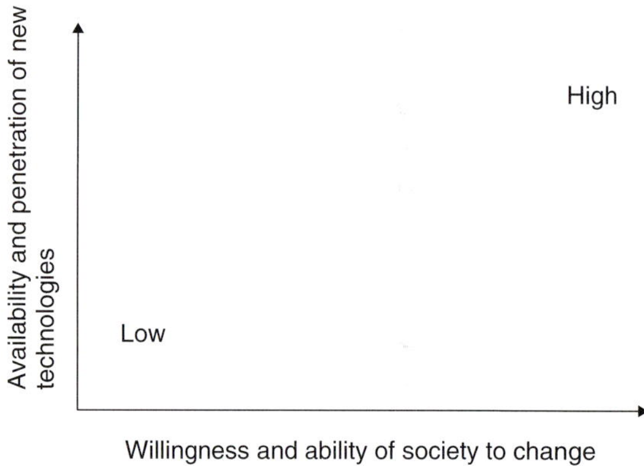

Figure 8.6   Response space. *Source*: Tompkins and Adger (2005).

adaptation. Furthermore, the willingness of a society to change is related to the awareness dimension of both adaptive and mitigative capacity as well by the action dimension of adaptive and mitigative capacity. In addition, availability and penetration of new technologies in a society is related to the ability dimension of adaptive and mitigative capacity, see Figure 8.7. In terms of the indicators listed under each determinant, those related to mitigative capacity are listed first and those related to adaptive capacity are listed at the bottom.

8.6.1   *Methodology*

Although the response capacity could be considered to be the product of adaptive and mitigative capacities, similarities between adaptive and mitigative capacity indices present challenges when using them to assess the response capacity of regions. Combining the two mathematically into one single response capacity indicator would stress the importance of those indicators that are used in both adaptive and mitigative capacity components. As we have no reliable information of the relative importance of individual indicators beyond the determinant level, the results would not be well-grounded.

The assessment of response capacity is done by comparing mitigative and adaptive capacities through a simple two-dimensional classification.

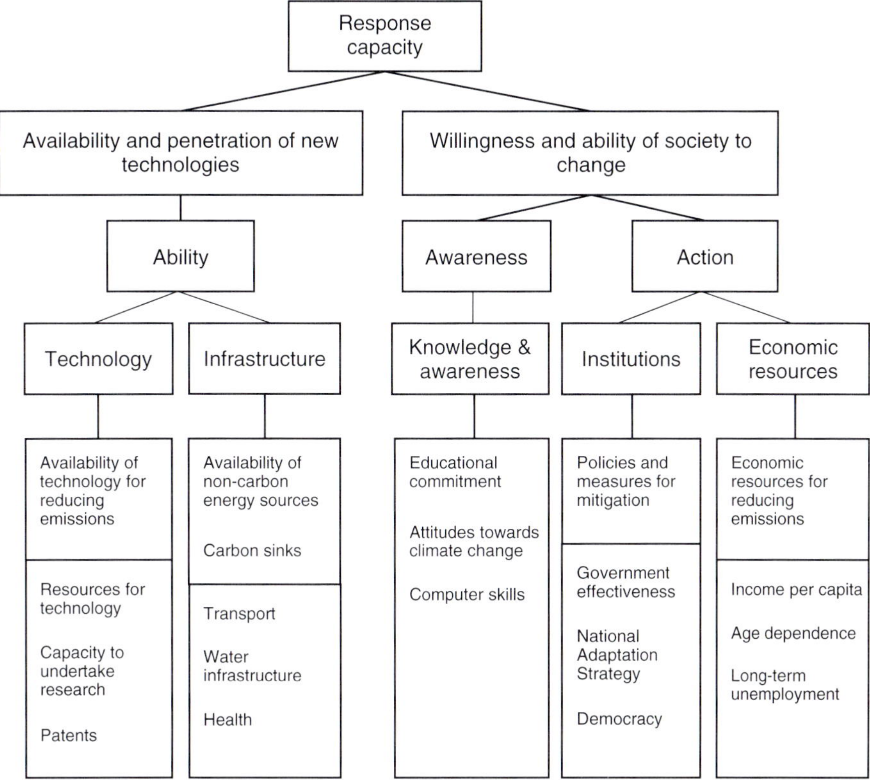

Figure 8.7   Response capacity combining adaptive and mitigative capacity.

The regions are divided into two classes by the median values of both indices. Integrating these creates four classes, high–high, high–low, low–high and low–low. This makes it possible to compare the differences between adaptive capacity and mitigative capacity and observe, in general terms, how the response capacity of regions varies across the European regions.

8.6.2   Map of response capacity

The map of response capacity of European regions integrates both adaptive and mitigative capacities (see Figure 8.8). This is due to the similarity of the two indices. They have many common indicators and merely adding the two together would stress those indicators.

As already discussed above, both mitigative and adaptive capacity are measured to a certain extent with the same set of indicators. Differences between mitigative and adaptive capacity arise mainly from the infrastructure determinants. For mitigative capacity, these consist of availability of non-carbon energy sources and types of land use, whereas for adaptive capacity, infrastructure indicators measure transport, water infrastructure and health. In addition, a different indicator is used in measuring the institutions determinant. For mitigation, the number of policies measured relates directly to mitigation, whilst for adaptation the indicator constructed measures the levels of the national adaptation strategy (NAS) (concern, recommendation, measure).

Overall, the map shows that there are regions that have high or low capacities in both adaptive and mitigative capacity, but also that there are regions within which either mitigative or adaptive capacity is lower than the other. Regions, which have high capacity to both mitigate and adapt, have carbon

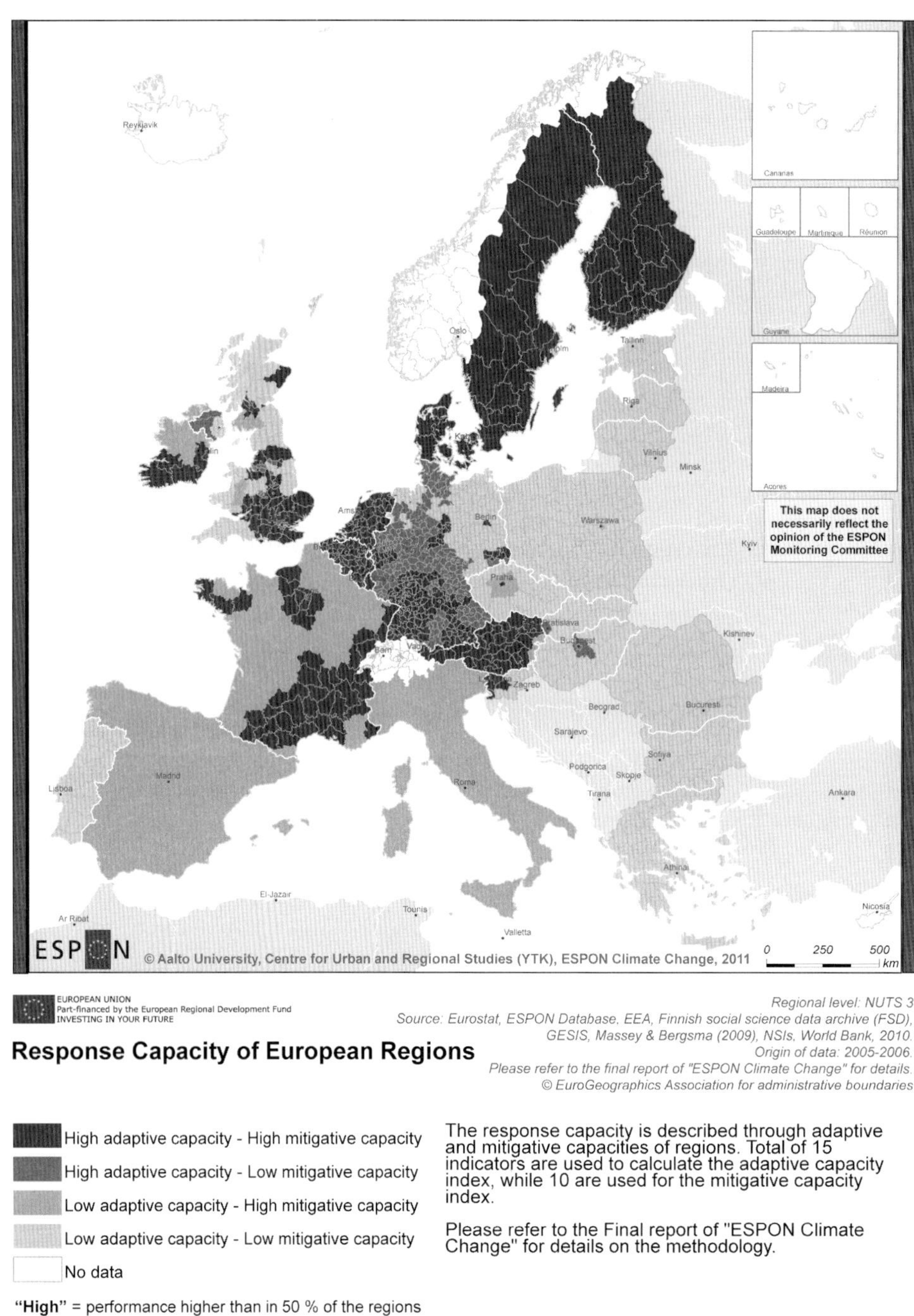

Figure 8.8   Response capacity of European regions. © ESPON 2013, IRPUD, ESPON Climate Project, 2011.

sinks and policies in place to reduce GHG emissions as well as infrastructure to deal with the impacts of climate change.

Regions that score low both on adaptive and mitigative capacity have less capacity to respond to climate change overall. Thus, these differences mainly arise from the fact that the countries within which the regions exist do not have national level policy measures to mitigate or adapt in place, since the policy indicators for both mitigative and adaptive capacity are measuring data on the national level. Furthermore, infrastructure in terms of lower capacity means lower potential for non-carbon energy sources and smaller areas of carbon sinks, or alternatively less infrastructure to deal with the impacts of climate change.

The differences between the types of regions also have implications for policy in terms of mitigation and adaptation. There are regions that have the capacity to mitigate and adapt to the impacts of climate change, whilst there are also regions which have a lower capacity to contribute to the mitigation of GHG emissions, and that also have lower capacity to adapt to the impacts. Similarly, there are regions that have a lower capacity to mitigate but higher adaptive capacity, or alternatively regions that have lower capacity to adapt but higher capacity to mitigate. In terms of policy implications, emphasis can be placed on increasing those dimensions that are the same for both capacities in order to improve the overall response capacity to climate change. Enhancing and increasing of mitigative and adaptive capacity is equally important, and can be in many cases complementary since it is recognised that similar capacities can underlie both actions.

8.7   Conclusion

This chapter discusses the mitigative as well as response capacity of European regions to deal with to climate change. In order to understand society's response capacity, literature on mitigative and response capacity is reviewed and a list of indicators to measure these concepts constructed. Mitigative

capacity consists of three dimensions and these dimensions are divided into determinants, which are knowledge and awareness, infrastructure, technology, economic resources and institutions. As a result, the chapter presents a map of European regions that illustrate these determinants.

Furthermore, the response capacity of European regions is reported, by combining both mitigative and adaptive indicators to reflect the ability of regions to respond to both reduction of greenhouse gases and adapt to the impacts of climate change.

The maps show how response and mitigative capacities vary between countries in Europe as well as within countries. Northern, Western and Central Europe have higher mitigative capacity than Southern Europe. Eastern Europe on the whole has a lower capacity than other parts of Europe. Within countries, mitigative capacity was higher in capital city regions in comparison with other regions in the countries. Only a few countries have uniform mitigative capacity across all regions. Adaptive capacity, partially being measured with the same indicators as mitigative capacity, shows similar trends in Europe.

References

Engle, N.L. (2011) Adaptive capacity and its assessment. *Global Environmental Change*, **21** (2), 647–656.

EUROSTAT (2009) Statistics Database-Population. (online) Available at: <http://epp.eurostat.ec.europa.eu /portal/page/portal/statistics/search_database> (accessed 4 March 2013).

EUROSTAT (2010) Statistics Database-Energy statistics. (online) Available at: <http://epp.eurostat.ec.europa.eu /portal/page/portal/statistics/search_database> (accessed 4 March 2013).

EUROSTAT (2012) *Indicators – Headline Indicators*. (online) Available at: <http://epp.eurostat.ec.europa.eu /portal/page/portal/sdi/indicators> (accessed 05 October 2012).

Hinkel, J. (2011) Indicators of vulnerability and adaptive capacity: Towards a clarification of the science–policy interface. *Global Environmental Change*, **21** (1), 198–208.

IEA (International Energy Agency) (2009) *World Energy Outlook*. International Energy Agency, Paris.

IPCC (2001) *Climate Change 2001: Impacts, Adaptation and Vulnerability*. Cambridge University Press, Cambridge.

IPCC (2007a) *Climate Change 2007: The Physical Science Basis. Contribution of Working Group I to the Fourth Assessment. Report of the Intergovernmental Panel on Climate Change.* Cambridge University Press, Cambridge.

IPCC (2007b) *Contribution of Working Group II to the Fourth Assessment Report of the Intergovernmental Panel on Climate Change.* Cambridge University Press, Cambridge.

Kirkinen, J., Martikainen, A., Holttinen, H., *et al.* (2005) *Impacts on the energy sector and adaptation of the electricity network business under a changing climate in Finland.* [pdf] Helsinki: Finnish Environment Institute Mimeographs. Available at: <http://www.ymparisto.fi/download.asp?contentid=45340&lan=EN> (accessed 05 October 2012).

Malone, E.L. and Engle, N.L. (2011) Evaluating regional vulnerability to climate change: purposes and methods. *Wiley Interdisciplinary Reviews: Climate Change*, **2** (3), 462–474.

Ribeiro, M., Losenno, C., Dworak, T., *et al.* (2009) *Design of Guidelines for the Elaboration of Regional Climate Change Adaptation Strategies.* Ecologic Institute, Vienna.

Richerzhagen, C. and Scholz, I. (2008) China's capacities for mitigating climate change. *World Development*, **36** (2), 308–324.

Rothman, D.S., Amelung, B. and Polomé, P. (2003) Estimating Non-Market Impacts of Climate Change and Climate Policy. (pdf) Available at: <http://www.oecd.org/dataoecd/6/30/2483779.pdf> (accessed 05 October 2012).

Rosenström, U. (2009) Sustaibnable development indicators - Much wanted, less used? Ph.D. thesis, University of Helsinki.

Schröter, D., Acosta-Michlik, L., Arnell, A.W., *et al.* (2004) *ATEAM Final report 2004: Section 5 and 6 and Annex 1 to 6.* Potsdam Institute for Climate Impact Research, Potsdam.

Tompkins, E.L. and Adger, W.N. (2005) Defining response capacity to enhance climate change policy. *Environmental Science & Policy*, **8** (6), 562–571.

UNFCCC (United Nations Framework Convention on Climate Change) (1998) *Kyoto Protocol to the United Nations Framework Convention on Climate Change.* United Nations, Washington.

UNFCCC (2010) *Greehouse gas emissions database.* (online) Available at: <http://unfccc.int/ghg_data/items/3800.php> (accessed 05 October 2012).

Vanhanen, M. (2010) *FSD1289 Measures of Democracy 1810–2010.* Finnish Social Science Data Archive, Tampere.

Winkler, H., Baumert, K., Blanchard, O., *et al.* (2007) What factors influence mitigation capacity? *Energy Policy*, **35** (1), 692–703.

Yohe, G. (2001) Mitigative capacity – the mirror image of adaptive capacity on the emissions side. *Climatic change*, **49** (3), 247–262.

Chapter 9

Overall impact and vulnerability to climate change in Europe

Johannes Lückenkötter, Christian Lindner and Stefan Greiving

TU Dortmund University, Institute of Spatial Planning (IRPUD), August-Schmidt-Strasse 10, 44227 Dortmund, Germany

Abstract

The previous chapters focused on the individual components of climate change vulnerability (exposure, impacts, adaptive capacity). Typically, many vulnerability assessments refrain from taking the next step of aggregating their individual results into a summary indicator of the overall impact or overall vulnerability to climate change. This neglects, however, that the number and particular selection of indicators used in any study already constitutes a selective view that may be guided by the intention to cover the *most important* aspects of a given domain, such as social climate change impacts. However, particularly from a policy point of view (which is often more concerned with the overall patterns and implications of scientific results instead of the fine details) it seems reasonable, if not inevitable, to aggregate the individual components and then critically discuss the resulting overall picture. This chapter hence consists of two sections. In the first section the environmental, physical, economic, social and cultural impacts are combined to an overall impact of climate change. This combination is based on a Delphi survey, which allowed assignment of particular weights for each dimension of impacts based on an assessment by European representatives of the ESPON Monitoring Committee. An alternative approach, using equal weights, is also presented and discussed. Furthermore, the social impacts are recalculated on the basis of projected population data, in order to examine the effects of considering

European Climate Vulnerabilities and Adaptation: A Spatial Planning Perspective, First Edition.
Edited by Philipp Schmidt-Thomé and Stefan Greiving.
© 2013 John Wiley & Sons, Ltd. Published 2013 by John Wiley & Sons, Ltd.

current (which is to date the common practice) versus future sensitivity. The second section focuses on vulnerability to climate change, as the final result of combining overall impact and overall adaptive capacity. Again, an alternative approach is shown, in which equal weights are used throughout all aggregation procedures. Finally, the vulnerability results of the ESPON Climate project are compared with the only other existing European integrated vulnerability assessment, which was conducted for the Regions 2020 report of the EU's General Directorate for Regional Policy.

9.1   Overall potential impact of climate change

According to the IPCC, climate change impacts are consequences of climate change on natural and human systems (McCarthy *et al.*, 2001). Potential impacts are defined as impacts that may occur given a projected change in climate and do not take into account adaptation. The potential impacts of climate change based on the A1B forcing scenario for the five thematic dimensions of the ESPON Climate project have been presented in Chapters 5 and 6.

Comparing the five thematic impact maps in Figure 9.1, it becomes apparent that the impacts can be grouped into impacts primarily caused by extreme events (e.g. flooding and heat waves) and those caused by slow changes of average climate conditions. The former group consists of potential physical, cultural and social impacts, which are mainly concentrated in coastal and mountainous areas as well as regions with major rivers or urban centres. The second group of climate change impacts relate to changes in average climate conditions. In contrast to the more robust sub-systems that are primarily affected by extreme events, the economic and environmental sub-systems are sensitive even to slow climatic changes, which affect almost all European regions.

9.1.1   *Aggregating the five impact dimensions*

When aggregating these environmental, physical, economic, social and cultural impacts a crucial question is whether they are treated as equally important or whether some impacts are considered as more important than others. The ESPON Climate project decided to deal with this normative question in a participatory and transparent way. As explained in Chapter 2, the project made use of a Delphi-based survey among the ESPON Monitoring Committee, which represented the European Commission, 27 European countries and four Partner States. Table 9.1 presents the outcomes of the Delphi survey, namely the average weights attributed to each dimension of impact (and adaptive capacity) after the final survey round (for further discussion of this method see Chapter 2).

Apparently the Committee members attached more importance to environmental and economic impacts, somewhat less importance to physical and social impacts and considered impacts on cultural assets as least important. The impact weights were then used for calculating the overall potential impact of climate change.

The resulting map of the aggregated impact of climate change in Europe is shown in Figure 9.2. The following spatial patterns can be clearly distinguished. Coastal regions are projected to be negatively affected, largely because their high concentrations of physical, economic, social and cultural assets would face increasing flood hazards. Southern European regions are expected to be negatively impacted because their hotter and drier climates severely worsen conditions for their populations, economies and natural environments. Highly negative impacts are therefore projected for Southern Europe's agglomerations and tourist resorts along the coasts. However, inland mountain regions that are dependent on the primary sector and winter and/or summer tourism would also be highly affected. In contrast, many Central, Eastern and Northern European regions would face virtually no

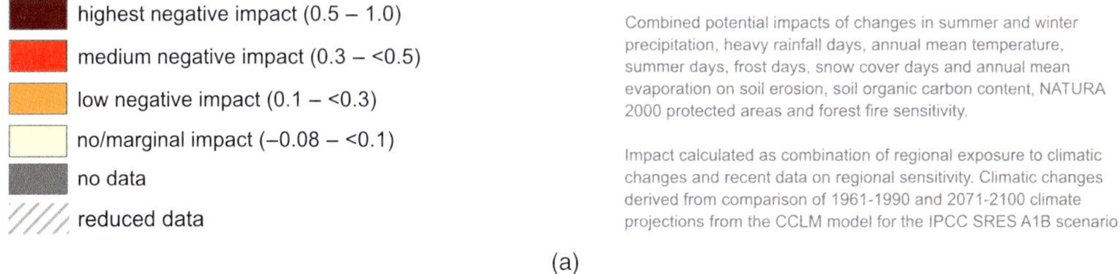

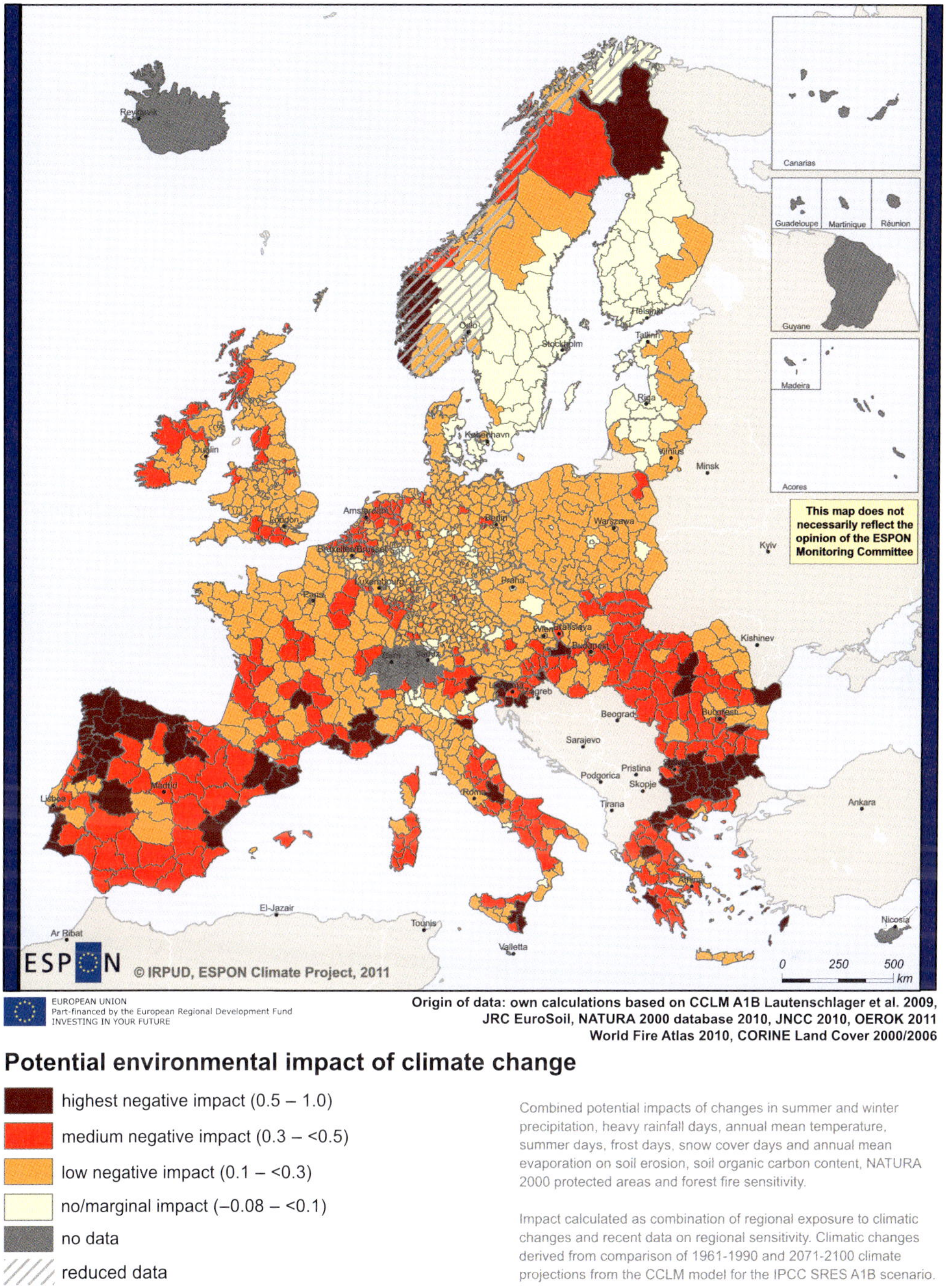

Origin of data: own calculations based on CCLM A1B Lautenschlager et al. 2009, JRC EuroSoil, NATURA 2000 database 2010, JNCC 2010, OEROK 2011 World Fire Atlas 2010, CORINE Land Cover 2000/2006

Potential environmental impact of climate change

- highest negative impact (0.5 – 1.0)
- medium negative impact (0.3 – <0.5)
- low negative impact (0.1 – <0.3)
- no/marginal impact (−0.08 – <0.1)
- no data
- reduced data

Combined potential impacts of changes in summer and winter precipitation, heavy rainfall days, annual mean temperature, summer days, frost days, snow cover days and annual mean evaporation on soil erosion, soil organic carbon content, NATURA 2000 protected areas and forest fire sensitivity.

Impact calculated as combination of regional exposure to climatic changes and recent data on regional sensitivity. Climatic changes derived from comparison of 1961-1990 and 2071-2100 climate projections from the CCLM model for the IPCC SRES A1B scenario.

(a)

Figure 9.1 Potential environmental, physical, economic and cultural impacts to climate change. © ESPON 2013, IRPUD, ESPON Climate Project, 2011.

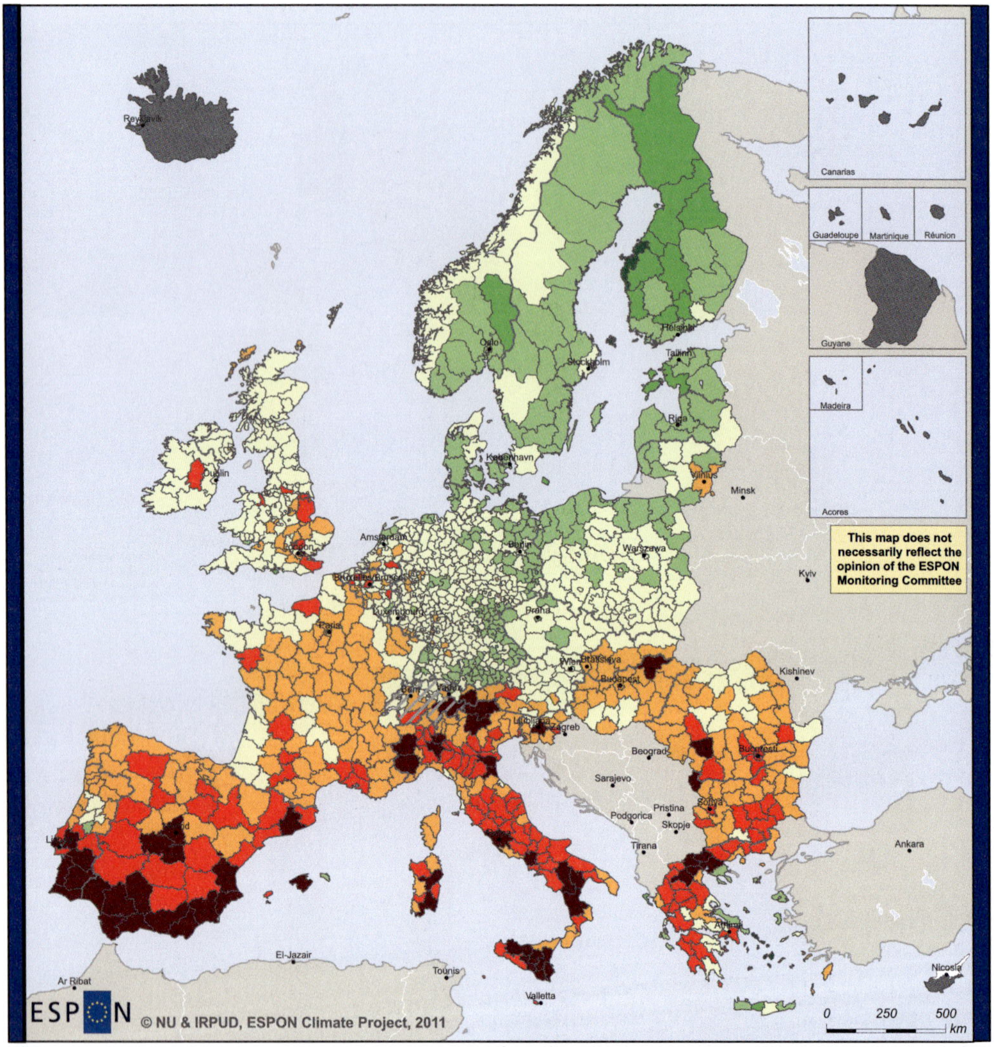

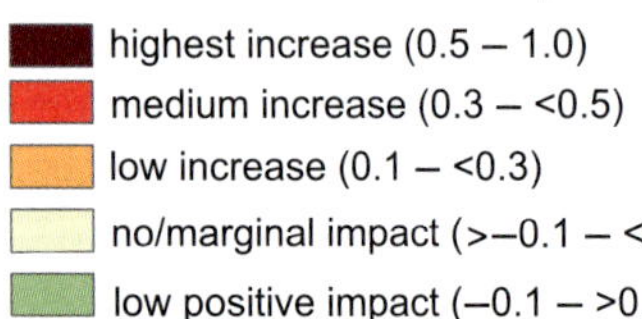

Potential economic impact of climate change

- highest increase (0.5 − 1.0)
- medium increase (0.3 − <0.5)
- low increase (0.1 − <0.3)
- no/marginal impact (>−0.1 − <0.1)
- low positive impact (−0.1 − >0.3)
- medium positive impact (−0.3 − −>0.5)
- high positive impact (−0.5 − −0.66)
- no data
- reduced data

Combined potential impacts of changes in annual mean evaporation, summer days, snow cover days, frost days, changes in inundation heights of a 100 year river flood event and a sea level rise adjusted 100 year coastal storm surge event on agriculture, forestry, summer and winter tourism, energy supply and demand.

Impact calculated as combination of regional exposure to climatic changes and recent data on regional sensitivity. Climatic changes derived from comparison of 1961-1990 and 2071-2100 climate projections from CCLM model for the IPCC SRES A1B scenario. Fluvial flooding changes based on LISFLOOD model. Regional coastal storm surge heights projected by DIVA were adjusted with 1 m sea level rise.

(b)

Figure 9.1   (*continued*)

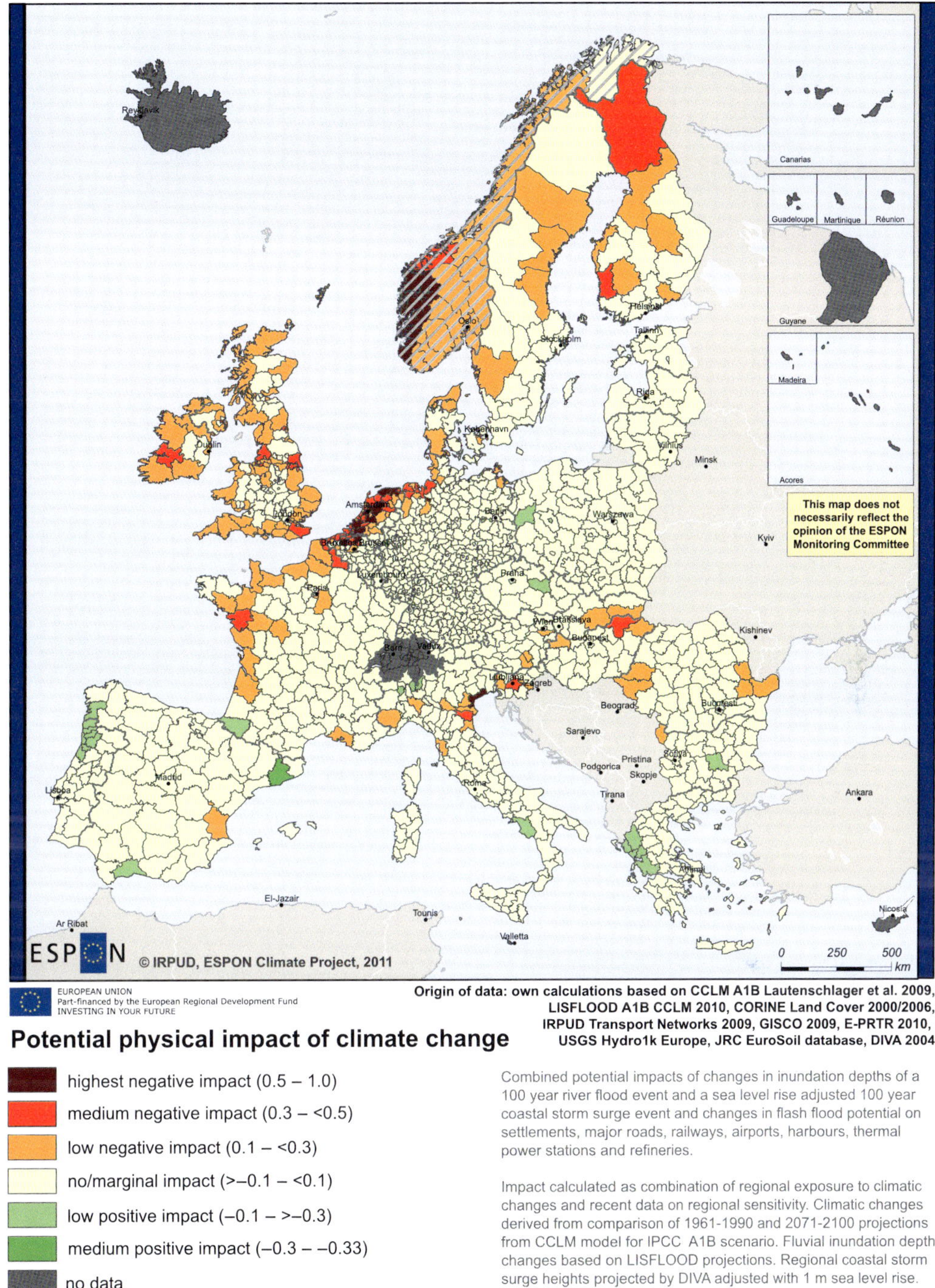

EUROPEAN UNION
Part-financed by the European Regional Development Fund
INVESTING IN YOUR FUTURE

Origin of data: own calculations based on CCLM A1B Lautenschlager et al. 2009,
LISFLOOD A1B CCLM 2010, CORINE Land Cover 2000/2006,
IRPUD Transport Networks 2009, GISCO 2009, E-PRTR 2010,
USGS Hydro1k Europe, JRC EuroSoil database, DIVA 2004

Potential physical impact of climate change

- highest negative impact (0.5 – 1.0)
- medium negative impact (0.3 – <0.5)
- low negative impact (0.1 – <0.3)
- no/marginal impact (>–0.1 – <0.1)
- low positive impact (–0.1 – >–0.3)
- medium positive impact (–0.3 – –0.33)
- no data
- reduced data

Combined potential impacts of changes in inundation depths of a
100 year river flood event and a sea level rise adjusted 100 year
coastal storm surge event and changes in flash flood potential on
settlements, major roads, railways, airports, harbours, thermal
power stations and refineries.

Impact calculated as combination of regional exposure to climatic
changes and recent data on regional sensitivity. Climatic changes
derived from comparison of 1961-1990 and 2071-2100 projections
from CCLM model for IPCC A1B scenario. Fluvial inundation depth
changes based on LISFLOOD projections. Regional coastal storm
surge heights projected by DIVA adjusted with 1 m sea level rise.

(c)

Figure 9.1   *(continued)*

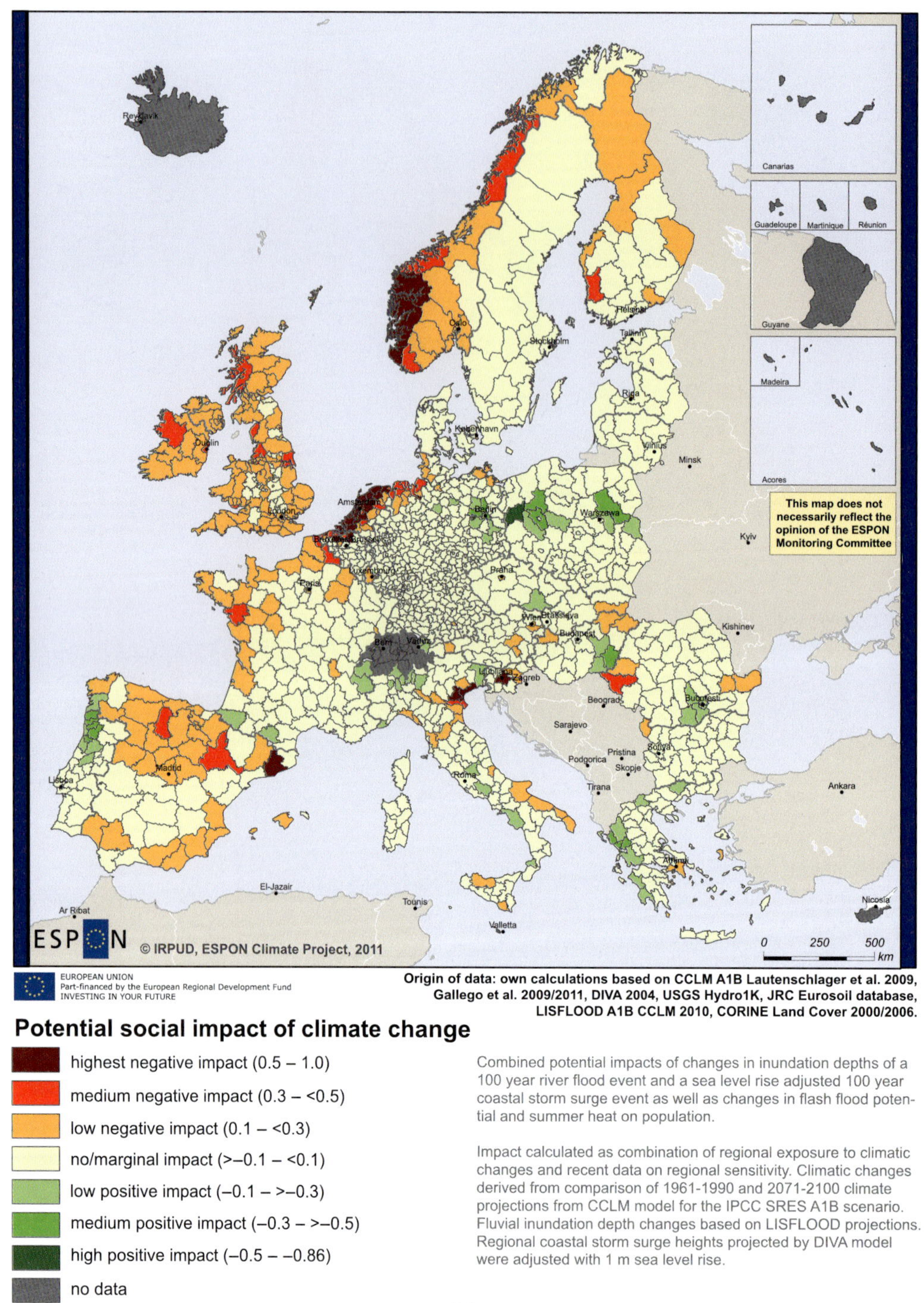

EUROPEAN UNION
Part-financed by the European Regional Development Fund
INVESTING IN YOUR FUTURE

Origin of data: own calculations based on CCLM A1B Lautenschlager et al. 2009,
Gallego et al. 2009/2011, DIVA 2004, USGS Hydro1K, JRC Eurosoil database,
LISFLOOD A1B CCLM 2010, CORINE Land Cover 2000/2006.

Potential social impact of climate change

- highest negative impact (0.5 – 1.0)
- medium negative impact (0.3 – <0.5)
- low negative impact (0.1 – <0.3)
- no/marginal impact (>–0.1 – <0.1)
- low positive impact (–0.1 – >–0.3)
- medium positive impact (–0.3 – >–0.5)
- high positive impact (–0.5 – –0.86)
- no data

Combined potential impacts of changes in inundation depths of a
100 year river flood event and a sea level rise adjusted 100 year
coastal storm surge event as well as changes in flash flood potential and summer heat on population.

Impact calculated as combination of regional exposure to climatic
changes and recent data on regional sensitivity. Climatic changes
derived from comparison of 1961-1990 and 2071-2100 climate
projections from CCLM model for the IPCC SRES A1B scenario.
Fluvial inundation depth changes based on LISFLOOD projections.
Regional coastal storm surge heights projected by DIVA model
were adjusted with 1 m sea level rise.

(d)

Figure 9.1   (*continued*)

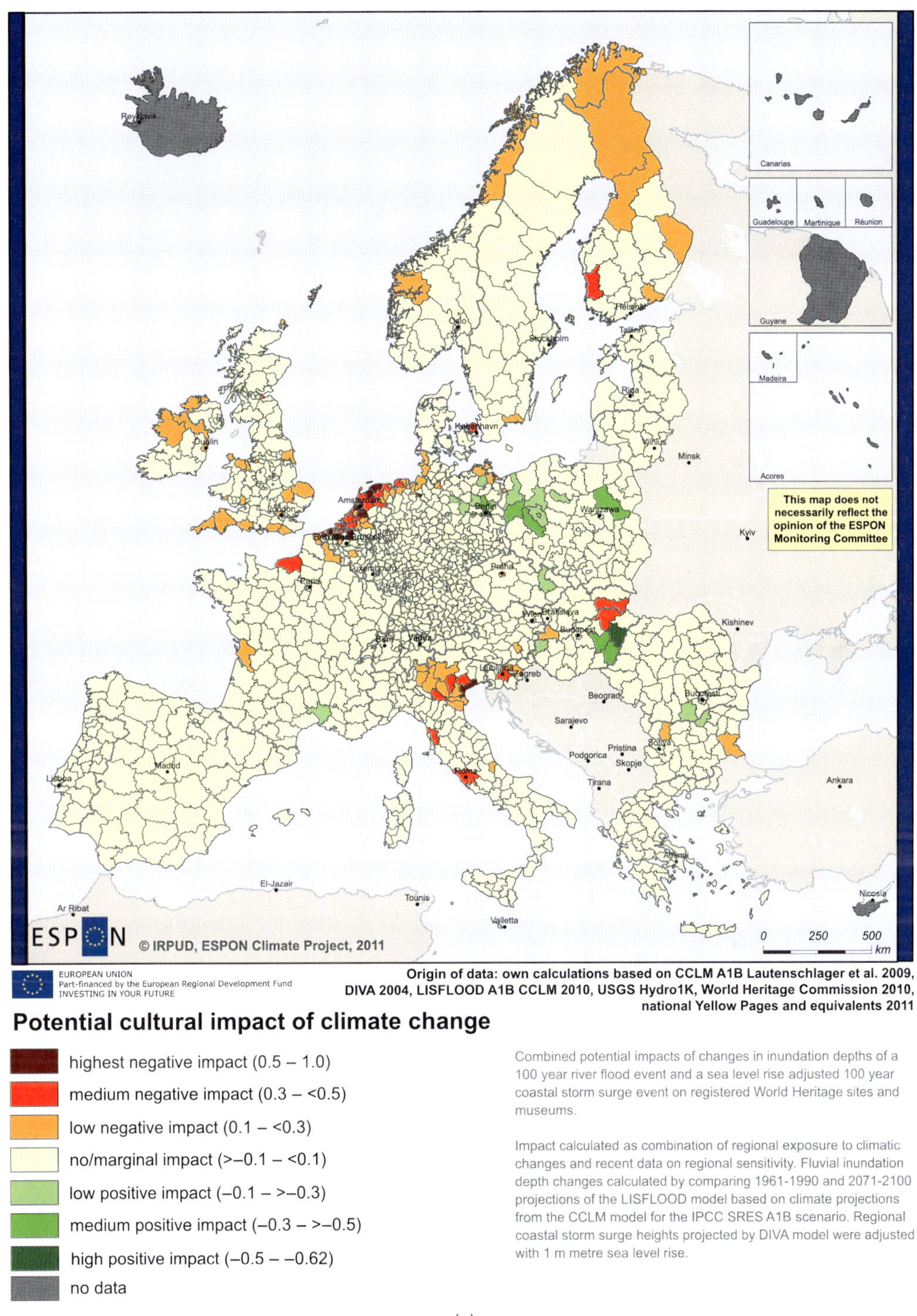

Origin of data: own calculations based on CCLM A1B Lautenschlager et al. 2009, DIVA 2004, LISFLOOD A1B CCLM 2010, USGS Hydro1K, World Heritage Commission 2010, national Yellow Pages and equivalents 2011

Potential cultural impact of climate change

- highest negative impact (0.5 – 1.0)
- medium negative impact (0.3 – <0.5)
- low negative impact (0.1 – <0.3)
- no/marginal impact (>–0.1 – <0.1)
- low positive impact (–0.1 – >–0.3)
- medium positive impact (–0.3 – >–0.5)
- high positive impact (–0.5 – –0.62)
- no data

Combined potential impacts of changes in inundation depths of a 100 year river flood event and a sea level rise adjusted 100 year coastal storm surge event on registered World Heritage sites and museums.

Impact calculated as combination of regional exposure to climatic changes and recent data on regional sensitivity. Fluvial inundation depth changes calculated by comparing 1961-1990 and 2071-2100 projections of the LISFLOOD model based on climate projections from the CCLM model for the IPCC SRES A1B scenario. Regional coastal storm surge heights projected by DIVA model were adjusted with 1 m metre sea level rise.

(e)

Figure 9.1   *(continued)*

Table 9.1 Weights for the sensitivity/impact and adaptive capacity dimensions based on Delphi survey.

Sensitivity/impact		Adaptive capacity	
Cultural sensitivity	0.1	Economic resources	0.21
Economic sensitivity	0.24	Knowledge and awareness	0.23
Environmental sensitivity	0.31	Infrastructure	0.16
Physical sensitivity	0.19	Institutions	0.17
Social sensitivity	0.16	Technology	0.23

negative impacts or would even experience positive impacts of climate change – mainly due to marginal climatic changes or to decreases in flood-related climate parameters.

9.1.2 *Alternative aggregation using equal weights*

Of course the weights used for the various impact dimensions represent the 'subjective' assessment of the ESPON Monitoring Committee. In contrast to their ratings one could just as well argue that social or even cultural impacts are equally important as environmental impacts. It is clear, however, that such an equal weighting of impact dimensions is just as subjective and in no way more 'objective'. Nevertheless, since most studies do not transparently address these weighting issues and simply apply equal weights, the ESPON Climate project also produced an aggregate impact map using equal weights (see Figure 9.3). The North–South gradient is still observable, but it is less accentuated than in the original impact map. Moreover, urban centres and mountain regions stand out more, because social, physical and cultural impacts, which are spatially more concentrated, gained in importance. Conversely, the influence of the more ubiquitous environmental and economic impacts was downgraded. Or, to look at it from an exposure point of view, one can say that the equal weight version gives more emphasis to impacts from extreme events (various flooding and heat events), whereas the version with the ESPON Monitoring Committee

weights gives more emphasis to 'creeping' changes of normal climatic conditions – and thus underlines the mostly negative environmental and economic impacts in southern parts of Europe more.

9.1.3 *'Pure' future impacts: an example using population projections*

Climate change impacts are about effects of climate change that may occur in the future. Hence the projected future climate would have to be related to projected future sensitivities (see Füssel 2007, 164). However, as explained in Chapter 2, future projections for the various sensitivity indicators used in the ESPON Climate project are not available (with few exceptions) – especially not for the time horizon 2071–2100. Therefore, the project had to relate future climatic changes to current sensitivities, which is the common practices in such analyses to date but in a way implies that climate change would happen immediately, affecting the current socio-economic and environmental system. In order to nevertheless gauge the effect that a truly dynamic assessment would have, that is, in which sensitivities are also changing as well, the project received support from the ESPON project DEMIFER, which dealt with demographic changes in Europe was able to provide population projections until the year 2100 (for NUTS 2 regions and broken down into age groups). Based on these projections ESPON Climate then recalculated social impacts for Europe.

Figure 9.4 juxtaposes the two social impact maps based on current and future population data, respectively. For many parts of Europe only marginal differences can be observed. Some regions in Spain have less negative impacts in the 'dynamic' population version, while the region around Madrid has more negative impacts. These changes are largely due to internal migration and ageing. More dramatic changes can be observed in Eastern and South-Eastern Europe, where DEMIFER projections anticipate huge population losses in some regions. Social sensitivity would therefore decrease and in

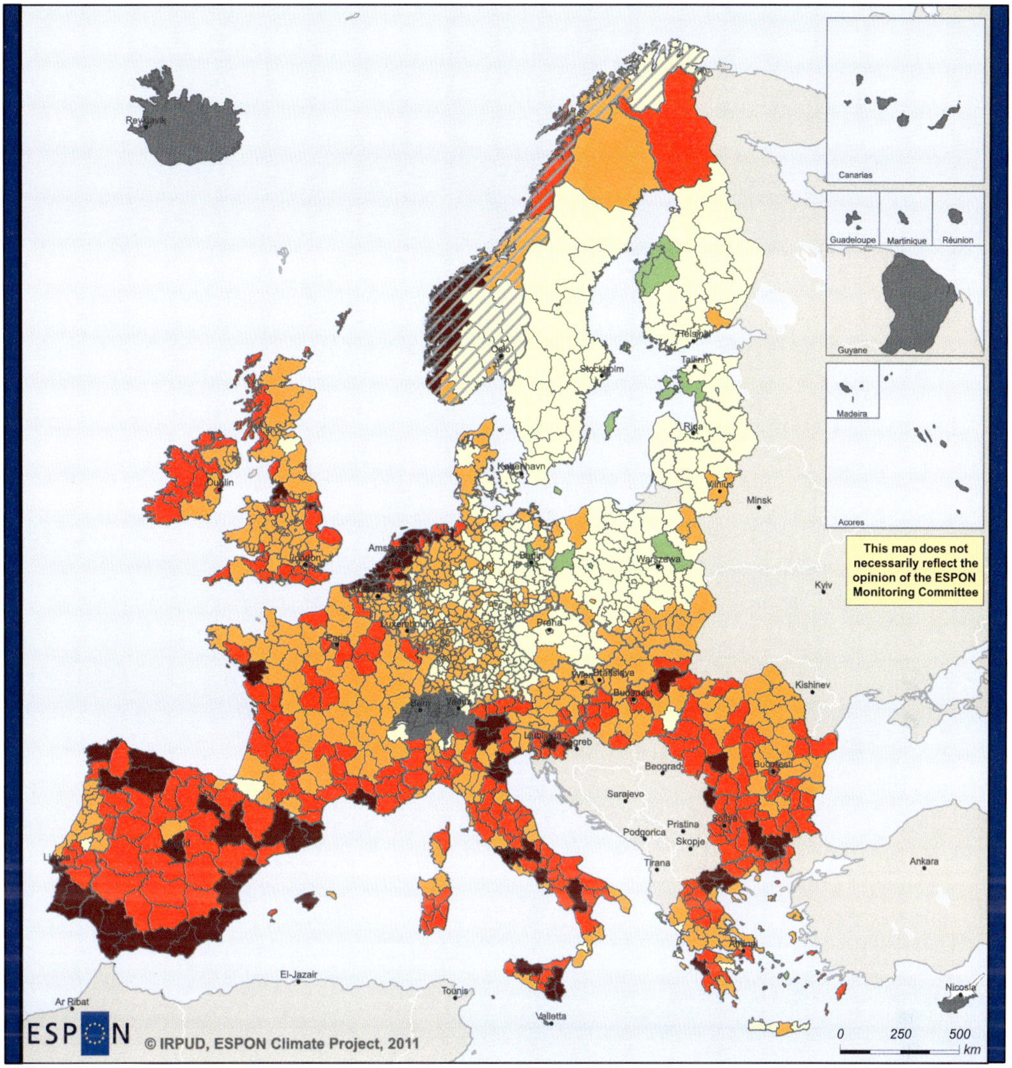

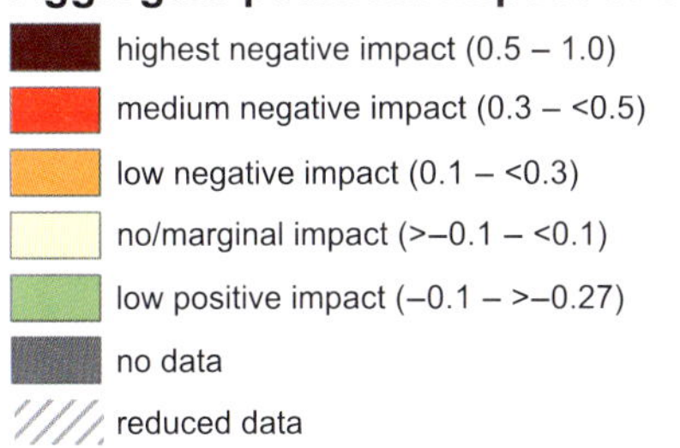

Aggregate potential impact of climate change

- highest negative impact (0.5 – 1.0)
- medium negative impact (0.3 – <0.5)
- low negative impact (0.1 – <0.3)
- no/marginal impact (>−0.1 – <0.1)
- low positive impact (−0.1 – >−0.27)
- no data
- reduced data

Weighted combination of physical (weight 0.19), environmental (0.31), social (0.16), economic (0.24) and cultural (0.1) potential impacts of climate change. Weights are based on a Delphi survey of the ESPON Monitoring Committee.

Impact calculated as combination of regional exposure to climatic changes and recent data on regional sensitivity. Climatic changes derived from comparison of 1961-1990 and 2071-2100 climate projections from the CCLM model for the IPCC SRES A1B scenario.

Figure 9.2 Aggregate potential of climate change in Europe. © ESPON 2013, IRPUD, ESPON Climate Project, 2011.

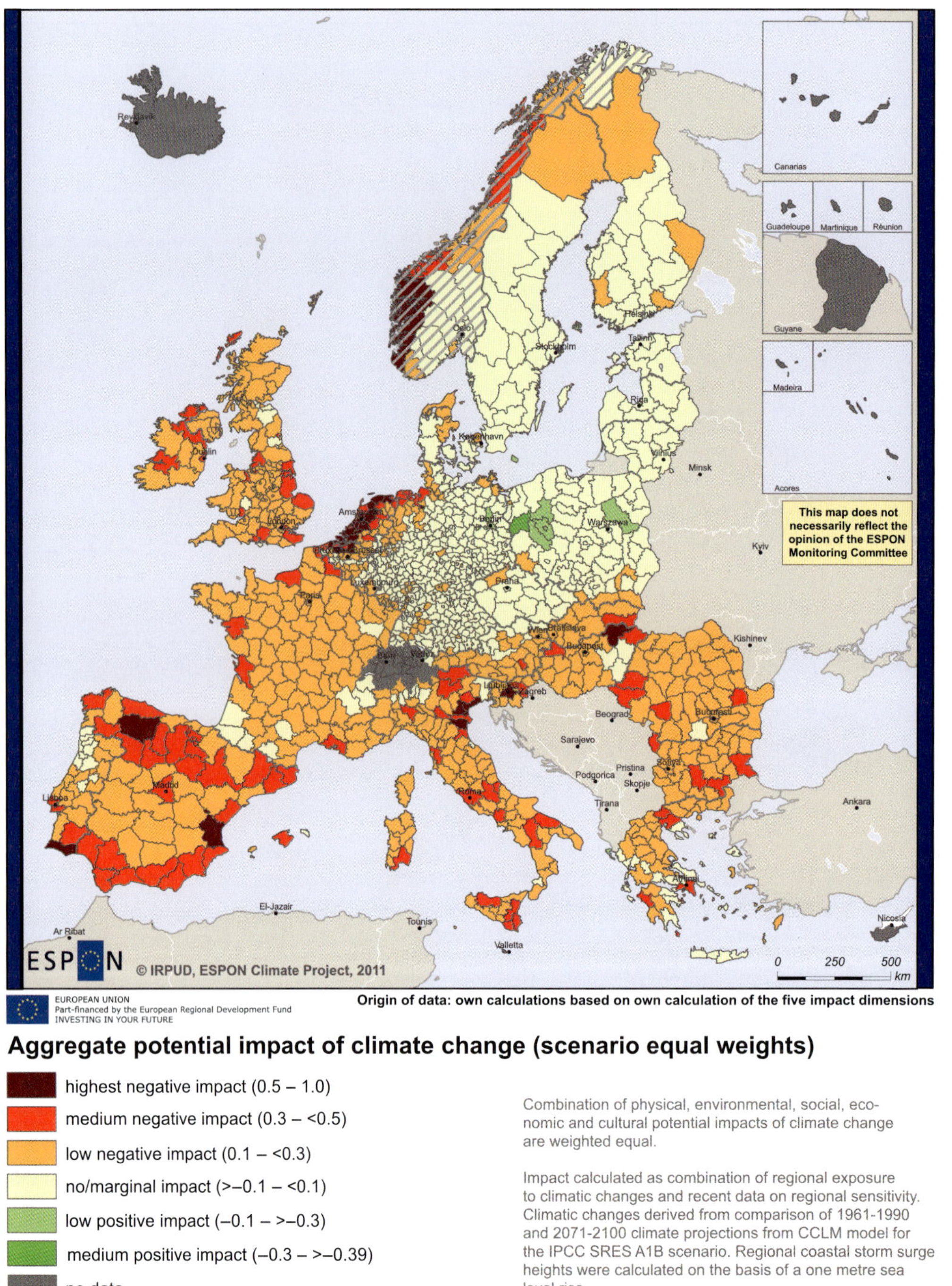

Figure 9.3   Aggregate potential of climate change using equal weights. © ESPON 2013, IRPUD, ESPON Climate Project, 2011.

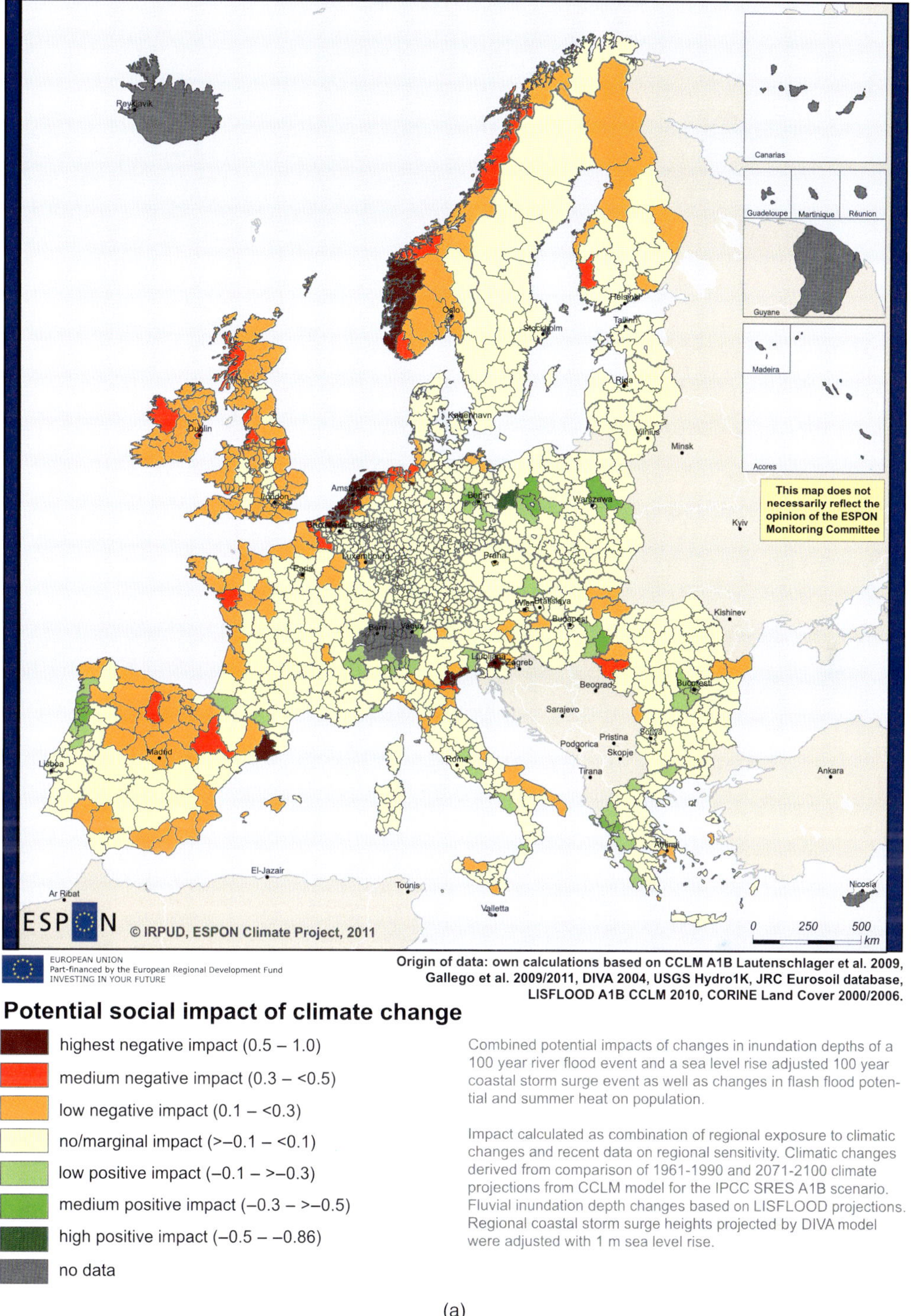

Potential social impact of climate change

Combined potential impacts of changes in inundation depths of a
100 year river flood event and a sea level rise adjusted 100 year
coastal storm surge event as well as changes in flash flood potential and summer heat on population.

Impact calculated as combination of regional exposure to climatic
changes and recent data on regional sensitivity. Climatic changes
derived from comparison of 1961-1990 and 2071-2100 climate
projections from CCLM model for the IPCC SRES A1B scenario.
Fluvial inundation depth changes based on LISFLOOD projections.
Regional coastal storm surge heights projected by DIVA model
were adjusted with 1 m sea level rise.

(a)

Figure 9.4   Social impacts: comparison of static versus dynamic sensitivities. © ESPON 2013, IRPUD, ESPON Climate Project, 2011.

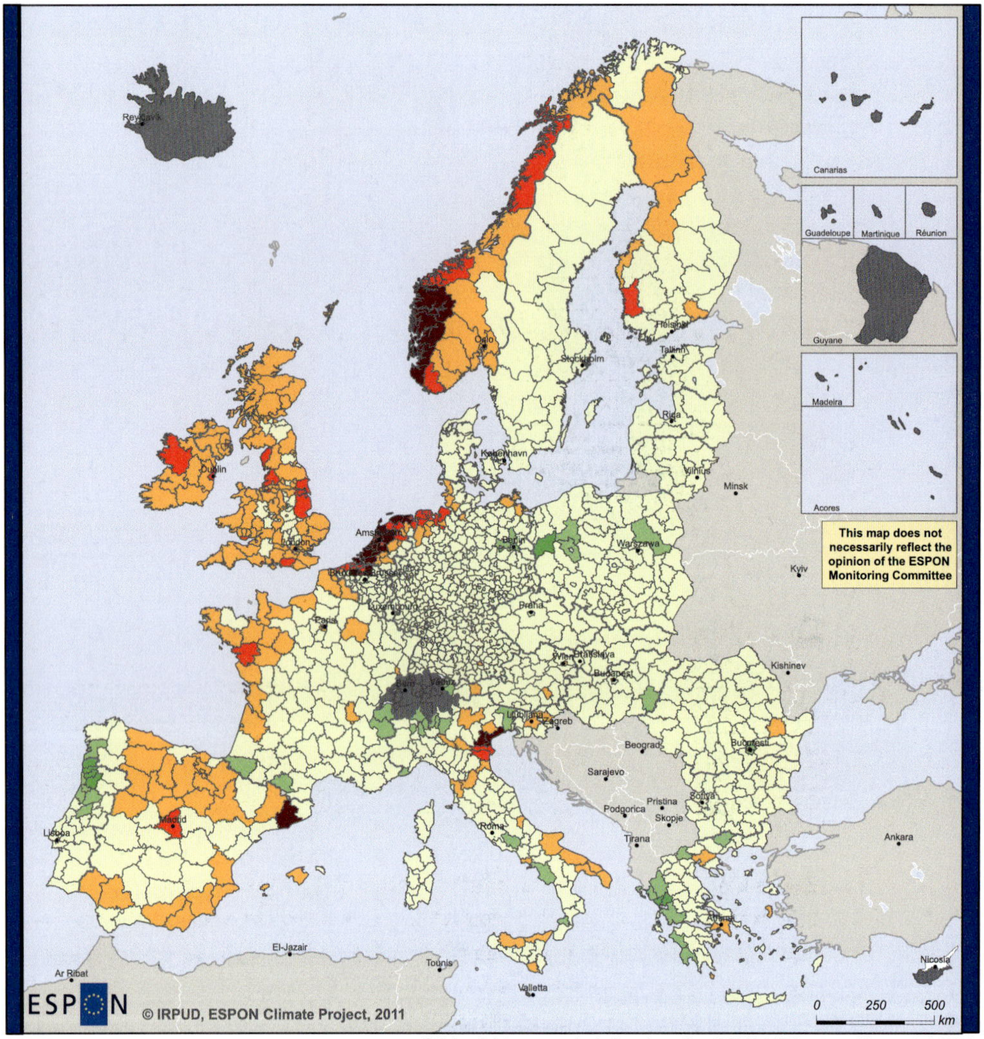

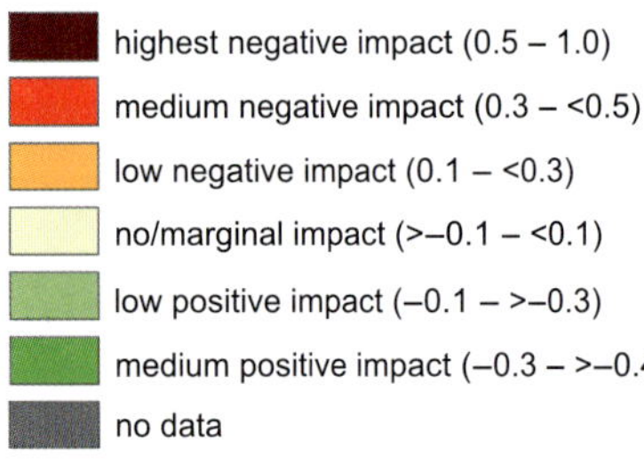

EUROPEAN UNION
Part-financed by the European Regional Development Fund
INVESTING IN YOUR FUTURE

Origin of data: own calculations based on CCLM A1B Lautenschlager et al. 2009, Gallego et al. 2009/2011, DIVA 2004, USGS Hydro1K, JRC Eurosoil database, LISFLOOD A1B CCLM 2010, CORINE Land Cover 2000/2006, ESPON DEMIFER Project.

Scenario: Combined social impact to climate change in 2100

- highest negative impact (0.5 – 1.0)
- medium negative impact (0.3 – <0.5)
- low negative impact (0.1 – <0.3)
- no/marginal impact (>–0.1 – <0.1)
- low positive impact (–0.1 – >–0.3)
- medium positive impact (–0.3 – >–0.49)
- no data

Combined potential impacts of changes in inundated heights of a 100 year river flood event and a sea level rise adjusted 100 year coastal storm surge event and changes in flash flood potential on population, using DEMIFER population projections for the year 2100.

Impact calculated as combination of regional exposure to climatic changes and recent data on regional sensitivity. Climatic changes derived from comparison of 1961-1990 and 2071-2100 climate projections from CCLM model for the IPCC SRES A1B scenario. Regional coastal storm surge heights were calculated on the basis of a one metre sea level rise.

(b)

Figure 9.4   (*continued*)

consequence so would the potential social impacts of climate change.

A shrinking and ageing population would actually also have effects on a region's adaptive capacity. Population losses usually are caused by or lead to declining economic activities and thus to a reduced economic adaptive capacity. Likewise, an older and more dependent population puts more pressure on, for example, health infrastructure (which in turn is harder to maintain in sparsely populated areas) and will most likely also lead to less knowledge and awareness of climate change and appropriate adaptation options.

9.2 Vulnerability to climate change

Vulnerability to climate change is a contested concept. This is partly due to the fact that it relates to possible harm from a possible future climate and thus cannot be observed and empirically tested. Hence it will always remain a theoretical concept (cf. Hinkel, 2011). No definition of climate change vulnerability can therefore *per se* claim superiority over other definitions. Also, there are many definitions of climate change vulnerability that are all based on different scientific approaches and methodologies to climate change research (see Füssel, 2007 for a systematic overview, as well as Thywissen, 2006; Adger, 2006; Eakin and Luers, 2006; and Hofmann, Hinkel and Wrobel, 2011). One of the more widely used definitions is the one developed by the IPCC for its Third Assessment Report. While also criticised as partly confusing or impractical (Downing *et al.*, 2001) it nevertheless 'consistently characterises the future (or dynamic) vulnerability of any natural or social system to global climate change' (Füssel, 2007, p. 164). As explained in Chapter 2, ESPON Climate therefore decided to follow the IPCC concept of vulnerability used in the Third and Fourth Assessment Report, where vulnerability is defined as 'the degree to which a system is susceptible to, and unable to cope with, adverse effects of climate change, including climate variability and extremes. Vulnerability is a function of the character, magnitude and rate of climate change and variation to which a system is exposed, its sensitivity and its adaptive capacity' (McCarthy *et al.*, 2001, p. 995). [1]

9.2.1 *Combining climate change impacts and adaptive capacity*

Accordingly, the vulnerability of a region in regard to climate change is determined by its potential climate change impacts (which is a product of exposure and sensitivity) and its capacity to adapt to climate change (see Chapter 2 for more methodological details). Figure 9.5 shows these two constituent elements and the resulting vulnerability of European regions in the form of maps. Note that for conceptual reasons the vulnerability value of a region can only range from 0 to 1, because vulnerability is a negative condition and there cannot be 'positive vulnerability'.

Looking at the impact and adaptive capacity maps one can observe an almost inverted pattern: most regions for which climate change impacts are expected to be the most severe (mainly in the South) are in fact least able to adapt to the potential impacts. Therefore, the vulnerability map seems to mirror the territorial pattern of potential impacts, but with an even more pronounced South–North gradient. This is due to the high adaptive capacity in Scandinavian and Western European regions, which partly compensates for the potential impacts projected for these regions. On the other hand, in the Mediterranean region and in South–East Europe, where medium to high negative impacts are expected, the ability to adapt to climate change is generally lower thus resulting in even higher levels of vulnerability. The overall most vulnerable types of regions in Europe are therefore: (i) Coastal regions in Southern Europe with high population and high dependency on summer tourism, (ii) mountain regions with high dependence on winter and summer tourism and

---

[1] Note that the IPCC also uses other definitions of vulnerability. For example in the SREX report vulnerability is simply defined as 'propensity or predisposition to be adversely affected' (IPCC 2012, p. 564) and that this definition was taken up in the draft of the Fifth IPCC Assessment Report.

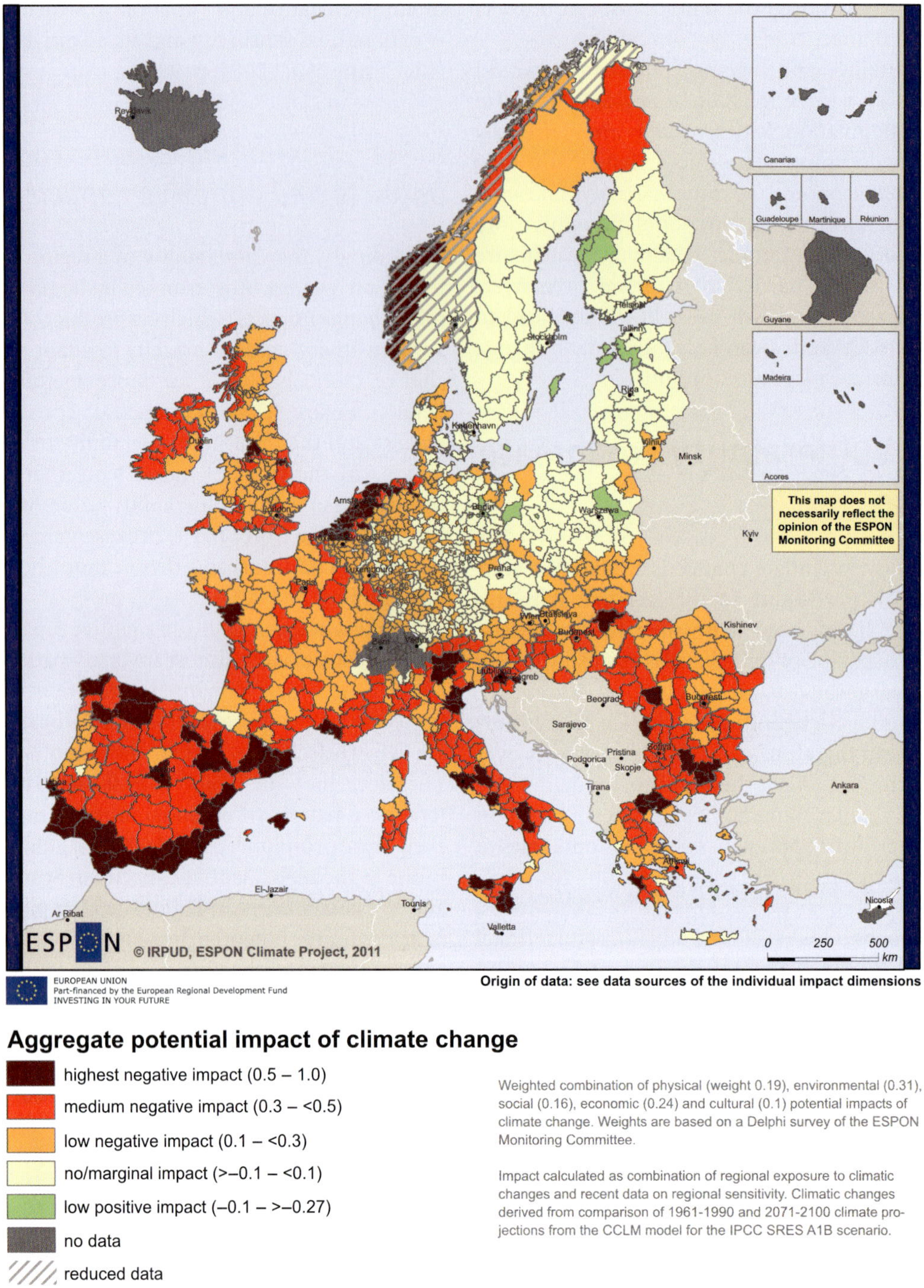

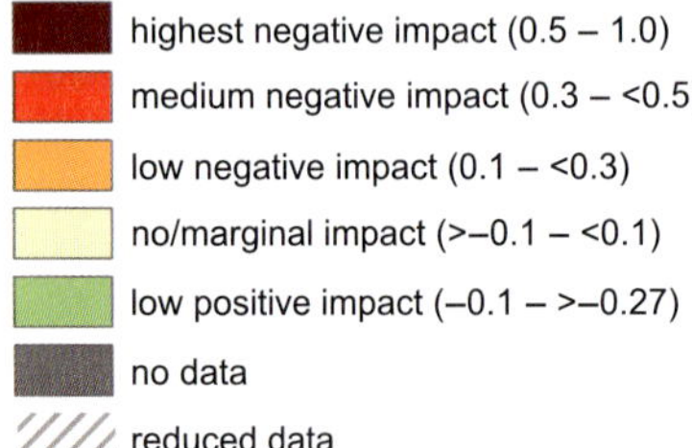

Aggregate potential impact of climate change

(a)

Figure 9.5 Potential aggregate impact, adaptive capacity and vulnerability. © ESPON 2013, IRPUD, ESPON Climate Project, 2011.

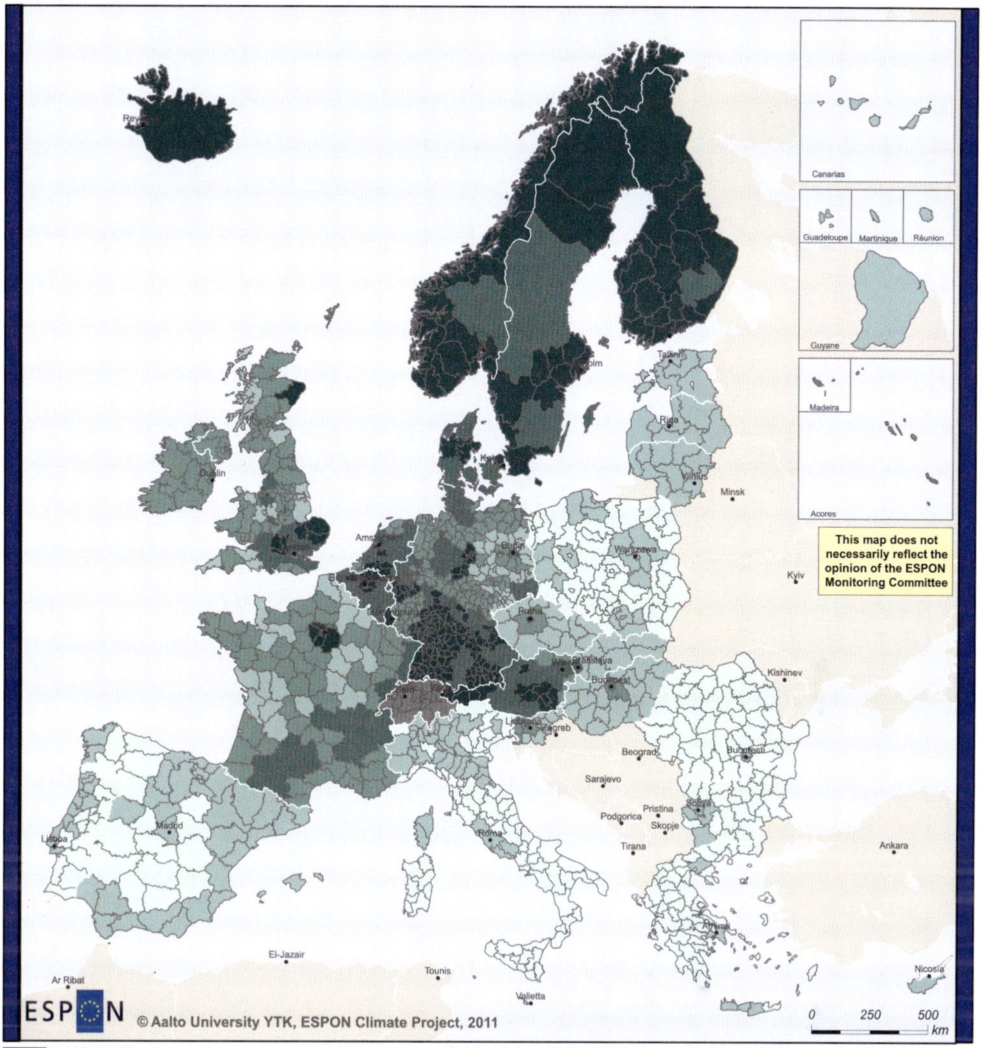

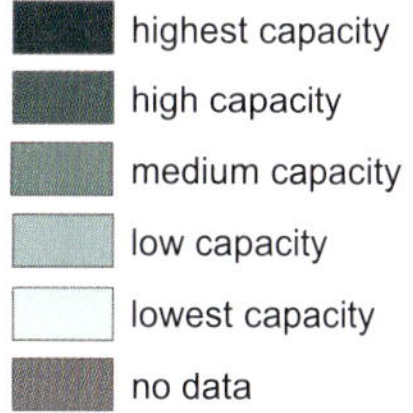

Overall capacity to adapt to climate change

Overall adaptive capacity towards climate change classified by quintiles.

The overall adaptive capacity was calculated as weighted combination of economic capacity (weight 0.21), infrastructural capacity (0.16), technological capacity (0.23), knowledge and awareness (0.23) and institutional capacity (0.17). Weights are based on a Delphi survey of the ESPON Monitoring Committee.

(b)

Figure 9.5   (*continued*)

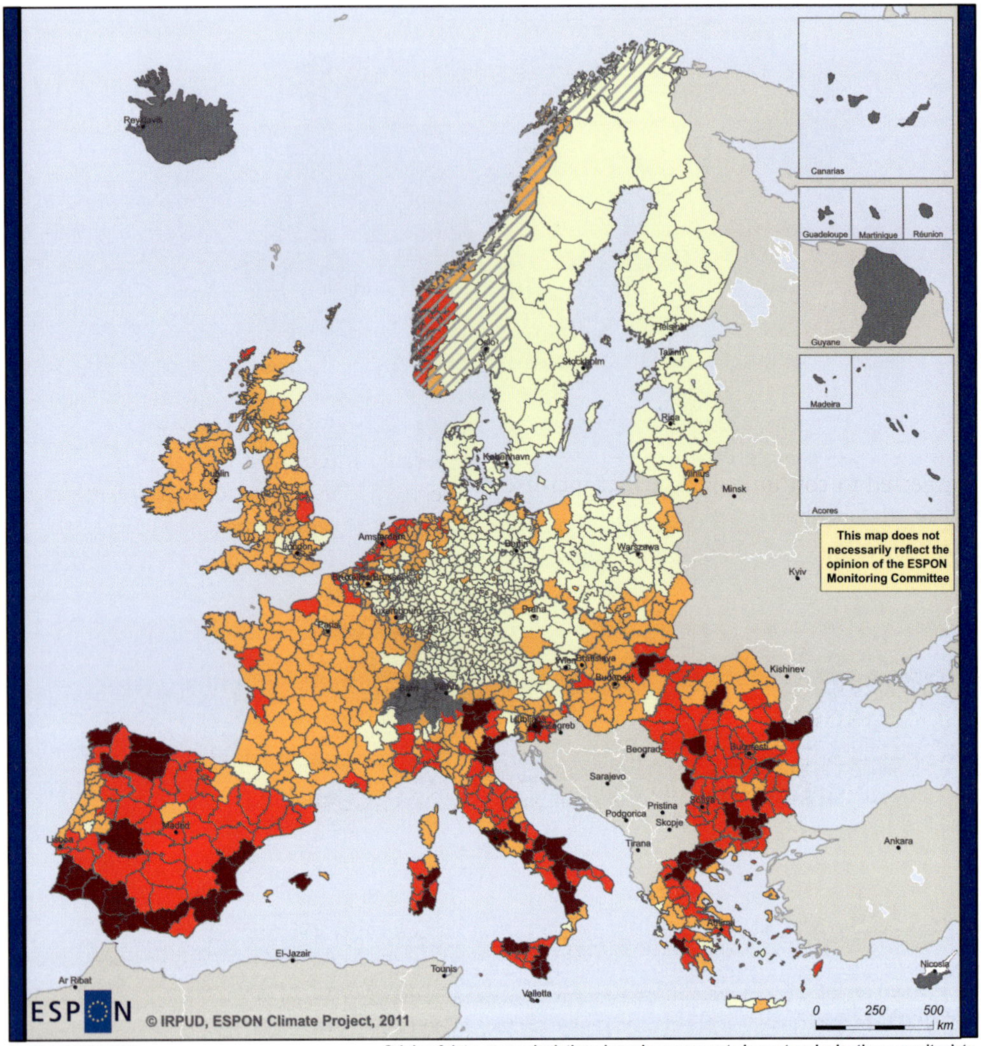

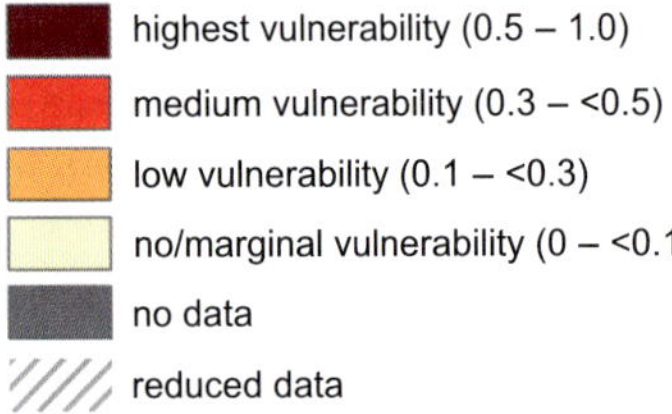

Potential vulnerability to climate change

- highest vulnerability (0.5 – 1.0)
- medium vulnerability (0.3 – <0.5)
- low vulnerability (0.1 – <0.3)
- no/marginal vulnerability (0 – <0.1)
- no data
- reduced data

Vulnerability calculated as the combination of regional potential impacts of climate change and regional capacity to adapt to climate change.

The potential impacts were calculated as a combination of regional exposure to climate change (difference between 1961-1990 and 2071-2100 climate projections of eight climatic variables of the CCLM model for the IPCC SRES A1B scenario as well as resulting inundation depth changes for a 100 year return flood event based on river flooding projections of the LISFLOOD model and coastal storm surge height projections of the DIVA model adjusted with a 1 m sea level rise) and most recent data on the weighted dimensions of physical, economic, social, environmental and cultural sensitivity to climate change. Adaptive capacity was calculated as a weighted combination of most recent data on economic, infrastructural, technological and institutional capacity as well as knowledge and awareness of climate change.

(c)

Figure 9.5   (*continued*)

(iii) agglomerations with high population density, where particularly the problem of urban heat might become viral.

These climate change vulnerability patterns run counter to the EU's aim of territorial cohesion. Climate change would trigger a deepening of the existing socio-economic imbalances between the economic core of Europe and its Southern and South-Eastern periphery. Most likely these imbalances will even increase in the future: particularly economic crises may reduce both the individual adaptive capacity of a member state as well as Europe's collective adaptive capacity as a whole. Furthermore, severe demographic changes such as massive outmigration and ageing in Eastern Europe that are projected to continue, will further increase regional climate change sensitivity and decrease adaptive capacity (e.g. an older regional population is more sensitive to heat and less capable to adapt to climate change).

The vulnerability patterns described above thus call for additional efforts in balancing and harmonising spatial differences to ensure a balanced and sustainable territorial development of the EU as a whole, strengthening its economic competitiveness and capacity for growth, while respecting the need to preserve its natural assets and ensuring social cohesion, as stated by the Green Paper on Territorial Cohesion (EC, 2008). Furthermore, territorially differentiated adaptation strategies seem to be needed particularly for coastal regions in the Mediterranean, but also the Alps. Moreover, agglomerations – especially in Southern Europe – are vulnerable for several reasons, but urban heat might be the most relevant from a long-term perspective as this poses not only risks for human health, but also leads to additional energy demand for cooling and, as a second-order effect, possibly to more frequent power failures.

9.2.2 Alternative vulnerability calculation using equal weights

Underlying the vulnerability patterns described above are the relative weights that the ESPON Monitoring Committee members assigned to the various impact and adaptive capacity dimensions. In order to make the effects of these normative decisions transparent, Figure 9.6 shows the ESPON Climate vulnerability map next to a version where all impact and adaptive capacity dimensions were weighted equally (which, again, constitutes some sort of weighting). The effects are quite significant. Overall the level of regional vulnerability is much less and the North–South gradient is also less pronounced. In the South of Europe, coastal regions and individual urban regions come out as vulnerability 'hot spots', whereas in the map on the left the vulnerability level of southern regions seems generally 'lifted' to a higher negative vulnerability category. The main reason for these differences stems from the different impact weights. The ESPON Monitoring Committee put more emphasis on the environmental and economic dimensions, whose negative impacts are affecting almost all regions in the South of Europe. In contrast, the equal weights scenario reduces these effects and thus highlights those regions with assets exposed to changes in extreme weather events such as all types of flooding and summer heat.

Neither of the two vulnerability maps can in principle claim to be 'more objective' than the other. However, from an uncertainty point of view the ESPON Monitoring Committee based vulnerability results could perhaps be more robust than the equal weights scenario. This is because general temperature and precipitation changes are much better understood and projected with greater certainty than the very complex and very locally specific projections for potential flash floods, river floods and sea-level rise induced coastal floods. A further aspect is that such flood events would usually only affect a small part of a particular region, whereas general climatic changes in air temperature and precipitation patterns affect all parts of every region. In combination these two aspects may actually make the map on the left more conservative and less uncertain than the more moderate-looking map on the right.

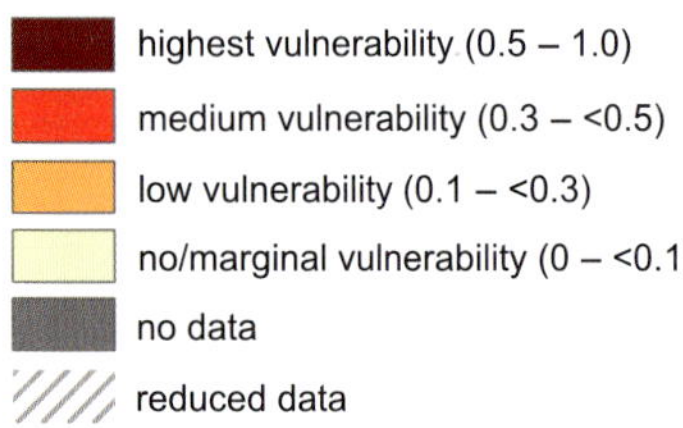

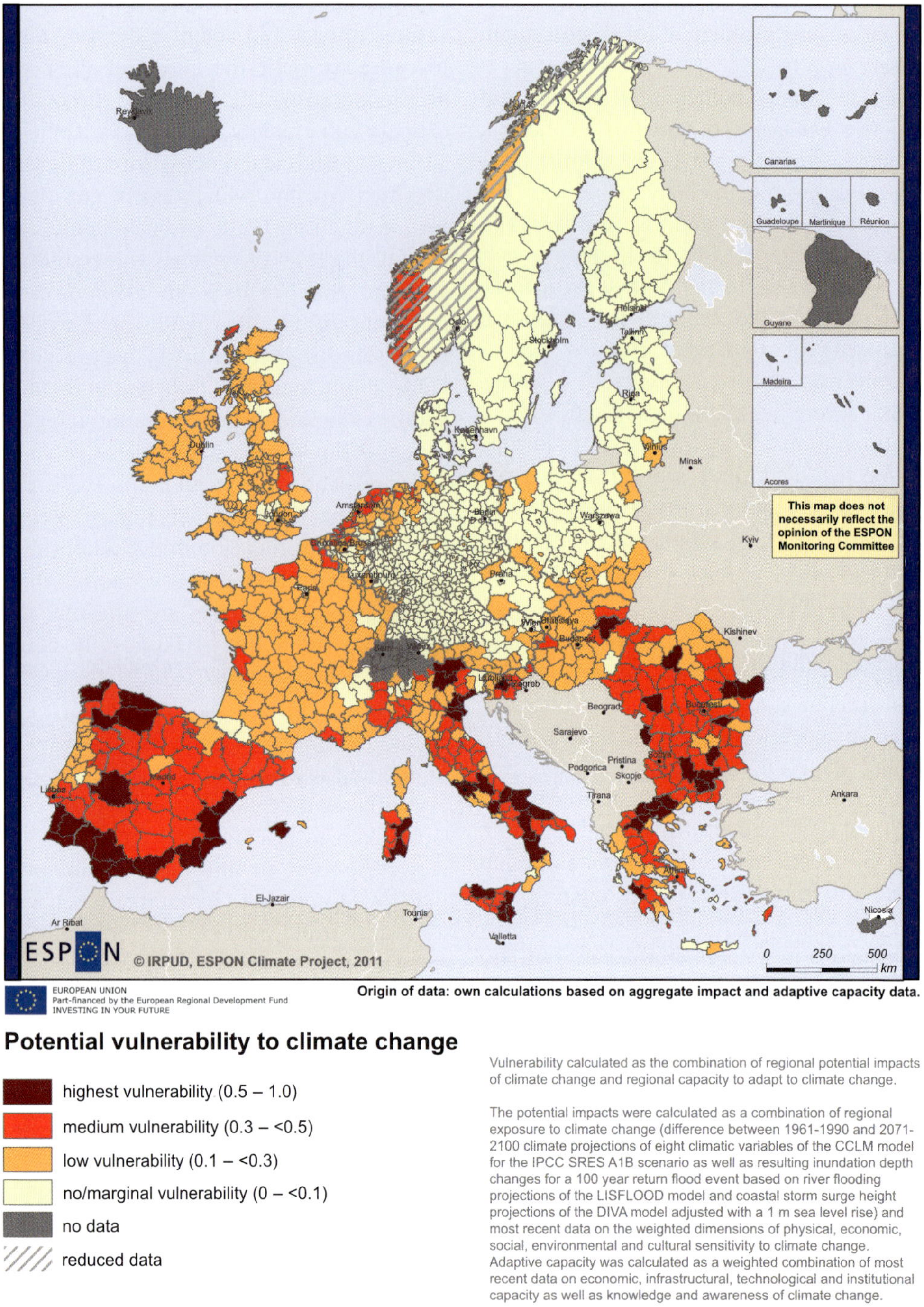

Potential vulnerability to climate change

Vulnerability calculated as the combination of regional potential impacts of climate change and regional capacity to adapt to climate change.

The potential impacts were calculated as a combination of regional exposure to climate change (difference between 1961-1990 and 2071-2100 climate projections of eight climatic variables of the CCLM model for the IPCC SRES A1B scenario as well as resulting inundation depth changes for a 100 year return flood event based on river flooding projections of the LISFLOOD model and coastal storm surge height projections of the DIVA model adjusted with a 1 m sea level rise) and most recent data on the weighted dimensions of physical, economic, social, environmental and cultural sensitivity to climate change. Adaptive capacity was calculated as a weighted combination of most recent data on economic, infrastructural, technological and institutional capacity as well as knowledge and awareness of climate change.

(a)

Figure 9.6   Comparison of ESPON climate change vulnerability versus equal weight vulnerability. © ESPON 2013, IRPUD, ESPON Climate Project, 2011.

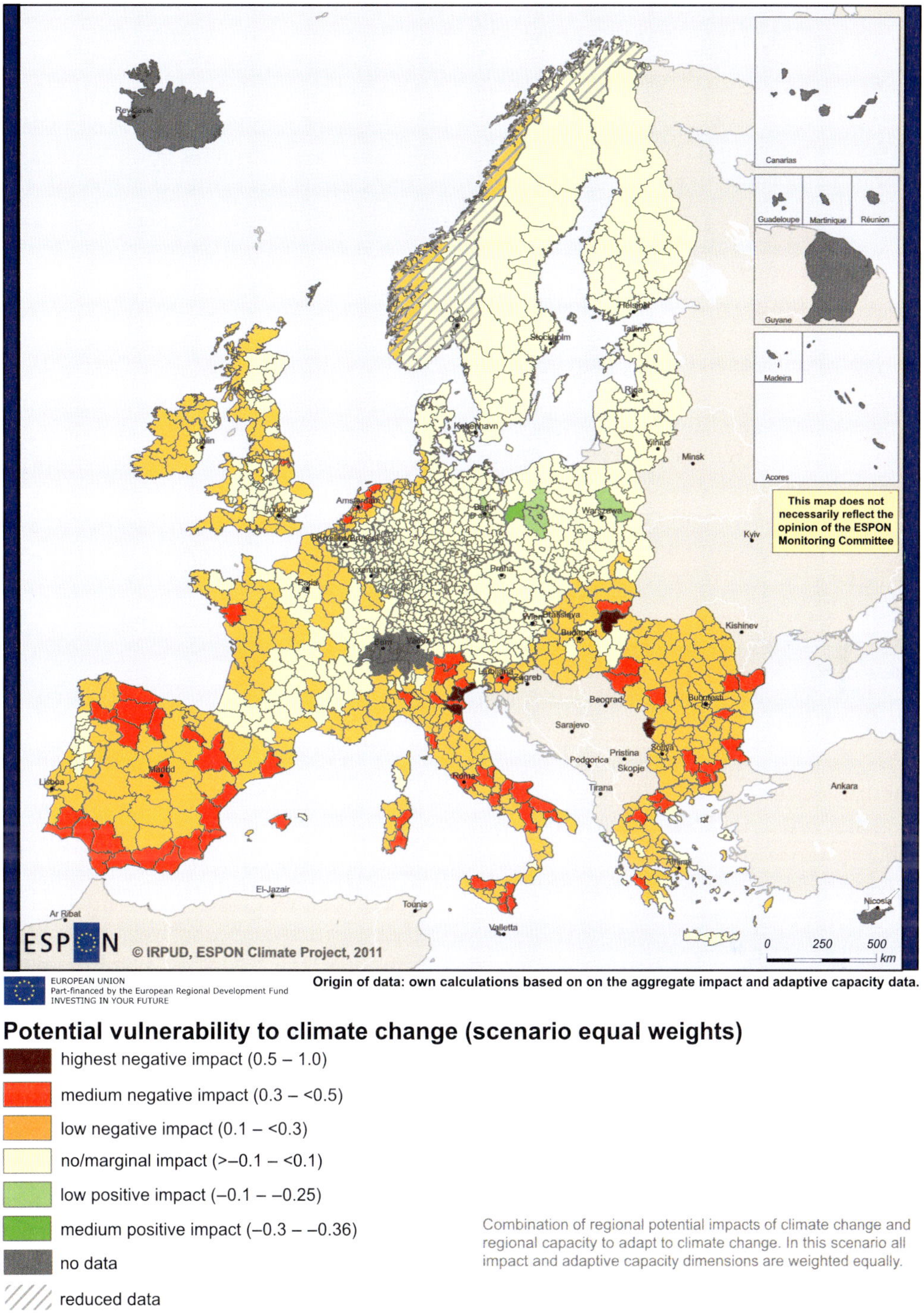

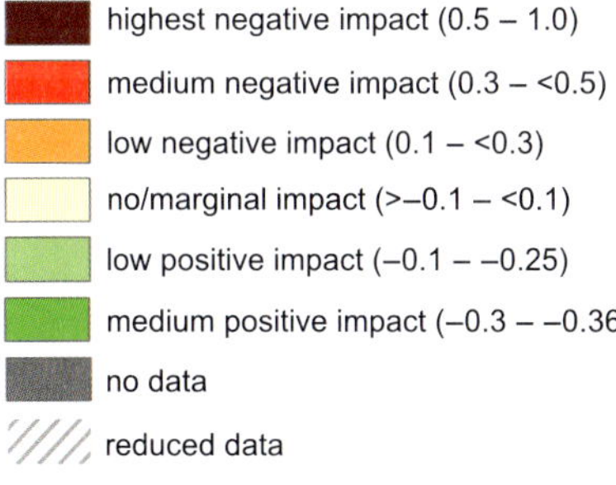

Origin of data: own calculations based on on the aggregate impact and adaptive capacity data.

Potential vulnerability to climate change (scenario equal weights)

- highest negative impact (0.5 – 1.0)
- medium negative impact (0.3 – <0.5)
- low negative impact (0.1 – <0.3)
- no/marginal impact (>–0.1 – <0.1)
- low positive impact (–0.1 – –0.25)
- medium positive impact (–0.3 – –0.36)
- no data
- reduced data

Combination of regional potential impacts of climate change and regional capacity to adapt to climate change. In this scenario all impact and adaptive capacity dimensions are weighted equally.

(b)

Figure 9.6   (*continued*)

9.2.3 Comparison with Regions 2020 climate change vulnerability index

While there are lots of studies on individual aspects of projected climate change in Europe, its potential impacts and the capacity to adapt to them, there is to date only one study at the European level that attempted a cross-sectoral, integrated vulnerability assessment like the one presented in this chapter. As part of its Regions 2020 report the European Commission's Directorate General for Regional Policy (DG Regio) commissioned a background paper on the challenges of climate change for Europe's regions (Römisch, 2009). With a few modifications this vulnerability assessment was integrated into the Regions 2020 background report on climate change (DG Regio, 2009), whose main findings already entered into the Regions 2020 main report (DG Regio, 2008) (Figure 9.7).

Methodologically, the Regions 2020 vulnerability index differs from the ESPON Climate vulnerability assessment as follows: (a) the Regions 2020 index is based very broadly on air temperature and precipitation changes compared with the ESPON assessment that uses eight climate change variables, (b) the Regions 2020 index uses the IPCC's A2 climate forcing scenario, whereas the ESPON study is based on the A1B scenario, (c) geographically the Regions 2020 index operates on the NUTS 2 level (provinces), the ESPON study on the NUTS 3 level (counties), (d) DG Regio makes use of five sensitivity indicators [2] compared with ESPON Climate's 19 indicators, (e) no adaptive capacity indicators were used in the Regions 2020 vulnerability index (the original study by Römisch at least had one such indicator, namely GDP per capita), while ESPON Climate uses 15 adaptive capacity indicators and, finally, (f) the Regions 2020 report combined their

indicators by using equal weights, whereas ESPON Climate weighted the various dimensions of their vulnerability assessment.

Not surprisingly, the results of the two vulnerability assessments differ. The overall spatial patterns are similar, in that regions in the South and South-East of Europe are the most vulnerable regions in both studies. However, the Regions 2020 results are not as fine-grained in these areas both in terms of geographical resolution and in terms of vulnerability categories. For France, the UK and the Benelux countries results are largely similar, but Germany, Poland, Denmark and Sweden are depicted as more vulnerable by the DG Regio study. It is difficult to identify which particular methodological differences account for each of these resulting variations.

9.3 Conclusions

This chapter has shown that the environmental, physical, economic, social and cultural impacts of climate change can be roughly grouped into those impact dimensions that are mainly affected by climate change induced extreme events and other dimensions mainly affected by slow changes of average climatic conditions. These differences also showed up in the overall climate change impact map and in particular when comparing the impact map based on weights from a Delphi survey to the impact map based on equal weights between impact dimensions. How the impact patterns might change if not current but projected future sensitivity data were generally available was demonstrated using the example of social impacts.

Following the first vulnerability concept of the IPCC (McCarthy *et al.*, 2001) the climate change impacts and related adaptive capacities were combined. The vulnerability results were then discussed in relation to the EU's territorial cohesion policy, which according to this vulnerability assessment would come under increasing pressure from climate change. The significant influence of normative underpinnings on the vulnerability results was demonstrated and discussed by comparing the Delphi-weighted with the equally weighted

---

[2] 'Vulnerability of areas to drought, change in population affected by 100 year return river floods under the A2 scenario, exposure of the agriculture, fisheries and tourism sector to climate change expressed by proportion of these sectors in regional GVA in regions where these sectors will be negatively vaffected, and exposure of densely populated areas to coastal erosion expressed in coastal populations living below 5 m elevation' (DG Regio, 2009a, p. 23).

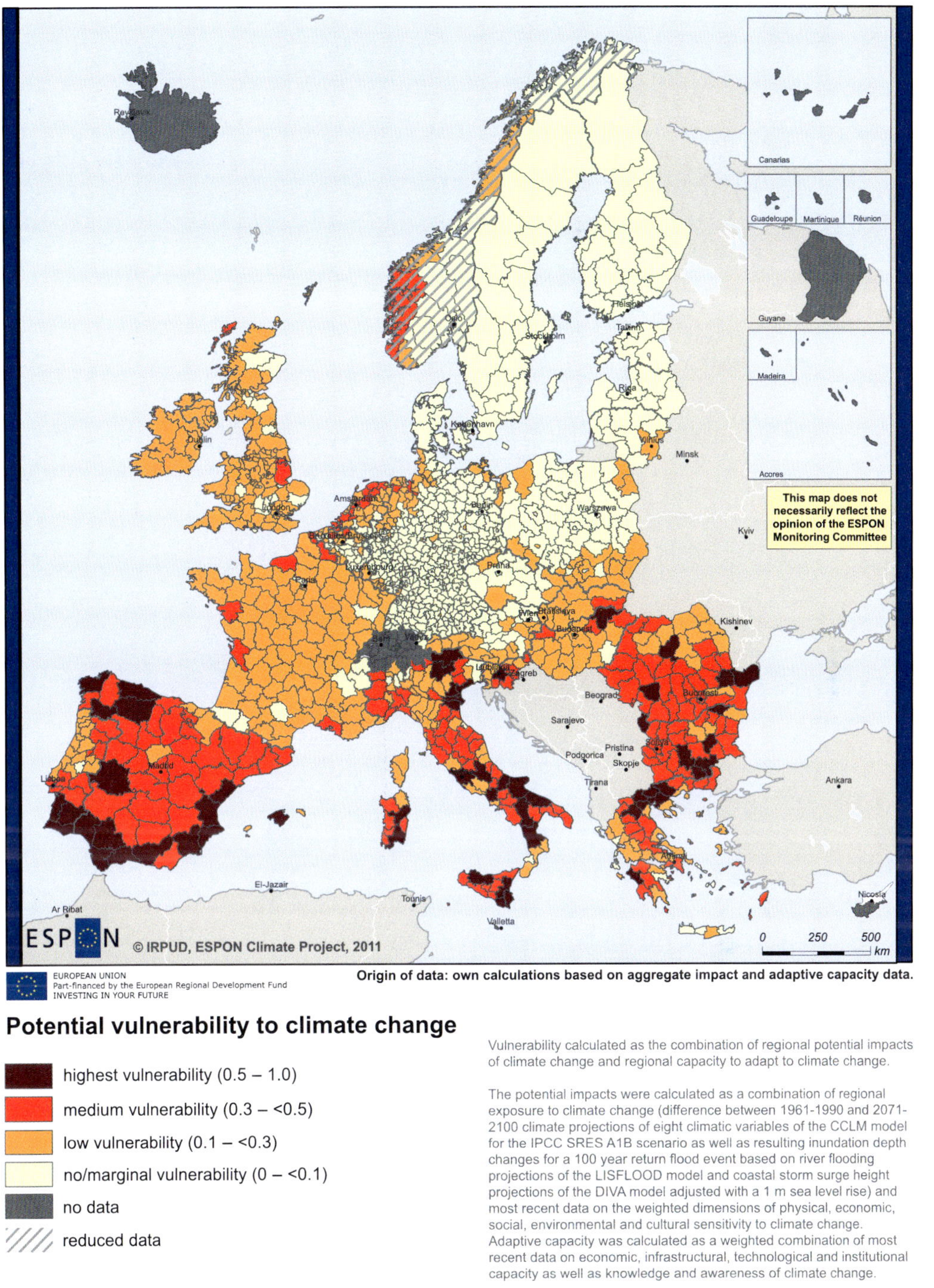

Potential vulnerability to climate change

Vulnerability calculated as the combination of regional potential impacts of climate change and regional capacity to adapt to climate change.

The potential impacts were calculated as a combination of regional exposure to climate change (difference between 1961-1990 and 2071-2100 climate projections of eight climatic variables of the CCLM model for the IPCC SRES A1B scenario as well as resulting inundation depth changes for a 100 year return flood event based on river flooding projections of the LISFLOOD model and coastal storm surge height projections of the DIVA model adjusted with a 1 m sea level rise) and most recent data on the weighted dimensions of physical, economic, social, environmental and cultural sensitivity to climate change. Adaptive capacity was calculated as a weighted combination of most recent data on economic, infrastructural, technological and institutional capacity as well as knowledge and awareness of climate change.

(a)

Figure 9.7 Comparison of ESPON Climate's vulnerability assessment and DG Regio's Regions 2020 climate change vulnerability index. © ESPON 2013, IRPUD, ESPON Climate Project, 2011.

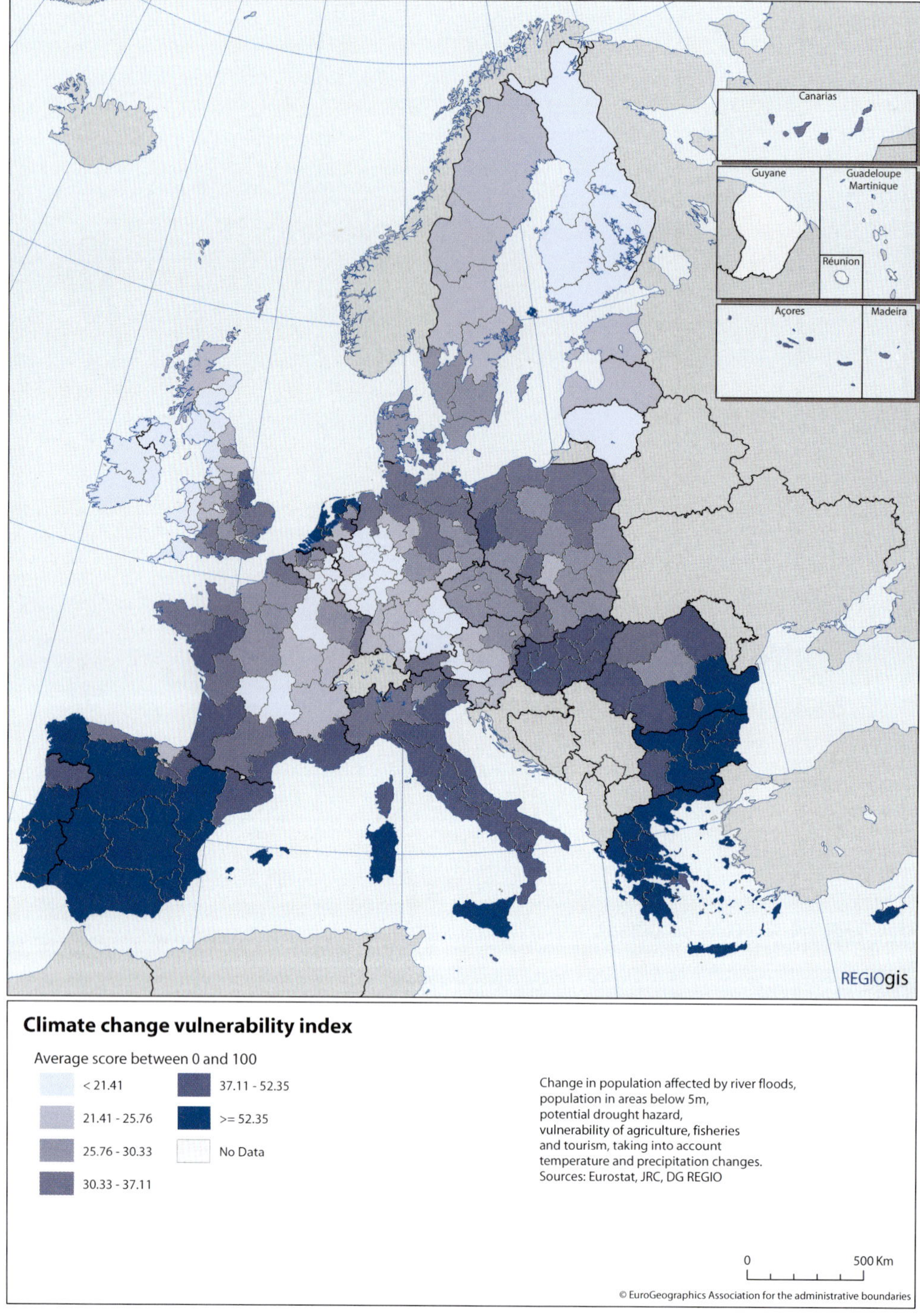

Figure 9.7 (*continued*). *Source*: Regions 2020. An assessment of future challenges for EU Regions, Commission staff working document, November 2008. © European Union, 1995-2013. © EuroGeographics Association for the administrative boundaries. http://ec.europa.eu/regional_policy/sources/docoffic/working/regions2020/pdf/regions2020_en.pdf.

vulnerability version. Finally, it was shown that ESPON Climate's vulnerability assessment goes far beyond the only other pan-European, integrated vulnerability study that currently exists – both in terms of underlying methodology and specificity of results. Therefore the methodology and results of this vulnerability assessment could prove to be a useful tool for European policy-making, but also for regional case studies on climate change. The following chapters will follow up on both of these directions and discuss in detail the regional and European implications of ESPON Climate's vulnerability assessment.

References

Adger, W.N. (2006) Vulnerability. *Global Environmental Change*, **16** (3), 268–281.

Directorate General Regional Policy (2008) Regions 2020: An Assessment of Future Challenges for EU Regions, Brussels.

Directorate General Regional Policy (2009) Regions 2020: The Climate Change Challenge for European Regions, Brussels.

Downing, T.E., Butterfield, R., Cohen, S., *et al.* (2001) Climate Change Vulnerability: Linking Impacts and Adaptation. Report to the Governing Council of the United Nations Programme. United Nations Environmental Programme, Nairobi, Kenya.

Eakin, H. and Luers, A.L. (2006) Assessing the vulnerability of social-environmental systems. *Annual Review of Environment and Resources*, **31**, 365–394.

European Commission (EC) (2008) Green Paper on Territorial Cohesion. http://ec.europa.eu/regional_policy/archive/consultation (accessed 5 March 2013).

ESPON Climate (2011) Climate Change and Territorial Effects on Regions and Local Economies: Scientific Report, Luxembourg.

Füssel, H.-M. (2007) Vulnerability: A generally applicable conceptual framework for climate change research. *Global Environmental Change*, **17**, 155–167.

IPCC (Intergovernmental Panel on Climate Change) (2012) *Managing the Risks of Extreme Events and Disasters to Advance Climate Change Adaptation. A Special Report of Working Groups I and II of the Intergovernmental Panel on Climate Change* (eds C.B. Field, V. Barros, T.F. Stocker, D. Qin, D.J. Dokken, K.L. Ebi, M.D. Mastrandrea, K.J. Mach, G.-K. Plattner, S.K. Allen, M. Tignor and P.M. Midgley), Cambridge University Press, Cambridge and New York.

Hinkel, J. (2011) Indicators of vulnerability and adaptive capacity: Towards a clarification of the science–policy interface. *Global Environmental Change*, **21**, 198–208.

Hofmann, M., Hinkel, J. and Wrobel, M. (2011) Classifying knowledge on climate change impacts, adaptation, and vulnerability in Europe for informing adaptation research and decision-making: a conceptual meta-analysis. *Global Environmental Change*, **21**, 1106–1116.

McCarthy, J.J., Canziani, O.F., Leary, N.A., Dokken, D.J., White, K.S. (eds) (2001) *Climate Change 2001: Impacts, Adaptation, and Vulnerability: Contribution of Working Group II to the Third Assessment Report of the Intergovernmental Panel on Climate Change*. Cambridge University Press, Cambridge.

Römisch, R. (2009) Perspectives of 2020 Regional Disparities and Future Challenges – A report to the Directorate-General for Regional Policy Unit Conception, Forward Studies, Impact Assessment. Available at: http://ec.europa.eu/regional_policy/sources/docgener/studies/ (accessed 5 March 2013).

Thywissen, K. (2006) Components of risk. A Comparative Glossary. SOURCE publication series of the United Nations University-Institute for Environment and Human Security (UNU-EHS) No. 2, Bonn.

Chapter 10

Role of case studies – methodological concept

Stefan Greiving[1] and Philipp Schmidt-Thomé[2]

[1]*TU Dortmund University, Institute for Spatial Planning (IRPUD), August-Schmidt-Strasse 10, 44227 Dortmund, Germany*
[2]*Geological Survey of Finland (GTK), P.O. Box 96, 02151 Espoo, Finland*

Abstract

Case studies of the ESPON Climate project require an integrated twofold approach: an analytical approach and an explorative approach. Seven case studies were chosen that cover all five types of climate change regions identified in the exposure cluster analysis, but also different legal–administrative families. These studies not only provide an in-depth analysis of climate change vulnerability in Europe, but also explore a wide variation of European regions and corresponding differences in climate change exposure, sensitivity, impact and adaptive capacity.

10.1   Case studies

The seven case studies of the ESPON Climate project serve to cross-check and extend the findings of the pan-European assessment. They provide in-depth regional analyses of climate change vulnerability (exposure, sensitivity, impact, adaptation). The studies cross-check the indicators and findings of the European-wide analysis with the results of the case study areas, but also explore the diversity of response approaches to climate change. Opposite to the European-wide analysis, the regional approaches also pay attention to qualitative information, gathered from interviews with stakeholders, in order to address cultural and institutional factors. Finally, they develop conclusions for the implementation of measures at the European level.

Thus, the case studies needed to integrate a twofold approach:

- An analytical approach coherent with the overall methodology of the project in order to ensure comparability between each other and connectibility with the overall analysis on the European scale;

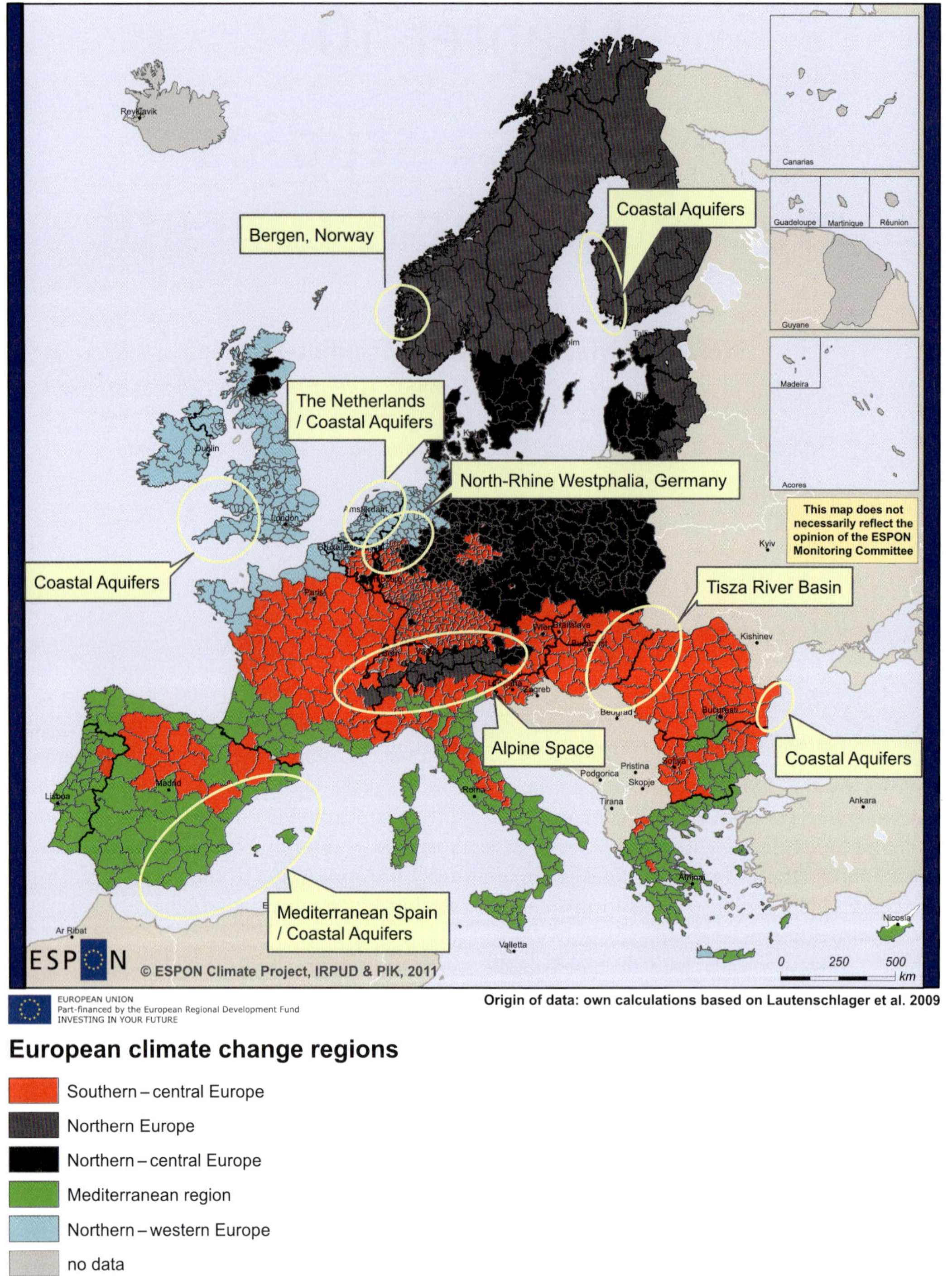

Figure 10.1 Case study locations within the major climate change regions. © ESPON 2013, IRPUD, ESPON Climate Project, 2011.

- an explorative approach focusing on aspects not covered in the European-wide analysis, such as understanding the cultural and institutional factors influencing climate change effects on different European regions, and aspects peculiar to the respective case study area which can best be captured by the case study approach.

In addition each case study explores certain dimensions of exposure, sensitivity and adaptation to climate change of particular relevance to it.

Figure 10.1 shows the locations of the seven ESPON Climate case studies by mapping them onto the typology of climate change regions.

However, a representation of the different climate change regions was not the only criterion for the selection of case studies. The legal frameworks as well as the political–administrative systems are fairly heterogeneous within Europe with regard to criteria such as centralism versus federalism (Newman and Thornley, 1996). They determine to a significant extent how adaptation responses are designed and by which institutions they are implemented (Greiving and Fleischhauer, 2012). Although the policies of the European Community shape national policies more and more, the specifics of the different Member States mainly characterise the design of national policies. In some of the EU Member States, for example, a new development is legally allowed when it is to conform to the regional plan. This so called regulatory function of spatial planning is known under the term 'conforming planning' in the international discourse on planning theory (Larsson, 2006; Rivolin, 2008). In most of the EU Member States,

Table 10.1 Case studies and selection criteria.

Case study area		Geographic coverage			Climate change regions
	ESPON three-level approach	**Macro-geographic regions**	**Geomor- phological character**	**Legal- administrative families**	
Alpine space	Trans-national	Central and Southern Europe	Mountain area	Germanic/ Napoleonic	Northern Europe Northern-Central Europe, Southern-Central Europe
Tisza river	Trans-national	Central and Eastern Europe	River basin	Eastern European	Northern-Central Europe, Southern-Central Europe
North Rhine-Westphalia	Regional	Germany (Western Europe)	River basin, hilly land	Germanic	North-Western Europe, Northern-Central Europe, Southern-Central Europe
Coastal Mediterranean Spain, Balearic Islands	Regional	Southern Europe	Coastal area	Napoleonic	Mediterranean region
Bergen	Local	Norway (Northern Europe)	Coastal area, mountain area	Scandinavian	Northern Europe
The Netherlands	National	Western Europe	Coastal area, river basin, lowlands	Napoleonic	North-Western Europe
Coastal Zone Aquifers	Trans-national	Finland, the Netherlands, United Kingdom, Spain, Romania	Coastal area, lowlands	Scandinavian, Eastern European, Napoleonic	All climate change types covered

and thus also in the majority of case studies, however, the so called development function dominates at the regional planning level, which is discussed under the term 'performing planning'. This planning type is characterised by non-binding programmatic and/or strategic statements. Potential projects are then evaluated against the question whether they support the implementation of the programme or strategy. Furthermore, there are – if at all – only partially binding effects for the subordinated local level. The research question in this respect is if the planning system has an influence on the role of spatial planning in national adaptation strategies.

Seven case studies were chosen that cover all five types of climate change regions identified in the exposure cluster analysis, but also different legal–administrative families. As seen in Table 10.1 their coverage ranges from local and regional to national and trans-national. Furthermore, they cover different macro-geographic regions of Europe, represent different geomorphological types (high and low mountains, lowlands, rivers, coasts) and all European legal–administrative families except for the British one. Thus the case studies not only provide an in-depth analysis of climate change vulnerability in Europe, but at the same time they explore a wide variation of European regions and corresponding differences in climate change exposure, sensitivity, impact and adaptive capacity.

Table 10.1 summarises the different selection criteria for the case studies.

References

Greiving, S. and Fleischhauer, M. (2012) National climate change adaptation strategies of European states from a spatial planning and development perspective. *European Planning Studies*, **20** (1), 27–47.

Larsson, G. (2006) *Spatial Planning Systems in Western Europe – An Overview*. IOS Press, Delft.

Newman, P. and Thornley, A. (1996) *Urban Planning in Europe – International Competition, National Systems and Planning Projects*. Routledge, London.

Rivolin, U.J. (2008) Conforming and performing planning systems in Europe: An unbearable cohabitation. *Planning, Practice & Research*, **23** (2), 167–186.

Chapter 11

Integrated assessment of vulnerability to climate change: the case study North Rhine–Westphalia

Anne Holsten, Carsten Walther, Olivia Roithmeier and Jürgen P. Kropp
Potsdam Institute for Climate Impact Research, P.O. Box 60 12 03, 14412 Potsdam, Germany

Abstract

This chapter presents an integrated climate change vulnerability assessment, including the physical, social, environmental and economic dimensions for municipalities of the German Federal State of North Rhine–Westphalia. Individual impacts of sectors are quantified by tailor-made approaches, for which relevant exposure variables are directly connected to specific sensitivity variables. A new framework for the aggregation of the components of vulnerability is presented, which is often neglected in such assessments. The aggregation across the impacts for the sectors is carried out by two alternative methods: the arithmetic mean of the values and a typological approach via a cluster analysis. The first approach enables the identification of which municipalities will have to deal with the highest impact burden, while at the same time it also allows tracking back decisive factors of the system to support target-oriented adaptation measures. The approach of typological categorisation provides an overview over the territorial distribution of impacts by identifying regions of similar impact characteristics. The results for both approaches show regional impact 'hotspots' in the metropolitan area, the foothills of the mountains and in the East of the state. Such comprehensive knowledge across sectors on this spatial scale is important for the development of regional adaptation strategies, which are currently being drafted

European Climate Vulnerabilities and Adaptation: A Spatial Planning Perspective, First Edition.
Edited by Philipp Schmidt-Thomé and Stefan Greiving.
© 2013 John Wiley & Sons, Ltd. Published 2013 by John Wiley & Sons, Ltd.

at national and state levels for Germany. However, various issues of these approaches, for example the subjectivity of selection of impacts and the aggregation across sectors, still remain unresolved and are discussed in this chapter. This stresses the need to consider the aim and methodological advantages and disadvantages of the applied methodologies before applying any vulnerability assessment.

11.1   Introduction

Spatially explicit vulnerability assessments regarding climate change have become common usage (Preston, Yuen and Westaway, 2011), especially with a sectoral focus (O'Brien *et al.*, 2004; Zebisch *et al.*, 2005; Ciscar *et al.*, 2011). They include sectors such as agriculture, forestry, human health or tourism. However, the implementation of such approaches is still challenging due to their interdisciplinary character, spatially and temporally heterogeneous processes and to the normative judgements involved (Hinkel, 2011; Preston, Yuen and Westaway, 2011). Moreover, existing methodologies are heterogeneous and lack standardised methodologies (Preston, Yuen and Westaway, 2011). Therefore, integrated assessments still remain rare, particularly regarding the consideration of both biophysical and socioeconomic determinants. Science with respect to vulnerability assessments is often under tension between the demands for an integration of sectors or dimensions and sector-specific analysis to support concrete decision making. Thus, novel approaches of vulnerability assessments are in demand, which integrate sectors, while at the same time conveying sector-specific information.

In this chapter we operationalise a climate change vulnerability assessment in a standardised and comparable way by means of tailor-made approaches for various sectors of the German Federal State North Rhine–Westphalia (NRW). The assessment of its vulnerability is of particular interest to decision makers, which is apparent from previous climate change-related studies financed by the state (Spekat, Gerstengarbe, Kreienkamp and Werner,

2006; Kropp *et al.*, 2006, 2009). Based on these results, an adaptation strategy at state level has been published in 2009 (MUNLV, 2009). However, a theoretical description of the processes of climate change effects is missing. Also, this strategy is still at an early stage and lacks further concretisation and prioritisation regarding specific adaptation measures and time horizons. Therefore, a more detailed and spatially explicit knowledge is necessary for a wide range of sectors. In spite of the regional application of this assessment we aim at standardising such approaches. Thus, our methodology is in principle transferable to other regions.

In the following we will first provide a description of the study region and an overview of past climatic changes and their effects in this region. We will then propose a methodology of an integrated vulnerability assessment for a selection of impact processes, which are defined by a system-specific sensitivity related to a relevant climatic stimuli. This approach is then applied to the municipalities in the case study region based on two aggregation methods to quantify the effects of climate change and two regional climate models, CCLM and REMO. Spatially explicit results of impacts and vulnerability are then discussed and the main conclusions of our approach are drawn.

11.2   Description of the study area

The German Federal State of North Rhine–Westphalia (NRW) is located in the West of Germany and comprises 396 municipalities (Figure 11.1). With a population of 17.8 million (2011) and an average population density of over 500 persons km^{-2}, NRW is the most populous and at the same time the most densely populated state in Germany. Regional characteristics are quite diverse in terms of climate and geomorphology as well as in socio-economic structure. Two main types of landscapes can be found in NRW, namely the North German Lowlands, with elevations just a few meters above sea level, and the North German

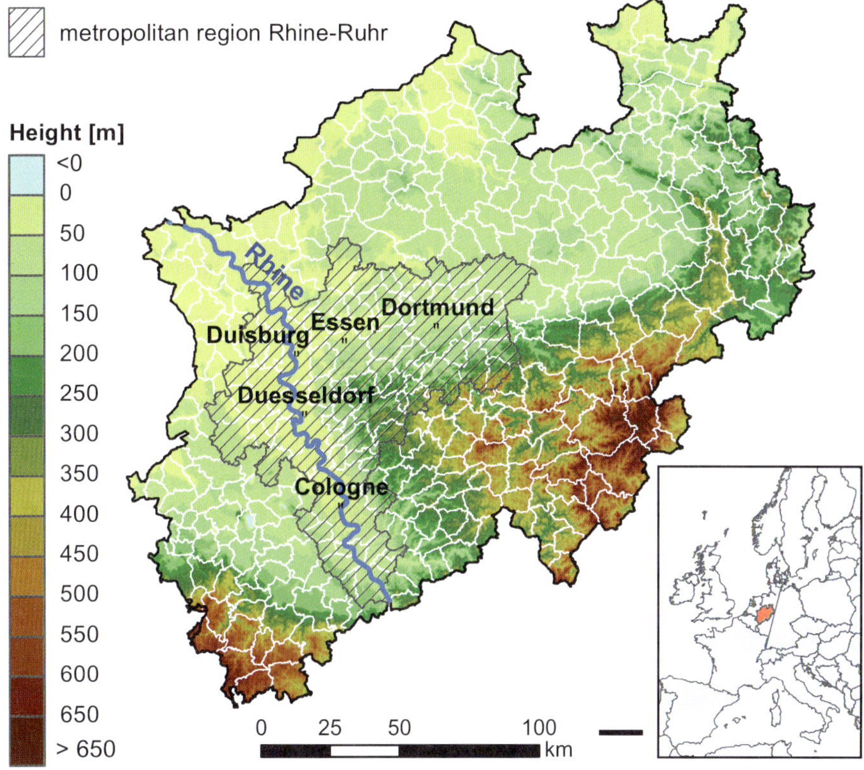

Figure 11.1 The study region North Rhine–Westphalia and its location within the European Union. The hatched area indicates the metropolitan Rhine–Ruhr region. Municipalities are delineated by white borders.

Low Mountain Range (Sauerland, Eifel Mountains) with elevations of up to 850 m. The lowlands comprise the Rhine–Ruhr Area, which is one of the largest metropolitan areas in Europe with a population of approximately 10 million and a very high population density of 2100 capita km^{-2}. On the contrary, in the mountain regions population density is rather low with 150 capita km^{-2}.

These main landscape types are also distinguishable regarding their climatic characteristics: annual mean temperature amounts to approximately 10 °C (1961–1990) in the lowlands and about 5 °C in the mountain regions. Yearly mean precipitation of up to 1500 mm has been recorded in higher elevations, while the Rhine Valley and the Bay of Cologne receive a mean sum of around 620 mm per year (Kropp *et al.*, 2009). Thus, while in summer it can become very hot in the Rhine–Ruhr Basin, the temperature

in the highlands and mountainous regions is more moderate. Demographic change towards an elderly population is apparent in NRW. In the year 2008 19% of the population were over 65 years old and an increase of up to 29% is expected until 2030 (IT.NRW, 2010a). As a former mining region, NRW was the turbine of economic growth after the Second World War (Storchmann, 2005). In the second half of the 20th century the region underwent a radical structural change from around 600 000 people employed in mining in the 1950s (Storchmann, 2005) to around 35 000 in 2009 (IT.NRW, 2010b). Currently more than 60% of the people are employed within the service sector. Despite the rather small share of employees within the energy supply sector (1%), NRW can be regarded as an energy state, producing one third of the overall electricity demand of Germany (Energy Agency NRW, 2007).

NRW contributes over 20 % to the overall German GDP (MWEBWV, 2010). Consequently, the investigation of climatic changes is especially important, as severe impacts in this region may reduce the overall economic performance of Germany.

11.3 Past climatic changes and their impacts in the study area

Climatic changes are already apparent in the region. Since the beginning of the 20th century until today, average air temperature rose from 8.4 to 9.6 °C (MUNLV, 2009). Thus, warming took place in NRW at a higher rate than the global average (Gerstengarbe, Badeck, Hattermann and Krysanova, 2004). Moreover, an acceleration of mean temperature rise between 1991 and 2000 compared with the 30 years before, has been observed (Spekat, Gerstengarbe, Kreienkamp and Werner, 2006). Generally a decrease in winter and an increase in summer temperature has been recorded. In the last decades the number of frost and ice days has declined considerably. In contrast, the number of summer days rose up to 20 days, particularly in the South of NRW, and hot days increased by up to eight days (Gerstengarbe, Badeck, Hattermann and Krysanova, 2004). In addition, the water temperatures of the river Rhine have increased during the last decades (Greis, 2007; MUNLV, 2009). For example, the warming trend has affected plant phenology in NRW, which can be described by a theoretical concept (Rybski, Holsten and Kropp, 2011). This shows that the onset dates of phenological events of plants in spring (e.g. flowering) has advanced in this region in recent decades. Moreover, the burden of heat waves causes additional threats to human health in the region (Lissner, Holsten, Walther and Kropp, 2012).

Average annual precipitation changed from around 790 mm towards the end of the 19th century to 910 mm today, which represents an increase of around 15%. This trend is mainly apparent from 1960 onwards, thenceforth, years with an annual precipitation of more than 1000 mm, in particular, occurred more often (MUNLV, 2009). Further, days with rainfall above 20 mm per day have increased since the 1950s, particularly during winter and in areas of high annual precipitation (Gerstengarbe, Badeck, Hattermann and Krysanova, 2004; aqua_plan GmbH, hydro & meteo GmbH and dr. papadakis GmbH, 2010). These events are likely to have led to flash floods, which occur in NRW at a rate above the average for Germany (Castro *et al.*, 2008).

A period of stronger winds was observed in the mid-1970s and at the beginning of the 1990s at Dusseldorf Airport (Kasperski, 2002), and the annual number of days with wind speeds above Beaufort 8 has increased by 40% in Dusseldorf between 1969 and 1999 (Otte, 1999).

11.4 Methodology of an integrated vulnerability assessment

11.4.1 *Vulnerability concept*

An array of definitions of vulnerability has evolved from different research disciplines, such as in the hazard, development and sustainability or climate change context (Fuchs, Kuhlicke, Meyer and May, 2011). We base our work on the common framework within the climate change context following IPCC (2001) and Füssel and Klein (2006): 'Vulnerability is the degree to which a system is susceptible to (. . .) adverse effects of climate change (. . .) as a function of the character, magnitude and rate of climate variation to which a system is exposed, its sensitivity and its adaptive capacity'. Thus, vulnerability V can be regarded as a function of the components adaptive capacity AC and impacts I, which in turn are expressed by the sensitivity S and exposure E:

$$V = f(I, AC), \text{with } I = f(E, S) \qquad (11.1)$$

Vulnerability frameworks still remain abstract and lack an indication regarding their aggregation procedure (Hinkel, 2011), albeit the relevance of the aggregation method for these assessments (Preston, Yuen and Westaway, 2011). In the following, we therefore propose a formalisation of a method to aggregate these components. Preston, Yuen and Westaway (2011) identify as a key challenge the lack of specificity in existing vulnerability assessments to state which system is vulnerable to which climatic stimuli. Therefore, system specific linkages between sensitivity and exposure variables are essential. Vulnerability can be described by the 'individual and collective physical, social, economic and environmental conditions' (UN/ISDR, 2004). The physical dimension refers to the built environment, including settlements and infrastructure, the social dimension considers the human wellbeing, the economic dimension represents economic activities such as agriculture or tourism and the environmental dimension refers to the natural environment. We integrate this into our vulnerability framework as a basis for our analysis, which comprises the components E, S, I and AC as well as the four dimensions (Figure 11.2).

These dimensions comprise, for example, human health, tourism, agriculture, forestry, settlements or protected areas, which in the following will be named 'sectors'. Thus, the dimensions and finally the total impacts cover the impacts over several specific sectors, which therefore represents an integrated vulnerability assessment. In the following analysis we quantify the different constituents of the vulnerability and present two approaches for their aggregation. For information on future climate change we apply results based on two regional dynamical climate models, CCLM (Lautenschlager *et al.*, 2009) and REMO (Jacob, Mahrenholz and Keup-Thiel, 2006) to allow for a comparison of our results between models.

11.4.2 *Selection of impact processes*

We restricted our analysis to ten different impacts, which are summarised in Table 11.1. They are described by their impact process, including the affected exposure unit, the relevant climatic stimuli, the relevance of the potential impact for NRW and the description of quantification of the sensitivity. We aimed at integrating climate-relevant impacts systematically and at the same time consider the largest range of sectors possible. However, for several impacts, sufficient data or suitable methods for quantification were lacking. For example, despite the climatic relevance of the energy sector regarding supply (Förster and Lilliestam, 2009) and demand (Olonscheck, Holsten and Kropp, 2011)

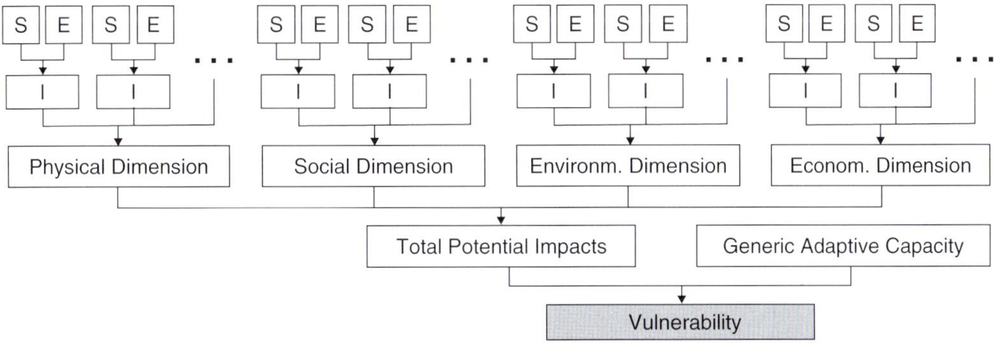

Figure 11.2 Schematic overview over the components and dimensions of the vulnerability analysis. Sensitivity indicators (*S*) are combined with relevant exposure indicators (*E*) expressing specific impacts (*I*). These are aggregated to the physical, social, environmental and economic dimension and to the total potential impacts. Together with the generic adaptive capacity they describe the vulnerability.

Table 11.1 Overview of considered impacts, which are expressed by sensitivities and relevant climatic stimuli considered regarding the physical, social, environmental and economic dimension. Positive relationships are marked by ↑ (e.g. impacts on humans increase with increasing heat days), negative feedback processes by ↓. For abbreviations see Table 11.2.

Exp. unit	Stimuli	Regional relevance and relation to exposure	Methodology of the sensitivity indicators
Physical			
Settlements	Flash floods (CHR ↑)	Settlements located in steep river catchments with a short time lag of the runoff water are prone to flash floods, which can be caused by heavy precipitation events (Collier, 2007; Castro *et al.*, 2008). Around 20% of flash floods in Germany occurred within the state NRW in the last decades, which in relation to its area lies above the average for Germany (Castro *et al.*, 2008). A concentration of past flash floods was found for the Rhine valley. We quantify the exposure to flash floods by changes in days with heavy precipitation.	Urban areas prone to flash floods can be quantified by a flow accumulation analysis (Castro *et al.*, 2008). We calculate the potential runoff by applying the Curve Number method (USDA, 1972; 2007), accounting for land use (ATKIS25 dataset) and hydrological soil types with the ArcCN tool (Zhan and Huang, 2004). We assigned Hydrological Soil Types (A = high infiltration capacity and low runoff to D = low/high, respectively) to regional soil types of NRW (BK50) using the criteria concerning minimum hydraulic conductivity, depth of least permeable layers and depth to water table (USDA, 2007). Flow accumulation is analysed with ArcGIS based on the potential runoff and the flow direction of the elevation model (DEM50). Considering the potential time lag (assuming a hypothetical 60 min event with 10 mm precipitation and considering the slope and runoff potential), the peak flow is quantified based on Castro *et al.* (2008). We then calculate the share of urban area affected by a high peak flow of $>1\,m^3$ on the total municipal area.
Settlements	Pluvial flooding (CHR ↑)	Besides areas of intense flow accumulation, landscape sinks, where water accumulates, are also threatened (Castro *et al.*, 2008; Grünewald *et al.*, 2009). It mainly causes drainage problems and thus economic damage (Jonkman, 2005), which is important in NRW due to its high urban density and sealed area (Held, 2000). Moreover, NRW is characterised by landscape sinks or depressions due to former lignite mining often without drainage systems (Drecker, Genske, Heinrich and Noll, 1995; Hydrotec, 2004; Grünewald *et al.*, 2009). In accordance with the previous indicator, we quantify the exposure to pluvial flooding with changes in days of extreme precipitation.	To analyse the sensitivity to pluvial flooding we combine the approaches of accumulated surface runoff and landscape sinks. Using ArcGIS, we identify sinks within the relief and calculate the amount of runoff necessary to completely fill each sink, dependent on the sink volume, drainage area and the surrounding topology. The potential runoff is calculated according to the methodology of flash floods described above. The volume of the sink, to be potentially flooded, is divided by the calculated runoff of the respective drainage area. Only settlement area (ATKIS25 data) within the sinks was considered. The value concerning the ratio between drainage area runoff and sink volume was summed for the municipalities and weighted by their area.
Social			
Humans	Heat (CHD ↑)	Extremely high temperatures are associated with significantly increased mortality and morbidity rates especially in older age groups (Kosatsky, 2005; Vandentorren and Empereur-Bissonnet, 2005). These severe impacts were also apparent in NRW during	The sensitivity of humans to heat is applied from Lissner, Holsten, Walther and Kropp (2012). This indicator comprises the sensitivity of the population, expressed by the share of population ≥65 years or older. The potential for an urban heat island (UHI) is represented by the degree of

Table 11.1   (*continued*)

Exp. unit	Stimuli	Regional relevance and relation to exposure	Methodology of the sensitivity indicators
		the extraordinary warm year of 2003 (Hellmeier, Stausberg and Hollmann, 2007). Thus, we represent the exposure by changes in heat days.	urbanity, expressed by the minimum value of either the population density or the share of sealed surface. A fuzzy logic algorithm is applied to the identified influence variables. The fuzzified variables sealed surface and population density were aggregated to indicate an UHI effect and then combined with the share of elderly population.
Environmental			
Protected areas	Drought (CWB ↓)	Protected areas experience large impacts of climate change, for example in form of distribution changes of species (Pompe *et al.*, 2008), phenological changes (Menzel *et al.*, 2006; Rybski, Holsten and Kropp, 2011) or extinction of species (Thuiller *et al.*, 2005). In Europe, the Natura 2000 network is of major importance for the conservation aims. Until 2080 over 60% of the plant and animal species listed in the Habitats directive could be driven out of the protected areas due to climate change (Araújo *et al.*, 2011). The climatic water balance has been identified as a key driver of the distribution of species (Svenning and Skov, 2006; Vohland and Cramer, 2009; Crimmins *et al.*, 2011).	We quantify the sensitivities of habitat types within protected areas to warmer and drier conditions, based on indicators developed by Petermann *et al.* (2007) comprising biogeographic information, namely the restriction of current area borders, ground water dependency, past trend in area decrease, restriction to high altitudes and neobiota influence for habitat types in Germany. We substituted their further proposed indicator 'qualitative risk' by the locally available 'conservation state' (Natura 2000 database), to indicate already existing pressures. We further include information on cold and wet-tolerant plants, especially those sensitive to warmer and drier conditions (Petermann *et al.*, 2007; Araújo *et al.*, 2011). These were defined by their temperature- (values 2–4) and moisture-tolerance values (values 7–9) (Ellenberg, 1992; BfN, 2011). Indicator values were assigned by equidistant categories of their share of the characteristic plant pool, from 0 to 100%. From the same dataset, the share of stenocious plants regarding temperature and moisture conditions (not indifferent) was calculated and classified analogously. The 10 subindicators are averaged for each habitat. The sensitivity of the protected area (data for 454 out of 518 areas) is calculated according to the area covered by its habitats. On the municipal level, the indicator is multiplied by the share of protected areas.
Soils	Water erosion (CHR ↑)	Erosion is especially relevant on temporarily uncovered agricultural soil, representing around 32% of the area in NRW. Climate, topography and cultivation processes already caused considerable damage through water erosion in NRW (Kehl, Everding, Botschek and Skowronek, 2005), which could be further aggravated by altered seasonal precipitation patterns	Soil erosion can be estimated by the Universal Soil Loss Equation (Schwertmann, Vogl and Kainz, 1990; Renard, Foster and Weesies, 1997) comprising sensitivity variables of soil erodibility, slope, slope length, cultivation and soil conservation. Slope length cannot be sufficiently calculated based on the available elevation data of 50 m resolution. Also, the inclusion of cultivation and soil conservation measures is restricted due to

(*continued overleaf*)

Table 11.1   (*continued*)

Exp. unit	Stimuli	Regional relevance and relation to exposure	Methodology of the sensitivity indicators
		(Sauerborn, Klein, Botschek and Skowronek, 1999). Soil water erosion is especially high during heavy rainfall events (Müller, 2003; Boardman, 2006), therefore we apply changes in days with heavy precipitation as the exposure.	lack of data. Thus, we apply this formula in a simplified way by considering the erodibility (K-factor, from the soil map BK50) and slope (S-factor, from the DEM50). We therefore describe the potential soil erosion sensitivity, which can be further influenced by agricultural activities. The variables are converted into a scale of 0–1 and multiplied according to the equation. The mean value on agricultural sites within the municipality is then multiplied by the share of agricultural area.
Lakes	Decrease in water volume (CWB ↓)	Lakes provide numerous services, for example, ground water renewal, water supply, recreation and habitats. The water level regime and lake level fluctuations are of key importance to their structure and functioning (Riis and Hawes, 2002; Coops, Beklioglu and Crisman, 2003); extreme fluctuations might exceed species adaptive capacity (Coops, Beklioglu and Crisman, 2003; Leira and Cantonati, 2008), driven by an imbalance in gains and losses of water. We therefore express the exposure by the climatic water balance.	Shallow lakes are especially sensitive to a decrease in water volume (Scheffer and van Nes, 2007). To therefore indicate the sensitivity of lakes to this exposure, we calculated the ratio of surface area (m^2) to volume (m^3) as two commonly available indicators. Municipalities, which do not comprise lakes or with a lack of data were assigned the lowest sensitivity. Anthropogenic interventions (i.e. water withdrawal, land use changes or agricultural activities) also affect the water balance and water quality of lakes (Bates, Kundzewicz, Wu and Palutikof, 2008), however, these are difficult to quantify. Thus, we focused on natural lakes (8) and lakes evolved through excavation (91), which are common especially in the Rhine valley due to gravel mining. The indicator is then multiplied by the share of lake area within the municipality.
Economic			
Winter tourism	Shortening of season (CSC ↓)	NRW comprises one of the largest winter sport region north of the Alps. With a gross annual turnover in the Sauerland area of around € 100 million (IFT, 2008) it is of high regional economic relevance. To ensure a profitable season, over 250 snow machines provide conditions for alpine tourism in NRW (WSA, 2011). The exposure is quantified by changes in days with snow cover.	The sensitivity of winter tourism to climate change is considered to be especially high for resorts in low mountain ranges (Sauter, Weitzenkamp and Schneider, 2009; Steiger, 2010). In NRW, all ski resorts are located in such area of comparatively low altitude, the Sauerland and Eifel mountains. Thus the size of the skiing area indicates the sensitivity, expressed by the length of ski runs within the municipality. Owing to more advantageous conditions, resorts within the core area identified by Roth *et al.* (2001) are considered to be half as sensitive as areas in the surroundings. Regarding the Eifel mountains, for Monschau only the number of lifts was available, thus the length of its ski runs was set to the average length for the remaining resorts.
Forestry	Windthrow (CSD ↑)	Storms are among the most important natural stressors for forests. In 2007, the storm "Kyrill" caused the highest insured	The sensitivity of forests to windthrow is applied from Klaus, Holsten, Hostert and Kropp (2011). It comprises forest characteristics, soil characteristics

Table 11.1   (*continued*)

Exp. unit	Stimuli	Regional relevance and relation to exposure	Methodology of the sensitivity indicators
		losses in Central Europe since at least 1990 (Munich Re, 2008). One third of the European and half of the German forest loss was recorded in NRW (MUNLV, 2010). Exposure is thus represented by changes in the number of storm days.	and topography. These variables have been aggregated by Klaus, Holsten, Hostert and Kropp (2011) by means of a logistic regression model validated for the storm event 'Kyrill'. We finally multiplied this indicator by the share of forests within the municipalities.
Forestry	Forest fires (CRH ↓)	Forest fire occurrence is relatively low in NRW. However, fire events have occurred in small numbers each year in the past according to the official fire statistics. During extremely hot summers, fire damage increased considerably. Forest fires in NRW show a stronger correlation with relative humidity than with temperature, precipitation and commonly applied meteorological forest fire indices (Holsten, Dominic, Costa and Kropp 2013). We, therefore apply changes in the humidity as the exposure variable.	A multitude of factors concerning climatic, environmental and human determinants play a role in the occurrence of forest fires (Syphard *et al.*, 2008; Costa, Thonicke, Poulter and Badeck, 2010; Reineking, Weibel, Conedera and Bugmann, 2010). A key factor is the fuel moisture, influenced by the soil moisture (Nelson, 2001; Bartsch, Balzter and George, 2009). Owing to a lack of spatially explicit data on fuel biomass, we apply soil moisture characteristics (potential available field capacity) under forests. Considering the observed fires in NRW from 1993 to 2009, burnt area was 1.55 times higher for needle leaved than for broad leaved forest, thus, we apply this ratio for the sensitivity (ATKIS25 dataset). Also humans influence fires, especially through the accessibility of forests (Chuvieco *et al.*, 2010; Costa, Thonicke, Poulter and Badeck, 2010; Reineking, Weibel, Conedera and Bugmann, 2010). We implement this by the distance to the nearest settlements (CORINE). This dataset was preferred over the regional ATKIS dataset to account for distances to objects outside of NRW. All three indicators are rescaled between 0 and 1, averaged with equal weight and multiplied by the share of forests within the municipalities.
Agriculture	Drought (CWB ↓)	Around half of the area of NRW is used for agriculture; two thirds of this area underlie crop production (LWK NRW, 2008) and is highly dependent on climatic conditions. In East Germany, future water deficit is expected to increase leading to droughts and production limitations (Schindler *et al.*, 2007). We therefore express the exposure by changes in the water balance.	The soil moisture regime is considered as one of the main determinants of constraining plant growth (Müller *et al.*, 2010). In Germany, soils most affected by yield decreases under climate change are characterised by low water retention capacities (Wechsung, Gerstengarbe, Lasch and Lüttger, 2008). We therefore express the sensitivity of agricultural soil in NRW to drought by the potential available field capacity under agricultural land (based on ATKIS data). The indicator is then multiplied by the share of agricultural area within the municipality.

and its strong economic importance for NRW, we could not consider this sector in our analysis due to a lack of a coherent database of its water use and other plant characteristics. Also, climate change is expected to increase river flooding, especially along the Rhine (Te Linde *et al.*, 2011). Yet, simulations of extreme flooding events in Germany involve considerable uncertainties, especially regarding high water discharge events on large time horizons (Huang, Krysanova, Oesterle and Hattermann, 2010). We therefore refrained from including this impact in our analysis.

The selection of the impact processes, that is, the omission due to lack of data, leads to some subjectivity regarding the final results. However, our approach is more focused on demonstrating an integration of different sectoral impacts with the aim of providing an integrated framework for a vulnerability assessment. Given a larger set of impact processes, the methodology could be applied in the same manner.

The indicators will be developed on a spatial scale of LAU2 (i.e. municipalities, see Figure 11.1) to provide more detailed spatial information than the European analysis (see Chapter 9), which focuses on NUTS 3 regions. Decision-making processes are generally within the scope of municipalities. They can be scaled up to larger administrative boundaries while being spatially resolved to delineate geographic characteristics of the area. For individual indicators the analysis is based on even more fine-scaled approaches, which are then aggregated to the administrative level.

While climatic changes are often apparent in the form of extreme events (Rahmstorf and Coumou, 2011; IPCC, 2012), incremental developments may result in extreme events in terms of natural or societal impacts (Glade, Felgentreff and Birkmann, 2010). We, therefore apply exposure variables as proxies for extreme events and for slower climatic changes. Thus, Table 11.1 identifies relevant climactic stimuli that are transferred to exposure variables prior to the aggregation to the impacts. To consider the direction of change, absolute exposure variables E_i are rescaled between -1 (decrease in climatic stimuli) and 1 (increase), based on the maximum absolute change in either direction for both models ($E_{i,\,\mathrm{CCLM}}$ and $E_{i,\,\mathrm{REMO}}$). Thus, relative changes in the exposure variables are given by:

$$E_{\mathrm{norm}} = E_i/|E_{\mathrm{max}}|$$
$$\text{with } E_{\mathrm{max}} := \max\{|E_{i,\mathrm{CCLM}}|, |E_{i,\mathrm{REMO}}|\} \quad (11.2)$$

For a graphical representation of this procedure, exemplified for changes in heavy precipitation days occurring over municipalities, see Figure 11.3. The range of values over the study area for these indicators is summarised in Table 11.2.

Sensitivity values are calculated according to the methodology summarized in Table 11.1. For quantifying sensitivities existing methods were applied when possible. For other sensitivity indicators, we develop new approaches. We focus on the relative vulnerability; therefore sensitivity values (e.g. sensitivity of forests to windthrow within a region) are first multiplied by the local relevance of each indicator (e.g. share of forest area within the region) prior to the rescaling procedure.

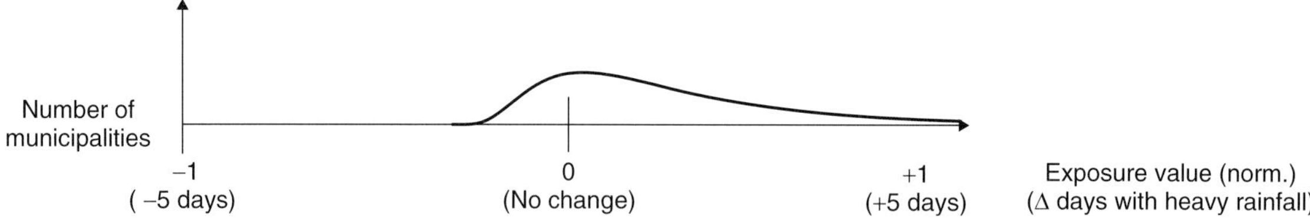

Figure 11.3 Rescaling procedure of exposure variables: schematic distribution of climatic changes over the municipalities, exemplary for changes in heavy precipitation days. Changes are displayed by their absolute value and their values after rescaling.

Table 11.2   Value ranges of projected changes in the selected climatic variables and corresponding rescaled exposure values. The first line refers to values of the model CCLM the second line to the model REMO, respectively.

Abbreviation	Name	Absolute range	Rescaled range
CWB	Change in climatic water balance (mm) (precipitation– evaporation)	−82.4 to 12.4 −69.7 to 48.6	−1 to 0.15 −0.85 to 0.59
CHD	Change in heat days with daily maximum temperature $\geq 30\,^\circ$C (No. of days)	6.2 to 25.0 5.5 to 16.2	0.25 to 1 0.22 to 0.65
CHP	Change in heavy precipitation days with daily precipitation ≥ 20 mm (No. of days)	0.2 to 2.2 −1.3 to 4.2	0.06 to 0.52 −0.31 to 1
CHR	Change in relative humidity (%)	−1.7 to −0.5 −1.3 to 0.3	−1 to 0.26 −0.77 to 0.19
CSC	Change in snow cover days (No. of days)	−16.7 to −1.8 −0.32 to −0.03	−52.0 to −22.2 −1 to −0.43
CSD	Change in storm days with daily maximum wind speed ≥ 20.5 m s^{-1} (No. of days)	3.4 to 6.9 0.49 to 1	0.14 to 5.6 0.02 to 0.66

Adaptive capacity is strongly dependent on the spatial scale (Adger and Vincent, 2005). Various studies have attempted to quantify adaptive capacity at the national or county level (e.g. Brooks, Adger and Kelly, 2005; Cutter and Burton, 2010). We apply generic macro-scale indicators according to the framework of Metzger and Schröter (2006) for European regions to our regional level. This means that the resulting generic index captures a cross-sectoral capacity of a region to adapt instead of reflecting the individual's ability. It therefore describes the context within which an individual could adapt.

Municipalities in NRW are of comparatively small extent, often comprising only one small- to medium-sized city. Therefore, indicator values that are spatially homogeneous at this scale (e.g. implementation level of national or state adaptation strategies) or indicators representing processes active beyond the municipal border (e.g. technological resource availability or traffic infrastructure) are not suitable here. We therefore concentrate on economic resources as well as knowledge and awareness (see Table 11.3). The former is described by the personal (household level) as well as the municipal financial situation. Regarding the public household, we considered the status of the financial budget

of the municipalities.[1] With respect to the private household, we consider the available income of households in 2009.[2] The knowledge of citizens is expressed by their educational school level as the share of population of principal residence with the highest school education level (secondary school or higher) in 2009.

There is a lack of data regarding the awareness of people in NRW concerning climate change-related adaptive capacity. Therefore we make use of proxy indicators of the initiative of the community (mainly driven by personal motivations) with respect to the engagement in three climate change or sustainability initiatives: (i) the 'Network of Municipal Climate Concepts' ('Netzwerk Kommunale

---

[1] Classification as follows: 1 = truly balanced, 0.8 = virtually balanced, 0.6 = approved reduction in common reserves without obligation of budget consolidation concept, 0.4 = approved budget consolidation concept, 0.2 = budget consolidation concept not approved, 0 = overindebted. For six municipalities, the decisions on the budget consolidation concept were still open at the end of 2009, thus we chose the classification of a non-approved concept. This is plausible, regarding the recent budget data of 2010, where these municipalities are mostly listed as overindebted or with a non-approved budget consolidation concept.

[2] According to the State Office for Information and Technology NRW, https://www.landesdatenbank.nrw.de

Table 11.3 Adaptive capacity indicators.

Economic resources	Private households	Available income of private households
	Municipality	Status of financial budget of municipality
Knowledge and awareness	Participation	Participation in climate change and sustainability initiatives on municipal level
	Education	Percentage of population with highest education level

Klimakonzepte'), which supports the development of integrated mitigation and adaptation concepts with potential funding of the Environmental Ministry NRW,[3] (ii) the European Energy Award with state funding for mitigation measures,[4] (iii) the Agenda 21 initiative to promote local sustainability within municipalities.[5] We add the values of all indicators expressing the level of initiatives and normalise them analogously to the sensitivity. We average the two indicators for economic resources and knowledge and awareness, respectively, by applying equal weights. Also the indicator of economic resources has been aggregated with equal weight with the indicator of knowledge and awareness to the final indicator of relative adaptive capacity.

In contrast to the exposure, we define sensitivity and adaptive capacity as dimensionless characteristics of the system, which are characterised by solely positive values. Thus, the rescaling of the sensitivity values S_j (and analogously adaptive capacity) based on minimum and maximum values within the data range (S_{min} and S_{max}) is given by $S_{norm} = (S_j - S_{min})/(S_{max} - S_{min})$. Thus, rescaled values of sensitivity and adaptive capacity range between 0 (low) and 1 (high).

11.4.3 Data

We used the climate scenarios modelled by the regional dynamical climate models, REMO (Jacob, Mahrenholz and Keup-Thiel, 2006) and CCLM

(model version 2.4.11) (Lautenschlager *et al.*, 2009) with a spatial resolution of 0.1° and 0.2°, respectively. We averaged all available runs for the period from 1960 to 2100 (scenario A1B) (Nakicenovic, Alcamo and Davis, 2000). According to these models and scenario A1B, the annual mean temperature for NRW will increase on average by 3–3.3 °C until 2071–2100 compared with 1961–1990. Comparing the same time periods, the annual sum of rainfall will increase by 0–1% (cf. Meinke *et al.*, 2010). We calculated absolute changes between the reference period 1961–1990 and the scenario period 2071–2100 for the climatic variables listed in Table 11.2. Data on lake characteristics and elevation (DEM, 50 m resolution) were provided by the Agency for Nature, Environment and Consumer Protection, NRW (LANUV). A regional soil map (BK50, 1:50000) was available from the Geological Survey NRW, with information on soil type and structure. Further, we made use of highly resolved land-use data (ATKIS25, Authoritative Topographic-Cartographic Information System, 1:25000), which we converted into a raster dataset in the same resolution as the DEM. Additionally, we applied CORINE Land Cover data (CLC 2006, Federal Environment Agency, DLR-DFD 2009) and socio-geographic data (population density, education and income level, sealed surface) from the Statistical Agency NRW. Information on Special Areas of Conservation were obtained from the EU Natura 2000 database. Regional characteristics of their habitat compositions were derived from LANUV NRW (LANUV, 2009). Data on forest area damaged during the storm event 'Kyrill' in 2007 for parts of the state were derived from the State Office for Forest and Timber NRW. Forest fire statistics were applied from the Federal Agency for Agriculture and Food (BLE) from 1993 to 2009. Data on

---

[3] Classification as follows: all participants = 1, municipalities chosen for further funding = 2, winners = 3.

[4] Data according to the Energy Agency of NRW, classification as follows: all participants = 1, municipalities with award = 2, municipalities with gold award = 3.

[5] According to the Agenda 21 Forum (Agenda 21 Treffpunkt, 2011), participants are assigned values of 1.

the length of ski runs for the Sauerland mountains were obtained from Roth *et al.* (2001), data for the Eifel from websites of its municipalities. The status of the financial budgets of municipalities was available from the Ministry of Home and Municipal Affairs NRW (MIK). Information regarding municipal initiatives was derived from the Energy Agency NRW (Energy Agency NRW, 2009), the Agenda 21 Forum (Agenda 21 Forum, 2005) and the Environmental Ministry NRW (MUNLV). Owing to a lack of projected data of sensitivity and adaptive capacity values up to 2100, they are expressed by their current status.

11.4.4   *Quantification of sector-specific impacts*

While most existing studies have focused on single components of vulnerability separately, only a few have given their combination much deeper thought (Hinkel, 2011). Two aggregation methods are common in vulnerability assessments: the arithmetic mean of the influencing factors or the multiplication. The latter implies that the inputs are perfectly substitutable, thus allowing for a compensation between them. This has been applied for a cross-sectoral analysis of climate change impacts in Germany (Rannow *et al.*, 2010). Regarding the calculation of climate change-related impacts from exposure and sensitivity, an aggregation algorithm seems suitable, which ensures that no climatic changes (i.e. zero exposure) always lead to zero impacts. This is visualized in Figure 11.4, where zero impacts are represented in grey colour according to an algorithm based on the arithmetic mean or multiplication.

Therefore the impacts within a dimension (physical, social, environmental or economic), for example, I_{phys}, are quantified based on the rescaled sensitivity $(S_{norm,k})$ and exposure values $(E_{norm,k})$ for all specific impacts within a dimension (with a maximum number of n), for example:

$$I_{phys} = 1/n * \sum_{k=1}^{k=n} S_{norm,k} * E_{norm,k} \qquad (11.3)$$

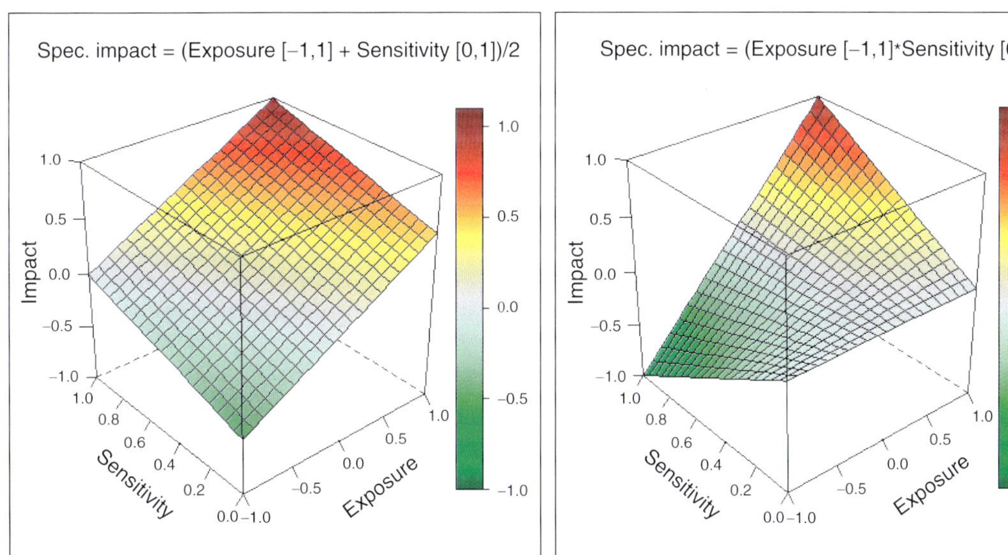

Figure 11.4   Schematic quantification of specific impacts as a function of exposure (−1 to +1) and sensitivity (0 to 1) based on the arithmetic mean (left) or multiplication (right). Resulting impacts values represent adverse (red) or beneficial effects (green). The multiplication process also entails overall lower absolute impact values. Also, the weighting of the input factors is not homogeneously distributed over the value range as with the arithmetic mean, rather, lower values have a higher influence on the final product.

The aggregation is carried out according to system-specific relationships between the exposure entity and the climatic stimuli (see also Figure 11.2).

11.4.5 Aggregation of sector-specific impacts by arithmetic mean

Quantitative aggregation via an arithmetic mean is often applied in vulnerability studies involving normative arguments (Hinkel, 2011). We apply this algorithm when aggregating the impacts of the four dimensions, for example I_{phys}, to the total impacts I_{total} using equal weights:

$$I_{\text{total}} = (I_{\text{phys}} + I_{\text{soc}} + I_{\text{env}} + I_{\text{econ}})/4 \qquad (11.4)$$

It has to be noted that prior to the aggregation to the four dimensions and from the four dimensions to the total impacts, we omit a rescaling procedure of the sector-specific or dimension-specific impact values. Thus, the magnitude of the ten considered impacts is maintained. Only at the final stage do we rescale again the resulting total impacts according to Equation 11.3, ranging from -1 (adverse effects) to 1 (beneficial effects). The consideration of beneficial effects of climate change, which are derived by diminishing of the exposure, is based on the assumption that impact processes between exposure and sensitivity work equally in both ways. Thus, we assume that increases in heavy rainfall days lead to adverse effects (e.g. flooding), while a reduction in these days by the same amount will attenuate the impacts equally.

Experience from vulnerability analysis has revealed a minor relevance of adaptive capacity indicators for stakeholders as they felt they could better estimate their adaptability for themselves (Hinkel, 2011). Moreover, it is still an under-researched topic and the relationship between climate impacts and adaptive capacity is not fully clear (Engle, 2011). We therefore refrain from the aggregation of adaptive capacity as applied for the exposure and sensitivity based on their multiplication. Instead we introduce a visual combination of the calculated values of impacts I and adaptive capacities AC for each municipality to express the vulnerability V (see Equation 11.1). Based on Metzger and Schröter (2006) we display the impact by hue and the adaptive capacity value by the transparency of the respective colour. Since adaptive capacity can potentially act in both ways, either reducing adverse or increasing positive effects, the transparency increases for adverse impacts from low to high adaptive capacity values and vice versa for positive impacts. This allows distinguishing between the impacts and adaptive capacity as determining factors of the vulnerability, which is of higher relevance for decision makers. Given the local knowledge regarding adaptive capacity from stakeholders within their own municipality, this component is more relevant to decision makers from the broader perspective. Thus, for a wider region, the potential to decrease vulnerability by increasing the local adaptive capacity can be identified.

Regarding this proposed approach of a vulnerability analysis, it has to be noted that the various steps of data rescaling lead to relative values of vulnerability. In other words, no absolute statements concerning the final degree of vulnerability (e.g. municipality X is vulnerable) are possible. However, relative statements for the study area can be made, (e.g. municipality X has a much higher vulnerability than municipality Y).

11.4.6 Typological aggregation of sector-specific impacts

The above presented approach of aggregating impacts via arithmetic mean aims at quantifying the full impact burden of regions over various sectors. However, even by including specific impacts with a clear direction according to Figure 11.4 (right), the subsequent aggregation represents a limitation as it allows for a compensation of impacts across sectors. Therefore we alternatively follow an additional approach by identifying regions of similar impact burden based on the calculated sector-specific impacts. These typologies of climate change impacts are developed by means of a cluster analysis, based on the specific indicators values calculated for all municipalities. The cluster mechanisms can be distinguished in hierarchical, partitioning and density-based methods (Handl, Knowles and Kell, 2005). Our analysis is mainly based on the second method, but used the hierarchical clustering to initialise the partitioning method. The cluster analysis is based on the sector-specific impacts of each municipality, which were rescaled between 0 and 1 prior to the calculations. To identify the most robust and therefore most representative number of clusters, several methods are applied: the variance ratio (Calinski and Harabasz, 1974), the consistency measure (Ben-Hur, Elisseeff and Guyon, 2002; Roth, Lange, Braun

and Buhmann, 2002) and the silhouette-width (Kaufman and Rousseeuw, 1990). Higher values of these measures indicate a better representation of the data regarding the respective number or clusters (for details on the method see Chapter 3).

11.5   Results

11.5.1   *Exposure*

The relative exposure differs between the regions in NRW with strongest increases in heat days in the Rhine Valley (Figure 11.5). Under both models,

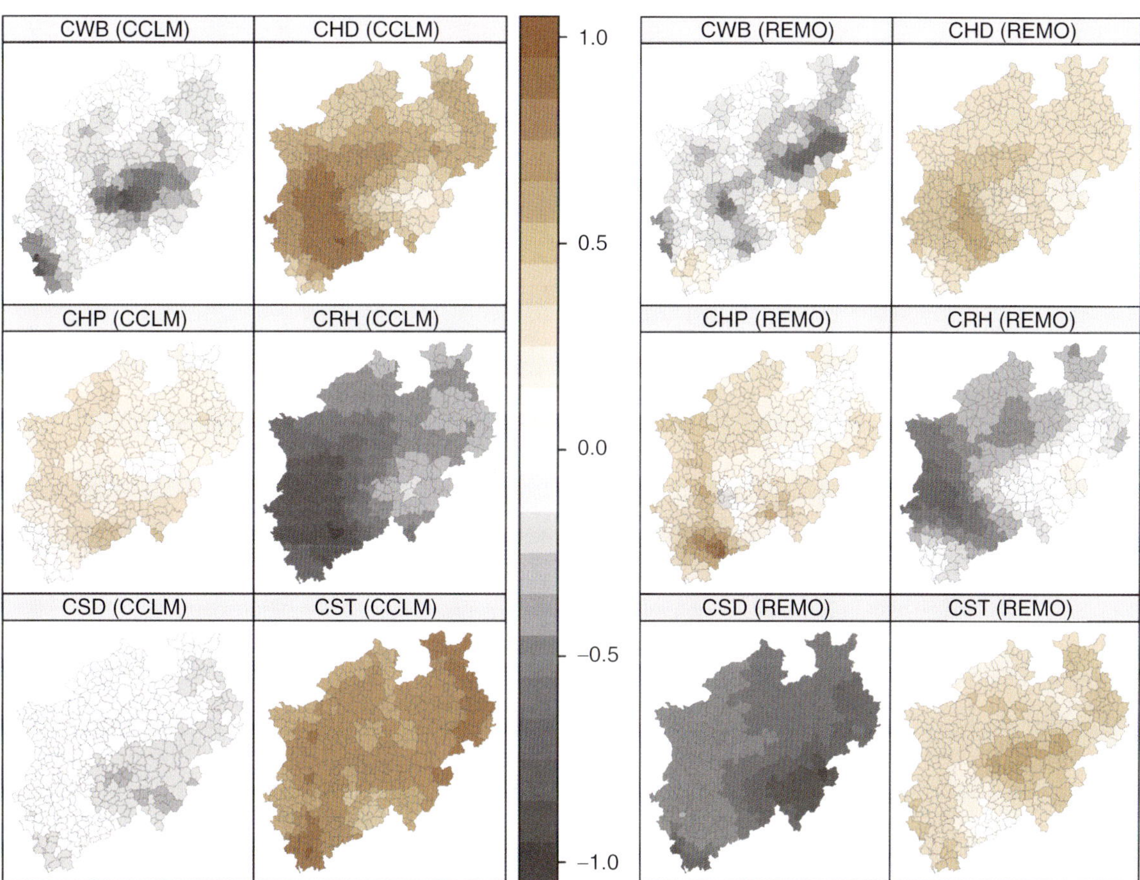

Figure 11.5   Rescaled exposure variables for the models CCLM (left) and REMO (right). Values are scaled to the data set of NRW for both models, ranging from −1 (decreases) to 1 (increases). The exposure is represented by changes in climatic variables between 1961–1990 and 2071–2100 under scenario A1B. For abbreviations see Table 11.2.

the mountainous areas exhibit the largest increases in storm days and the strongest reduction in snow conditions, however with much higher changes for the REMO model. Both models deviate considerably in the projection of hydrologic variables, both in the magnitude of change and in the direction. For example, according to the REMO model, the Western region of Sauerland experiences wetter conditions in future, whereas CCLM projects drier conditions.

11.5.2 Sensitivity

The spatial pattern of the sensitivity values shows great variations across the sectors (Figure 11.6). While silvicultural sensitivities are highest in the mountains, the sensitivity toward heat is most severe in metropolitan areas. Regarding the urban flooding processes, regional concentrations of high values are discernible in the Rhine Valley as well as at the foothills of the mountains. The strongest

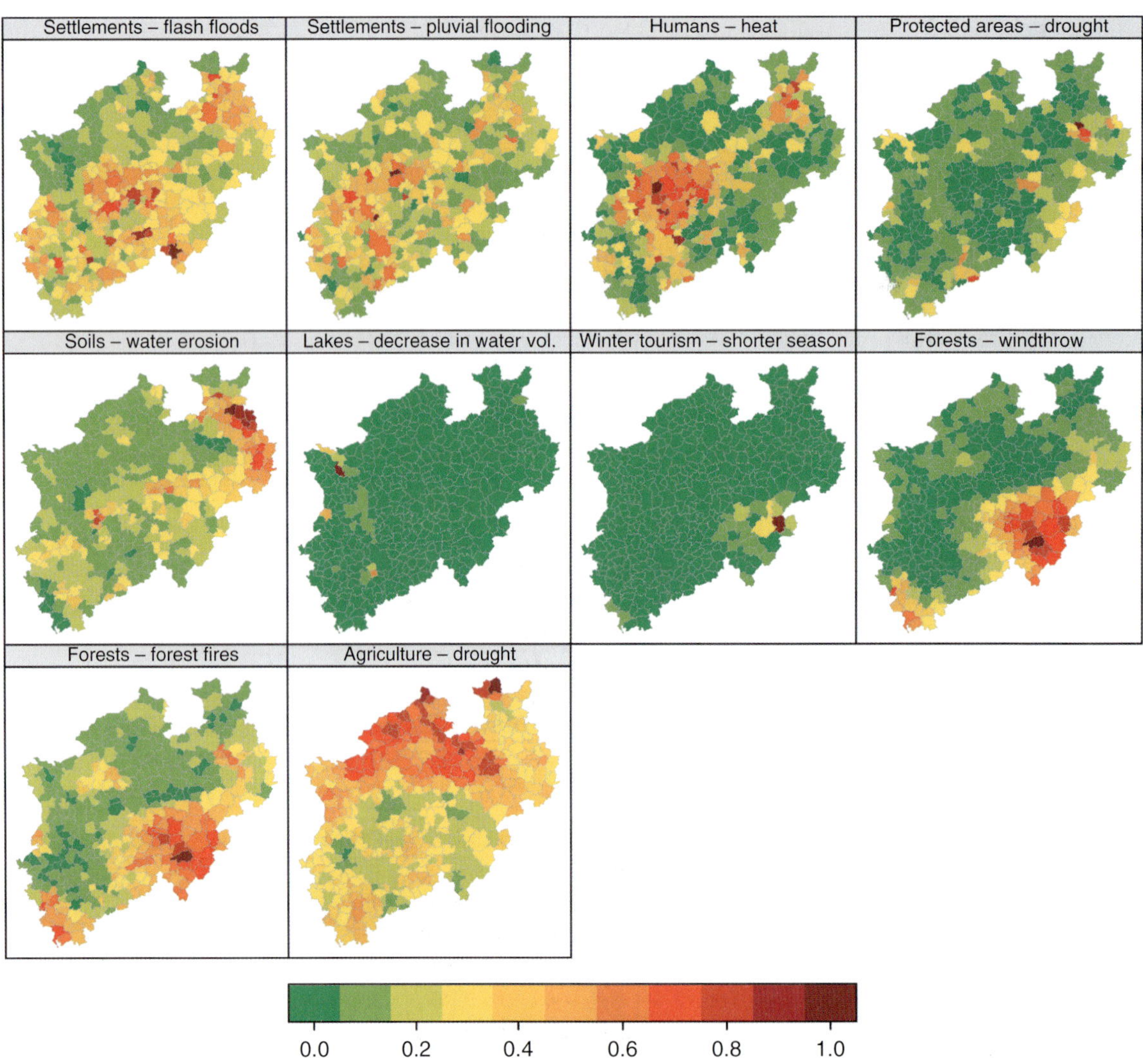

Figure 11.6 Sensitivity values ranging from 0 (low) to 1 (high). Values are scaled to the data set of municipalities in NRW.

susceptibility to erosion is found at the foothills of the Egge mountains in the East and highest sensitivity with respect to the agriculture in the lower lying Münsterland. Most sensitive protected areas and lakes are spatially scattered due to their sparse occurrence. The municipality of Winterberg clearly stands out as the most sensitive regarding winter tourism. These results are based on current sensitivities. When considering changes in the system characteristics over time, the described pattern changes. For example, a detailed study assessing heat wave impacts under climate change in NRW identified that future impacts are alleviated in the cities of Cologne and Düsseldorf due to a relatively low proprotion of an elderly population in the future, which are especially sensitive to heat (Lissner, Holsten, Walther and Kropp, 2012).

11.5.3   *Sector-specific impacts*

Climate change may have positive impacts in parts of the Rhine Valley according to the model REMO, projecting a reduction in heavy rainfall days, which influences the impacts with regard to flash floods and pluvial flooding (see Figure 11.7). The metropolitan area within the Rhine Valley (see Figure 11.1) is strongly affected by an increase in heat days when considering both climate models. Here, population density and sealed surfaces lead to local heat islands, thus increasing the impacts. Protected areas and lakes are characterised by relatively low impacts. This can be partly explained by the low relative sensitivity of the habitats within the protected areas and the lakes, mainly due to the small share of area of these exposure units within the municipalities. While a large spatial differentiation of sensitivity values is apparent regarding soil erosion, areas with strong increases in heavy precipitation events do not overlap areas of high sensitivity. Thus, the impact with regard to soil erosion is diminished. For forestry, strong impacts are prevalent in the mountains. These comprise both a large share of forest in the municipalities and a dominance of needle-leaved trees, which are especially sensitive to windthrow and forest fires. While storms are

projected to increase most in these mountains for both models, changes in relative humidity differ between the models, ranging from general decreases for CCLM to slight increases in the Eastern Sauerland mountains for REMO. Therefore, in areas of strong sensitivity to forest fire (Figure 11.6), potential impacts are alleviated by generally wetter conditions under the model. However, regarding the windthrow, high sensitivity values in the mountains coincide with strong increases in storms, which exacerbates the potential impact. Agriculture, however, shows the strongest relative impacts in the Eastern Westphalian Bay, with larger agricultural areas and soils of a lower water retention capacity. While only small changes in the climatic water balance occur over this region under the CCLM model, REMO simulates stronger decreases. Winter tourism is most affected in the higher elevated areas of Sauerland, with the strongest dependency on this sector. Most intense changes in snow cover are projected for the REMO model, compared with CCLM.

These results of sector-specific impacts form the basis of the following aggregation approaches: a quantitative aggregation of the impacts across the four dimensions via an arithmetic mean as well as a typological categorisation by a cluster analysis to identify regions of similar impact characteristics.

11.5.4   *Aggregation of impacts by arithmetic mean*

The sector-specific impacts (Figure 11.7) were first aggregated to the four dimensions. The physical impacts are strongest in the foothills of the mountains and in the Rhine Valley for both models (Figure 11.8). Environmental impacts exhibit lower values than the other dimensions over large parts of the state mainly due to a lower share of the respective exposure unit within the municipalities. Also for this dimension, regions of high sensitivity differ spatially from regions of strong climatic changes of relevant stimuli. Economic impacts are characterised by a rather heterogeneous picture regarding the sector-specific impacts of forestry, agriculture and tourism. While both impacts on the forestry sector

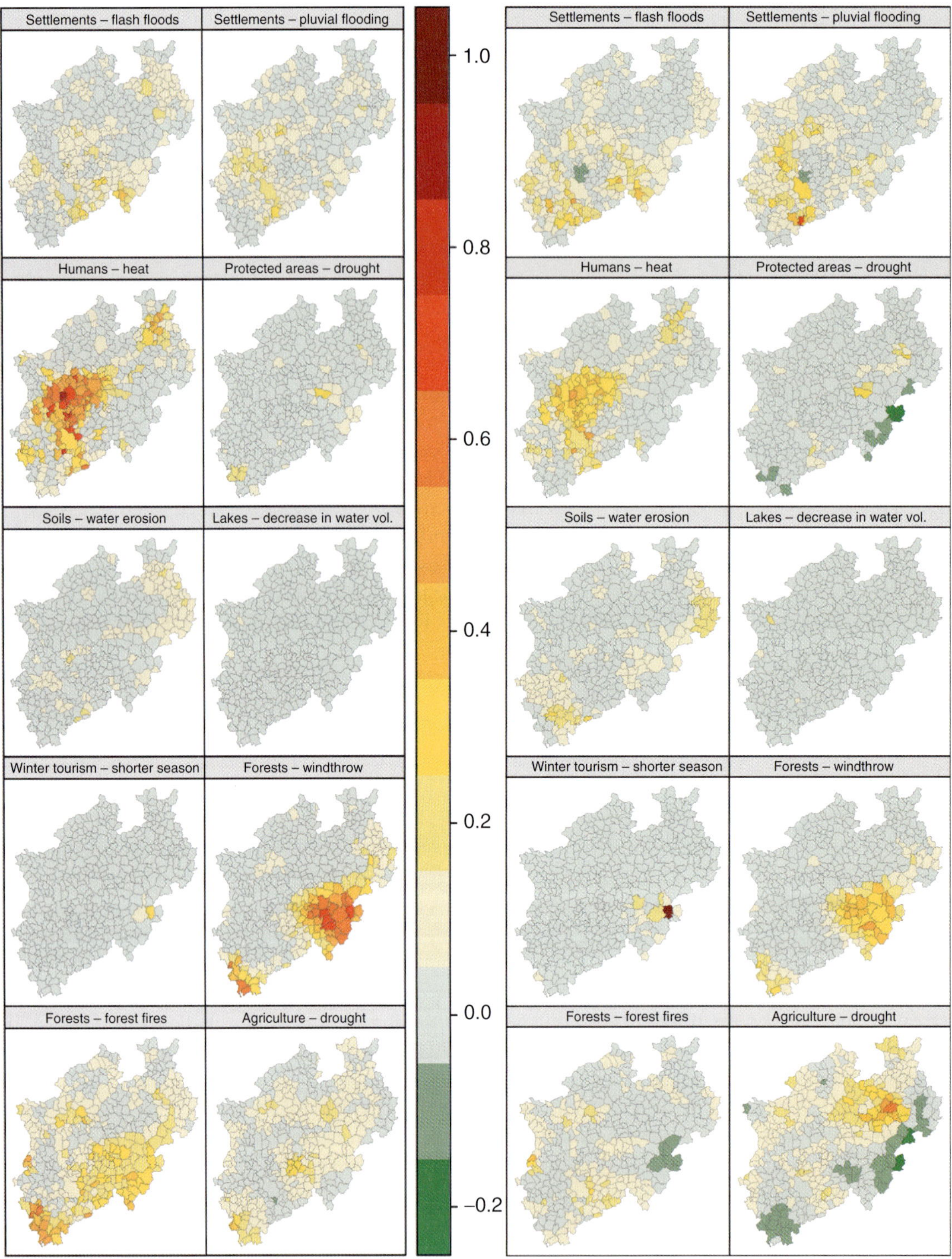

Figure 11.7 Sector-specific impacts according to the climate model CCLM (left) and REMO (right). The underlying exposure is represented by changes in climatic variables between 1961 and 1990 and 2071 and 2100 under scenario A1B.

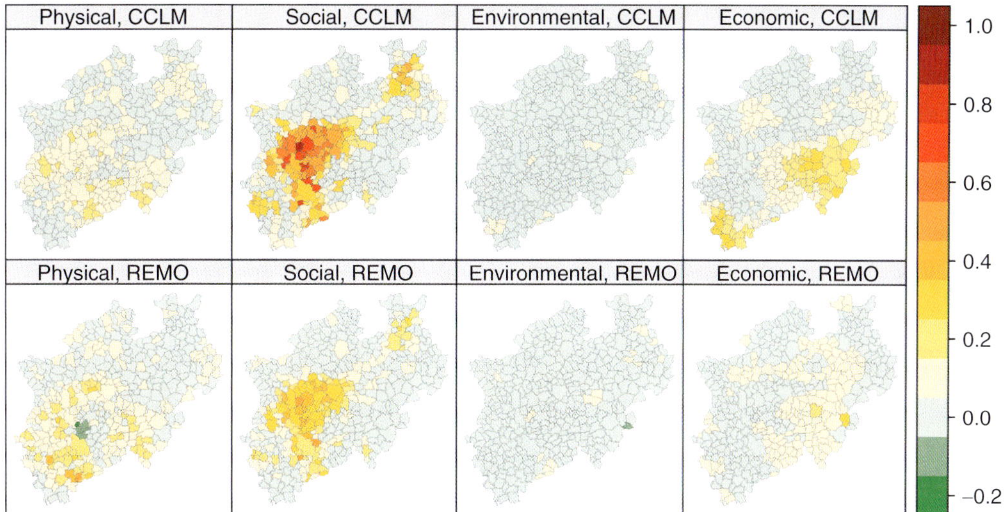

Figure 11.8   Aggregated potential impacts based on the climate models CCLM (top) and REMO (bottom) for the physical, social, environmental and economic dimension. Values from 0–1 represent adverse impacts, below 0 beneficial impacts. Underlying exposure and sensitivity variables have been scaled to the data set of both models. Exposure is represented by changes in climatic variables between 1961 and 1990 and 2071 and 2100 under scenario A1B.

(windthrow and forest fire) as well as the impact on the winter tourism sector are more prevalent in the mountains under both climate models, the agricultural is more affected in the Eastern Westphalian Bay, especially according to the REMO model. Overall, clearly the social impacts stand out. On the one hand, this is due to a spatial overlap of high sensitivity and exposure values. On the other hand, it is also due to the aggregation methodology, where heat wave impacts are the single determinant indicator for the social dimension, whereas other dimensions encompass more impacts. For example, the aggregation of impacts within the economic dimension leads to a lowering of the value for the municipality of Winterberg, with very high potential negative consequences regarding winter tourism, but beneficial impacts for the agricultural sector under the REMO model. For the REMO model, some regions with beneficial climate change impacts in the physical and environmental dimension stand out. These are due to changes in the exposure, that is, a projected decrease in days with heavy precipitation in the central Rhine Valley and an increase in the

climatic water balance in the Eastern part of the Sauerland region.

The total relative impacts (integrating all four dimensions) range from no changes to adverse effects of climate change over the state for both models (Figure 11.9). These are strongest for the upper Rhine Valley, especially in the densely populated metropolitan area. Also, the foothills of the mountains exhibit strong impacts, especially the Western part of the Sauerland and Northern part of the Eifel mountains. As a third affected region, the municipalities in the East of NRW stand out with higher potential impacts regarding the physical and social dimension as well as with respect to the forestry sector. Despite the differences in the projected climate data for both models, the total impacts are similar in their spatial pattern.

These aggregated impacts have been further overlayed by the generic adaptive capacity, which is displayed using a specific colour code. According to both of the applied models, the overall relative vulnerability to climate change, comprising the total impacts and the generic adaptive capacity, is low for

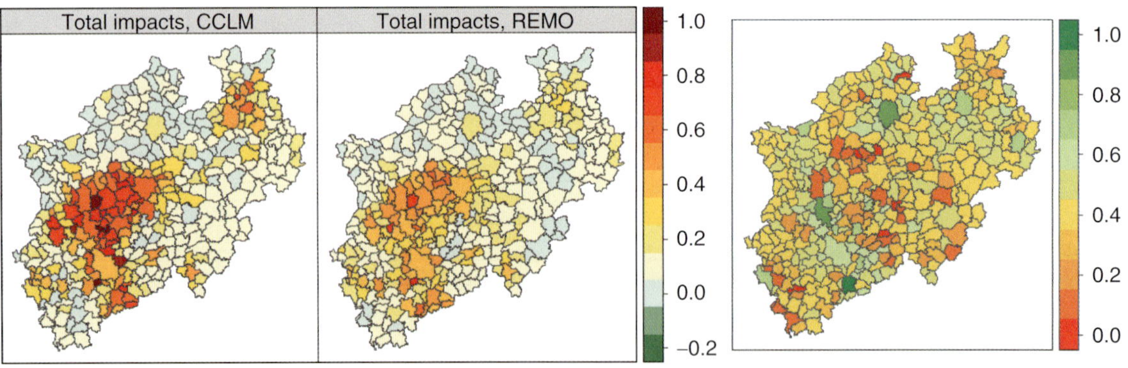

Figure 11.9 Total potential impacts (left) based on the climate models CCLM and REMO and generic adaptive capacity (right). Impact values from 0 to 1 represent negative impacts, below 0 positive impacts. Values of adaptive capacity range from 0 = low to 1 = high. Values are scaled to the data set of both models. The exposure included in the impacts is represented by changes in climatic variables between 1961 and 1990 and 2071 and 2100 under scenario A1B.

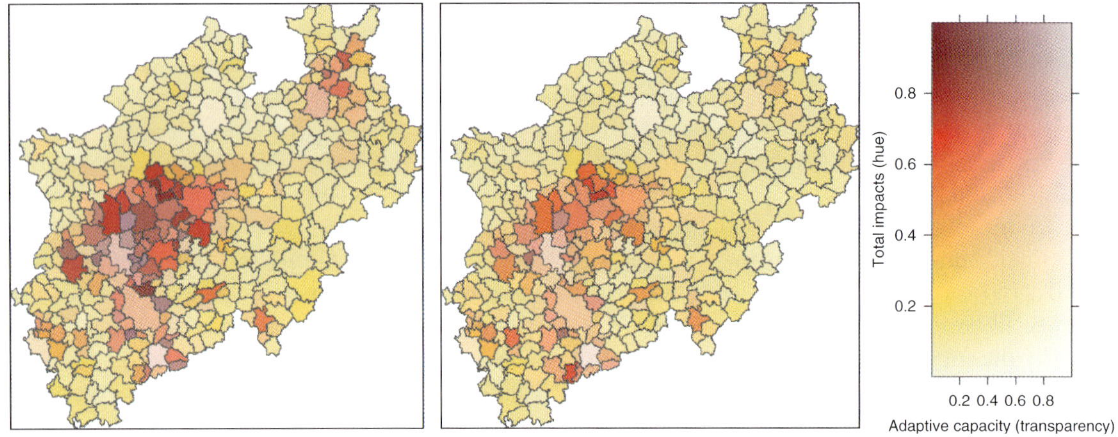

Figure 11.10 Visualization of the overall vulnerability based on aggregated impacts and the generic adaptive capacity for CCLM (left) and REMO (right). A high adaptive capacity reduces negative impacts (hue from yellow to red), which is visualised by changes in the level of transparency. The underlying exposure is represented by changes in climatic variables between 1961 and 1990 and 2071 and 2100 under scenario A1B.

large parts of the lowlands (Figure 11.10). By and large, most vulnerable municipalities lie within the metropolitan area, the mountainous areas as well as their foothills, similar to the spatial distribution of the impacts. However, the pattern of vulnerability is more heterogeneous, which is caused by the spatially strongly distributed values of the adaptive capacity. This effect is most apparent in the densely populated metropolitan area, where municipalities display overall high impacts under both models.

However, our results show a strong adaptive capacity for several of its municipalities (e.g. Bonn or Dusseldorf), while others are characterised by very low capacities (e.g. Duisburg), mainly due to a strained financial situation. The overall pattern of vulnerability is fairly similar between the two models applied. By including the adaptive capacity, climate change effects can be alleviated, resulting in lower values of the vulnerability. This is the case for parts of the Rhine area, while high adverse impacts combined with a low adaptive capacity still result in high vulnerabilities in the Ruhr area.

11.5.5 *Typological categorisation of impacts*

As a second approach, regions of similar impact burden have been identified through a cluster analysis based on the calculated sector-specific impacts as shown in Figure 11.7. The number of clusters ensuring a stable representation of the data has been tested using several methods. Regarding the impacts based on climate data of the CCLM and REMO model, the method of the variance ratio and the silhouette index point to three clusters (Figure 11.11a–d). This number is also represented by a local maxima regarding the consistency measure for the CCLM model (Figure 11.11e). The median consistency measure based on results of the REMO model indicates an optimal number of four clusters, however, with a very small deviation to the three cluster option (Figure 11.11f). Also for the latter, the median value is higher. In order to compare our results, we therefore chose to represent our impact results based on three clusters for both climate models.

These clusters are characterised by different combinations of the sector-specific impacts (Figure 11.12) and different spatial distributions within the state NRW (Figure 11.13). For both models, the first cluster (brown colour) shows a strong impact regarding storm causing windthrow in forests, especially in the mountainous regions of NRW. These include the Sauerland and Eifel Mountains as well as parts of the Egge Mountains in the East. The second cluster is especially marked by a high social impact with respect to heat waves and is located in the Rhine Valley and metropolitan region. Also, the region around the city of Bielefeld is assigned to this cluster. The third cluster is characterised by a more balanced pattern of impacts, with relatively lower values regarding heat impacts and impacts on forests but higher values with respect to environmental and agricultural impacts. This cluster is more dominant in the low-lying Westphalian Bay and the foothills of the mountains. A slightly higher number of municipalities are allocated to cluster 1 according to the CCLM model compared with the REMO model. However, overall, the identified clusters are fairly robust regarding the chosen climate model.

When comparing the pattern of these clusters to the adaptive capacity of the municipalities (see Figure 11.9, right), a spatial overlap is not directly apparent. Rather, the adaptive capacity varies significantly between municipalities. However, the upper Rhine region including Bonn, Cologne or Düsseldorf as well as the city of Bielefeld in the Northeast exhibit higher adaptive capacities. At the same time, they are all located within the second cluster, which is marked by high social impacts. Yet, the remaining area of this cluster, mainly the Ruhr region shows a low adaptive capacity. Low values of adaptive capacity are also found for several municipalities belonging to the first cluster of the mountainous regions, characterised by stronger impacts regarding the forestry sector. The third cluster, located in the region of Münsterland and parts of the Bay of Cologne show, by and large, a medium level of adaptive capacity, with the exception of the city of Münster.

11.6   Discussion

The approaches presented allow for comparative and integrated assessments of climate change vulnerabilities, while enabling a sector-specific perspective of climate change effects. This fulfils the needs of decision-makers. We demonstrate the approach through a multisectoral, regional case study in the German Federal State of North Rhine–Westphalia, which exhibits a strong spatial heterogeneity, while being of special relevance for the German economy. Sensitivity is quantified by means of tailor-made approaches for specific sectors for biophysical and socioeconomic dimensions. This is then related directly to relevant exposure indicators, defined as relative changes of climate variables between the

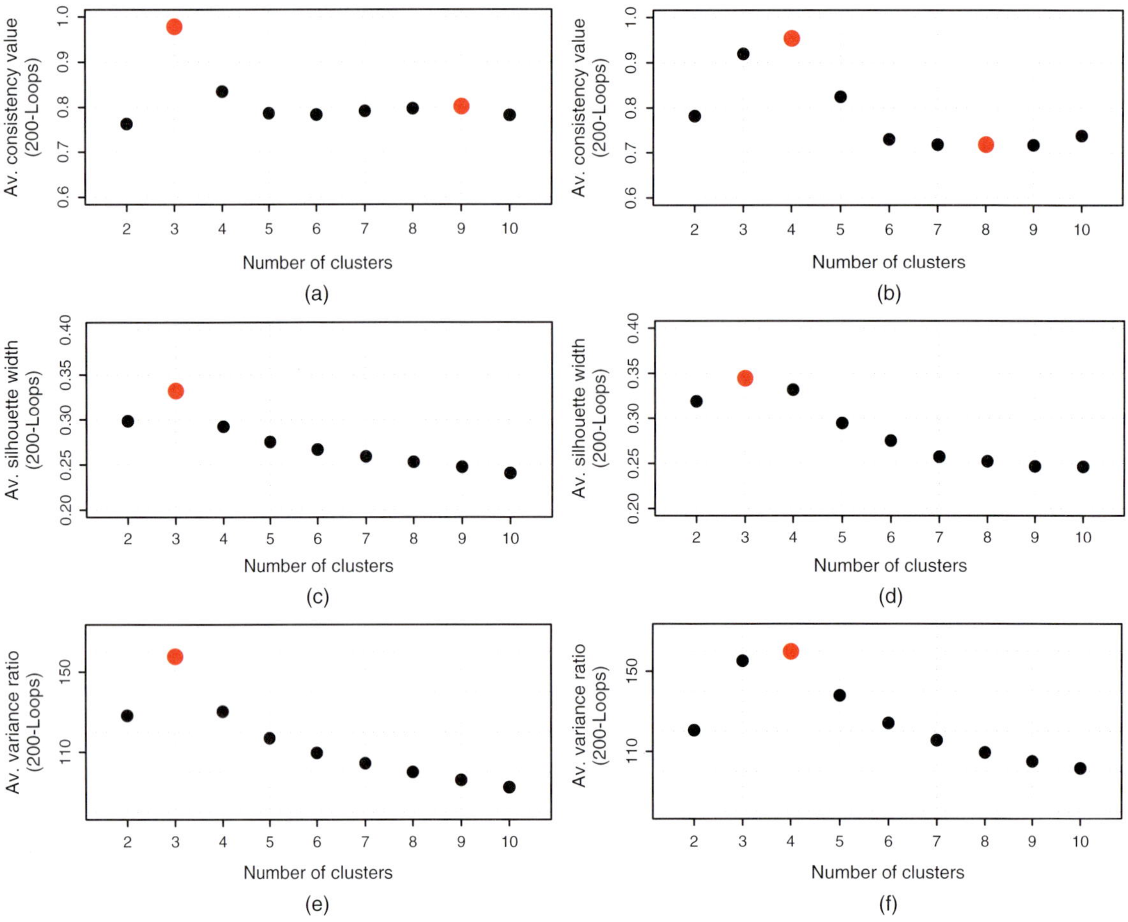

Figure 11.11   Consistency measure (a, b), ratio of the between cluster variance and the inner cluster variance (c, d) and silhouette width (e, f) for cluster numbers 2–15 based on the models CCLM (left) and REMO (right). The values represent averages for 200 repetitions of the cluster-specific measures. Local maxima regarding the optimal number of clusters are marked by red points.

past and future based on two regional climate models (CCLM and REMO). This consideration of direct linkages between the exposure unit and specific climatic stimuli has been often neglected in previous vulnerability analysis (Preston, Yuen and Westaway, 2011). Further, by multiplying sensitivity and exposure to quantify impacts, we ensure that regions experiencing no climatic changes are indeed characterised by zero impacts at the level of sector-specific impacts. Nevertheless, by this approach, zero impacts in a region could either derive from zero exposure but also from zero sensitivity,

as the minimum value within the study area. In the latter case, climatic impacts still might occur in the region.

The applied quantitative methodology of exposure and impacts shares common ground with the 'vulnerability cube' proposed by Lin and Morefield (2011), who classify vulnerability by means of axes expressing specific indicators in a multidimensional cube. However, they restrict their concept to visualisation, whereas we involve a mathematical function of sensitivity, exposure and impacts, which can be visualised in a three-dimensional space.

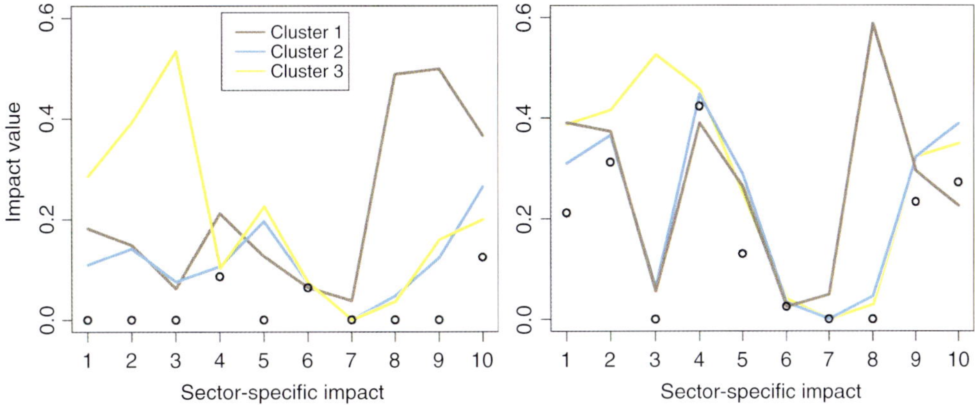

Figure 11.12   Values for the identified cluster centres regarding the ten sector-specific impacts in NRW, according to the model CCLM (left) and REMO (right). The colours of the clusters are identical to those used in Figure 11.13. Note that for the cluster analysis, each input variable was rescaled between 0 and 1, original impact values of zero are marked by black circles (e.g. for all clusters an increase in vulnerability to heat is apparent). Indicators are numbered according to the impacts described for their exposure unit in Table 11.1: 1 = settlements flash floods, 2 = settlements pluvial flooding, 3 = humans heat, 4 = protected areas drought, 5 = soils water erosion, 6 = lakes decrease in water volume, 7 = winter tourism shortening of season, 8 = forests windthrow, 9 = forest fires, 10 = agriculture drought. The underlying exposure is represented by changes in climatic variables between 1961 and 1990 and 2071 and 2100 under scenario A1B.

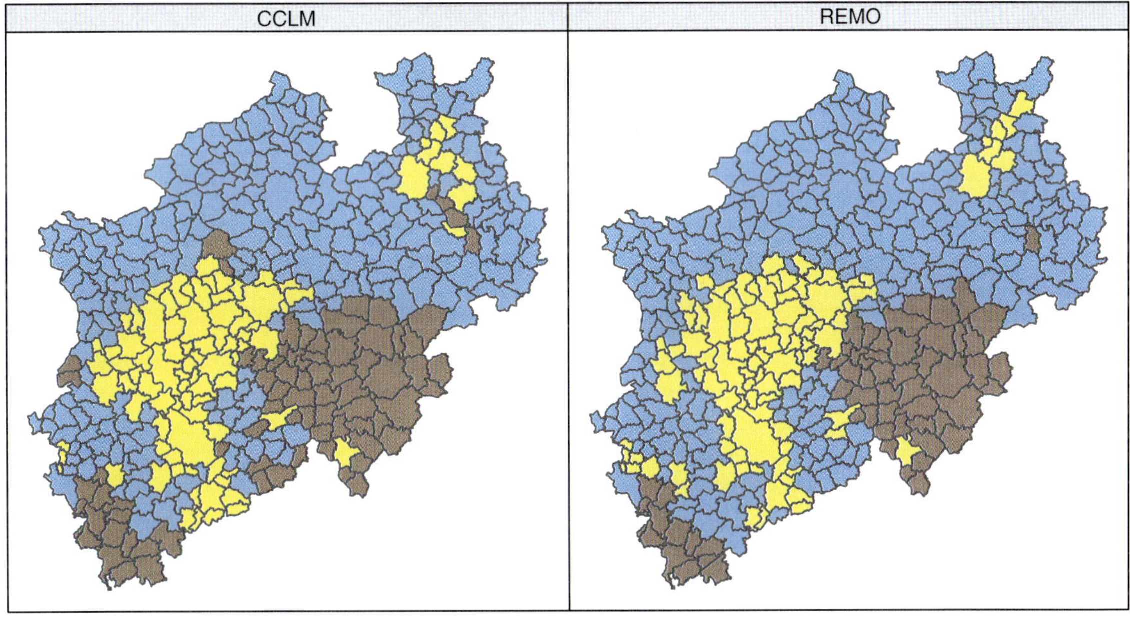

Figure 11.13   Spatial distribution of the three identified clusters based on the model CCLM (left) and REMO (right). The cluster colours are identical to those used in Figure 11.12. The underlying exposure is represented by changes in climatic variables between 1961 and 1990 and 2071 and 2100 under scenario A1B.

In general, a consensus exists, regarding the meaning of the components of vulnerability (Costa and Kropp, 2012), yet their aggregation is often not discussed (Hinkel, 2011). We presented two alternative approaches – an aggregation via an arithmetic mean and a typological categorisation – to aggregate sector-specific impacts for the municipalities of NRW. After reducing complexity through aggregation, the first method enables a cross-sectoral view on the spatial distribution of vulnerability. At the same time, it also allows back tracking of decisive factors of the system to support target-oriented adaptation measures. However, while for sector-specific impacts regions of zero impacts (e.g. due to no climatic changes) are still clearly identifiable, our concept allows for a compensation between impact or sectors, for example between the environmental and economic dimension. This could be refined by considering different weighting factors within the aggregation. However, these differ between regions, presumably even within our study region and between different stakeholder groups that were questioned.

The alternative typological categorisation aimed at identifying regional similarities regarding sector-specific impacts affecting the municipalities through a cluster analysis. In contrast to the first approach, it hinders the representation of the full impact burden across sectors. However, sector-specific impacts can still be inferred from the respective clusters, and these are not compensated across the sectors. Moreover, it also provides a form of aggregation of the results, enabling a quick overview of the region.

While we applied direct linkages between sensitivities and exposure variables, we express the adaptive capacity in a generic manner. This included cross-sectoral features such as financial resources and education level. Given a more comprehensive database, adaptive capacity could also be integrated in our concept in a system specific way, for example, capacity of citizens to adapt to heat waves or sector-specific institutional characteristics. This would then fully complement the integrated approach of our vulnerability assessment.

We have concentrated on the spatial scale of municipalities. In cases where more-fine-scaled data were available, this information was scaled up to these administrative boundaries. Such subnational spatial level supports the comparability of regions and aggregates regional process and patterns for decision-makers. However, up-scaling also entails the levelling out of local information (Fekete, Damm and Birkmann, 2010). By focusing on municipalities, individual or household impacts are not represented, nor information on larger spatial scales. Also, for some sectors, the regional spatial boundaries at which key decisions are taken (e.g. forest districts) differ from the universally applied municipal boundaries in our study.

Our methodologies are in general transferable to other regions, but the selection of impacts processes should be adapted to the specific regional relevance. This step is crucial and involved some subjectivity regarding the results. We considered a wide range of regionally relevant and climate-dependent sectors, however, which was limited by data availability. Yet, given a better database, the approach could also be extended for a wider range of sectors. Further information could also be derived from stakeholder involvement, especially regarding the quantification of adaptive capacity. This would also alleviate the potential bias of an assessment toward the selection of impacts, which are quantified with existing data sources. Further, it has to be stressed that the results of this case study express relative vulnerabilities, which only allow for a comparative interpretation of the values within the study area. For some sectors, absolute vulnerabilities or impacts could be determined. This has been achieved by (Klaus, Holsten, Hostert and Kropp, 2011) for the quantification of windthrow impacts in forests of NRW, where sensitivity was directly related to actual past damage occurring during a severe winter storm. However, such data were not available for the full range of sectors analysed. The advantages and disadvantages of the presented approaches are summarised in Table 11.4.

The application of the quantitative approach to the region of NRW showed regional impact 'hotspots' in the metropolitan area, the foothills of the mountains and in the East. A higher potential impact of climate change has also been found by Rannow

Table 11.4   Advantages and disadvantages of the presented arithmetic mean aggregation approach and the typological categorisation approach.

Approach	Advantages	Disadvantages
Arithmetic mean aggregation	Quantification of aggregated impact burden across sectors	Subjectivity due to selection of impact processes
	Integration of biophysical and socioeconomic dimension	Subjectivity due to weighting factors between impacts or sectors
	Transparent formalisation of procedure	Approach allows for a compensation of positive and negative impacts across sectors
	Clear relation between system-specific sensitivities and relevant exposure variables	Interpretation of results is limited to relative comparison within the respective study area
	Decisive factors can be traced back to sector-specific impacts, sensitivities or exposures	
Typological aggregation	Impacts between sectors are not compensated	Total impact burden not quantified
	Regional typologies are identified	Subjectivity due to weighting factors between impacts or sectors
	Individual impact values can be inferred from the clusters	Formalisation of procedure less transparent
	Clear relation between system-specific sensitivities and relevant exposure variables	Interpretation of results is limited to relative comparison within the respective study area
	Decisive factors can be traced back to sector-specific impacts, sensitivities or exposures	Coarse resolution of results

et al. (2010) for the Rhine Valley in NRW. However, they conclude a clearer gradient from higher impacts in the West to lower ones in the East of NRW. While the underlying climate projection (REMO, scenario A1B) is comparable, they assume a substitution between low climatic changes and a high sensitivity and apply a set of considered impacts, which deviates considerably from ours. By applying a cluster analysis, Kropp *et al.* (2006) have also identified most vulnerable areas regarding heat waves in the Rhine–Ruhr region, and in the mountainous regions with respect to the forestry sector.

According to our framework, high specific impacts accrue from both high sensitivities and high exposure coinciding spatially. Regarding our results, this is the case for impacts of heat wave on humans in the metropolitan area and of storms on forest stands in the mountains. Further adaptation measures focusing on these impacts could thus reduce the consequence of climate change considerably. Thereby synergies across sectors should be prioritised, which are possible to identify based on our multisectoral approach. For example, the conversion of coniferous dominated forests in the mountains could both reduce the impacts regarding windthrow and forest fires. As a possible adaptation option to heat wave impacts, sealed surfaces especially in the Rhine Valley could be reduced, which at the same time, may diminish impacts from floods.

The typological approach has yielded spatial classifications of impacts characteristics. These were also found to overlap partly with the results of the adaptive capacity. A cluster has been identified, which show strong heat wave related impacts. While parts of this region exhibit a high adaptive capacity (e.g. the upper Rhine Valley), the reduction of impacts could be hindered by a very low adaptive capacity, mainly due to a constrained financial situation. The mountains have also been identified comprising a specific pattern of impacts, with strongest values for the forestry sector, especially regarding windthrow. At the same time, several municipalities in this region show a low adaptive capacity.

11.7 Conclusion

While sectoral impact assessments have become common usage in the climate change field, integrated approaches are still scarce. This information, however, is of importance to inform and prioritise adaptation processes and is requested by decision makers (e.g. Patt, Klein and Delavegaleinert, 2005; Preston, Yuen and Westaway, 2011). Integrated assessments can provide a basis for policy decisions for the overall allocation of adaptation funds by supporting an overview over the relevant impacts in a region. Also cross-sectoral interactions, for example benefits and conflicts, can be identified to avoid adaptation measures, which might represent maladaptation for another sector. However, given the sectoral structure of regional administrations, assessments providing detailed information at this level are also helpful. Thus, to initiate informed adaptation, knowledge on several levels is needed. In light of the National German Adaptation Strategy enacted in 2008, each of the Federal States is demanded to develop regional adaptation strategies. A start towards the planning and implementation of adaptation measures for NRW was made with the state's strategy, published the following year by the Environmental Ministry (MUNLV, 2009). While a qualitative overview over potential impacts of sectors is provided, it still lacks a comprehensive quantitative approach. Our cross-sectoral analysis fills this knowledge gap and supports the concretisation and prioritisation regarding specific adaptation measures. Thus, on the one hand, those regions have been identified which will have to deal with the highest impact burden and therefore have the highest need for adaptation. On the other hand, detailed information on the concrete sectoral impacts and underlying cause-and-effect chains has enabled efficient and purposeful adaptation. This knowledge on expected impacts in other sectors

in the same location is also important to avoid maladaptation. Additionally, the approach of typological categorisation provides an overview over the territorial distribution of impacts by identifying regions of similar impact characteristics. However, various issues of these approaches, for example the subjectivity of selection of impacts and the aggregation across sectors still remain unresolved. This stresses the need to consider the aim and methodological advantages and disadvantages of the applied methodologies before applying any vulnerability assessment.

Acknowledgements

Parts of this study were supported the European Spatial Planning and Observation Network (ESPON) within the ESPON programme and by the Ministry for Climate Protection, Environment, Agriculture, Nature Conservation and Consumer Protection North Rhine–Westphalia (MUNLV). We wish to thank the project team for fruitful discussions on the methodology, our colleagues Ylva Hauf for methodological support and Tabea Lissner for commenting on the manuscript.

References

Adger, W.N. and Vincent, K. (2005) Uncertainty in adaptive capacity. *Comptes Rendus Geoscience*, **337** (4), 399–410.

Agenda 21 Forum (2005) *Lokale Agenda 21 und Schule. Beschlüsse in NRW.* (online) Available at: <http://www.agenda21-treffpunkt.de/lokal/stadt/abc_stadt/nrw.htm> (accessed 4 October 2012).

aqua_plan GmbH, hydro & meteo GmbH & Co. KG and dr. papadakis GmbH (2010) *ExUS NRW- Extremwertstatistische Untersuchungen von Starkregen in Nordrhein-Westfalen (ExUS) – Veränderungen in Dauer, Intensität und Raum auf Basis beobachteter Ereignisse und Auswirkugen auf die Eintretenswahrscheinlichkeit.* Landesamt für Natur, Umwelt und Verbraucherschutz Nordrhein-Westfalen, Recklinghausen.

Araújo, M.B., Alagador, D., Cabeza, M., Nogués-Bravo, D. and Thuiller, W. (2011) Climate change threatens European conservation areas. *Ecology Letters*, **14** (5), 484–92.

Bartsch, A., Balzter, H. and George, C. (2009) The influence of regional surface soil moisture anomalies on forest fires in Siberia observed from satellites. *Environmental Research Letters*, **4** (4), 1–5.

Bates, B.C., Kundzewicz, Z.W., Wu, S. and Palutikof, J.P. (2008) *Climate Change and Water – IPCC Technical Paper VI*. Intergovernmental Panel on Climate Change, Geneva.

Ben-Hur, A., Elisseeff, A. and Guyon, I. (2002) A stability based method for discovering structure in clustered data. *Pacific Symposium on Biocomputing*, **7**, 6–17.

BfN (Bundesamt für Naturschutz) (2011) *FloraWeb – Daten und Informationen zu Wildpflanzen und zur Vegetation Deutschlands.* (online) Available at: <http://www .floraweb.de/> (accessed 31 July 2009).

Boardman, J. (2006) Soil erosion science: Reflections on the limitations of current approaches. *Catena*, **68** (2–3), 73–86.

Brooks, N., Adger, W.N. and Kelly, P.M. (2005) The determinants of vulnerability and adaptive capacity at the national level and the implications for adaptation. *Global Environmental Change-Human and Policy Dimensions*, **15** (2), 151–163.

Calinski, T. and Harabasz, J. (1974) A dendrite method for cluster analysis. *Communications in Statistics Theory and Methods*, **3** (1), 1–27.

Castro, D., Einfalt, T., Frerichs, S., *et al.* (2008) *Vorhersage und Management von Sturzfluten in urbanen Gebieten (URBAS). Schlussbericht.* Hydrotec Ingenieurgesellschaft für Wasser und Umwelt mbH, Fachhochschule Aachen, FB Architektur, Deutscher Wetterdienst, Meteorologisches Observatorium Hohenpeißenberg, Aachen.

Chuvieco, E., Aguado, I., Yebra, M., *et al.* (2010) Development of a framework for fire risk assessment using remote sensing and geographic information system technologies. *Ecological Modelling*, **221** (1), 46–58.

Ciscar, J.-C., Iglesias, A., Feyen, L., *et al.* (2011) Physical and economic consequences of climate change in Europe. *Proceedings of the National Academy of Sciences of the United States of America*, **108** (7), 2678–2683.

Collier, C. (2007) Flash flood forecasting: What are the limits of predictability? *Quarterly Journal of the Royal Meteorological Society*, **133** (622), 3–23.

Coops, H., Beklioglu, M. and Crisman, T.L. (2003) The role of water-level fluctuations in shallow lake ecosystems – workshop conclusions. *Hydrobiologica*, **506–509**, 23–27.

Costa, L. and Kropp, J.P. (2012) Linking components of vulnerability in theoretic frameworks and case studies. *Sustainability Science*, 1–9.

Costa, L., Thonicke, K., Poulter, B. and Badeck, F. (2010) Sensitivity of Portuguese forest fires to climatic, human and landscape factors: Differences between extreme fire

years and long term averages. *Regional Environmental Change*, **11** (3), 543–551.

Crimmins, S.M., Dobrowski, S.Z., Greenberg, J.A., *et al.* (2011) Changes in climatic water balance drive downhill shifts in plant species' optimum elevations. *Science*, **331** (6015), 324–327.

Cutter, S. and Burton, C. (2010) Disaster resilience indicators for benchmarking baseline conditions. *Homeland Security and Emergency Management*, **7** (1), 1–22.

Drecker, P., Genske, D., Heinrich, K. and Noll, H. (1995) Subsidence and wetland development in the Ruhr district of Germany. *IAHS Publications-Series of Proceedings and Reports – International Association of Hydrological Sciences*, **234**, 413–422.

Ellenberg, H. (1992) *Zeigerwerte von Pflanzen in Mitteleuropa.* Verlag Erich Goltze, Göttingen.

Energy Agency NRW (2007) *North Rhine-Westphalia. Economic Structure and Climate Protection. At First Glance.* EnergieAgentur.NRW, Dusseldorf.

Energy Agency NRW (2009) *NRW – Kommunen, Gemeinden und Kreise. Beteiligung am European Energy Award.* (pdf) Available at: <http://www.energieagentur.nrw.de /european-energy-award/willkommen-beim-european- energy-award-5808.asp> (accessed 6 December 2009).

Engle, N.L. (2011) Adaptive capacity and its assessment. *Global Environmental Change*, **21** (2), 647–656.

Fekete, A., Damm, M. and Birkmann, J. (2010) Scales as a challenge for vulnerability assessment. *Natural Hazards*, **55** (3), 729–747.

Förster, H. and Lilliestam, J. (2009) Modeling thermoelectric power generation in view of climate change. *Regional Environmental Change*, **10** (4), 327–338.

Fuchs, S., Kuhlicke, C. and Meyer, V. (2011) Editorial for the special issue: vulnerability to natural hazards: the challenge of integration. *Natural Hazards*, **58** (2), 609–619.

Füssel, H. and Klein, R. (2006) Climate change vulnerability assessments: an evolution of conceptual thinking. *Climatic Change*, **75** (3), 301–329.

Gerstengarbe, F.-W., Badeck, F., Hattermann, F. and Krysanova, V. (2004) *Erstellung regionaler Klimaszenarien für Nordrhein-Westfalen. Bericht zum Werkvertrag.* (pdf) Available at: <http://www.lanuv.nrw.de/klima/pdf /klimastudie_nrw.pdf> (ccessed 4 October 2012).

Glade, T., Felgentreff, C. and Birkmann, J. (2010) Editorial for the special issue: extreme events and vulnerability in environment and society. *Natural Hazards*, **55** (3), 571–576.

Greis, S. (2007) *Gewässertemperaturentwicklung und industrieller Kühlwasserbedarf*. Diploma thesis, Julius-Maximilians-University, Würzburg.

Grünewald, U., Schümberg, S.B., Wöllecke, G., *et al.* (2009) *Gutachten – Zu Entstehung und Verlauf des extremen*

Niederschlag-Abfluss-Ereignisses am 26.07.2008 im Stadtgebiet von Dortmund einschließlich der Untersuchung der Funktionsfähigkeit von wasserwirtschaftlichen Anlagen und Einrichtungen der Stadt Emscherg. Cottbus, Karlsruhe, BTU Cottbus, ARCADIS Consult GmbH, Köln.

Handl, J., Knowles, J. and Kell, D.B. (2005) Computational cluster validation in post-genomic data analysis. *Bioinformatics*, **21** (15), 3201–3212.

Held, T. (2000) Regenwasserversickerung in urbanen Landschaften. Probleme und Lösungsansätze in der Planung am Beispiel der Emscherregion. *Zeitschrift für Angewandte Geographie*, **4**, 20–26.

Hellmeier, W., Stausberg, J. and Hollmann, B. (2007) Untersuchungen in NRW zu Auswirkungen der Hitzewelle 2003 auf die kurzzeitige Mortalität. Tech. rep., LöGD. Materialien 'Umwelt und Gesundheit'.

Hinkel, J. (2011) Indicators of vulnerability and adaptive capacity: Towards a clarification of the science-policy interface. *Global Environmental Change*, **21** (1), 198–208.

Holsten, A., Dominic, A.R., Costa, L. and Kropp, J.P. (2013) Evaluation of the performance of meteorological forest fire indices for German Federal States. *Forest Ecology and Management*, **287** (1), 123–131.

Huang, S., Krysanova, V., Oesterle, H. and Hattermann, F.F. (2010) Simulation of spatiotemporal dynamics of water fluxes in Germany under climate change. *Hydrological Processes*, **24** (23), 3289–3306.

Hydrotec (2004) *Hochwasser-Aktionsplan Emscher. Grundlagen, Überflutungsgebiete, Schaden.* Emschergenossenschaft/Hydrotec, Aachen.

IFT (2008) *Masterplan Wintersportarena Sauerland/Siegerland-Wittgenstein.* IFT – Freizeit- und Tourismusberatung GmbH, Köln.

IPCC (2001) Annex B: Glossary of Terms, in (eds R.T. Watson and the Core Writing Team, 2001) *Climate Change 2001: Synthesis Report – Contribution of Working Groups I, II and III to the Third Assessment Report of the Intergovernmental Panel on Climate Change*, Cambridge University Press, Cambridge, New York, pp. 365–389.

IPCC (2012) Summary for Policymakers, in *IPCC, 2012. Managing the Risks of Extreme Events and Disasters to Advance Climate Change Adaptation – A Special Report of Working Groups I and II of the Intergovernmental Panel on Climate Change*, Cambridge University Press, Cambridge, New York.

IT.NRW (2010a) *Bevölkerungsvorrausberechung. Landesdatenbank NRW.* (online) Available at: <https://www.landesdatenbank.nrw.de> (accessed 1 February 2011).

IT.NRW (2010b) *Verarbeitendes Gewerbe sowie Bergbau und Gewinnung von Steinen und Erden in NRW: Betriebe, Beschäftigte, Lohn- und Gehaltsumme sowie Umsatz,* Zentraler Statistischer Auskunftsdienst, Dusseldorf.

Jacob, D., Mahrenholz, P. and Keup-Thiel, E. (2006) *REMO A1B Scenario Run 'REMO UBA A1B D3' and REMO Climate off the 20th century run 'REMO UBA C20 D3', Datastream 3, UBA project, CERA-DB,* World Data Center for Climate, Hamburg.

Jonkman, S. (2005) Global perspectives on loss of human life caused by floods. *Natural Hazards*, **34** (2), 151–175.

Kasperski, M. (2002) A new wind zone map of Germany. *Journal of Wind Engineering and Industrial Aerodynamics*, **90**, 1271–1287.

Kaufman, L. and Rousseeuw, P.J. (1990) *Finding Groups in Data,* John Wiley & Sons, Inc., New York.

Kehl, M., Everding, C., Botschek, J. and Skowronek, A. (2005) Erosion processes and erodibility of cultivated soils in North Rhine-Westphalia under artificial rain – I. Site characteristics and results of laboratory experiments. *Journal of Plant Nutrition and Soil Science*, **168** (1), 34–44.

Klaus, M., Holsten, A., Hostert, P. and Kropp, J.P. (2011) Integrated methodology to assess windthrow impacts on forest stands under climate change. *Forest Ecology and Management*, **261** (11), 1799–1810.

Kosatsky, T. (2005) The 2003 European HeatWaves. *Eurosurveillance*, **10** (7–9), 148–149.

Kropp, J.P., Block, A., Reusswig, F., *et al.* (2006) Semiquantitative Assessment of Regional Climate Vulnerability: The North-Rhine Westphalia Study. *Climatic Change*, **76** (3–4), 265–290.

Kropp, J., Holsten, A., Lissner, T., *et al.* (2009) *Klimawandel in Nordrhein-Westfalen. Regionale Abschätzung der Anfälligkeit ausgewählter Sektoren – Abschlussbericht,* Potsdam-Institut für Klimafolgenforschung, Potsdam.

LANUV (Landesamt für Natur, Umwelt und Verbraucherschutz NRW) (2009) *Kartieranleitungen in Nordrhein-Westfalen.* (online) Available at: <http://www.naturschutzinformationen-nrw.de> (accessed 31 July 2009).

Lautenschlager, M., Keuler, K., Wunram, C., *et al.* (2009) *Climate Simulation with CLM, Climate of the 20th Century run no. 1, 2 and 3 and Scenario A1B runs no.1,2, Data Stream 3. Tech. rep., European region MPI-M/MaD,* World Data Center for Climate, Hamburg.

Leira, M. and Cantonati, M. (2008) Effects of water-level fluctuations on lakes: an annotated bibliography. *Hydrobiologia*, **613**, 171–184.

Lin, B.B. and Morefield, P.E. (2011) The vulnverability cube: A multi-dimensional framework for assessing relative vulnerability. *Environmental Management*, **48** (3), 631–643.

Lissner, T., Holsten, A.,Walther, C. and Kropp, J.P. (2012) Towards sectoral and standardised vulnerability assessments: the example of heatwave impacts on human health. *Climatic Change*, **112** (3–4), 687–708.

LWK NRW (Landwirtschaftskammer Nordrhein-Westfalen) (2008) *Zahlen zur Landwirtschaft in Nordrhein-Westfalen 2008*, Landwirtschaftskammer Nordrhein-Westfalen, Düsseldorf.

Meinke, I., Gerstner, E.-M., von Storch, H., *et al.* (2010) Regionaler Klimaatlas Deutschland der Helmholtz-Gemeinschaft informiert im Internet über möglichen künftigen Klimawandel. *DMG Mitteilungen*, **2**, 5–7.

Menzel, A., Sparks, T.H., Estrella, N., *et al.* (2006) European phenological response to climate change matches the warming pattern. *Global Change Biology*, **12** (10), 1969–1976.

Metzger, M.J. and Schröter, D. (2006) Towards a spatially explicit and quantitative vulnerability assessment of environmental change in Europe. *Regional Environmental Change*, **6** (4), 201–216.

Müller, I. (2003) Erosionsschutzmaßnahmen in der Landwirtschaft – Demonstrationsvorhaben, in *LUA 2004. Jahresbericht 2003*, Landesumweltamt NRW, Essen, pp. 73–75.

Müller, L., Schindler, U., Mirschel, W., *et al.* (2010) Assessing the productivity function of soils. A review. *Agronomy for Sustainable Development*, **30** (3), 601–614.

Munich Re (2008) *Highs and Lows. Weather Risks in Central Europe*, Münchener Rückversicherungs-Gesellschaft, Munich.

MUNLV (2009) *Anpassung an den Klimawandel. Eine Strategie für Nordrhein-Westfalen*, Ministerium für Klimaschutz, Umwelt, Landwirtschaft, Natur- und Verbraucherschutz NRW, Düsseldorf.

MUNLV (2010) *Abschlussbericht der Landesregierung von Nordrhein-Westfalen zu den Folgen des Sturmereignisses 'Kyrill' vom 18./19. Januar 2007*, Ministerium für Klimaschutz, Umwelt, Landwirtschaft, Natur- und Verbraucherschutz NRW, Münster.

MWEBWV (2010) *Basisdaten zum Standort NRW*. (pdf) Ministerium für Wirtschaft, Energie, Bauen, Wohnen und Verkehr des Landes Nordrhein-Westfalen, Düsseldorf. Available at: <http://www.mwme.nrw.de/witelegramm/basisdaten/basisdaten_14\ _2_08.pdf> (accessed 1 July 2011).

Nakicenovic, N., Alcamo, J. and Davis, G. (2000) *Special Report on Emissions Scenarios (SRES) – Working Group III of the Intergovernmental Panel on Climate Change (IPCC)*, Cambridge University Press, Cambridge.

Nelson, R.M. (2001) Water relations of forest fuels, in (eds E.A. Johnson and K. Miyanishi) *Forest Fires: Behavior and Ecological Effects*. Academic Press, San Diego, pp. 79–150.

O'Brien, K.L., Leichenko, R., Kelkar, U., *et al.* (2004) Mapping vulnerability to multiple stressors: climate change and globalization in India. *Global Environmental Change, Part A* **14** (4), 303–313.

Olonscheck, M., Holsten, A. and Kropp, J.P. (2011) Heating and cooling energy demand and related emissions of the German residential building stock under climate change. *Energy Policy*, **39** (9), 4795–4806.

Otte, U. (1999) Häufigkeit von Sturmböen in den letzten Jahren, in *DWD, 1999. Klimastatusbericht 1999*. Deutscher Wetterdienst, Offenbach, pp. 1–9.

Patt, A., Klein, R. and Delavegaleinert, A. (2005) Taking the uncertainty in climate-change vulnerability assessment seriously. *Comptes Rendus Geosciences*, **337** (4), 411–424.

Petermann, J., Balzer, S., Ellwanger, G., *et al.* (2007) Klimawandel – Herausforderung für das europaweite Schutzgebietssystem Natura 2000, in (eds S. Balzer, M. Dieterich, B. Beinlich) *Natura 2000 und Klimaänderungen*. Vol. 46, Bundesamt für Naturschutz, Bonn, Bad Godesberg.

Pompe, S., Hanspach, J., Badeck, F., *et al.* (2008) Climate and land use change impacts on plant distributions in Germany. *Biology Letters*, **4** (5), 564–567.

Preston, B.L., Yuen, E.J. and Westaway, R.M. (2011) Putting vulnerability to climate change on the map: a review of approaches, benefits, and risks. *Sustainability Science*, **6** (2), 177–202.

Rahmstorf, S. and Coumou, D. (2011) Increase of extreme events in a warming world. *Proceedings of the National Academy of Sciences of the United States of America*, **108** (44), 17905–17909.

Rannow, S., Loibl, W., Greiving, S., *et al.* (2010) Potential impacts of climate change in Germany – Identifying regional priorities for adaptation activities in spatial planning. *Landscape and Urban Planning*, **98** (3–4), 160–171.

Reineking, B., Weibel, P., Conedera, M. and Bugmann, H. (2010) Environmental determinants of lightning- v. human-induced forest fire ignitions differ in a temperate mountain region of Switzerland. *International Journal of Wildland Fire*, **19** (5), 541–557.

Renard, K.G., Foster, G.R. and Weesies, G.A. (1997) *Predicting Soil Erosion by Water. A Guide to Conservation Planning with the Revised Universal Soil Loss Equation (RUSLE)*, Vol. 703, U.S. Deptartment of Agriculture, Washington DC.

Riis, T. and Hawes, I. (2002) Relationships between water level fluctuations and vegetation diversity in shallow water of New Zealand lakes. *Aquatic Botany*, **74** (2), 133–148.

Roth, R., Türk, S., Armbruster, F., *et al.* (2001) *Masterplan Wintersport im Sauerland und Siegerland-Wittgenstein. Räumliche Entwicklungskonzeption und Marktanalyse. Tech. Rep. 6*, Institut für Natursport und Ökologie der Deutschen Sporthochschule Köln, ift Freizeit- und Tourismusberatung GmbH, Köln.

Roth, V., Lange, T., Braun, M. and Buhmann, J. (2002) A resampling approach to cluster validation, in (eds W. Härdle and B. Rönz) 2002 *Proc. Conf. Computational Statistics*, Physika-Verlag, Heidelberg, pp. 123–128.

Rybski, D., Holsten, A. and Kropp, J.P. (2011) Towards a unified characterisation of phenological phases: Fluctuations and correlations with temperature. *Physica A – Statistical Mechanics and Its Applications*, **390** (4), 680–688.

Sauerborn, P., Klein, A., Botschek, J. and Skowronek, A. (1999) Future rainfall erosivity derived from large-scale climate models – methods and scenarios for a humid region. *Geoderma*, **93** (3–4), 269–276.

Sauter, T., Weitzenkamp, B. and Schneider, C. (2009) Spatio-temporal prediction of snow cover in the Black Forest mountain range using remote sensing and a recurrent neural network. *International Journal of Climatology*, **30** (15), 2330–2341.

Scheffer, M. and van Nes, E.H. (2007) Shallow lakes theory revisited: various alternative regimes driven by climate, nutrients, depth and lake size. *Hydrobiologia*, **584**, 455–466.

Schindler, U., Steidl, J., Müller, L., *et al.* (2007) Drought risk to agricultural land in Northeast and Central Germany. *Journal of Plant Nutrition and Soil Science*, **170** (3), 357–362.

Schwertmann, U., Vogl, W. and Kainz, M. (1990) *Bodenerosion durch Wasser. Vorhersage des Abtrags und Bewertung von Gegenmaßnahmen.* 2nd edn., Eugen Ulmer GmbH & Co., Stuttgart.

Spekat, A., Gerstengarbe, F.-W., Kreienkamp, F. and Werner, P.C. (2006) *Fortschreibung der Klimaszenarien für Nordrhein-Westfalen. Bericht im Auftrag der Landesanstalt für Ökologie, Bodenordnung und Forsten Nordrhein-Westfalen. Tech. Rep. 2*, Climate & Environment Consulting Potsdam GmbH, Potsdam.

Steiger, R. (2010) Climate change impact assessment in winter tourism, in (eds A. Matzarakis, H. Mayer and F. Chmielewski) 2010 *Proceedings of the 7th Conference on Biometeorology*, Meteorologisches Institut der Universität Freiburg, Freiburg, pp. 410–415.

Storchmann, K. (2005) The rise and fall of German hard coal subsidies. *Energy Policy*, **33** (11), 1469–1492.

Svenning, J.-C. and Skov, F. (2006) Potential impact of climate change on the Northern nemoral forest herb flora of Europe. *Biodiversity and Conservation*, **15** (10), 3341–3356.

Syphard, A.D., Radeloff, V.C., Keuler, N.S., *et al.* (2008) Predicting spatial patterns of fire on a southern California landscape. *International Journal of Wildland Fire*, **17** (5), 602–613.

Te Linde, A.H., Bubeck, P., Dekkers, J.E.C., *et al.* (2011) Future flood risk estimates along the river Rhine. *Natural Hazards and Earth System Sciences*, **11** (2), 459–473.

Thuiller, W., Lavorel, S., Araujo, M.B., *et al.* (2005) Climate change threats to plant diversity in Europe. *Proceedings of the National Academy of Sciences of the United States of America*, **102** (23), 8245–8250.

UN/ISDR (2004) *Living with Risk. A Global Review of Disaster Reduction Initiatives*, United Nations, Geneva.

USDA (1972) *National Engineering Handbook. Section 4 Hydrology. Vol. Hydrology*, U.S. Department of Agriculture, Soil Conservation Service, Washington DC.

USDA (2007) Hydrologic Soil Groups, in *National Engineering Handbook, Part 630 Hydrology*, United States Department of Agriculture, Natural Resources Conservation Service, Washington DC, Chapter 7, p. 14.

Vandentorren, S. and Empereur-Bissonnet, P. (2005) Health Impact of the 2003 Heat-Wave in France, in (eds W. Kirch, B. Menne and R. Bertollini) *Extreme Weather Events and Public Health Responses*, World Health Organization, Geneva, pp. 81–88.

Vohland, K. and Cramer, W. (2009) Auswirkungen des Klimawandels auf gefährdete Biotoptypen und Schutzgebiete. *Jahrbuch für Naturschutz und Landschaftspflege*, **57**, 22–27.

Wechsung, F., Gerstengarbe, F.-W., Lasch, P. and Lüttger, A. (2008) *Die Ertragsfähigkeit ostdeutscher Ackerflächen unter Klimawandel. PIK Report Nr. 112*, Potsdam Institute for Climate Impact Research, Potsdam.

WSA (2011) *Wintersport und -spaß. Wintersport-Arena Sauerland.* (online) Available at: <http://www.wintersport-arena.de/de/wintersport-arena/region.php> (accessed 31 July 2011).

Zebisch, M., Grothmann, T., Schröter, D., *et al.* (2005) *Climate Change in Germany – Vulnerability and Adaption of Climate sensitive Sectors*, Umweltbundesamt, Nr. 10/2005, Dessau, Germany.

Zhan, X.Y. and Huang, M.L. (2004) ArcCN-Runoff: an ArcGIS tool for generating curve number and runoff maps. *Environmental Modelling & Software*, **19** (10), 875–879.

Chapter 12

Climate change impacts on the Hungarian, Romanian and Slovak Territories of the Tisza catchment area

Krisztián Schneller[1], Gábor Bálint[2], Alina Chicoş[2], Mária Csete[3],
Jan Dzurdzenik[4], Annamária Göncz[1], Alexandru-Ionut Petrisor[5],
Tesliar Jaroslav[4] and Tamás Pálvölgyi[3]

[1]VÁTI Nonprofit Ltd. Spatial Planning Department, Gellérthegy u. 30–32, 1016 Budapest, Hungary
[2]VITUKI Environmental and Water Management Research Institute, Kvassay 1, 1095 Budapest, Hungary
[3]BME (Budapest University of Technology and Economics), Műegyetem rkp., 1111 Budapest, Hungary
[4]Agency for the Support of Regional Development Košice, n.o., Strojárenská 3, 040 01 Košice, Slovakia
[5]URBAN-INCERC, Soseaua Pantelimon, nr. 266, Sector 2, 021652 Bucharest, Romania

Abstract

The river Tisza has the largest catchment area among the tributaries of the river Danube. It covers nearly $160\,000\,km^2$ and has about 14 million inhabitants. Extreme weather phenomena are already a serious problem in the region. According to the forecasts, the frequency of extreme weather events in the context of droughts and excess water flooding is expected to increase as a result of climate change. In the course of the vulnerability assessment the impacts of changing frequency of extreme weather (floods, drought and excess water inundation) events were analysed using the uniform methodology.

In the case study, the result values of exposure indicators of COSMO CLM were compared with results of regional analysis. According to the results of the comparison, the quantitative change of summer and winter precipitation were taken as exposure indicators form the COSMO CLM study.

European Climate Vulnerabilities and Adaptation: A Spatial Planning Perspective, First Edition.
Edited by Philipp Schmidt-Thomé and Stefan Greiving.

Sensitivity was analysed by means of two indices in the economic dimension, while in the physical dimension the impacts were assessed directly by four indices, where the analysis were based on the maps of LISFLOOD model. Indicators of adaptive capacity characterise the social and economic as well as infrastructure conditions, showing how the regions are capable of coping with unfavourable changes. Vulnerability was calculated on the basis of potential aggregated impact and aggregated adaptive capacity.

Based on the results of impact analysis in the physical dimension the most negative impacts are to be expected on the upstream (mountainous and hilly regions) section of the Tisza and its tributaries, although most of the current potential flood prone areas are found on the downstream section (on the plain) of the Tisza and tributaries. In terms of economic impacts the picture is more divers, but the highest increase is predicted mainly on the lowland and hilly part of the Tisza basin.

The geographical distribution of the aggregated adaptive capacity indices shows a diverse picture. In Slovakia, adaptive capacity is medium or high, while in Hungary all degrees of adaptive ability can be experienced. The regions of Romania are characterised by low and very low values except for Arad, Timiş and Cluj counties.

The ultimate result of the vulnerability analysis proved the results obtained from the two partial analyses (aggregated impact and adaptive capacity), namely, that in the water catchment area of the Tisza the most vulnerable counties are around the Apuseni Mountains in Romanian.

12.1   Introduction

The river Tisza has the largest catchment area among the tributaries of the river Danube. It covers nearly $160\,000\,\mathrm{km}^2$ and has about 14 million inhabitants. Extreme weather phenomena are already a serious problem in the region. According to the forecasts, the frequency of extreme weather events in the context of droughts and excess waters flooding (undrained runoff) is expected to increase as a result of climate change. Regarding the impact of climate change on discharges there are lots of uncertainties, which are described in a separate chapter of the case study based on the national and regional level researches.

In the course of the vulnerability assessment the impacts of changing frequency of extreme weather (floods, drought and excess water inundation) events were analysed using the uniform methodology.

The adaptability to the unfavourable impacts of the more and more extreme weather (warming and drying climate, excess water inundation and flood) can be enhanced in the region by means of: adapting land use structures; more effective possibilities of water retention and discharge regulation; promoting the policies in support of the above, with special regard to the distribution of domestic and EU resources of sustainable agriculture, forest and environmental management as well as water management and flood control; joint elaboration of transnational plans of water and land management.

12.2   Brief characteristics of Tisza River Basin

The Tisza River Basin is the largest sub-basin in the Danube River Basin, covering nearly $160\,000\,\mathrm{km}^2$ (20% of the Danube Basin), it is home to approximately 14 million people (Figure 12.1). The greatest part of the basin is dominated by the mountains (Carpathians). The South-Western and middle parts of the catchment area are flat plains. Most of the catchment area is covered by sediments (limestone in the mountains and sand and loess on the plain) and in some parts, for instance in the inner range of the Carpathian Mountains, there are also volcanic origin mountains. The geographical differences are the determinants for land use: in the mountainous areas the forests and on the plains arable land is the dominant land-use type. Forests cover 27% of the catchment area: on the high mountains coniferous woods and on the mountains of medium height deciduous trees are dominant. The majority of grasslands (pastures and meadows) are also on

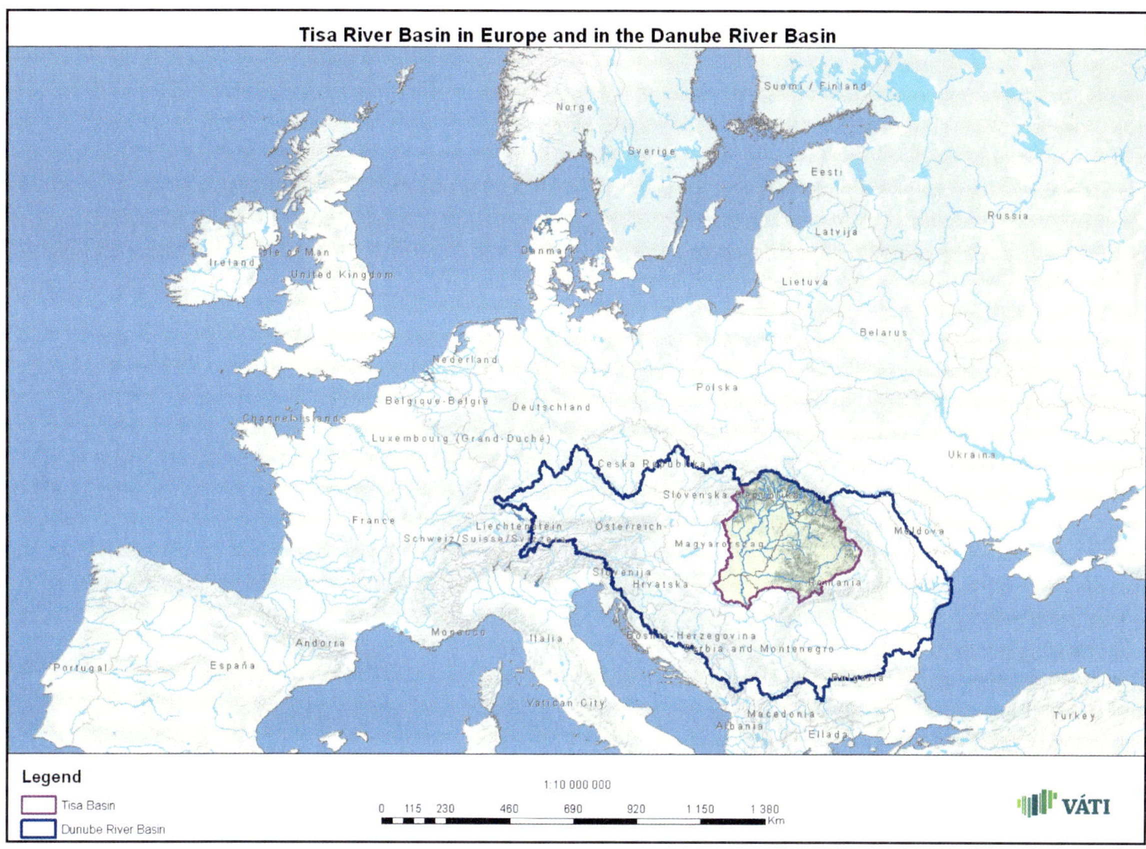

Figure 12.1   The location of Tisza basin in Europe. *Source*: ArcGIS Online; ICPDR (International Commission for the Protection of Danube River).

mountains. The plains are predominantly plough lands. Arable land covers 35% of the catchment area, the greatest part is in Hungary (ICPDR, 2007).

In the catchment area of the River Tisza there are only two big cities (Cluj and Timisoara) with over 300 000 inhabitants and 26 cities are in the range of 50 000–300 000. The municipalities with populations of 10 000–50 000 are, for the most part, on the Great Plain of Hungary. The majority of the municipalities of the Tisza river catchment area have less than 10 000 inhabitants. There are a large number of very small communities with populations of less than 1000. The majority of these are in the hilly and mountainous areas (SEE-TICAD, 2011).

The climate in the river basin is exposed to Atlantic to Mediterranean and continental effects. The differences are particularly significant in terms of precipitation. In mountainous areas the annual average precipitation is over 1000 mm, and in the lowlands it is actually below 500 mm. Rainfall in the Carpathian Mountains can be substantial and sudden. Extensive runoff, floodplain deforestation and river canalisation reduce the ability of the catchment to attenuate the flood wave. When heavy rains occur, flooding threatens human lives as water levels rise quickly without sufficient retention capacity (ICPDR, 2007).

In the water catchment area of the River Tisza, exposure to droughts, inland waters and floods is already a serious threat. Droughts are hazards mainly for agriculture, and they have an unfavourable impact on the habitats that are valuable from the point of view of nature conservation. Excess water flooding causes damage for agriculture, and is also dangerous for the settlements. It is not rare that excess water flooding and droughts hit the same territory in the same year. Exposure to floods is characteristic for the entire region. For the hilly parts, short floods of great intensity are characteristic, while on the plains floods last longer (Farago, Lang and Csete, 2010).

12.3 Climate change impacts on Tisza River Basin

12.3.1 *Expected climate change impacts on the Hungarian part of the basin*

According to national climate research, climate change will bring about severe draughts, more frequent and more intensive extreme weather events all over the Carpathian Basin (Farago, Lang and Csete, 2010). Considering that in this area 28 years out of 100 are arid, and where a sequence of dry years is frequent, whilst in other years floods, excess water inundation and frost damage may also occur, a prospective warming and drying climate raises serious issues concerning domestic food security (Csete, 2008). In critical years, the price of imported foodstuffs sharply increases. One of the most important natural resources of Hungary is soil. It is not only the source of raw material, but also the largest reservoir of the entire country. Soil quality may be adversely affected by climate change (Pálvölgyi *et al.*, 2010; Pálvölgyi and Czira, 2011). The prevention of the reduction and degradation of agricultural land is therefore of utmost importance.

The most serious issue of water management is that a great proportion of the territory of the Hungarian part of the basin is endangered by floods. Furthermore, one third of the Hungarian lowlands (most of them are located in the Tisza Basin) is moderately or heavily endangered by excess water inundation. Research indicates that the most vulnerable aspect of water management is the reduction

of the available water resources (Csete and Szendrő, 2011). The research report of the Scientific Expert Group on Climate Change (SEG, 2007) claims that as regards the impact of climate change on biodiversity, Hungary is one of the most vulnerable countries in Europe. A serious problem is the large-scale spread of invasive species caused by the improvement of survival capability due to increasing percentage of summer droughts and warmer winters (Sipkay, Kiss, Vadadi-Fulop and Hufnagel, 2009).

12.3.2 *Expected climate change impacts on Romanian part of the catchment area*

Agriculture is among the most affected economic sectors in Romania because of its dependence on weather conditions. Changes in temperature and precipitation patterns will modify the vegetation periods and involve shifts in the relative ratios of woodland and pasture land. Floods represent the most widespread hazard in the Tisza River Basin, with numerous losses of human lives and a high-proportion of material damage. The rapid and violent floods cause losses of human lives and important material damage because the water level increases within a very short period of time (Romanian Ministry of the Environment, 2005b).

According to the Romanian Ministry of the Environment (2007), floods have a high frequency in spring (30–50%) and a lower in autumn (10–20%) and winter (15–30%). Romania is also characterised by a high biodiversity. However, under the present conditions, particularly the effect of climate change, many plants, animals and habitats are endangered and the landscape modifications are the first sign of environmental deterioration. Romania has, among the 27 member states of the EU, the highest biogeographical diversity (with five biogeographical regions out of the 11 at the European level) and most of these areas are in a good conservation status.

The health status of millions of people is projected to be affected in the Tisza River Basin due to climate change effects. Some of these expected changes could produce, in conjunction with extreme local socioeconomic conditions (extreme poverty, social exclusion) serious effects, affecting food supply and therefore increasing malnutrition, mortality and diseases; they may create injuries to humans through extreme weather events, increase the frequency of respiratory and gastro-intestinal diseases related to climate change and alter the spatial distribution of some infectious diseases (IWGCCH, 2010).

12.3.3　Expected climate change impacts on the Slovakian part of the river basin

The expected climate change will have serious consequences for Slovakia in the long term but some effects, such as an increasing frequency of extreme weather events, is already a reality today. During the last decade the extreme situations have increased. Heavy rains and flash floods, which have suddenly occurred, caused not only economic damage in settlements and infrastructure, but have taken human lives too. Changing agro-climatic conditions will have an effect requiring changing the varieties of cereals species. Forests will be exposed to the growing extreme weather conditions, for example, droughts that will cause increased risk of forest fires, or wet periods with significant soil moisture increase, which will undermine forest stability (e.g. windthrow damage in the past). The condition of the forests is deteriorating because of the predominance of the spruce tree in the forest ecosystems, which is prone to the spread of the bark beetle. According the state water management policy of the Slovak Republic, this expected climate change will have a significant impact on total runoff as well as on its distribution within the year. Flood disasters in Slovakia are expected to accelerate this trend and will increase by 19% up until 2100 (Aaheim *et al.*, 2008; AEA, 2007).

The tourism sector can partly benefit from the projected climate change, in general. In the lowlands the importance of the use of water resources will grow for sports and recreation, including spa resorts. An example of the negative impact are when the water reservoirs suffer from the occurrence of high levels of toxic cyan bacteria, which, apart from the chemical nature of the water quality, is also related to an increase in the temperatures of the backwaters. No significant negative impact is expected in the mountain areas on active outdoor sports such as climbing, cycling and hiking. However, changes to the landscape, including a change in the compositions of forests and change of agriculture crops, the impact of long-term droughts in combination with sudden floods and other extreme weather events, will necessarily have an impact on the attractiveness of the region, quality of available services and thus it will inevitably also be reflected in the visitors rates.

Most negative impact is expected for the ski resorts. According to the prognosis of the Slovak Hydro-Meteorological Institute (Pecho and Fasko, 2010) the levels of snow cover during the winter will not be lower than 1100 m. With very few exceptions the ski resorts of Eastern Slovakia are located below this level and experience from recent winters suggests that ski-resorts which are located even higher are facing difficulties with the snow cover. Along with insufficient snowfall it is envisioned that water levels will also drop, which will mean that production of artificial snow becomes very expensive and requires using more of the already inadequate supply of energy. Furthermore, development of ski tourism is typically linked to spa and wellness tourism, which will, with a high probability, also lead to a negative impact on this segment of tourism.

12.3.4　Climate change impact on river floods in the Tisza Basin based on the CLAVIER project

Although sporadic investigations targeted parts of the catchment areas (Bálint *et al.*, 1995; Bálint *et al.*, 1996; Bálint and Gauzer, 1998; Nováky, 2002) the first river basin wide simulations by Radvánszky and Jacob (2008) produced results only recently and with complex analyses tools within the CLAVIER project (2009).

Detailed studies within the CLAVIER project applied the VITUKI NHFS modelling system (Gauzer and Bartha, 2001) and the Vidra–Consul Model (Becjer and Serban, 1990) to route meteorological input (the ECHAM5 driven–REMO50 scenario) to produce discharge series at selected gauging stations. Both tools are semi-distributed: the NHFS model calculates a 0.1° grid-wise water balance followed by sub-basin wise runoff concentration and channel routing performed for defined river reaches; the Vidra–Consul spatial structure is based on small runoff producing sub-basins linked by routing modules.

Hydrological impact and water management implications of climate projections in the Tisza Basin concluded that the changes in simulated river discharge under the investigated climate scenarios show a general pattern of change in the discharge regime towards higher runoff in winter and early spring and lower runoff in summer. This pattern can be seen at all stations and can be explained by less precipitation in summer, more precipitation in winter and spring and higher temperatures resulting in an earlier snowmelt. Regimes with a significant snowmelt component, such as the upper Tisza, also show a shift and reduction in the runoff peak in late spring.

Based on the hydrological simulations, the analysis of the mean seasonal and annual discharge simulated time series show the following impact of the A1B climate change scenario (ECHAM5 driven–REMO50):

- slight increase in Upper Tisza and tributaries (less than 5%);
- no change or general decrease of the mean annual discharges, down to 15% for Central Tisza and the southern basins;
- the mean seasonal discharges are increasing only in the winter season, in all the basins but with significant variations (3–42%).

For the spring–autumn period, the hydrological simulations indicate a decrease in the mean discharges for most of the catchments in the southern basins of between −15 and −20%.

In order to estimate the A1B climate change scenario impact on the hydrological regime, the variations of the mean monthly simulated discharges for selected 30 year periods (1961–1990 as the reference period and 2001–2030, 2011–2040, 2021–2050 as representative periods for the future) were also analysed.

The results in most cases indicate a slight decrease in the annual mean flow throughout the region with significant spatial variability and even some increase for the high elevation zones in the Upper Tisza sub-catchment areas. The decrease in spring runoff is compensated by the flow resulting from the thaw during late winter. An important conclusion of the CLAVIER project was related to the indication of significant variability between different sub-basins: increasing trends in the Upper Tisza and a decrease over the catchment areas of the southern tributaries.

12.3.5 *Summary*

Based on the literature review on the lowland part of the river basin, an increase in the drought periods will have serious consequences for agriculture and habitats; on the hilly and mountainous parts the increasing flood situation will cause greater damage in the built-up environment. In the European comparison, the Tisza basin belongs to the areas that are of the most backward and underdeveloped from an economic point of view, which makes adaptation to extreme weather conditions even more difficult.

12.4 Vulnerability assessment of the Tisza River case study area

Based on the previously introduced Hungarian, Romanian and Slovak national literature, the following described exposure data validation and the availability of sensitivity data, the focus of the vulnerability assessment of Tisza case study was based on:

- the increasing drought and excess water flooding (undrained runoff) impact on agriculture;
- the impact of changes in river flooding on the built-up environment.

The vulnerability study was carried out (with small and large changes) using the uniform methodology of the project (see Figure 12.3). Some dimensions (e.g. environmental dimension) were not assessed because of data availability problems (accessibility of habitat type maps). The smallest area units of the analysis were NUTS 3 regions. The case study area of the River Tisza extends to ten NUTS regions in Hungary, 13 in Romania and three in Slovakia[1] (Figure 12.2).

In the case study, the results for the values of exposure indicators of the COSMO-CLM were compared with the results of some regional analyses. According to the results of the comparison, the quantitative *changes of summer and winter precipitation* were taken as exposure indicators from the COSMO-CLM study. For these indices, the uncertainties are minimal and the results from the different pieces of research are almost similar.

Sensitivity was analysed by means of two indices in the economic dimension, while in the physical dimension the impacts were assessed directly by four indices, where the analysis was based on the maps of the LISFLOOD model (recalculated by TU Dortmund).[2] The impact assessment is based on the aggregation of the six indices of economic and physical dimensions.

Indicators of *adaptive capacity* characterise the social and economic as well as infrastructure conditions, showing how they are capable of coping with unfavourable changes. The calculation of aggregated adaptive capacity are based on six indices of four determinants (knowledge and awareness, technology, infrastructure and economic resources).

Vulnerability was calculated on the basis of potential aggregated impact and aggregated adaptive capacity.

12.4.1 Qualitative validation of COSMO-CLM exposure indicators in the context of the Tisza River case study

In order to assess the uncertainty of climate change vulnerability in the Tisza River case study area it is essential to verify the exposure indicators (Table 12.1). The main objective of the evaluation of exposure indicators is to provide a qualitative comparison between the exposure indicators used in the Tisza River case study and the relevant model outputs. Exposure indicators for the Tisza River case study were provided by the Potsdam Institute for Climate Impact Research (PIK) and based on the latest outputs of the COSMO-CLM model (or CCLM). The model runs have been conducted in conjunction with the global coupled atmosphere–ocean model ECHAM5/MPI-OM and based on the SRES scenario A1B. The change indicators always relate the reference time frame (1961–1990) to the climate conditions within the projected periods as calculated by the CCLM model (e.g. 2071–2100). All of the seven exposure indicators provided by PIK have been compared qualitatively with the relevant results from regional climate change research (Szépszó and Horányi, 2008; Lautenschlager *et al.*, 2009; Bartholy, Pongrácz, Torma, and Pieczka, 2009; CLAVIER project[3]).

The main results of the qualitative validation are as the follows.

1. **Change in annual mean temperature**   According to our comparative analysis the annual mean temperature change is less pronounced than seasonal changes. This may be important in the case of flood-related and heat-wave related vulnerability assessments. The quasi homogeneous

---

[1]NUTS 3 regions (counties) in Hungary: Bács-Kiskun, Békés, Borsod–Abaúj–Zemplén, Csongrád, Hajdú–Bihar, Heves, Jász–Nagykun–Szolnok, Nógrád, Pest and Szabolcs–Szatmár–Bereg.
NUTS 3 regions (counties) in Romania: Alba, Arad, Bihor, Bistriţa–Năsăud, Cluj, Harghita, Hunedoara, Maramureş, Mures, Sălaj, Satu–Mare, Sibiu and Timiş.
NUTS 3 regions (kraj) in the Slovak Republic: Banská Bystrica Region, Košice Region and Prešov Region.
[2]In the vulnerability assessment the results of the recalculated LISFLOOD model were applied because the regional researches do not provide data regarding the changing flood-prone areas.

[3]www.clavier-eu.org

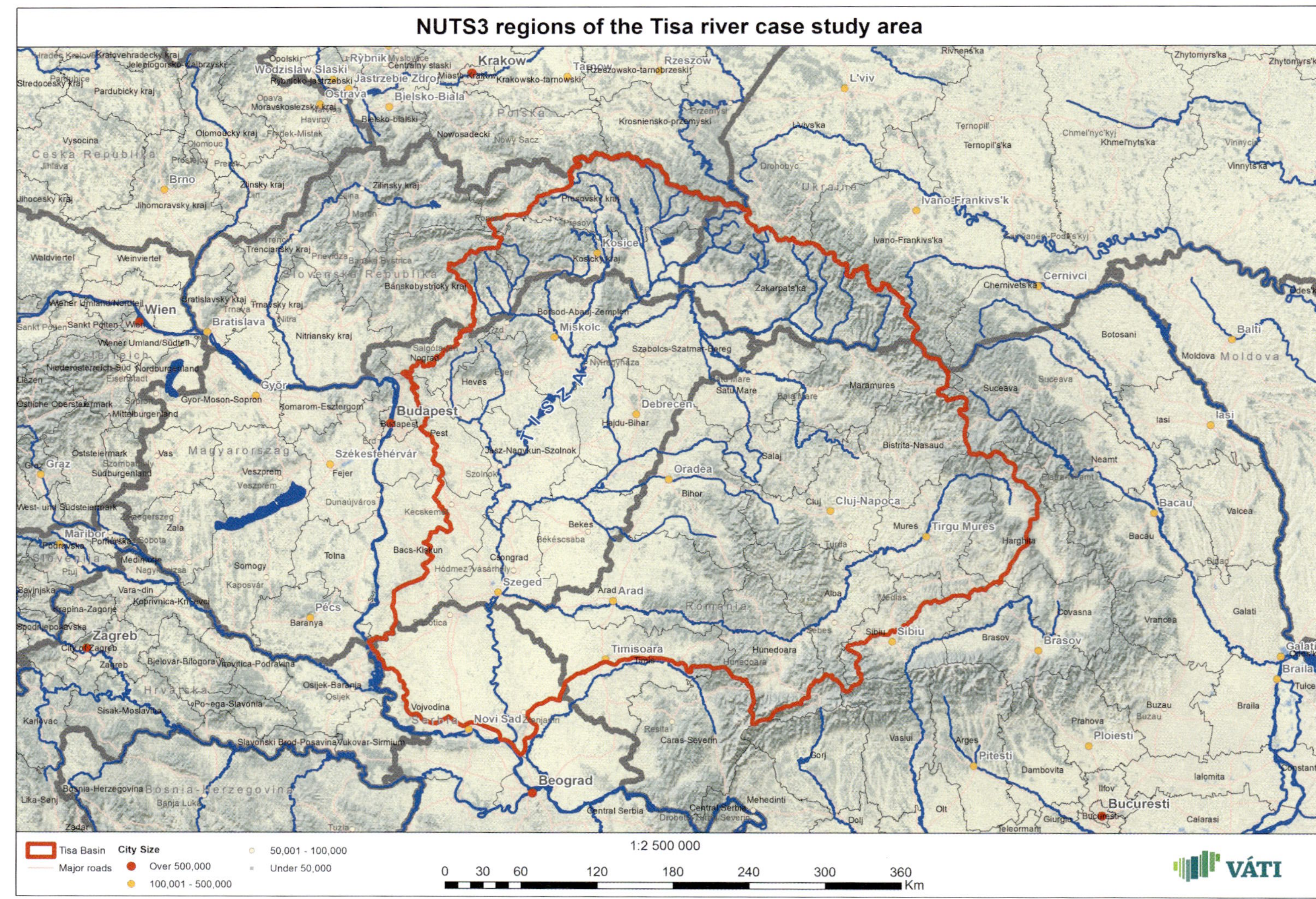

Figure 12.2   Tisza river case study area. *Source*: ArcGIS Online; ICPDR (International Commission for the Protection of Danube River).

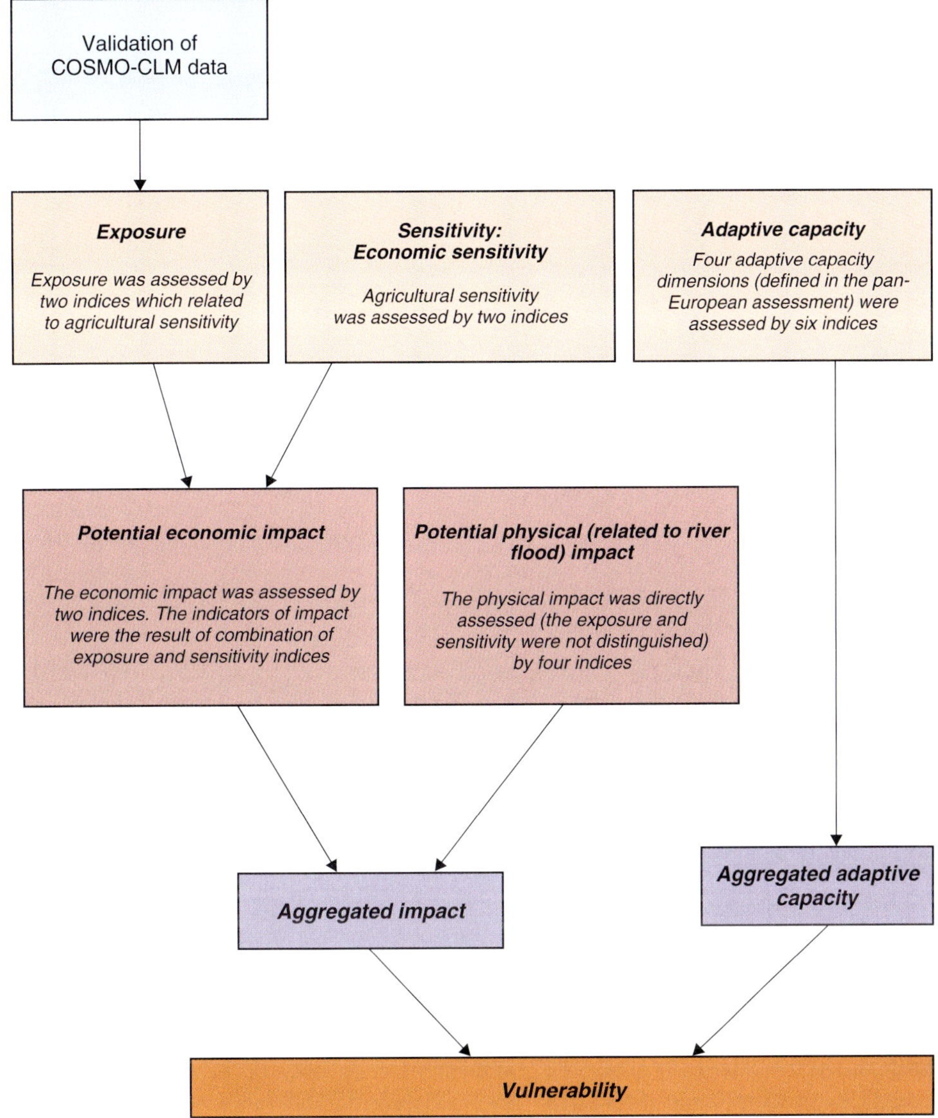

Figure 12.3   Overview of the vulnerability assessment in the Tisza River Basin.

spatial structure of warming does not allow a proper classification into five classes.[4]

2. **Change in annual mean number of frost days** The decrease in number of frost days is a characteristic indicator of regional climate change in the Tisza region, which coincides well with the results derived from the literature mentioned above.

3. **Change in annual mean number of summer days** The plain spatial pattern in this exposure indicator will present difficulties when classifying into five classes.

4. **Relative change in annual mean precipitation in winter months** This is a characteristic indicator

---

[4]During the vulnerability assessment, indicator values (among other exposure indicators) were classified into five classes following the equal interval method.

Table 12.1 Indicator characteristics of the 26 NUTS 3 regions in the Tisza case study area.

Indicator characteristics	COSMO-CLM exposure indicators							
	Change in annual mean temperature	Decrease in number of frost days	Increase in number of summer days	Increase in winter precipitation	Decrease in summer precipitation	Change in days with heavy rainfall	Relative change in evaporation	Decrease in snow cover days
Maximum value	3.69	55	36	20.3	42.3	1.01	−5.2	42
Minimum value	3.38	41	28	6.6	25	0.01	+4.4	3
Average	3.57	49	34	13.6	35.4	0.55	−0.3	18
Variation	0.076	3.9	1.6	3.9	3.8	0.28	2.1	11.9
Representativity	+	++	+++	+++	++	++	+	+
Problem with classification	!	–	!	–	–	!	!	–
Magnitude of change	++	+++	+++	+	++	≈ 0	≈ 0	++
Coincidence with ref. run	+++	+++	NA	++	+	NA	+	NA

Notes: + low/poor; ++ medium; +++ high; ! significant problem with classification.

of climate change in the Tisza region, which is in medium coincidence with the results derived from the literature (Szépszó and Horányi, 2008).

5. **Change in annual mean number of days with heavy rainfall**   This shows insignificant changes (less than one day) in the Tisza region. Therefore this indicator will not contribute to increased vulnerability in the case study area.

6. **Relative change in annual mean evaporation**   This indicator shows minor (probably insignificant) changes in the Tisza region. Therefore this indicator will not significantly contribute to a vulnerability increase in the case study area.

7. **Change in annual mean number of days with snow cover**   This is a characteristic indicator of regional climate change in the Tisza region, which is in medium coincidence with the results derived from the literature (Szépszó and Horányi, 2008).

12.4.2   *Exposure analysis of the Tisza Basin*

In the analysis, the quantitative changes in the summer and winter precipitation were examined as exposure indicators of the COSMO-CLM study (Figures 12.4 and 12.5). The considerable reduction in the volume of summer precipitation on the European level predicts a probable growth in the frequency of droughts, while the increase in the volume of winter precipitation suggests an unfavourable increase in the volume and duration of river flooding and excess water flooding.

12.4.3   *Sensitivity assessment*

In the case of sensitivity, only one dimension was assessed. The main reason for this was that some important sensitivity data were not accessible or available (e.g. habitat data for assessing the environmental dimension, monuments on flood prone areas for assessing cultural sensitivity, etc.). On the other hand, regarding the physical dimension (impacts of changes in river flooding on built-up environments) the sensitivity and exposure parts of the assessment were not separated.

As a result of data collection, two indices describe the economic sensitivity dimension (see Table 12.2).

Concerning economic sensitivity, the focus was on the sensitivity of agriculture, which can be well characterised by the water management properties of the various soil types. The soil with unfavourable

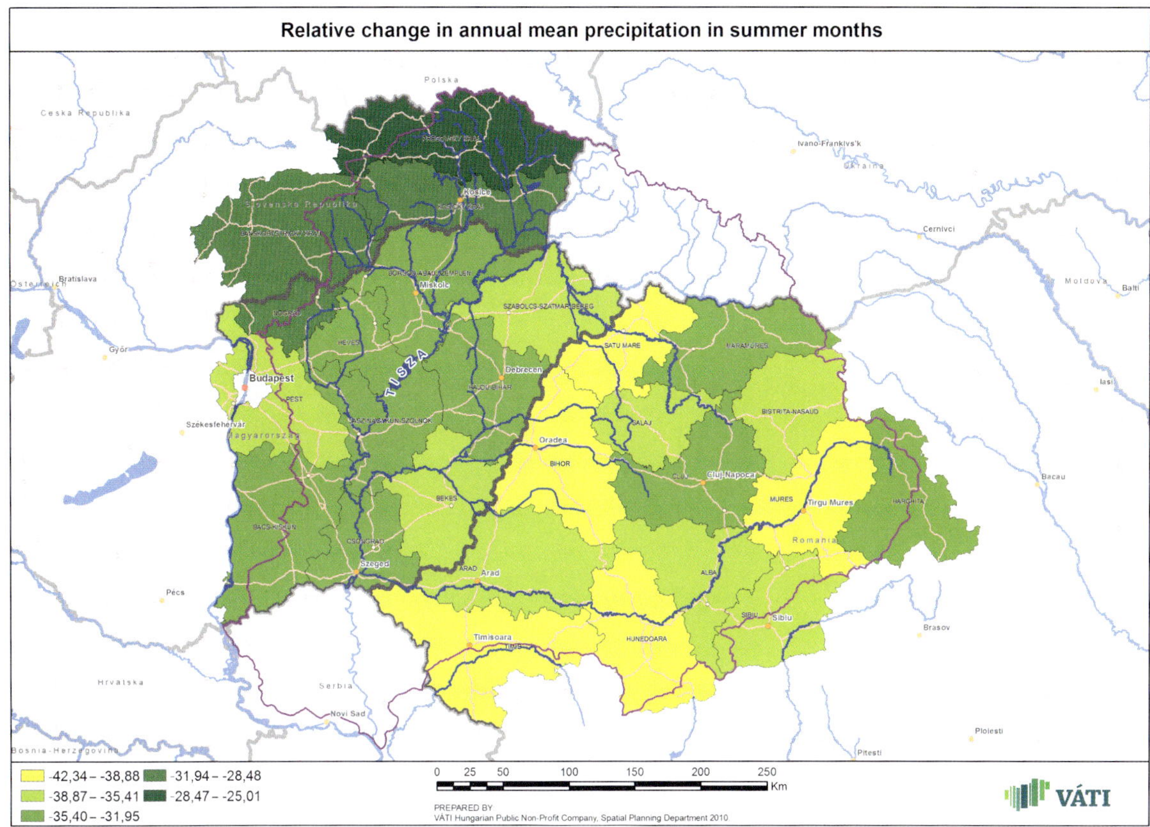

Figure 12.4   Relative change in annual mean precipitation in the summer months (in %). Based on data from COSMO-CLM, 2010; NUTS-3: TU Dortmund.

water management properties may considerably strengthen and intensify the climate impacts. The indicators of sensitivity to drought and excess water are based on the soil map of arable land from FAO-UNESCO. On the basis of the FAO soil map (soil type and texture) we developed five categories of sensitivity to excess water flooding and drought (Figures 12.6 and 12.7).

12.4.4   *Potential impact assessment*

The impact assessments were performed in physical and economic dimensions, using six indices (Table 12.3).

In terms of the physical dimension, the calculation of potential impacts is based on maps of changing potential flood-prone areas using the recalculated LISFLOOD model (Figure 12.8). The exposure and sensitivity indices were not analysed separately in the course of the analysis, but the potential impacts were assessed directly according to the definition in Table 12.3. In the physical dimension, in addition to the impacts on the settlements, also the impacts on transport networks (existing and planned) have been analysed. The results obtained for each of the indices have been classified according to a five-grade scale. As the change is duplex, in the classification of negative and positive impacts were distinguished.

In terms of the economic impacts, the indicator values were the results of a combination of the appropriate exposure and sensitivity indicators.

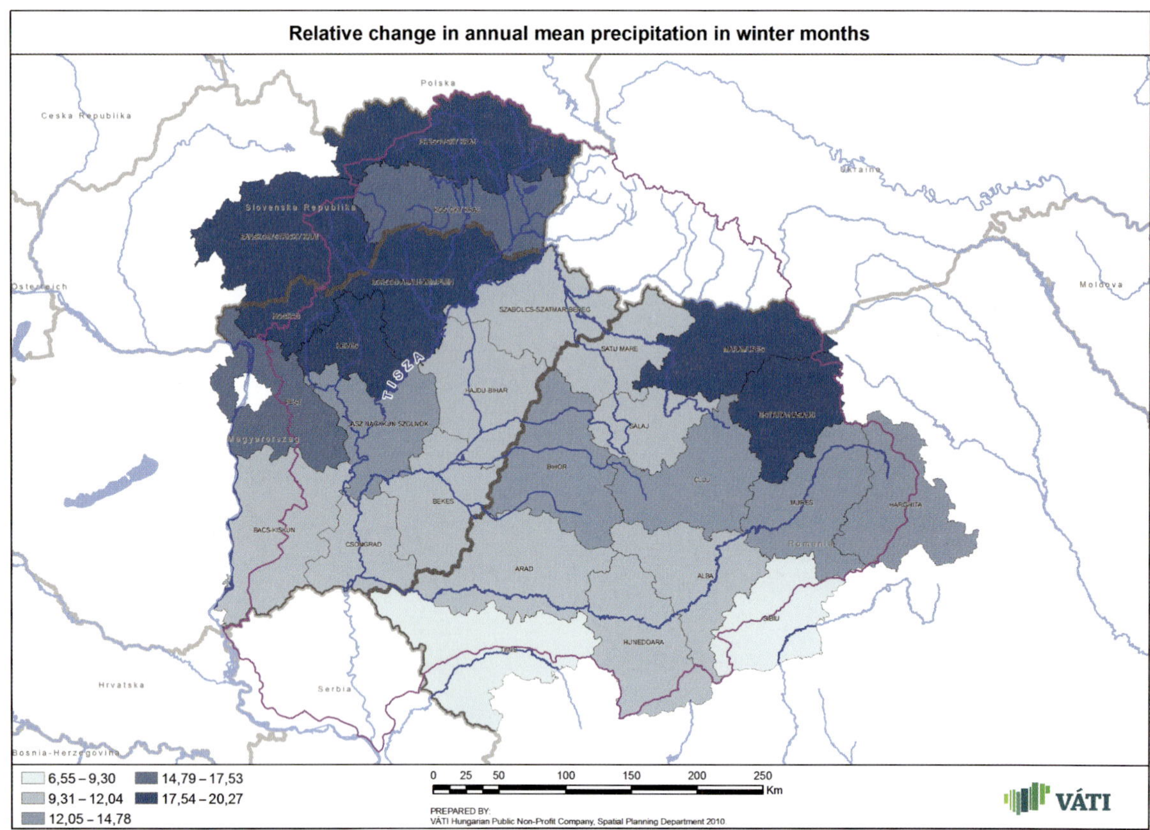

Figure 12.5   Relative change in annual mean precipitation in winter months (in %). Based on data from COSMO-CLM, 2010; NUTS-3: TU Dortmund.

Table 12.2   Sensitivity indicators of the Tisza River case study area.

Sensitivity dimension	Sensitivity indicator	Calculation method
Physical sensitivity	Regarding the physical dimension, the sensitivity indicators were not separately assessed (because it is not necessary), but the impacts were directly analysed according to the definition in Table 12.3	
Economic sensitivity (Sensitivity of agriculture)	*Soil properties influencing crop production sensitivity to drying climate*	The indicator is based on the categorisation and evaluation of FAO-UNESCO soil map
	Soil properties influencing crop production sensitivity to excess water	The indicator is based on the categorisation and evaluation of FAO-UNESCO soil map

Notes: *italic* – indicators defined in the Tisza River case study.

During the combination we used the same weight (0.5) for exposure and sensitivity. The results obtained for each of the two indices have been classified according to a five-grade scale. The weighting of dimensions was made on the basis of expert opinions.

Low negative combined physical impact is predicted mainly in the mountainous and hilly regions, especially in the Košice and Prešov regions (Figure 12.9). The negatively impacted settlement area is also significant in Szabolcs–Szatmár–Bereg

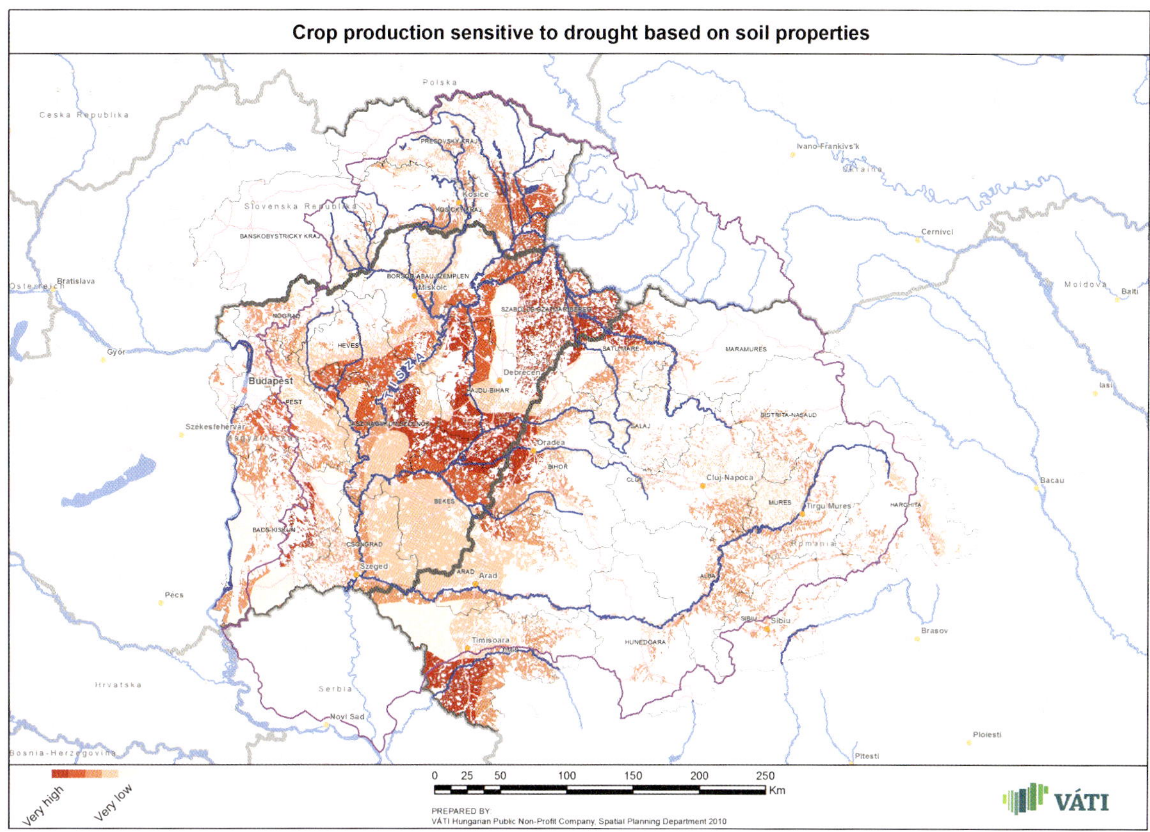

Figure 12.6 Sensitivity of crop production to drought (based on soil properties). Based on data from CORINE, 2006; FAO Digital Soil Map of the World, 2007.

megye (county). Very high positive impact, on the other hand, can be expected on the plain, and more specifically, in Békés and Hajdú–Bihar counties due to the decrease in the potential for flood prone areas.

Very high negative combined economic impact is predicted mainly in the NUTS 3 regions of the lowland part of Tisza River Basin and around the Apuseni Mountain, especially in the following regions: Borsod–Abaúj–Zemplén, Bihor, Bistriţa–Năsăud, Hajdú–Bihar, Heves, Jász–Nagykun–Szolnok, Mures and Satu Mare counties and in the Košice region. The lowest negative impact, on the other hand, can be expected in Arad, Bács–Kiskun, Csongrád and in Harghita counties (Figure 12.10).

The aggregated impact was assessed by combining and weighting the two impact dimensions. The physical dimension was given a weight of 0.4 and the economic dimension a weight of 0.6.

The aggregated impact varies in the case study area. Very high negative impact is in Bistriţa–Năsăud, Mures and Satu Mare counties and in the Košice region. On the other hand, very low positive impact is in Csongrád county (Figure 12.11).

12.4.5 *Adaptive capacity of the regions of Tisza Basin*

Adaptive capacity was characterised by the six indices of the four determinant adaptive characteristics. They are shown in Table 12.4.

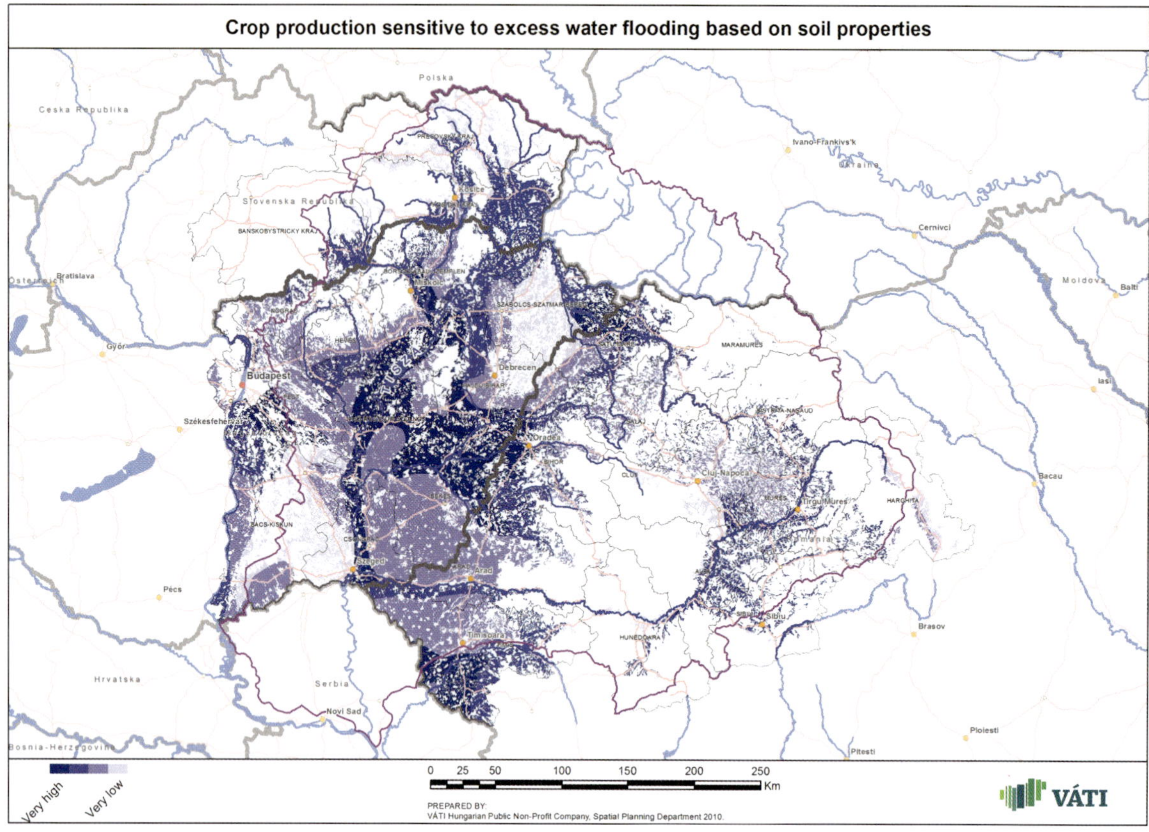

Figure 12.7   Sensitivity of crop production to excess water flooding (based on soil properties). Based on data from CORINE, 2006; FAO Digital Soil Map of the World, 2007.

Aggregated adaptive capacity was estimated by weighting the main factors (see Table 12.4) and then by the joint evaluation of the determinants, according to the uniform methodology.

The geographical distribution of the aggregated adaptive capacity indices shows a diverse picture (Figure 12.12). In the NUTS 3 regions of Slovakia, adaptive capacity is medium or high, while in Hungary all degrees of the ability of adaptation can be experienced: Pest and Csongrád counties have very favourable values, Nógrád and Békés counties have low and very low adaptive capacity values. In the case of Pest, the effect of Budapest should be highlighted, while in case of Hajdú–Bihar and Csongrád the universities must be mentioned. Contrerly Nógrád and Békés regions are peripheral rural areas. The NUTS 3 regions of Romania are characterised by low and very low values. Good adaptive capacities have been defined only for Arad, Timiş and Cluj counties.

Adaptive capacity in the water catchment area of the River Tisza is determined by several factors, which cannot really be measured at the moment. These factors are related to droughts and flooding. As regards floods, adaptive capacity is greatly influenced by the technical skills and knowledge of the people responsible for water management, and especially for hazard abatement and for defining the strategic issues of the elimination of damage caused by water (e.g. assessment of the possibility of river discharge on the protected side). It is also difficult to translate the county-level resources and institutions of regional and town planning into figures.

Table 12.3 Impact indicators.

Sensitivity/impact dimension	Name of impact	Definition
Physical dimension	Change of potential impact of 100 year river flood event on settlements	Settlement area in increasing or decreasing flood-prone areas/settlement area of NUTS 3 region
	Change of potential impact of 100 year river flood event on high-speed roads	Length of high-speed roads in increasing or decreasing flood-prone areas/length of high-speed roads on NUTS 3 level
	Change of potential impact of 100 year river flood event on main roads	Length of main roads in increasing or decreasing flood-prone areas/length of main roads on NUTS 3 level
	Change of potential impact of 100 year river flood event on main railway tracks	Length of main railway tracks in increasing or decreasing flood-prone areas/length of main railway tracks on NUTS 3 level
Economic dimension	*Potential impacts of decreasing summer precipitation on crop production*	Combination of exposure indicator *'Decrease in annual mean precipitation in summer months'* and sensitivity indicator *'Soil properties in terms of crop production sensitivity to drying climate'*
	Potential impacts of increasing winter precipitation on excess water flooding and crop production	Combination of exposure indicator *'Increase in annual mean precipitation in winter months'* and sensitivity indicator *'Soil properties in terms of crop production sensitivity to excess water flooding'*

Notes: *italic* – indicators defined in the Tisza River case study.

Similarly, the role of transnational cooperation can be assessed best on a national (NUTS 0) level.

12.4.6 *Vulnerability*

Vulnerability can be calculated on the basis of expected impact and aggregated adaptive capacity, using the same weight for both factors.

The geographical distribution of vulnerability also shows a diverse picture (Figure 12.13). The highest vulnerability is predicted in Romania in Bistriţa–Năsăud, Mures and Satu Mare. Contrarily only marginal effects are probable in Hungary in Csongrád county.

The findings of the vulnerability assessment – more or less – strengthen the results of national and regional research. Namely, the drying climate will have serious consequences for agriculture in the lowland and hilly parts of the region, while on the hilly and mountainous parts the increasing flood situation will cause greater damage in the built-up environments. The results of the impact assessment in terms of floods also satisfies the findings

of the regional study that is introduced (CLAVIER project).

12.5 Existing climate change adaptation strategies in the case study area

12.5.1 *Climate change adaptation strategies in Hungary*

The *National Climate Change Strategy* of Hungary 2008–2025 was elaborated in 2007 (NÉS, 2007). Detailed measures for strategy implementation are contained in the National Climate Change Programme, which covers two years. Sectors involved in the National Climate Change Strategy are: environmental protection, human environment and human health, water management, agriculture and forestry, spatial and urban planning as well as development and built-up environments. In the strategy

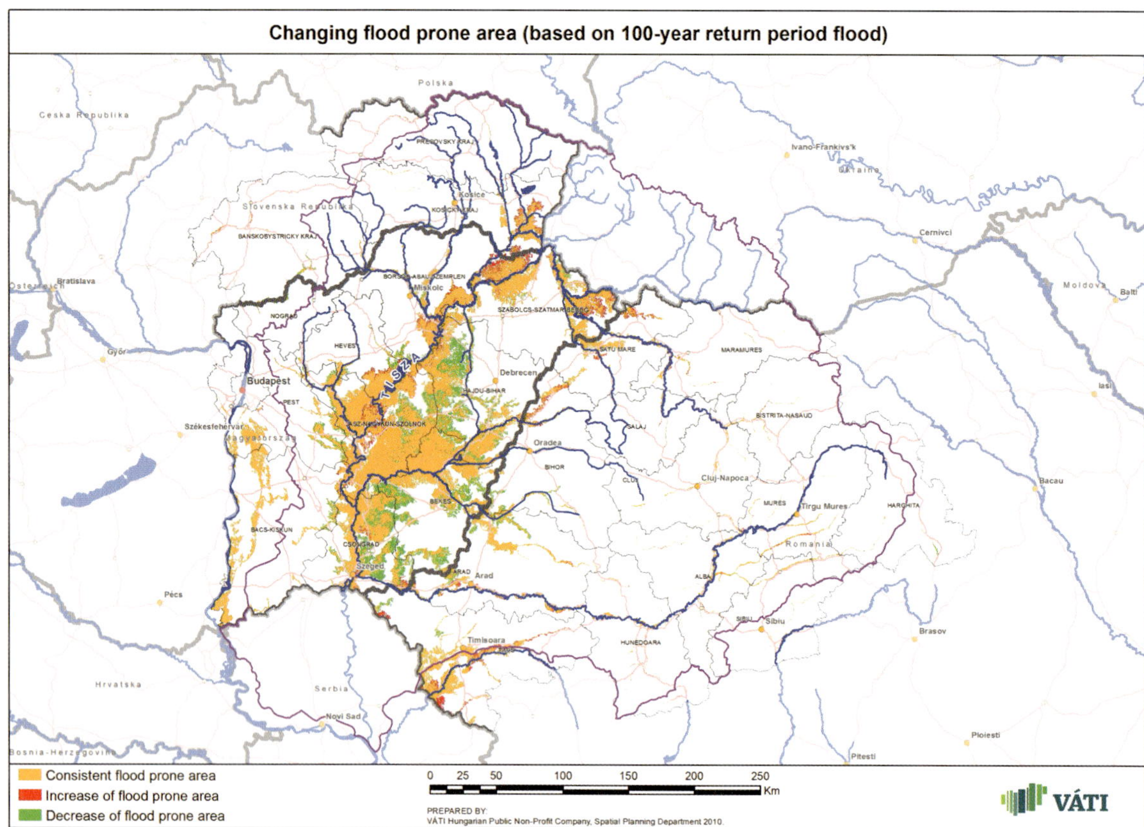

Figure 12.8   Changing flood-prone areas along the River Tisza and its main tributaries. Based on data from LISFLOOD project, recalculation: TU Dortmund IRPUD, 2010.

it is emphasised that in climate politics mitigation and adaptation measures should be planned, implemented and supervised in a harmonised way, to achieve mutual enhancement. Targets are set for the reduction of emissions in four sectors: energy, transport, agriculture and waste management. The main tasks are to facilitate energy efficiency in the community and public sectors, to promote the use of renewable energy resources, to control the energy demand of manufacturing and to re-use heat produced elsewhere, to reduce the electric energy consumption of manufacturing technologies, to transform the structure of transport and, thereby, reduction of its energy consumption. The adaptation measures are not addressed to the same activities as those specified for emission reduction.

The new Vásárhelyi Plan deserves special attention due to its scale, complexity and innovative operations.

Following multi-disciplinary research and a wide spread consultation process, long lasting consensus has already been attained in the 1990s in three main aspects:

- The issue of flood control in the region is not limited to Hungary. Coordinated action needs to be undertaken by all affected countries of the catchment area. This is of particular importance for Hungary, as the rivers flow from beyond the national borders.
- The responsibility for flood control is not limited to the water management sector. It needs

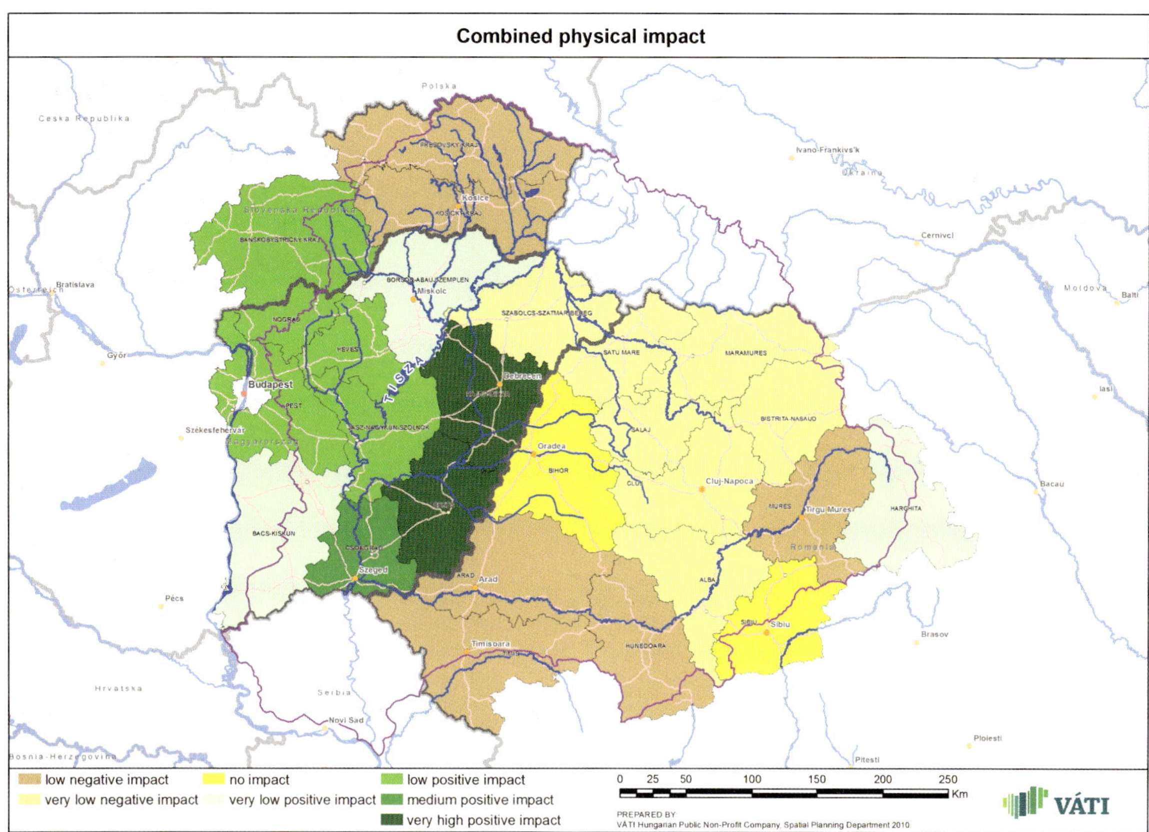

Figure 12.9   Combined physical impacts. Based on data from LISFLOOD A1B, COSMO-CLM 2010; See TICAD project 2009–2011.

an integrated approach, involving the participation of ecology and nature protection, agriculture, transport and particularly regional and urban development.

- In Hungary a major cause of floods and their devastating impact is the sudden and fast flow and equally sudden and fast growth of the volume of water in the rivers, which in turn leads to more sediments being deposited on the river bed. The river beds are thus built-up. Therefore flood control can no longer be effective by means of constructing higher and higher dykes. Far more effective and sustainable is to create (or re-install) natural reservoirs in the low-lying sections of the river valley.

Numerous relevant studies and comprehensive plans are available. The title of this large-scale planned intervention is the 'Vásárhelyi Development Plan' referring to the name of the engineer who first elaborated the river-regulation plan of Tisza during the 19th century. The Vásárhelyi Development Plan focuses on the establishment of clusters of natural reservoirs at several sites along the Hungarian section of the River Tisza. The purpose is to channel and deposit the surplus water during flood periods. To regulate and support the implementation of the Vásárhelyi Plan, regulation measures were approved: the Government Decision of the new Vásárhelyi Plan 1022/2003 (III.27.) and the ACT LXVII (2004) 'Implementation of Vásárhelyi Plan'.

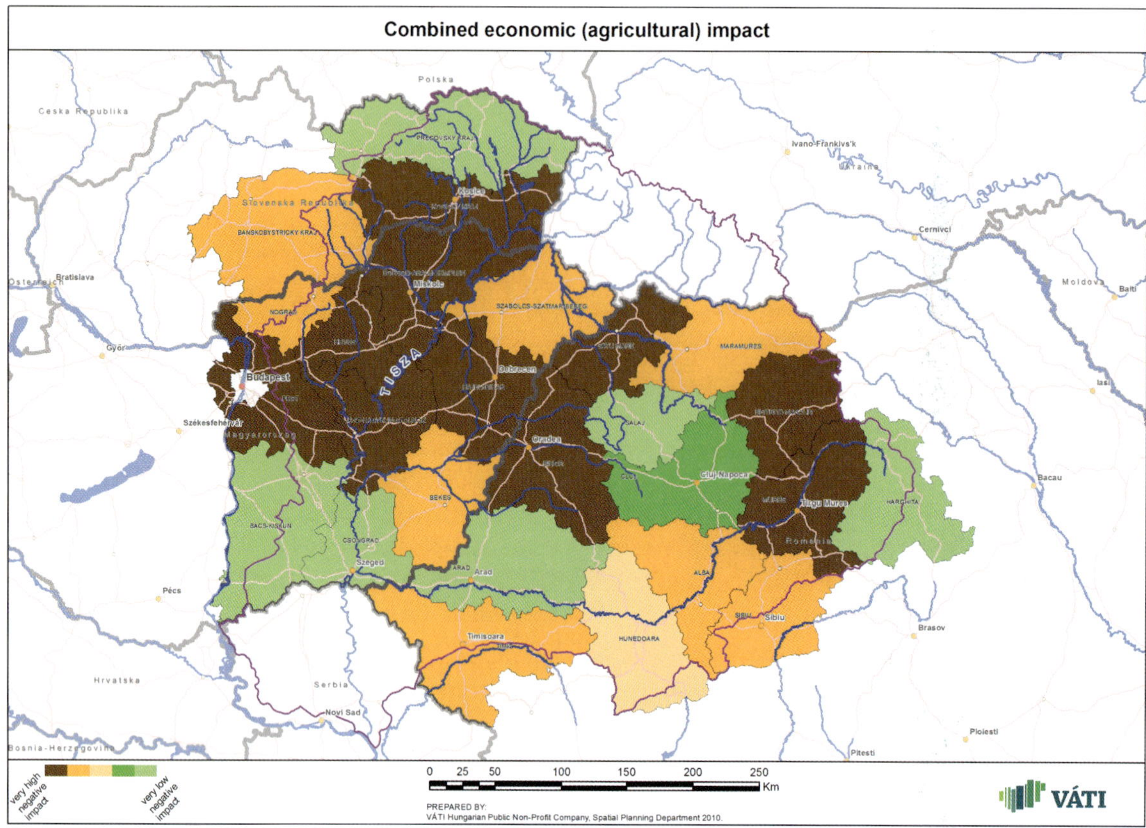

Figure 12.10 Combined economic (agricultural) impacts.

12.5.2 *Climate change adaptation of Romania*

In Romania, four legal documents tackle the issue of clime change. The *National Plan of Action on Climate Change 2005–2007* (Romanian Ministry of the Environment, 2005a) consists of a collection of 'fiches' for each required action, including, for example, strengthening the administrative capacity and institutional cooperation, identifying decision making instruments, promoting renewable energies, efficient use of energy, closure of waste deposits, introduction of integrated land-use systems (land use and its changes), increasing public awareness, improvement of public access to information and public participation.

In the *National Strategy on Climate Change 2005–2007* (Romanian Ministry of the Environment, 2005b), the general objectives (fulfilment of Kyoto obligations, setting and implementation of objectives and measures for adapting to climate change), are broken down into more specific objectives; the United Nations Framework Convention on Climate Change, the Kyoto Protocol and the European policy are discussed with special regard to the position of Romania; actions are presented for estimating greenhouse gas (GHG) emissions in Romania and the progress is assessed towards fulfilling the Kyoto objectives. The conclusions emphasised the need for action with respect to climate change towards the goals of sustainable development.

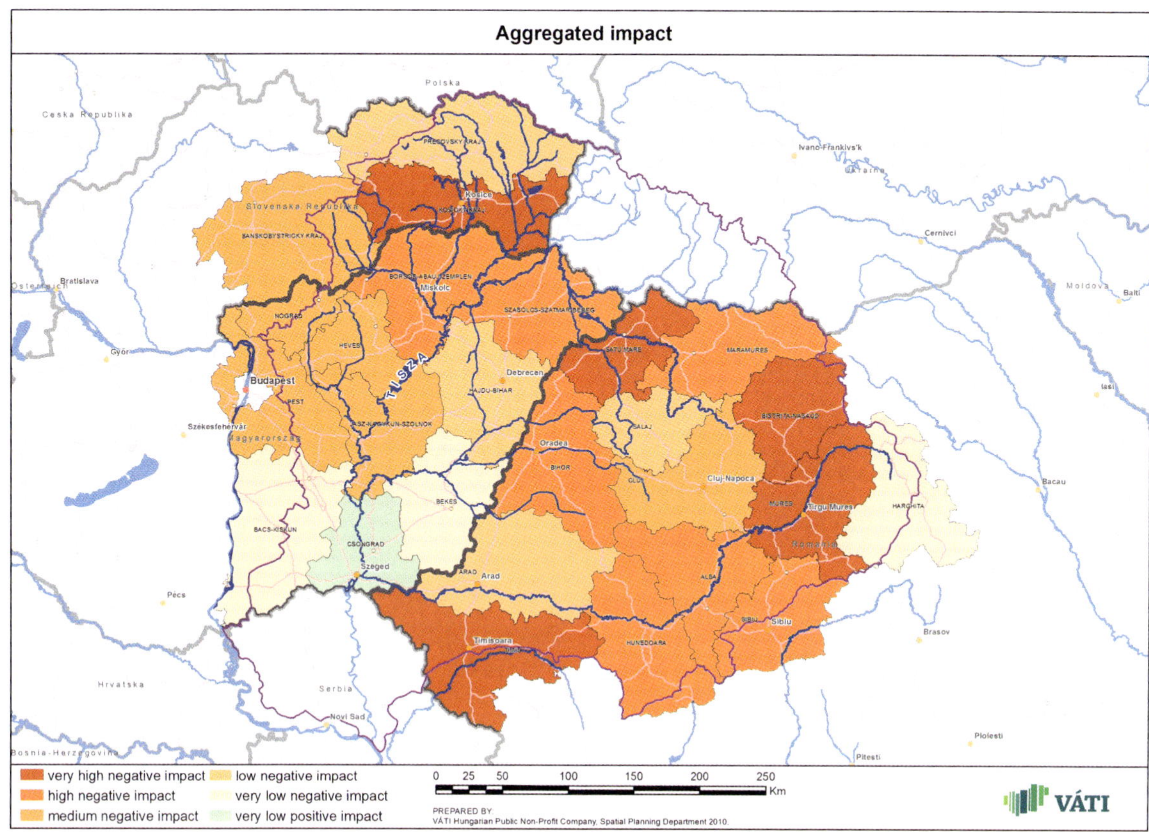

Figure 12.11   Aggregated impact.

Table 12.4   Indicators of adaptive capacity.

Adaptive capacity determinant	Adaptive capacity indicator	Calculation methodology	Weight of determinant
Knowledge and awareness	*Population with higher education*	Population with higher education (aged >24) total population	0.1
Technology	*Scientist and engineers in research and development*	Number of scientist and engineers/1 000 000 inhabitants	0.2
Infrastructure	*Irrigated area*	Territory of irrigated area/territory of agricultural area	0.3
	Share of Natura 2000 area	Territory of Natura 2000 sites/total territory of NUTS 3 area	
Economic resources	*Income per capita*	GDP/inhabitants	0.4
	Employment rate	Number employees/population between 15 and 64	

Notes: *italic* – indicators defined in the Tisza River case study.

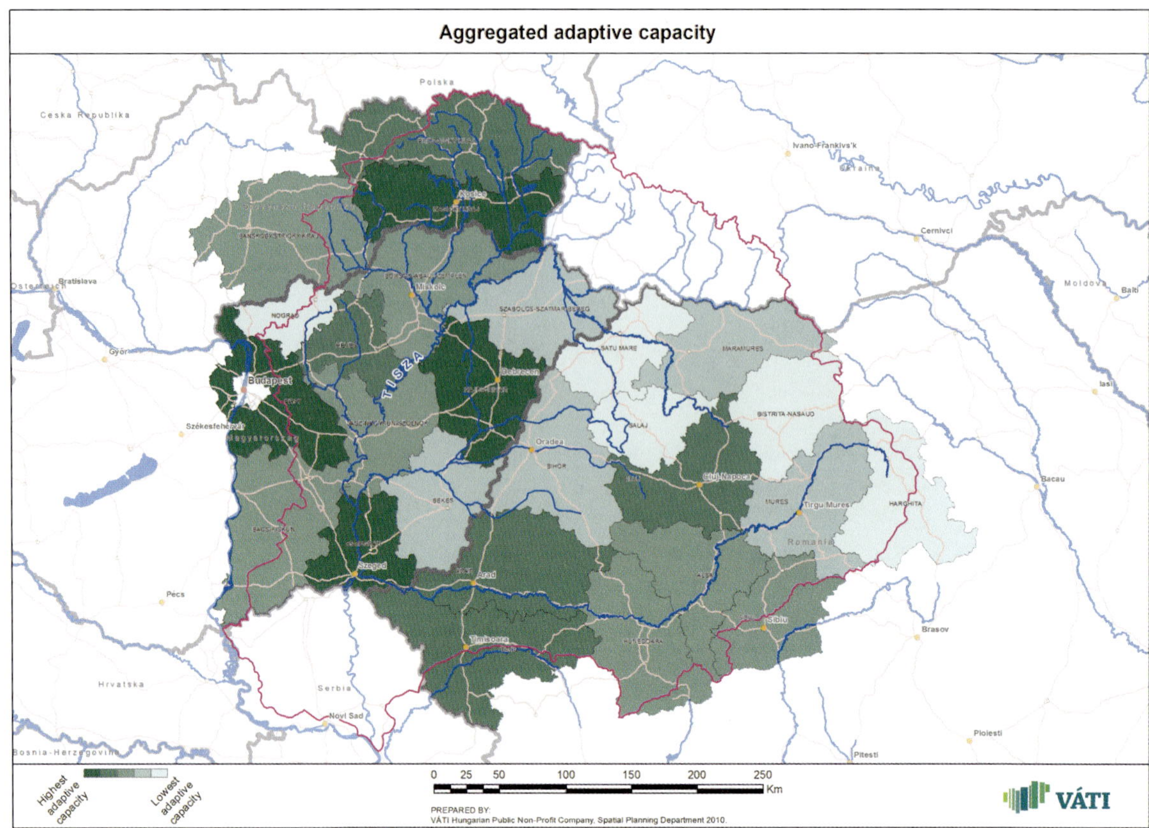

Figure 12.12   Aggregated adaptive capacity.

The *National Strategy on Long-Term Reduction of the Effects of Drought, Prevention and Combat of the Degradation of Lands and Desertification* (Romanian Ministry of Agriculture, 2008b) presents the necessity and opportunity for updating the 2000 document in the following subjects: land management, rehabilitation of degraded lands, water and soil management, horticulture, pastures and husbandry, forestry, regeneration of forests, conservation of biodiversity and landscapes, rural communities, research, monitoring, regional development and energetic security. The second part consists of a presentation of the actions and measures.

Short-term measures include, among others, change of land use practices, ecological restoration, development of adequate infrastructure, education and awareness, research and monitoring, capacity building, revision of legislation, correlation with other programs and EU institutions, average-term measures referring to research, monitoring and emergency services and long-term measures comprising water management, research and decentralised services.

The *Guide on Adaptation to Climate Change Effects* (Romanian Ministry of the Environment, 2008a) describes different climate change scenarios at global, European and national levels, and then discusses in detail their impact on and the vulnerability of different sectors: agriculture, biodiversity, water resources, forests, infrastructure, construction and urban planning, transportation, tourism, energy, industry, health, recreational activities and insurance.

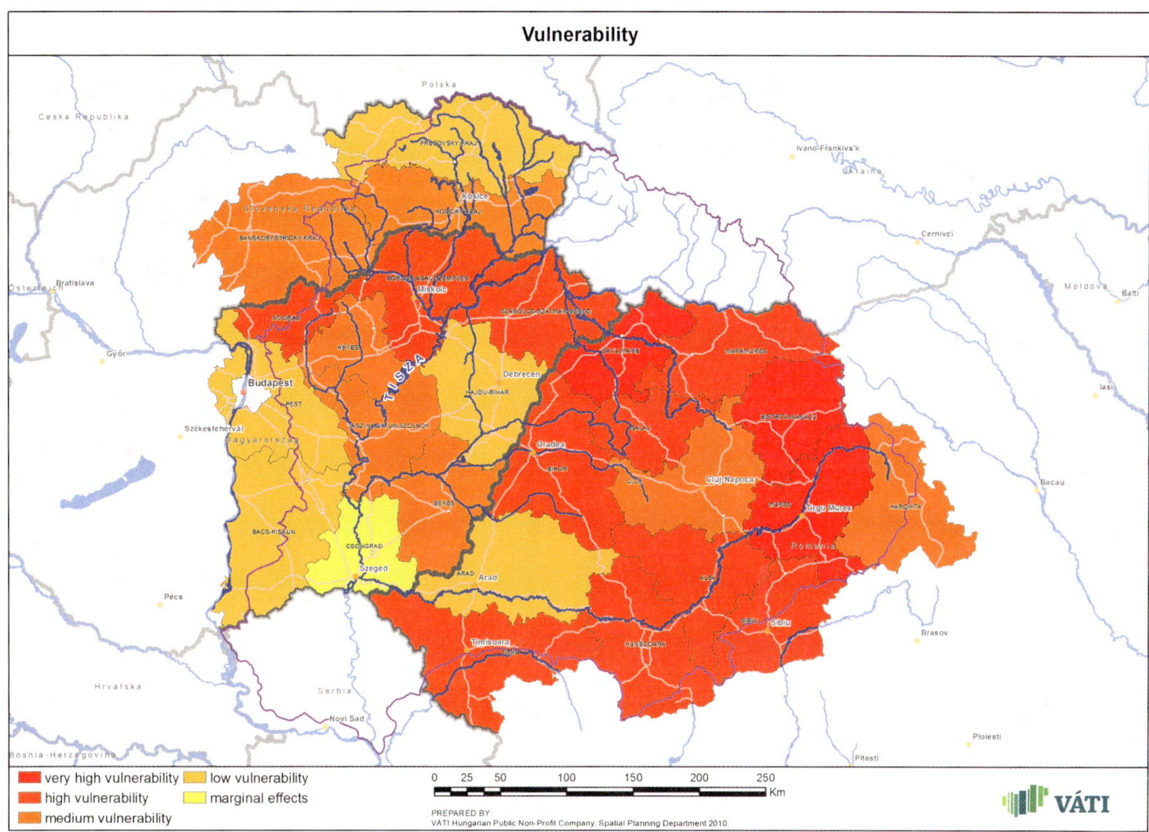

Figure 12.13   Spatial distribution of vulnerability.

12.5.3 *Climate change adaptation strategies of the Slovak Republic*

A *Slovak national climate strategy* has not been developed so far and it lacks a comprehensive approach to the implementation of the EU's legislative package on climate and energy at the national level. Some partial targets are in place – such as reducing CO_2 emissions by 8% from 2008 to 2012 as compared with those of 1990, in accordance with the Kyoto Protocol. A national climate strategy will be elaborated during year 2012, with coordination of works being covered by the Ministry of Transport, Construction and Regional Development of the Slovak Republic (SR).

In 2001 the document *Proposal of Adaptation Measures to Climate Change in Agriculture of SR* was elaborated and approved by Ministry of Agriculture of the SR. Measures are divided into nine groups including, for example, crop production, irrigation, forestry and water management.

However, the climate change issue is addressed in the *National Strategy for Sustainable Development* of the Slovak Republic (adopted by the Government of SR in 2001). To implement the priorities of the Strategy, the *Action Plan for Sustainable Development* for the years 2005–2010 has been developed. The roles of spatial plans in adaptation strategy are the following: spatial planning in the Slovak Republic is a relatively complex set of instruments and methods at the national, regional and local levels, with the emphasis on applying the decision-making power by the self-government and executive authorities at these levels.

12.6 Policy recommendation and implications

In summary, the results of our case study indicate that the area will have a warmer climate, more severe droughts and more frequent and intense extreme weather events. A probable growth in the frequency of droughts is expected due to the considerable reduction in the volume of summer precipitation, and also an increase in the volume and duration of excess water flooding could occur due to an increase in the volume of winter precipitation. The modification of river flood hazards seems to be related to the changing frequency of floods endangering protection structures and/or exceeding their crest.

However, the possible effects of these findings, as well as their importance, vary across the three case study countries – Hungary, Romania and Slovakia. Hungary exhibited damage due to the droughts, river floods, excess water inundations and frosts during the same year, and is likely to suffer from these effects in the future. Extreme weather events such as storms, river floods and droughts are expected in Romania at higher frequencies in the future. Slovakian forests, covering 41% of its territory, will be exposed to increased extreme weather conditions (droughts and wet periods with significant levels of soil moisture).

These phenomena are likely to influence mostly agriculture, resulting into lower productivity if no adaptation measures are taken due to the modifications in vegetation periods and crossover boundaries between forests and pasture land induced by the changes in temperature and precipitation patterns (Slovak Ministry of Agriculture and Development). The importance of these effects varies from one country to another; the example of Romania is noteworthy, where most of the agriculture

is subsistence farming rather than market-oriented activity. Furthermore, along with flooding periods, extended periods of extreme drought produce serious economic losses in agriculture, but also in transportation, energy provision, water management and households.

The overall findings and pictures derived from the case study are, in some cases, different from the ones obtained when assessing the European situation. The differences are, in general, due two causes:

1. **The multi-scale approach** The Tisza River case study could be considered a product obtained by 'zooming' into the overall pattern. While the maps showing the Tisza River area show interior differences, these are lost when the range of values is assessed for the entire European continent, since the values for Tisza area cluster in either the upper or lower portion of the overall range.

2. **The cooperation between the countries involved** In the Tisza River case study, when considering the cooperation between the countries involved, methodological differences were revealed, not only in the use of indicators, but also in the methodology used to compute a specific indicator. Nevertheless, when differences were visible to an extent which suggested that an entire country had values higher or lower than the others, a common methodology was used.

The findings summarised in the paragraphs above suggest that there is need for a multi-level strategy. While the European recommendations should be at the core of such strategies, the national, regional and local differences can only be addressed by special policies and strategies addressed at these particular spatial levels.

The main policy recommendations based on the results of the study can be grouped into the following three sections.

- **Spatial planning**
  - Vulnerability should be assessed in greater depth and extended to cover the entire Tisza River Basin area (Ukraine and Serbia in addition to Hungary, Romania and Slovakia).

- Data on vulnerability should be generated or obtained, and indicators developed.
- A new common adaptation strategy, comprising an Action Plan, should be elaborated and embedded into the national spatial development frameworks.
- A common spatial development master plan should be elaborated and an impact study of the interventions affecting land use prepared.
- **Agricultural policy**
  - Subsidies are to be provided for the aquatic habitats, areas affected by excess water flooding. Landscape management of flood plains is to be in the forefront of agro-environmental management.
  - Recommendations are to be elaborated for promoting the various opportunities of planting and breeding drought resistant species.
  - The farming communities are to be educated to respond more efficiently to climate change.
- **Water management**
  - The synergies between river flood protection and protection from excess water flooding on the one hand and spatial and rural development policies on the other hand should be exploited.
  - River discharge on the protected side (from flood) and the landscape management of flood plains need to be harmonised.

It is very important to define regional possibilities for climate change *mitigation* by introducing measures that help to reduce the GHG production. Although the Tisza River Basin region has prevailingly an agricultural character, the industrial plants are not exceptional here. A broad range of policy instruments needs to be brought into action in order to achieve the emission reductions necessary to deal with continuing global warming. A number of measures have been agreed at an international level. While international instruments are essential elements, they can also be supplemented by measures at the local and regional levels. These include mainly Corporate Social Responsibility programmes and action, support of local municipalities and non-governmental initiatives focused on climate change adaptation and mitigation. Intelligent urban planning, which incorporates the climate change factors into the consideration, is one of the most important existing policies to reduce the emission of greenhouse gases in the transport sector.

In addition, there is a clear need to increase the awareness of politicians in the addressed countries toward climate change issues, so that each country could start developing its own instruments, national and regional, on a spatial premise (delimit the areas at risk and develop specific policies for each of them, working together with the local administrations). This is, after all, an implementation of the principle 'Think globally, act locally'.

References

Aheim, A., Berkhout, F., McEvoy, D. *et al.* (2008) Adaptation to Climate Change: Why is it needed and how can it be implemented? CEPS Policy Brief No. 161, ADAM/CEPS, Brussels.

Atomic Energy Authority (AEA) (2007) Adaptation to Climate Change in the Agricultural Sector, AGRI-2006-G4-05, AEA Energy & Environment and Universidad de Politécnica de Madrid.

Bálint G., Gauzer B., Dobi I., and Mika J. (1995) On hydrological aspects of climate changes based on diurnal simulations, in *Drought in the Carpathian Region*, 3–5 May, 1995, Budapest -Alsógöd, Hungary, pp. 65–77.

Bálint G., Dobi I. and Mika J. *(1996)* Runoff simulation assuming global warming scenarios. 18th (XVIII) Conference of the Danube Countries on Hydrological Forecasting and Hydrological Bases of water Management, Schriftenreihe zur Wasserwirtschaft, Vol. 19/1, Technische Universität Graz, Graz, Austria, 1996.

Bálint, G. Gnandt, B. Hunyady, A. *et al.* (2009) Flood characteristics, *in CLAVIER Final Report*, Max Planck Institute for Meteorology, Hamburg.

Bartholy, J., Pongrácz, R., Torma, C. and Pieczka, I. (2009) Analysis of regional climate change modeling experiments for the Carpathian Basin. *International Journal of Global Warming*, **1** (1–3), 238–252.

Becker, A. and Serban, P. (1990) *Hydrological Models for Water-resource System Design and Operation. Operational Hydrology*, Report No. 35, WMO-No. 740, WMO, Geneva, Switzerland.

CLAVIER (Climate Change and Variability Impact on Central and Eastern Europe) (2009) *WP3b: EXTREMES – Results of WP3b: Extremes.* (online)

Available at: <http://www.clavier-eu.org/?q=node/876#top> (accessed December 2009).

Csete, M. (2008) The synergistic effects of adaptation to climate change on local, sub regional level, in *Proceedings of the Global Conference on Global Warming*, Istanbul, Turkey, 6–10 July 2008, pp.1244–1249.

Csete, M. and Szendrő, G. (2011) Implementation of local sustainability with regard to climate change – Hungarian case study, in *Proceedings of the 17th International Sustainable Development Research Conference*, New York, USA, 8–10 May 2011, pp. 83–85.

EEA (European Environmental Agency) (2006) Corine Land Cover. Available at http://www.eea.europa.eu/data-and-maps/data/corine-land-cover-2006-clc2006-100-m-version-12-2009 (accessed December 2010).

FAO (Food and Agriculture Organisation) (2007) Digital Soil Map of the World, http://www.fao.org/geonetwork/srv/en/metadata.show?id=14116 (accessed December 2010).

Farago, T., Lang, I. and Csete, L. (2010) *VAHAVA Report: Climate Change and Hungary: Mitigating the Hazard and Preparing for the Impacts*, VAHAVA Project, Budapest.

Government of Hungary (2003) Government Decision of new Vásárhelyi Plan 1022/2003, (III.27.)

Government of Hungary (2004) Act LXVII, "Implementation of Vásárhelyi Plan".

Gauzer B. and Bartha P. (2001) árvíziszimulációsvizsgálatok a Tisza Tokaj-Szeged közöttiszakaszán (Flood simulation for the Tokaj – Szeged reach of the Tisza River) VízügyiKözlemények, XXXIII. évf., 4. füzet., Hungarian Statistical Office (2007) Statistical Yearbook, pp. 512–537.

ICPDR (International Commission for the Protection of the Danube River) (2007) Tisza River Basin overview, in *ICPDR* , 2008. *Analysis of the Tisza River Basin 2007 – Initial step toward the Tisza River Basin Management Plan – 2009*. ICPDR, Vienna, pp. 12–19.

IWGCCH (The Interagency Working Group on Climate Change and Health) (2010) *A Human Health Perspective on Climate Change – A Report Outlining the Research Needs on the Human Health Effects of Climate Change*, Environmental Health Perspectives and the National Institute of Environmental Health Sciences, Research Triangle Park, NC.

Lautenschlager, M., Keuler, K., Wunram, C., *et al.* (2009) Climate Simulation with CLM, Climate of the 20th Century (run no. 1 and 2) and Scenarios A1B and B1 (run no. 1, 2 and 3), Data Stream 3: European region MPI-M/MaD. World Data Center for Climate.

NÉS (Hungarian Ministry for Environment and Water) (2007) *Nemzeti Éghajlatváltozási stratégia (National Climate Change Strategy) 2008–2025*. (pdf) Available at: <http://klima.kvvm.hu/documents/14/nes_080219.pdf> (accessed 30 November 2011).

Nováky, B. (2002) Impacts of climate change on water management, in *Strategic Issues of the Hungarian Water resources Management* (in Hungarian with English Summary), (ed. L Somlyódy), Academy of Science of Hungary, ISBN 963 508 333 5, Budapest, pp. 75–106.

Pálvölgyi T., Czira, T., Dobozi, E., *et al.* (2010) Methodology and results of climate change related vulnerability assessment on the level of micro-regions. *"Klíma-21" füzetek*, **62**, 88–101.

Pálvölgyi, T. and Czira, T. (2011) Climate change vulnerability in the level of micro-regions, in (eds M. Bulla and P. Tamás), *Vulnerability and Adaptation: Chances for Resilience*, MTA Research Institute for Sociology, Budapest, pp. 237–253.

Pecho, J. and Fasko, P. (2010) *Current Scientific information: Príde Slovensko o svoj sneh?* (online) Available at: <www.shmu.sk/sk/?page=1744&id=&ocid=156> (accessed 5 December 2011).

Radvánszky, B. and Jacob, D. (2008) A Tisza vízgyűjtő területének várható klímaváltozása és annak hatása a Tisza vízhozamára regionális klímamodell (REMO) és a lefolyási modell (HD) alkalmazásával. *Hidrológiai Közlöny*, **3**, 33–42.

Romanian Ministry of the Environment (2005a) National Plan of Action on Climate Change 2005–2007 (in Romanian), Romanian Ministry of the Environment, Bucharest.

Romanian Ministry of the Environment (2005b) National Strategy on Climate Change 2005–2007 (in Romanian), Romanian Ministry of the Environment, Bucharest.

Romanian Ministry of the Environment (2007) WATMAN – Information System for Integrated Water Management. Description of the project (in Romanian), Bucharest, p. 9.

Romanian Ministry of the Environment (2008a) Guide on Adaptation to Climate Change Effects (in Romanian), Romanian Ministry of the Environment, Bucharest.

Romanian Ministry of Agriculture (2008b) National Strategy on Long-Term Reduction of the Effects of Drought, Prevention and Combat of the Degradation of Lands and Desertification (in Romanian), Romanian Ministry of Agriculture, Bucharest.

SEG (Scientific Expert Group on Climate Change) (2007) *Confronting Climate Change: Avoiding the Unmanageable and Managing the Unavoidable*. Sigma Xi, United Nations Foundation, Durham, Washington DC.

Sipkay, C.S., Kiss, K.T., Vadadi-Fulop, C.S. and Hufnagel, L. (2009) Trends in research on the possible effects of climate change concerning aquatic ecosystems with special emphasis on the modeling approach. *Applied Ecology and Environmental Research*, **7** (2), 171–198.

Slovak Ministry of Agriculture and Rural Development (2001) *Proposal of adaptation measures to climate change in agriculture of SR* (in Slovak). Government Office of the Slovak Republic, Bratislava, 11.04.2001 / No. of the document: 2445/2001/ 49p.

Szépszó, G. and Horányi, A. (2008) Transient simulation of the REMO regional climate model and its evaluation over Hungary. *Quarterly Journal of the Hungarian Meteorological Service*, **112** (3–4), 203–231.

VÁTI Hungarian Nonprofit Ltd. for Regional Development and Town Planning (2011) Tisa Catchment Area Development – Synthesis of the National Analyses. (pdf) VÁTI Hungarian Nonprofit Ltd. for Regional Development and Town Planning, Budapest. Available at: <http://www.see-ticad.eu/index.php?option=com _docman&Itemid=150&lang=en> (accessed December 2010).

Chapter 13

Tourism, climate change and water resources: coastal Mediterranean Spain as an example

David Saurí[1], Jorge Olcina[2], José Fernando Vera[3], Javier Martín-Vide[4], Hug March[1,*], Anna Serra-Llobet[1] and Emilio Padilla[5]

[1]*Autonomous University of Barcelona, Geography Department, 08193 Bellaterra, Spain*
[2]*University of Alicante, Department of Regional Geographical Analysis, 03080 Alicante, Spain*
[3]*University of Alicante, Institute for Tourism Research, 03080 Alicante, Spain*
[4]*University of Barcelona, Department of Physical Geography, 08001 Barcelona, Spain*
[5]*Autonomous University of Barcelona, Department of Applied Economics, 08193 Bellaterra, Spain*

Abstract

Tourism is an activity that is particularly exposed to the effects of climate change. Some tourist modalities (beach, winter sports, etc.)* may show a high vulnerability to global warming and associated processes. Increased temperatures and reduced rainfall might generate a significant reduction in water resources in coastal Mediterranean areas. In this sense, the Spanish Mediterranean coast could become a laboratory for the analysis of the effects of the loss of water resources and their influence on tourism, which represents the main source of income in this part of Spain. This chapter analyses the different levels of exposure and vulnerability to climate change and the reduction of water resources in the tourist areas of the Spanish Mediterranean coast. It also analyses the implementation of mitigation and adaptation measures to climate change (use of alternative water resources such as wastewater and desalination) that can prevent the loss of competitiveness for the tourism destinations in the forthcoming decades.

*Current address Hug March: Open University of Catalonia, Internet Interdisciplinary Institute, 08018 Barcelona, Spain

European Climate Vulnerabilities and Adaptation: A Spatial Planning Perspective, First Edition.
Edited by Philipp Schmidt-Thomé and Stefan Greiving.
© 2013 John Wiley & Sons, Ltd. Published 2013 by John Wiley & Sons, Ltd.

13.1   Introduction

The Mediterranean coast, including the Balearic Islands, is the most important tourist area of Spain and a key pillar of the Spanish economy (Anton Clavé, Rullan Salamanca and Vera Rebollo, 2011). Climate is a fundamental constituent, and is perhaps the key influencing factor in explaining the attractiveness of this area for domestic and international tourists (Amelung, Nichols and Viner, 2007; JRC, 2009). Therefore, climate change must necessarily be an issue of concern for the tourist sector and, more generally, for the well-being of the local and foreign populations (Dubois and Ceron, 2006). In one sense this is not an unknown situation, as one of the most important challenges faced by Mediterranean societies during the last millennia was how to adapt to erratic climate patterns that are best represented by extremes (drought and flood) than by averages. Furthermore, this adaptation needed to tackle a fundamental constraint in this area: under natural conditions water availability tends to be at its lowest in summer when human and non-human (agricultural) demand is at its highest (Olcina Cantos, 2009).

According to the Fourth Assessment Report of the IPCC (2007), average temperatures in the Mediterranean basin may increase substantially during the 21st century. In turn, precipitation may decrease thus limiting the amount of water available. Moreover, economic activities such as tourism may be affected by heat waves and flood episodes as well as by storm surges that damage beaches and may force the construction of costly infrastructures. Of all these potential impacts of climate change, we have chosen to focus on water resources because of their strategic importance for tourism. Water is fundamental for life and economic activity, and tourism is no exception. Furthermore, water availability in quantity and quality remains essential in any tourist package. To give just one example on how water can be critical in this respect, we may recall how the German tourism industry abandoned *en masse* the tourist centre of Benidorm because of the water shortages experienced in this city during 1977 and 1978 (Gil Olcina, 2010). Water is also a lens through which we may examine different adaptation strategies to

climate change (changing consumption habits, 'production' of new resources, etc.). Finally, water can also help us to unravel how adaptation strategies may depend on the ways by which certain economic sectors such as tourism tend to develop spatially (Vera Rebollo, 2006; Rico Amorós, 2007; Baños Castiñeira, Vera Rebollo and Díez Santo, 2010). In other words, how adaptation to water stress induced by climate change varies according to the spatial imprint of different forms of tourist development.

The case study presented in this chapter explores three inter-related themes on the likely impact of climate change on the Spanish tourist industry. First, temperature and precipitation patterns are evaluated in order to unravel future trends regarding the availability of water, especially whether extreme events (droughts and floods) are to be expected. Preliminary results indicate that a reduction in precipitation is likely, especially during the warmest months of the year. The second theme concerns an analysis of the recent development of urban land uses related to tourism (hotels, apartments, villas, campsites, etc.). This analysis is very relevant because exposure and vulnerability to water scarcity change according to each different modality of tourism ('mass' beach hotels versus golf-based resorts, for instance) and because, in turn, each modality adjusts or adapts to these new climatic conditions in different ways. Benidorm, for instance, shows that water consumption per capita in houses and villas may be higher than water consumption in hotels, and much higher than water consumption in campsites (Rico Amorós, Olcina Cantos and Sauri, 2009). The third theme focuses on existing and potential water resources in the study area classified according to their origin. We have assessed the role of water desalination or water re-use as adaptive tools to become 'climate independent'. Finally, our vulnerability assessment will consider both the potential impacts induced by climate related water scarcity, and the adaptation measures taken. We include also the most important results of a survey on public perception of climate change, water resources and impacts in the selected tourist areas in order to contrast the findings of the vulnerability assessment with the views of residents on this matter.

13.2 The study area: a brief presentation

Our study area includes the Spanish Mediterranean Coast from the French border in the North to Gibraltar in the South and also the Balearic Islands. It is the most important tourist area of Spain with some 28 million visitors staying in hotels in 2008. Most tourists prefer beach destinations but certain large cities, such as Valencia and especially Barcelona, are also important poles of tourist attraction (EXCELTUR, 2005; Vera Rebollo and Baños Castiñeira, 2010).

We have divided this area into a number of smaller areas or 'tourist zones' according to the nomenclature of the Spanish Statistical Institute (INE), which provides aggregated data at this scale (see Figure 13.1). In total we have selected 16 'tourist zones', which include 128 municipalities. For the sensitivity analysis, and except for the aggregated data at the scale of the tourist zone, most of the data gathered for the case study have been collected at the municipal level. It has not been always possible to obtain data for the same year, and for some indicators, such as per capita water consumption, not all areas are represented because of the impossibility to obtain water consumption data at the desired scale.

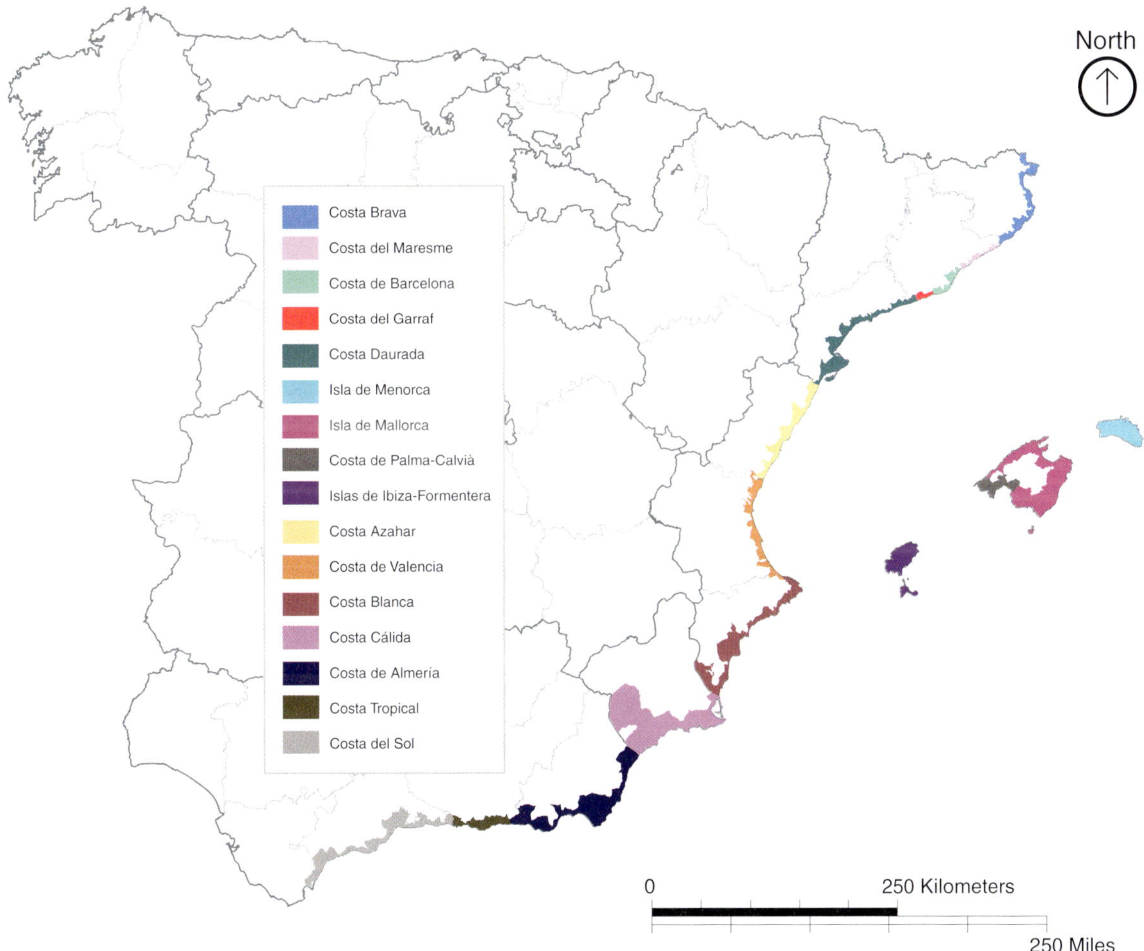

Figure 13.1   Study area and selected tourist zones.

It has not been possible to obtain data on municipal incomes and for this particular variable we have used regional figures at the NUTS 2 level.

The study area hosts the largest tourist enclaves of the Mediterranean, combining worldwide centres of mass tourism (Mallorca, Lloret de Mar, Benidorm, Costa del Sol) with the more recent expansion areas of the so-called residential tourism oriented towards private apartments and houses (Vera Rebollo and Ivars, 2003). Between 2001 and 2007, heavily investments in real estate, by foreigners and Spanish nationals, concentrated strongly in this area. Climate was of course an important factor in the attraction, but buyers also looked at advantages of location, security, promotion systems and commercial networks. The progression of residential tourism can be inferred from data on the total number of places of tourist residential accommodations: Valencia, with 2.6 million of places, is leading Spain in this type of accommodation, followed by Catalonia, with 1.9 million and Andalusia with 1.7 million. Among the factors driving this new tourist model, low-cost airlines have become a crucial factor (Vera Rebollo and Ivars, 2009). Along with vacation homes, a specific aspect of residential tourism is the settlement of European retirees in Mallorca, Alicante or the Costa del Sol (Schröder and Widmann, 2007).

13.3   Methodological outline

Our case study on Mediterranean Spain will use the concepts of exposure, sensitivity, adaptive capacity and vulnerability as defined by the IPCC. The combination of exposure and sensitivity produces impacts, while the combination of impacts and adaptive capacity yields vulnerability as a result. Hence, the main goal of the chapter is to characterise the vulnerability of the tourist areas of Coastal Mediterranean Spain (including the Balearic Islands) to water stress originating from climate change. Next, we expand this methodology and characterise the concepts introduced above.

Total values for each component and subcomponent of vulnerability are obtained after the joint consideration of the value of the specific component or subcomponent, and the weighted value of this component or subcomponent obtained by means of a Delphi Panel constituted by ESPON Committee members. Table 13.1 indicates the weighted values for the major components and subcomponents of the vulnerability matrix.

The next step was to select the dimensions relevant for our case study, and the variables that could represent best these dimensions taking into account data availability and quality at the desired spatial scales. Figure 13.2 shows the variables chosen for each dimension of the vulnerability assessment exercise for our study areas of coastal Mediterranean Spain.

Accordingly, the methodological approach followed in this chapter can be divided into four stages.

1. Definition of the variables used to calculate each dimension and compilation of relevant data (summarised in Table 13.2).
2. Assignment of a weight to each dimension according to the standard values established in the ESPON Climate project and the availability of data. We have assumed an error in the global calculation of the sensitivity due to the lack of pertinent data regarding cultural sensitivity (given a value of '0'), and the lack of data regarding water consumption for the Costa de Azahar tourist area (variable used to calculate environmental sensitivity).
3. Establishment of different ranks for each variable to calculate the value of each dimension for each 'tourist zone'. The **impact** of climate change in the Mediterranean coast of Spain has been calculated taking into account two dimensions: *exposure* and *sensitivity*. While *exposure* has been defined according to the expected decrease in available water resources, *sensitivity* has been calculated as a sum of economic sensitivity (jobs in hotel industry), environmental sensitivity (urban sprawl plus water consumption), physical sensitivity (number of hotels and apartment

Table 13.1   Weighted values of the different dimensions of vulnerability to climate change.

Dimension	Standard deviation	Average	Final value
Cultural sensitivity	4.340 890 85	10.52	**10**
Economic sensitivity	7.123 435 50	24.08	**24**
Environmental sensitivity	9.895 285 07	31.00	**31**
Physical sensitivity	6.849 574 20	18.60	**19**
Social sensitivity	6.720 615 05	15.80	**16**
Sum		100.00	100
Exposure	13.462 912 00	44.00	**44**
Sensitivity	13.462 912 00	56.00	**56**
Sum		100.00	100
Economic	8.122 397 02	20.84	**21**
Knowledge and awareness	8.509 798 27	23.00	**23**
Infrastructure	6.178 457 20	16.44	**16**
Institutions	7.757 362 09	16.48	**17**
Technology	8.135 723 69	23.24	**23**
Sum		100.00	100
Adaptive capacity	14.914 534 30	50.12	**50**
Impact	14.914 534 30	49.88	**50**
Sum		100.00	100

Source: Adapted from Greiving (2011).

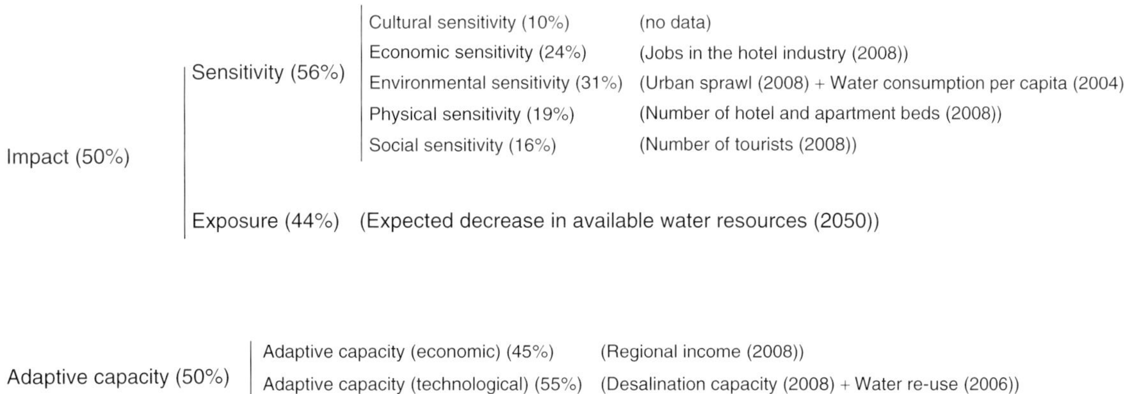

Figure 13.2   Dimensions included in the vulnerability index with the respective weights and the variables used. Adapted from Greiving (2011).

Table 13.2   Variables used for the integrated vulnerability assessment.

Tourist area	Population (2009)	Population growth 2000–2009 (%)	Number of beds in hotels and apartments (2008)	Number of tourists in hotels and apartments (2008)	Average water consumption (litres/person/day) (2004)	Jobs in hotels (2008)	Average income (€) (2008)	Desalination capacity (hm³) (2009)	Treated wastewater re-used (%) (2009)	Low density urbanism (% of urban land) (2006)
Costa Brava	98 249	58.62	69 548	1 697 051	573	6028	27 600	0	82.6	82
Maresme	426 565	23.49	19 865	1 003 349	343.5	3708	27 600	20	1.2	75
Costa Barcelona	2 251 600	7.39	53 787	5 696 838	204	9942	27 600	60	44.4	7
Garraf	27 668	42.27	4651	245 726	436.6	936	27 600	0	0	80
Costa Daurada	198 692	35.07	58 287	1 421 651	474.3	4664	26 900	0	44.2	83
Menorca	93 915	29.15	45 475	534 225	323	3973	24 600	5	66.8	82
Mallorca	69 818	44.20	106 580	1 720 239	676.1	11 604	25 400	5	31.3	66
Palma-Calvià	464 729	22.30	111 254	2 990 224	301.2	14 598	25 400	50	1	60
Ibiza and Formetera	139 114	45.06	84 048	1 296 110	261.8	7486	25 900	15	59.8	92
Costa Azahar	373 984	33.12	41 897	809 603	Not available	2766	24 500	0	27	66
Costa Valencia	894 228	12.04	31 487	1 972 231	430.7	3754	21 800	5	63.9	5
Costa Blanca	471 266	25.32	75 453	2 533 753	277.4	7943	19 600	40	87.1	64
Costa Cálida	370 276	30.85	20 839	568 261	343	2040	19 400	120	52.3	54
Costa Almería	465 954	35.93	39 911	1 212 034	257.9	3619	21 200	100	56.3	28
Costa Tropical	115 848	23.95	7112	231 453	333.8	829	16 900	0	39.7	60
Costa del Sol	1 209 680	26.05	110 924	4 451 018	424.8	15 212	18 200	100	29.9	70
TOTALS	7 671 586	30.92	881 118	28 383 766	377.4	99 102	23 800	520	41.1	61

Sources: MMA (2009), CORINE (2012), EUROSTAT (2012), INE (2012a,b) Water supply companies of the study area.

beds) and social sensitivity (number of tourists). **Adaptive capacity** has been calculated taking into account two dimensions: *economic* and *technological* adaptive capacity, with relative weights of 45 and 55%. Finally, the **vulnerability** is the result of the sum of the impact and the adaptive capacity.

4. Mapping of each dimension. The legend of the summary maps (exposure, sensitivity, impact, adaptive capacity and final vulnerability) is set to represent values between 0 and 1 that indicate qualitative assessments (i.e. 'low', 'medium' and 'high').

13.4   Vulnerability assessment

13.4.1   *Exposure*

According to the IPCC and to the Spanish Meteorological Agency (AEMeT), climate change is likely to produce a general negative effect on the existing water resources of the Spanish Mediterranean Coast. Models indicate that the increase in average temperature (and therefore of evapotranspiration) and especially the reduction in precipitation (up to 20% towards the end of the 21st century) may add important stresses to the population and economic activities, especially agriculture and tourism. The decline in precipitation would be more intense in the southern parts of the study area. Presently (IPCC, 2007), Spain enjoys a resource availability of water per person per year estimated at 3000 m^3, for a demand of 2000 m^3/person/year. At the end of the 21st century, the IPCC estimates that water availability would be reduced to just 450 m^3/person/year, for demands moderately higher than today. It is likely that this reduction will affect the Spanish Mediterranean Coast in particular.

The aim of this section is to provide some figures regarding the future availability of water resources for the tourist sector in our study area. These figures give an indication of likely physical scarcity conditions, which will have to be integrated with sensitivity and adaptive capacity to produce a vulnerability assessment of the Spanish Mediterranean Coast towards water scarcity situations induced by climate change. Our indicators in this case will be those climatic stimuli that are relevant for the estimation of the amount of surface water and groundwater potentially available for supplying the tourist demand of our study area. The climatic stimuli selected are precipitation, temperature and evaporation.

In 2009, the Spanish Meteorological Agency published the report: 'Proyecciones regionalizadas de cambio climático generadas por el proyecto EN-SEMBLES para un escenario de emisiones medio (A1B)', which presented future climate scenarios generated from some European research projects on climate, particularly PRUDENCE (AEMeT, 2009). These projections included data on maximum and minimum temperatures and on precipitation patterns, using ten regional models developed through the downscaling of five global models, under two emissions scenarios, A2 and B2. All these projections yielded rising temperatures throughout the 21st century for the Iberian Peninsula. However, our study area shows temperature increases significantly more moderate than those of the interior of the Iberian Peninsula, because of the influence of maritime effects in regulating thermal conditions.

For the period 2011–2040, emission scenarios show little impact on the increasing temperature, while in the last period (2071–2100) the differences are substantial, with increases more marked for A2 than for B2. On the other hand, there is also a significant pattern within the year (seasonality), with increases much more pronounced in summer than in winter and early spring. Increases in the Spanish Mediterranean Coast for 2071–2100, compared with 1961–1990, are approximately 1.5 to 3 °C in March for the Catalan, Valencia and Cadiz (Andalusia) areas, and more than 7 °C in August in the Balearic Islands and the Catalan Coast. Given these figures, a significant increase in evapotranspiration, very noticeable in summer, is expected. Precipitation presents more mixed results. During the first half of the 21st century there is no clear signal in any direction. However, in the second half of the century average precipitation in Peninsular Spain may decrease between 15 and 30% with respect to the values of 1961–1990.

A more recent update (July 2011) of the Spanish Meteorological Agency climate projections for temperature and precipitation extreme indexes under the intermediate A1B emission scenario, shows increases of the duration of heat waves and the number of hot days in all of our study area throughout the current century. By 2100 the duration of heat waves will increase by almost 30% and up to 40% in the Balearic Islands. Meanwhile, the increases in the number of hot days will reach values between 30 and 40% higher in Catalonia, Valencia, and Murcia, 40% higher in Andalusia and even above this figure in the Balearic Islands. The projections of heavy rainfall suffer from large uncertainties, with no clear trends, except in Catalonia, where a weak increase in heavy rainfall is expected. On average the duration of the typical dry period will increase about 10% by 2100 in the different regions (higher in Andalusia) although the uncertainty is also large. Finally, the number of rainy days shows decreases close to 20% by the end of the century, with the smaller percentages in Murcia and the Balearic Islands and the largest in Andalusia. In summary, drier and hotter extreme conditions are expected in the study area during the next decades.

Several assessments have attempted to predict possible reductions of water flows derived from climate change. Thus, the Spanish Plan for Adaptation to Climate Change of 2006 calculated a reduction of between 5 and 11% for the different river basins in the Spanish Mediterranean Coast towards the middle of the century (MMA, 2006). These figures were included in the review of river basin management plans undertaken by the Ministry of Environment. The results of the application of hydrological models, such as SIMPA, by the Centre for Hydrographic Studies (CEDEX), to climate data also indicate overall reductions in water availability. More specifically, the average runoff in Spain will decrease from 75 to 25 mm at the end of this century under A2 and B2 climate change scenarios.

Some regional agencies have also generated climate and hydrological forecasts. For example, a report by the Catalan Water Agency foresees a reduction in water resources from 495 hm^3 to 420 hm^3 (15%) in Catalonia throughout this century with a more noticeable decrease in the second half (Cunillera *et al.*, 2009). For the central sector of the Spanish Mediterranean Coast, occupied by the Jucar and Segura basins, Quereda, Monton and Escrig (2009) have developed two possible scenarios of water resources for the year 2030 that consider changes in runoff. The first scenario envisages an increase of 1 °C in mean annual temperature and no change in precipitation patterns. Under these climatic conditions, the Jucar and Segura basins would experience a reduction of 12% (some 550 hm^3/year) of their water resources currently estimated in 4596 hm^3/year. The second scenario implies a temperature rise of 1 °C and a reduction of 5% in average rainfall. This reduction would be translated into a runoff of only 57.6 mm, that is, 23% lower than the actual runoff of 74.3 mm. In turn, the water available would fall 22% or over 1000 hm^3/year.

Under both scenarios, but especially the second, the availability of water resources would cover at best 60% of the estimated water demand in 2020, which is estimated at some 7000 hm^3/year according to the Jucar and Segura river basin plans. For Andalusia and according to Corominas (2008) an 8% reduction in available water resources would represent a decrease of water for irrigation of 12% (up to 16% with unregulated irrigation water). These values are expected to double in 2050. Furthermore, the years of severe drought will increase by 7% for the horizon 2027, and 40% in 2050.

Until more detailed figures on water availability reduction may be generated by modelling, it appears reasonable to assume a gradual trend in which water in rivers and aquifers in our study area may decrease by: 5% in the period 2011–2020; 10% for the following two decades; 15% for the decades 2041–2050 and 2051–2060; and at least 20% for the remaining decades of the 21st century. Generally, reductions in runoff will have a latitudinal component. Reductions in summer may double annual reductions while reductions in winter and perhaps in autumn may be minimal, especially in the more Northern basins. For the purposes of this chapter, and taking into account all the estimations presented above, we assume the latitudinal decrease in the availability of water resources appearing in all sources and we quantify this decrease to be: of 5% in the tourist areas of Catalonia and the Balearic Islands (low decrease); of 10% in the tourist areas of Valencia, except for the Costa Blanca (medium decrease); and of 15% in the tourist areas of Costa Blanca, Costa Cálida and Andalusia (high decrease) (see Figure 13.3).

13.4.2   *Sensitivity*

In our case study sensitivity indicators attempt to assess the effects for the tourist sector of a decrease in water availability induced by climate change. We suggest five indicators that fall within the categories of physical sensitivity, environmental sensitivity,

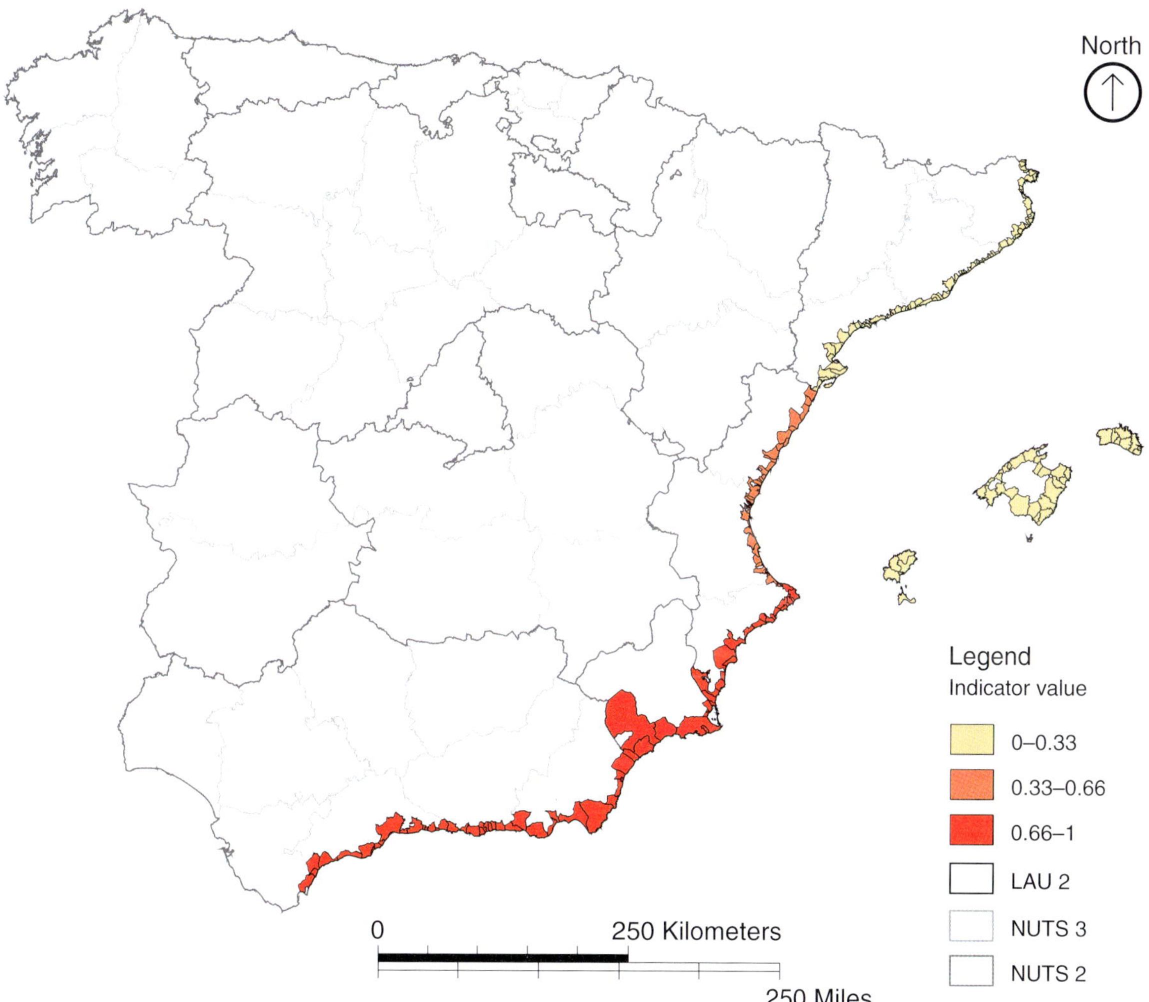

Figure 13.3   Decrease in available water resources around 2050 with respect to average values for 1961–1990. *Source:* Own elaboration.

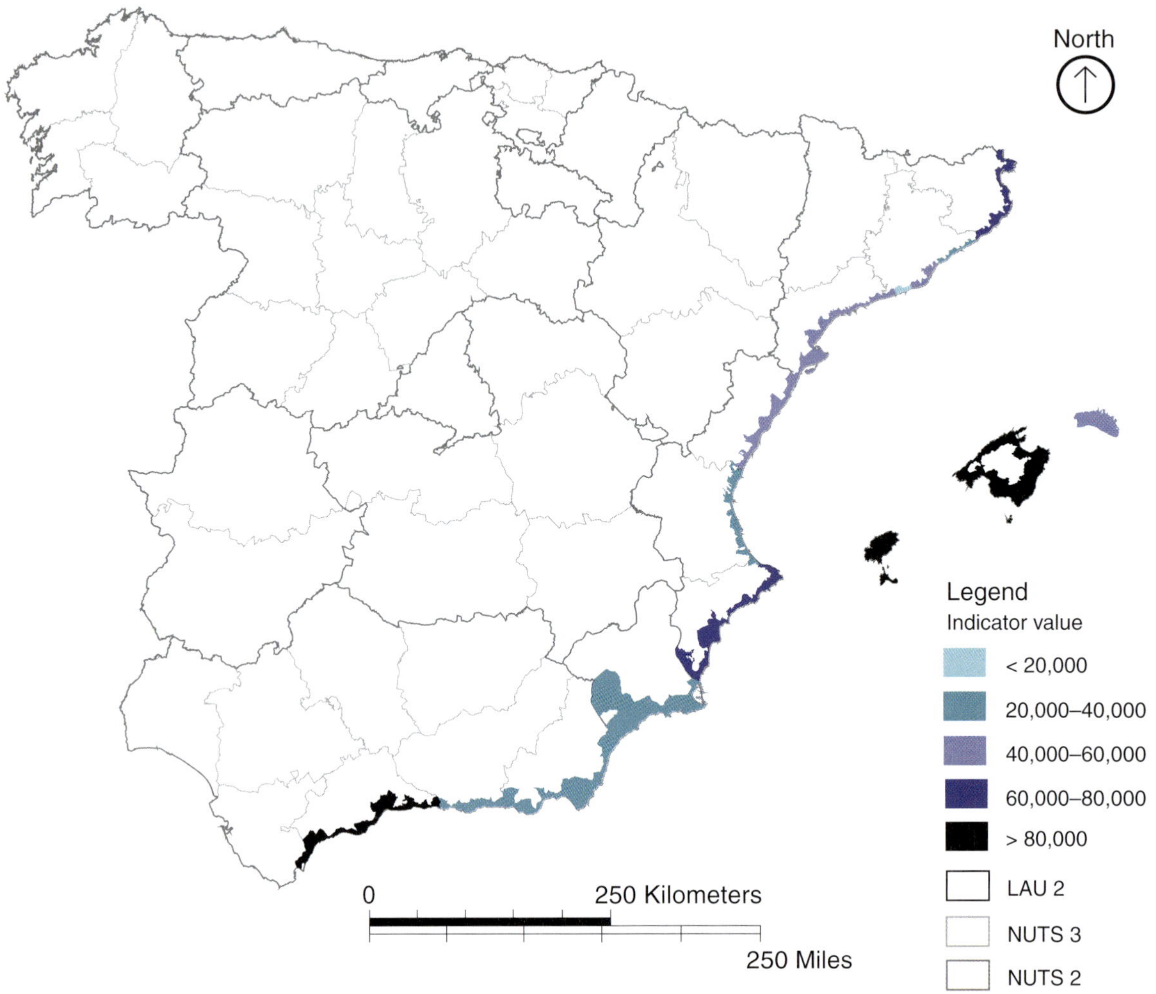

Figure 13.4   Number of beds in hotels and apartments (2008) (**Physical sensitivity**). *Source*: Own elaboration.

social sensitivity and economic sensitivity. With respect to what concerns physical sensitivity, we have chosen one indicator, and that is tourist infrastructure (measured as the total number of hotel and apartment beds in each tourist area). Environmental sensitivity is assessed through the variables urban sprawl (measured as the percentage of disperse urban land over the total urban land) and water consumption (water consumption per capita per day for each tourist area). The indicator chosen for social sensitivity is the total number of tourists in hotels that could be potentially affected by water shortages. Finally, economic sensitivity is measured by the number of jobs in hotels potentially threatened by the decline in tourist numbers that would follow a period or periods of water shortages.

In 2009, more than 9 million people were concentrated in the Spanish Mediterranean Coast (about a quarter of the total population of Spain). The regions of Catalonia, Valencia, Balearic Islands and Murcia, whose more active economic areas are located within our study area, represented nearly 40% of the Spanish GDP in 2008. Considering the

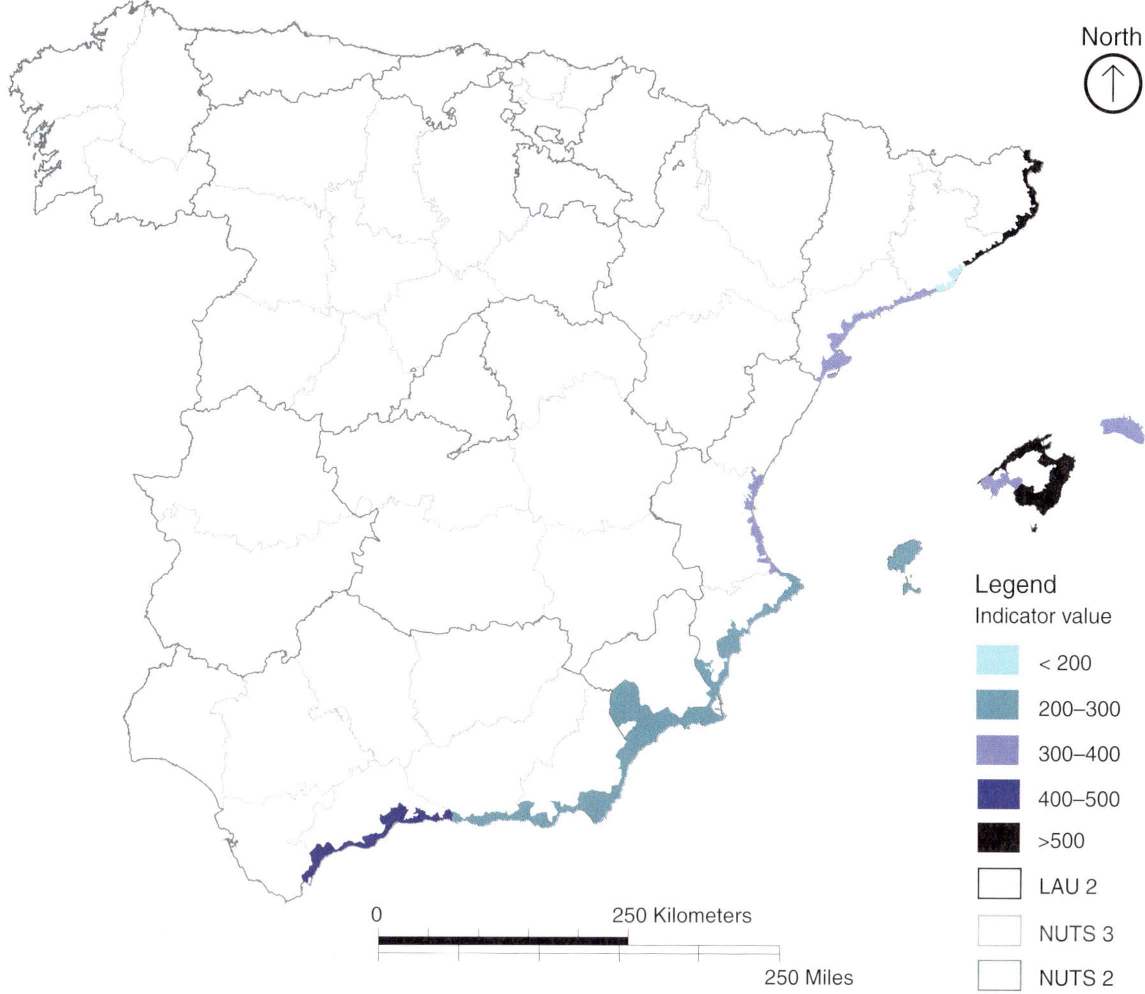

Figure 13.5   Water consumption in the different tourist areas (litres/person/day) (2004) (**Environmental sensitivity**). *Source*: Own elaboration.

population change, the study area observed a 31% increase between 2000 and 2009, mostly attributable to immigration. Population growth in the study area was double that of Spain as a whole, and is between six and seven times higher than that of Europe in the same period.

Physical sensitivity is expressed as tourist infrastructure (number of beds in hotels and apartments). There are some 860 000 hotel and apartment beds in our study area, ranging from more than 217 000 in the Balearic Islands to slightly above 5000 in the Garraf area (south of Barcelona) or in the Costa

Tropical of Granada. Most areas approach or exceed 75 000 beds and a few (Costa del Sol, for instance) exceed 100 000 beds (Figure 13.4). **Social sensitivity** is reflected through the indicator number of tourists in the different tourist areas. Our study area holds the highest tourist concentration of the entire Mediterranean Basin with more than 26 million tourists staying in hotels in 2008. The most important single destination is Barcelona with 5.6 million visitors, but taken together, the Balearic destinations hosted over 6 million tourists in 2008. Regarding **environmental sensitivity**, and with the exception

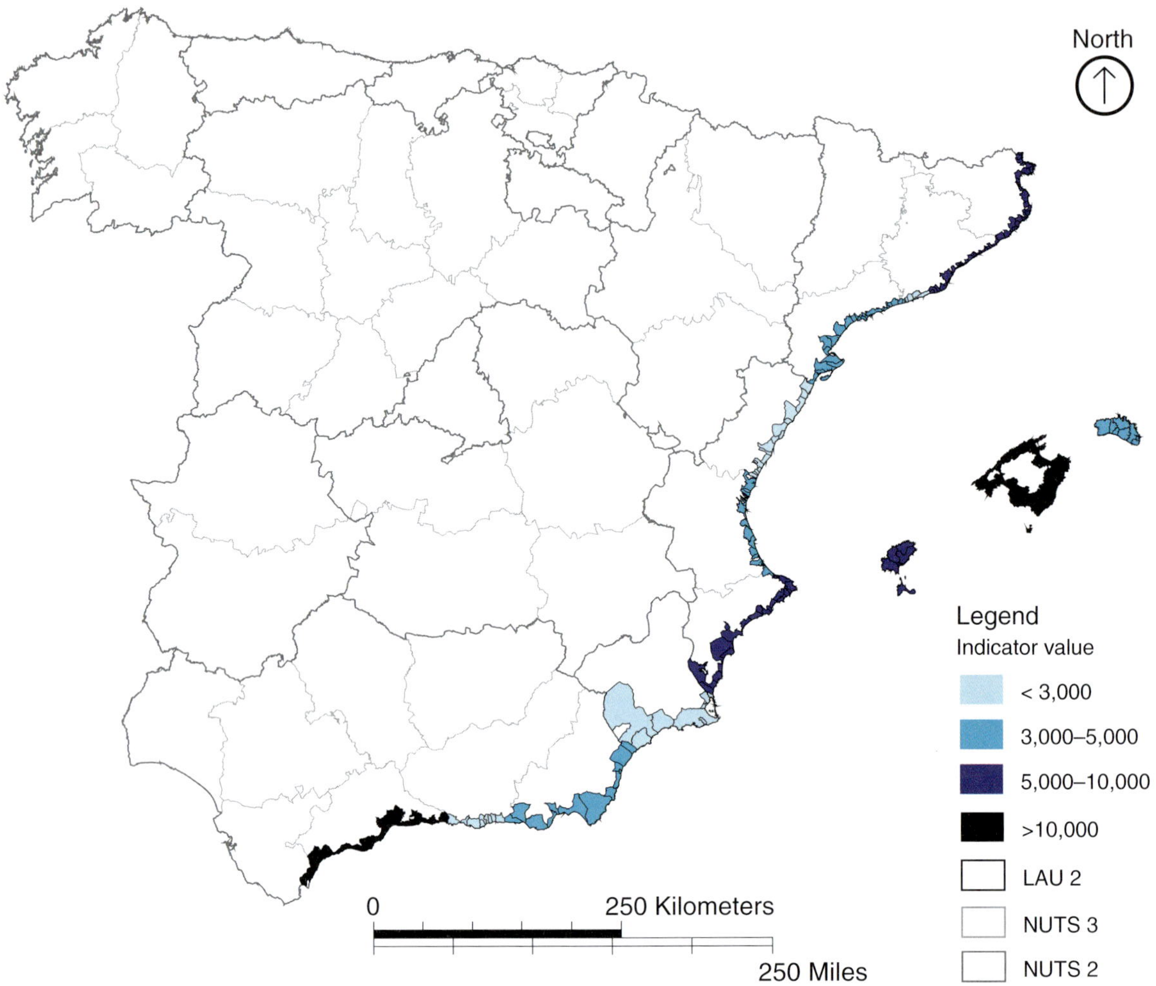

Figure 13.6 Jobs in the hotel industry (2008) (**Economic sensitivity**). *Source*: Own elaboration.

of the consolidated metropolises of Barcelona and Valencia, urban expansion in our study area during the recent decades has mainly taken a sprawled form. Tourist areas such as the Costa Brava, the Costa del Sol and much of the Balearic Islands maintain a disperse pattern of settlement with higher water consumptions per capita than the more compact areas (Figure 13.5). Sprawled areas could therefore show a greater sensitivity to water shortages than compact areas because of larger and more varied water uses. Finally, and regarding **economic sensitivity** as measured through the number of jobs in the hotel industry, there are almost 100 000 jobs in our study area, with important concentrations in the Balearic Islands (over 35% of the total), Costa del Sol, and Barcelona (Figure 13.6). Eventually, in Figure 13.7, we present the combined sensitivity to climate induced water shortages.

13.4.3 Impacts

Impacts result from the combination between exposure and sensitivity (Figure 13.8). In this case, high impacts are only recorded for the Costa del Sol tourist area where high exposure is coupled

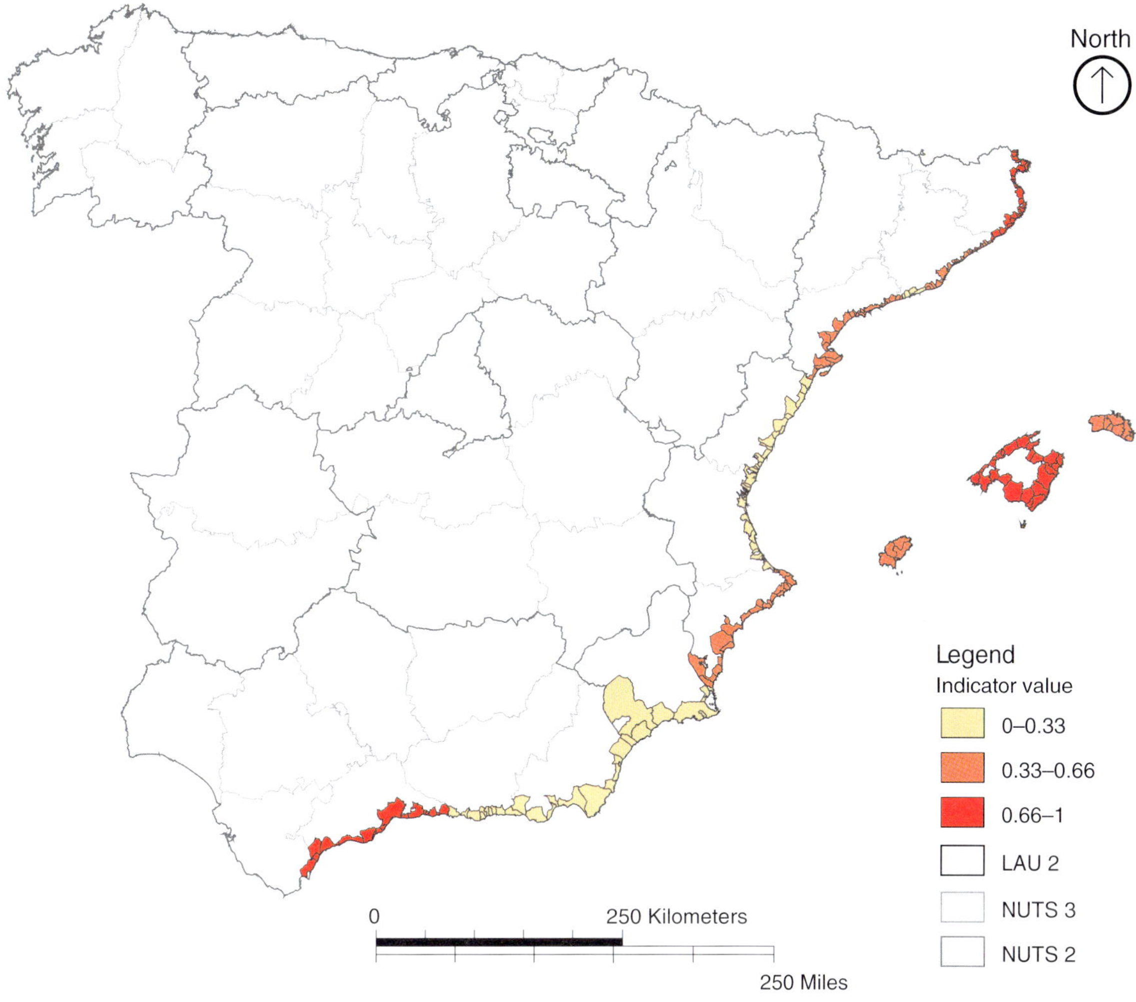

Figure 13.7   Combined sensitivity to climate induced water shortages. *Source*: Own elaboration.

with high sensitivity, as this is one of the most important tourist areas of Spain. The situations for the two Majorcan areas and of the Costa Brava area in the medium category are considered to be high sensitivities since both areas are also at the top of the Spanish tourist industry.

13.4.4   *Adaptive capacity*

Adaptative capacity of the tourist areas to climate change and particularly to the effects of increasing water scarcity will depend strongly on the orientation taken by water planning and management in these areas as well as on the resources available to develop new water management options (e.g. desalination plants). The Spanish Mediterranean Coast has suffered several episodes of water stress with important consequences on the tourist economy. We can recall here the cases of Benidorm in the late 1970s, of Majorca in the mid-1990s, or Barcelona in 2007–2008. Each of these episodes was solved by increasing and diversifying the sources for the supply of water, most notably through the construction of desalination plants in the more

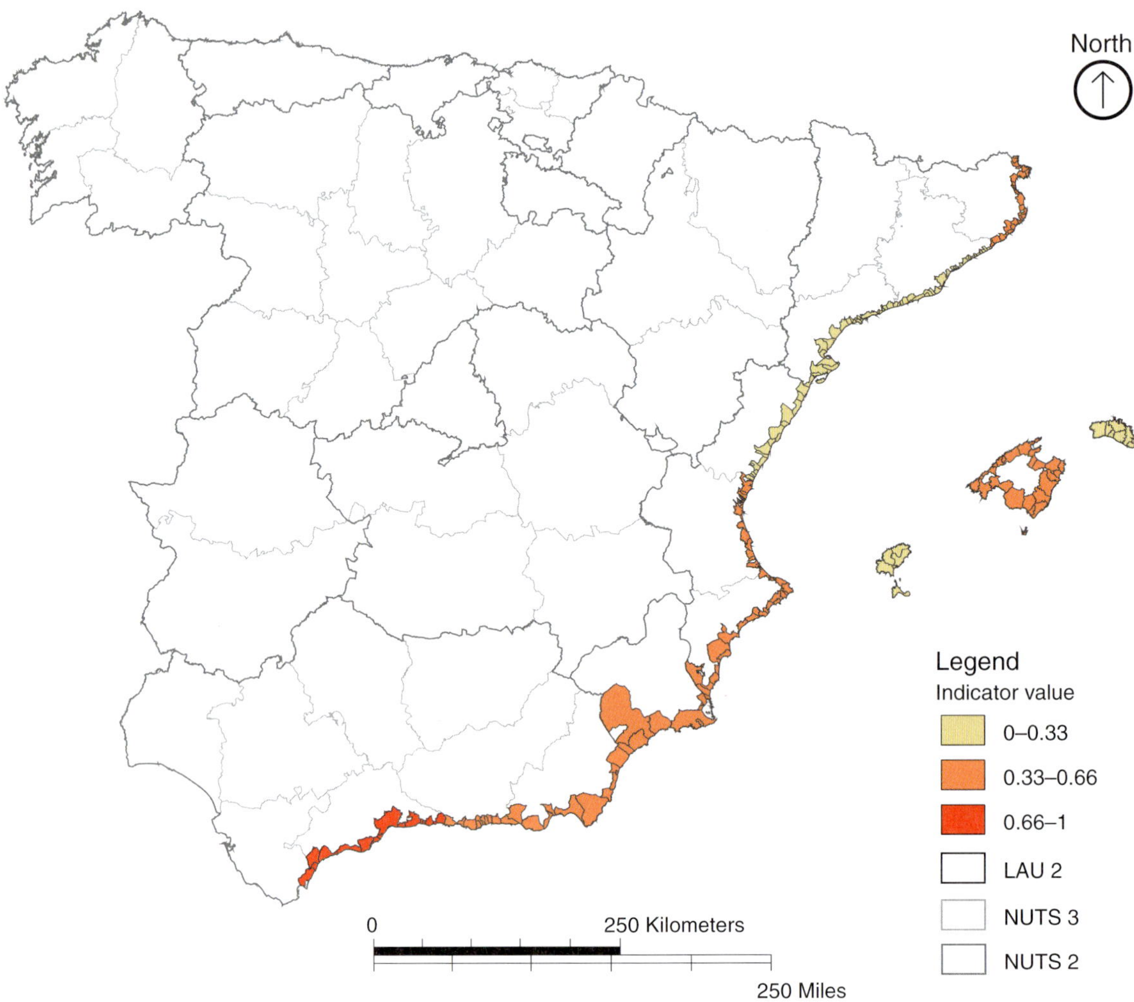

Figure 13.8   Impacts of climate induced water shortages in tourist areas selected. *Source*: Own elaboration.

recent crises. In fact, desalination is projected to expand considerably in the Mediterranean basin during the coming years. The advantage of desalination is that it is climate independent and thus immune to droughts, although vulnerable to energy availability and price. Treated wastewater may also become an important resource for non-drinking water uses, but adaptive capacity can include also actions addressed to improve consumption habits among tourists, such as the installation of efficient water fixtures in hotels or the design of landscaping adapted to local climatic conditions.

In this respect, and with the exception of hotels that may have some environmental certification, efficient water technologies in hotels are relatively scarce.

In our study area, adaptive capacity can also be evaluated through the evolution of daily per capita water consumption, which shows a decreasing trend during the period 2000–2008 in all cases for which there are available data. This decreasing trend may involve a large variety of factors that are not easy to differentiate but that probably include more efficient distribution networks, metering

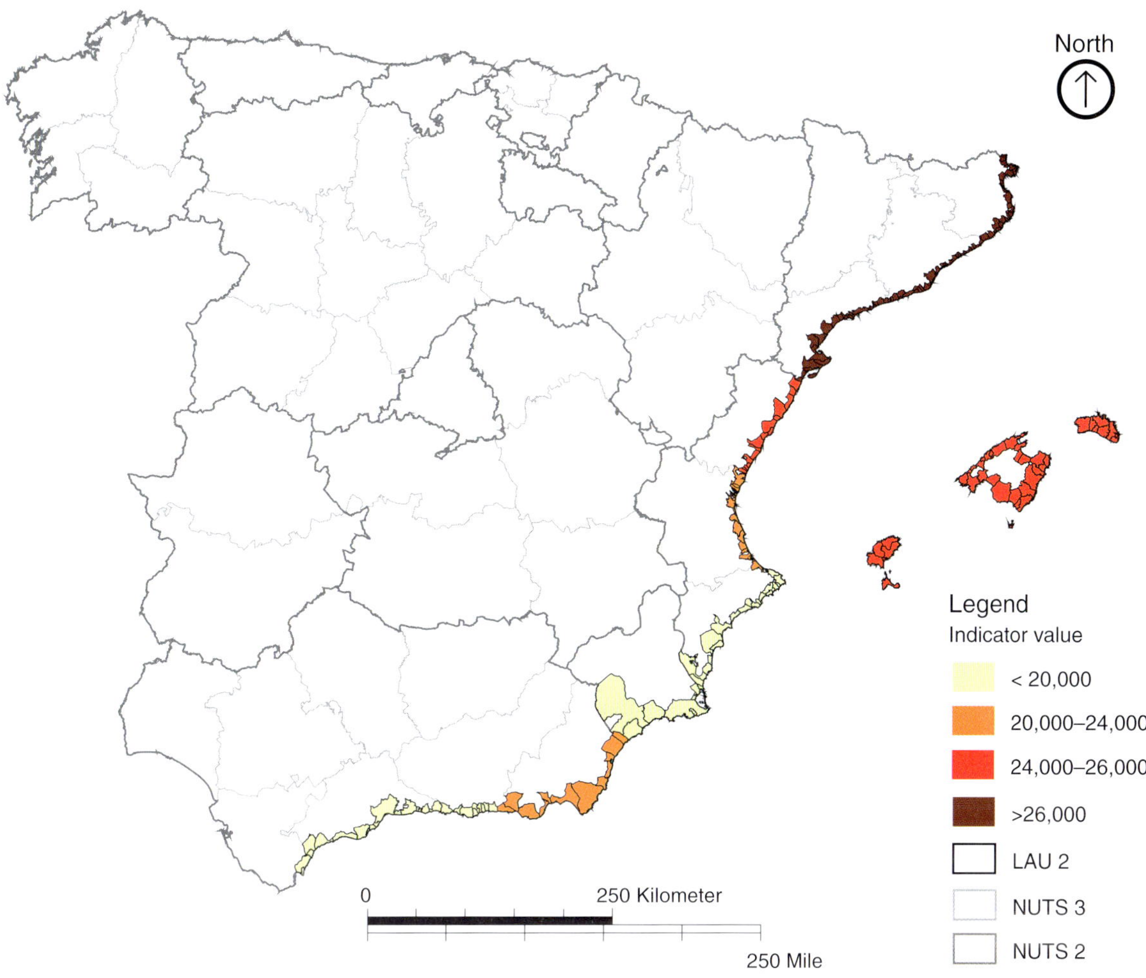

Figure 13.9   Average regional income (€) (2008) (**Economic adaptive capacity**). *Source*: Own elaboration.

and perhaps also better conservation attitudes by citizens, especially during times of drought, and structural demographic trends, such as the higher proportion of immigrants and the elderly in certain metropolitan cores, for example those of Barcelona and Valencia. However, we have considered this variable in the final calculations because of the paucity of data available.

Desalination capacity appears to be oversized in certain cases (Almería and Costa Cálida) and undersized in others (Valencia). The use of treated wastewater represents another interesting form of technological adaptation, and one that has received considerable attention in some of the tourist zones examined. In many areas more than 50% of treated wastewater may be available for certain uses (golf irrigation, for instance) and, in some cases (the Costa Brava and the Costa Blanca) it approaches 90%. Regarding income per capita, which is a standard proxy for economic adaptive capacity, this shows a clear North–South gradient whereby only the tourist areas of Catalonia and the Balearic Islands show incomes higher than the European average. At the other extreme, the Southern areas of Murcia and Andalusia remain well below this average (Figure 13.9).

The highest values for adaptive capacity are recorded for those zones combining relatively high scores in new water technologies with high incomes, whereas the lowest values indicate relatively low incomes and medium to low technological developments (Figure 13.10).

13.4.5 *Vulnerability*

Figure 13.11 shows the final results of our vulnerability assessment to climate induced water shortages for the different tourist areas of the Spanish Mediterranean Coast (see also Table 13.3). It can be seen how the highest values are found in the Costa del Sol, the Costa Tropical and the Costa Blanca areas. The Costa del Sol and the Costa Tropical areas observe high exposures and a relatively low adaptive capacity, mostly because of their relatively low incomes. Furthermore, and given its strong tourist orientation, the Costa del Sol ranks high in sensitivity to impacts. In the Costa Blanca, high exposure is accompanied by a high sensitivity and a relatively low adaptive capacity, again especially in what concerns income. However, the important

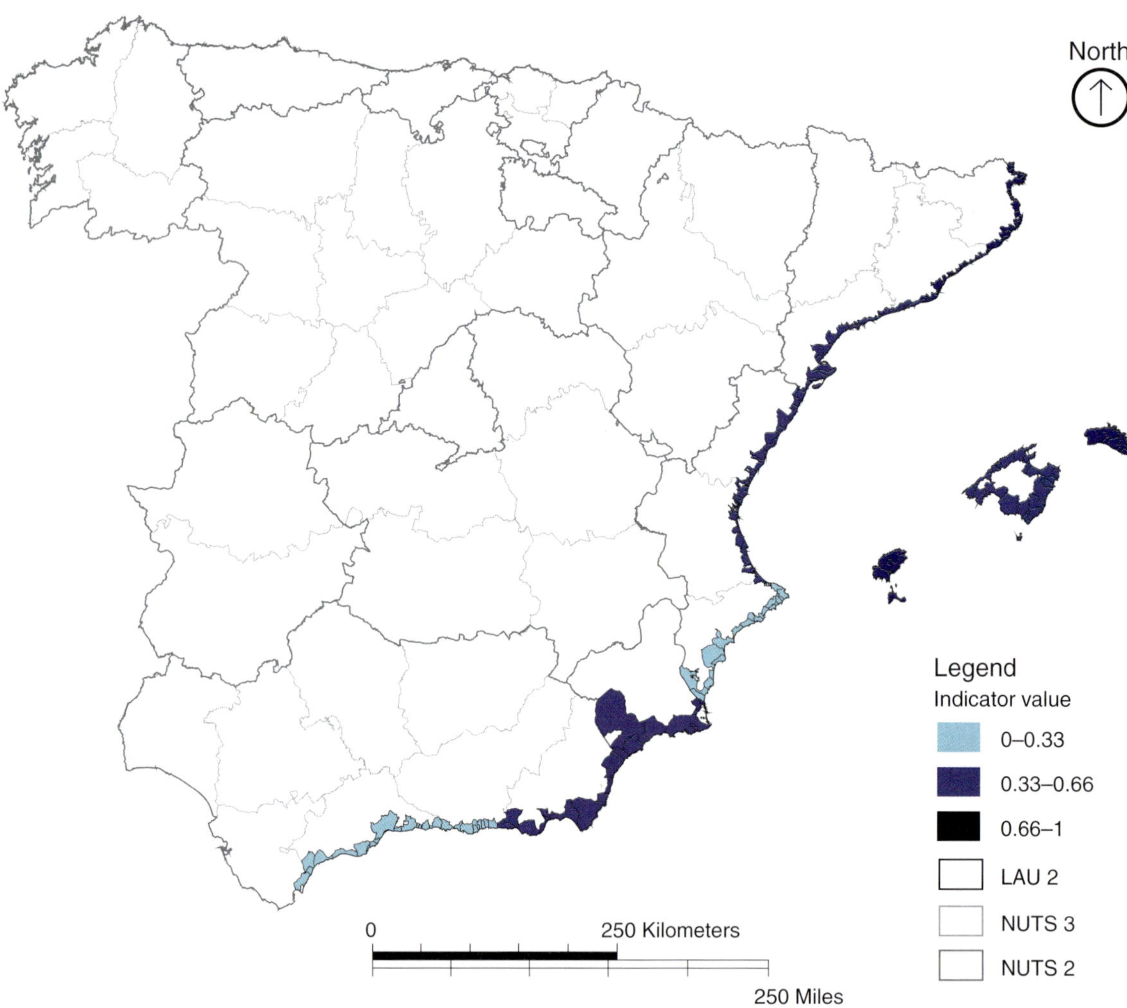

Figure 13.10 Adaptive capacity of the different tourist areas. *Source*: Own elaboration.

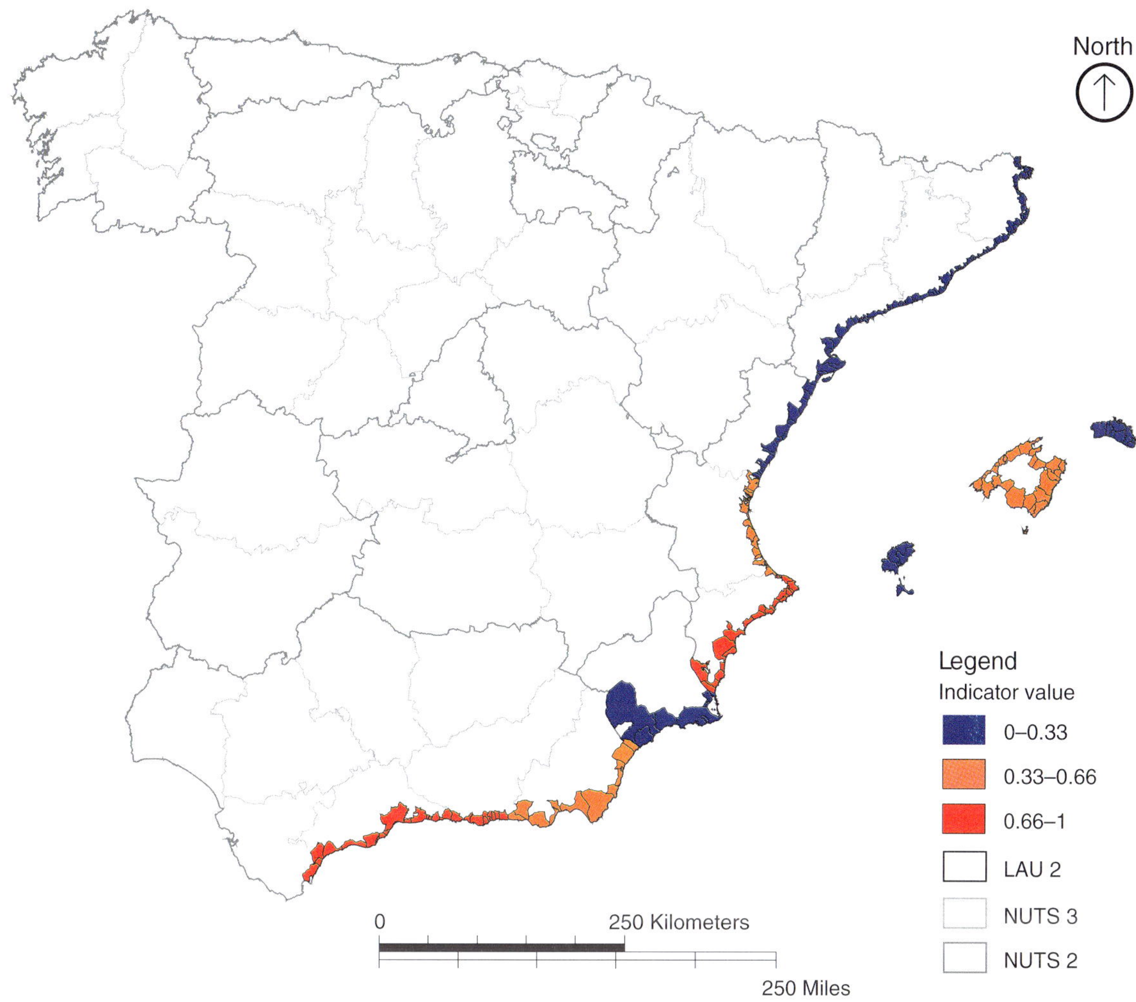

Figure 13.11   Aggregated vulnerability to climate induced water shortages of tourist areas in the Spanish Mediterranean Coast. *Source*: Own elaboration.

presence of desalination and especially of water re-use must also be noted. Medium vulnerability has been found for the tourist areas of the island of Mallorca, for Valencia and for the Costa de Almería and Costa Cálida areas. In these cases, high sensitivities (especially for Mallorca) are combined with medium exposures and medium to high adaptive capacities (note the large desalination capacity of the Costa de Almería area, which is able to offset the low values attained in income). Finally, vulnerability is at its lowest in the tourist areas of Catalonia, and in the islands of Menorca and Ibiza. For this group, low exposure combines with medium to high sensitivity and medium to high adaptive capacity, thus resulting thus in a low vulnerability score.

13.5   Response strategies and policy development

Strategies to manage water scarcity due to climate change in tourist areas should ideally be based

Table 13.3 Qualitative summary of components of the vulnerability assessment to climate induced water shortages in the different tourist areas of the Spanish Mediterranean Coast.

Area	Exposure	Sensitivity	Impact	Adaptive capacity	Vulnerability
Costa Brava	low	high	medium	high	low
Maresme	low	low	low	medium	low
Barcelona	low	medium	low	high	low
Garraf	low	low	low	medium	low
Costa Daurada	low	medium	low	medium	low
Menorca	low	medium	low	high	low
Mallorca	low	high	medium	medium	medium
Palma-Calvià	low	high	medium	medium	medium
Ibiza-Formentera	low	medium	low	high	low
Costa Azahar	medium	low	low	medium	low
Valencia	medium	low	medium	medium	medium
Costa Blanca	high	medium	medium	low	high
Costa Cálida	high	low	medium	medium	medium
Costa Almería	high	low	medium	medium	medium
Costa Tropical	high	low	medium	low	high
Costa del Sol	high	high	high	low	high

on previous vulnerability analyses. One important matter is whether the different European regions will have the capacity to access resources that are to a greater or lesser extent climate independent, such as desalted water or treated wastewater, or to resources traded with other large users, most especially irrigation. However, as indicated earlier, adaptive capacity may include other measures, such as an overall reduction in per capita use in the tourist industry after the implementation of water saving technologies in hotels or of landscaping practices that avoid species with high water requirements.

Regarding the case of the study area, responses to vulnerability to water shortages induced by climate change must also be seen in the light of the National and Regional Mitigation and Adaptation Strategies currently being developed in Spain. The Spanish National State and the different *Comunidades Autónomas* (regional governments) have all issued mitigation and adaptation documents (the latter still very incipient for certain areas). The National Adaptation Strategy of 2006 implied the elaboration of vulnerability assessments for 15 different natural and socioeconomic sectors, four of which (water resources, coastal areas, tourism and urbanism and

the built-up environment) are especially relevant for our case study. Spatial planning, particularly at local and regional levels, is central to the reduction of vulnerabilities in these four areas. Land use regulations, especially those limiting the extent of urban sprawl, can be understood as mitigation tools (for example, in managing urban and tourist growth through urban planning tools, the proliferation of gardens planted with high water consumption species, swimming pools or golf courses may be put under control) or as adaptation (land use regulations may also consider the design of gardens and golf courses using species adapted to the climate). Nevertheless, it is likely that most of the adaptive actions addressed to redress the impact of climate change on the existing water resources in coastal Mediterranean Spain will likely follow the course of new water producing technologies, such as those outlined above (Olcina Cantos and Moltó Mantero, 2010). As to the issues of governance, the requirements of the Water Framework Directive regarding public participation in water planning and management will surely be critical in designing future adaptations of the water sector to climate change. In this sense, one of the possible lessons to be learned from our case study

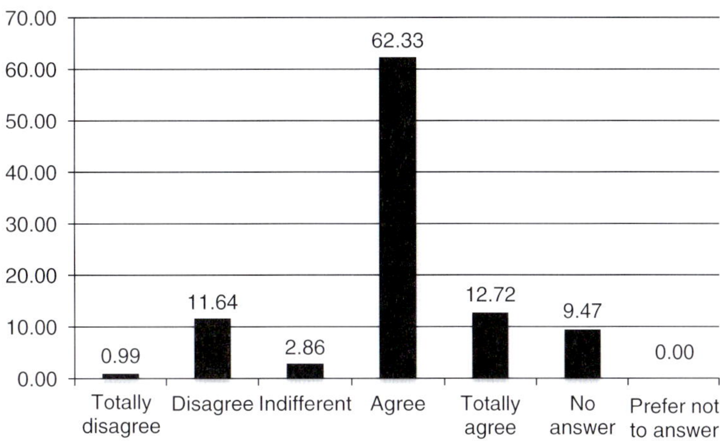

Figure 13.12 'Your residence area will be affected by climate change' (%).

is that new forms of governance (for example, joint management of local or regional water cycles by agricultural and urban interests and exchanges of water rights of different qualities) may have to be developed in order to avoid excessive dependency on new and costly water technologies.

13.6 Further outlook on the issue: public perception of the relationships between climate change, water resources and tourism in the tourist zones

In this final section we will present the main results of a survey on the public perception of the relationships between climate change, water and tourism in the study area. Our sample consisted of 1006 telephone interviews, stratified according to permanent population in the study areas. Interviews took place in September 2011 and involved a total of ten questions formulated through statements to which respondents had to react. Responses were coded using a five-point Likert scale: 'totally agree', 'agree', 'indifferent', 'disagree' and 'totally disagree'. The options 'no answer' and 'prefer not to answer'

were also available. Some of the more important results are summarised in Figures 13.12, 13.13, 13.14 and 13.15. First, it is important to stress that over three quarters of the sample (sum of those answering 'agree' or 'totally agree') consider that climate change will affect their area of residence. Furthermore, over 60% of the respondents agreed or totally agreed with the statement that their area of residence would receive less rain in the future, and only 24% believed otherwise. A staggering 72% agreed or totally agreed with the statement that less water would be available in the future. However, when asked about the repercussions of these trends in their tourist sector, most respondents (over 50%) were confident that climate and precipitation would not damage tourist activities in their areas, versus 36% that believed the contrary. While the use of desalted and treated wastewater could contribute in easing water deficits, overall 53% of respondents felt pessimistic about the future in that climate change would reduce the quality of life in their area of residence.

13.7 Conclusions

The vulnerability assessment to possible water shortages induced by climate change in the Spanish Mediterranean Coast developed in this chapter shows a distinct spatial pattern according to the

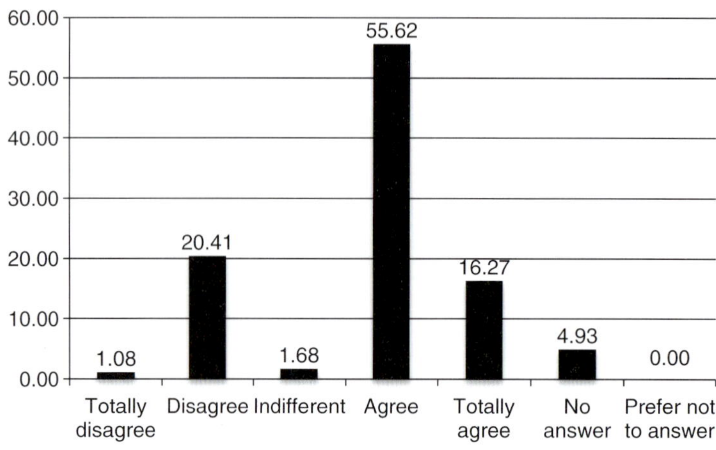

Figure 13.13   'Climate change will reduce water availability for tourist consumption in your area of residence' (%).

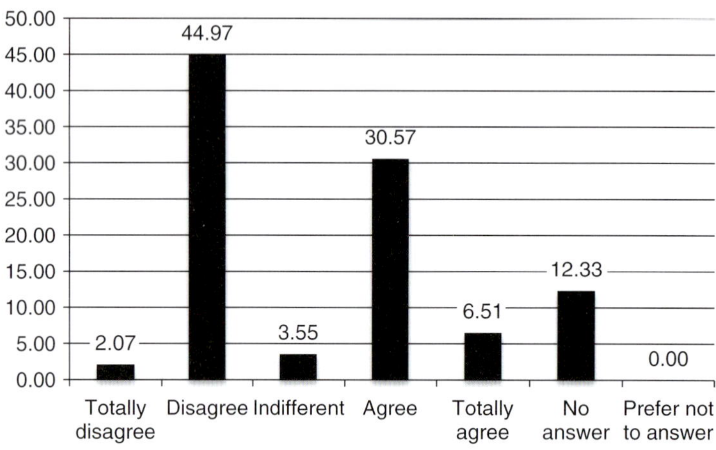

Figure 13.14   'Climate change will decrease tourist employment in your area of residence' (%).

combined dimensions of exposure, sensitivity and adaptive capacity, all weighted according to ESPON Climate experts. Generally, vulnerability tends to increase from North to South, mainly because of increasing exposure and decrease in adaptive capacity (especially in what concerns income) along this gradient. One extreme case is the Costa del Sol (one of the most important not only of Spain but of the entire Mediterranean) where scores for exposure, sensitivity and adaptive capacity combine to produce the highest vulnerability of the study area. At the opposite end, certain areas of Catalonia observe low vulnerabilities after a combination of low exposure and high adaptive capacity. Another interesting case are the Balearic Islands, which rank low in exposure but medium to high in sensitivity, thus indicating the strategic importance of tourism for their economies. Adaptive capacity, however, is in principle high enough to offset sensitivity and the resulting vulnerability is low.

The variables selected and the method chosen may be useful for other tourist areas of the Mediterranean coast. Generally, one could assume an increase in the vulnerability of Mediterranean

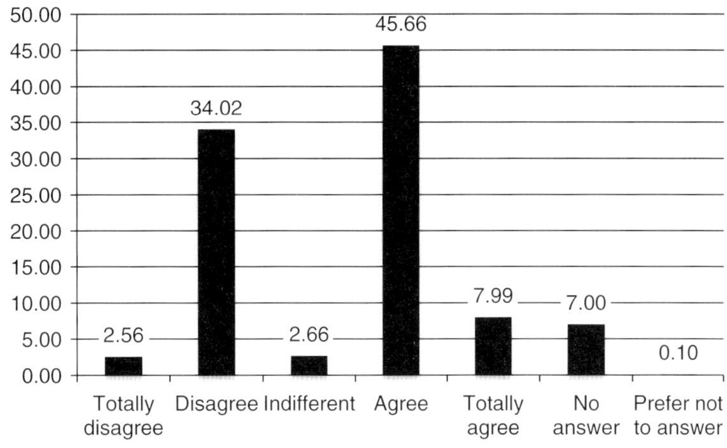

Figure 13.15 'Climate change will reduce life quality in your area of residence' (%).

tourist areas along a gradient West–East due to increasing exposure, perhaps medium to high sensitivity (due to the enormous growth of the tourist industry in certain areas such as the Balkans or the Eastern coasts) and low to medium adaptive capacities, which may change in the future if alternatives such as desalination (already present and growing in many Mediterranean countries) may be successfully implemented. However, sound adaptive capacities should move towards better water demand management in the tourist areas (to a certain extent this is only possible through the management of urban growth). As seen in the case of Spain, this alternative is still in its infancy. Another interesting possibility for increasing the adaptive capacity of Mediterranean tourist areas would be to negotiate with farmers for water either saved by the modernisation of irrigation channels or through exchanges of water of different qualities between agriculture and tourism. Finally, public perception of the matter illuminates a possible contradiction: prospects in terms of water resources may look grim but tourism will (presumably) find ways to overcome the problem.

Acknowledgements

We would like to thank all our colleagues in ESPON Climate, and especially Stefan Greiving and Philipp Schmidt-Thomé, for a very stimulating work atmosphere throughout the project. We also acknowledge the helpful comments provided by referees on previous versions of the chapter, and Anna Marín and Alfons Parcerisas for their input concerning the maps.

References

AEMeT (2009) *Proyecciones regionalizadas de cambio climático generadas por el proyecto ENSEMBLES para un escenario de emisiones medio (A1B)*. (pdf) Madrid: Ministerio de Medio Ambiente. Available at: <http://www.aemet.es/documentos/es/elclima/cambio_climat/escenarios/Informe_Escenarios.pdf> (accessed 8 October 2012).

Amelung, B., Nichols, S. and Viner, D. (2007) Implications of global climate change on tourist flows and seasonability. *Journal of Travel Research*, **45** (3), 285–296.

Anton Clavé, S., Rullan Salamanca, O. and Vera Rebollo, J.F. (2011) Mass tourism development on the Mediterranean coast. *Tourism Geographies: An International Journal of Tourism Space, Place and Environment*, **13** (3), 495–501.

Baños Castiñeira, C.J., Vera Rebollo, J.F. and Díez Santo, D. (2010) El abastecimiento de agua en los espacios y destinos turísticos de Alicante y Murcia. *Investigaciones Geográficas*, **51**, 81–105.

CORINE (Coordination of Information on the Environment) (2012) Corine Land Cover 2006. (online) Available at: <http://www.eea.europa.eu/data-and-maps/data

/corine-land-cover-2006-raster-2> (accessed 10 October 2012).

Corominas, J. (2008) *Los nuevos Planes Hidrológicos de las Cuencas Andaluzas*. (pdf) Madrid: Fundación Botín. Available at: <http://www.fundacionbotin.org/buscador_resultado.htm?q=Los%20nuevos%20planes%20de%20Cuenca%20seg%C3%BAn%20la%20Directiva%20Marco%20del%20Agua> (accessed 10 March 2012).

Cunillera, J., Mas, J., Manzano, A., Prat, N., Munné, A. and Sauri, D. (eds.) (2009) *Aigua i Canvi Climàtic*. Barcelona: Agència Catalana de l'Aigua.

Dubois, G. and Ceron, J.P. (2006) Tourism and climate change: Proposals for a research agenda. *Journal of Sustainable Tourism*, **14** (4), 399–415.

EUROSTAT (2012) *GDP at regional level*. (online) Available at: <http://epp.eurostat.ec.europa.eu/statistics_explained/index.php/GDP_at_regional_level> (accessed 10 October 2012).

EXCELTUR (2005) *Impactos sobre el entorno, la economía y el empleo de los distintos modelos de desarrollo turístico del litoral mediterráneo español, Baleares y Canarias. Executive Summary*. Madrid: Deloitte and Area de Estudios e Investigaciones de Exceltur.

Gil Olcina, A. (2010) Optimización de recursos hídricos y armonización de sus usos: el Consorcio de Aguas de la Marina Baja. *Investigaciones Geográficas*, **51**, 165–183.

Greiving, S. (ed.) (2011) *ESPON Climate–Climate Change and Territorial Effects on Regions and Local Economies in Europe. Final Report*. Brussels: European Commission, ESPON.

INE (Instituto Nacional de Estadística) (2012a) *Cifras de Población y Censos Demográficos*. (online) Available at: <http://www.ine.es/inebmenu/mnu_cifraspob.htm> (accessed 10 October 2012).

INE (Instituto Nacional de Estadística) (2012b) *Hostelería y Turismo*. (online) Available at: <http://www.ine.es/inebmenu/indice.htm#36> (accessed 10 October 2012).

IPCC (2007) *Climate Change 2007: Synthesis Report. Contribution of Working Groups I, II and III to the Fourth Assessment Report of the Intergovernmental Panel on Climate Change*. (online) Geneva: IPCC. Available at: <http://www.ipcc.ch/publications_and_data/publications_ipcc_fourth_assessment_report_synthesis_report.htm> (accessed 10 December 2011).

JRC (Joint Research Centre) (2009) *The PESETA project. Impact on climate change in Europe*. (online) Available at: <http://peseta.jrc.ec.europa.eu/> (accessed 10 December 2011).

MMA (Ministerio de Medio Ambiente) (2006) *Plan Nacional de Adaptación al Cambio Climático*. (pdf) Madrid: MMA. Available at: <http://www.magrama.gob.es/es/cambio-climatico/temas/impactos-vulnerabilidad-y-adaptacion/pna_v3_tcm7-12445.pdf> (accessed 8 October 2012).

MMA (Ministerio de Medio Ambiente) (2009) *Desalación*. (online) Available at: <http://servicios2.marm.es/sia/visualizacion/lda/recursos/noconvencionales_desalacion.jsp> (accessed 8 October 2012).

Olcina Cantos, J. (2009) Cambio climático y riesgos climáticos en España. *Investigaciones Geográficas*, **49**, 197–220.

Olcina Cantos, J. and Moltó Mantero, E. (2010) Recursos de agua no convencionales en España. Estado de la cuestión, 2010. *Investigaciones Geográficas*, **51**, 131–163.

Quereda, J., Monton, E. and Escrig, J. (2009) *Evaluación del cambio climático y de su impacto sobre los recursos hídricos en la cuenca del Júcar*. Valencia: Generalitat Valenciana, Fundación Agua y Progreso.

Rico Amorós, A.M. (2007) Tipologías de consumo de agua en abastecimientos urbano-turísticos de la Comunidad Valenciana. *Investigaciones Geográficas*, **42**, 5–34.

Rico Amorós, A.M., Olcina Cantos, J. and Sauri, D. (2009) Tourist land use patterns and water demand: evidence from the Western Mediterranean. *Land Use Policy*, **26**, 493–501.

Schröder, A. and Widmann, T. (2007) Demographic change and its impact on the travel industries: Oldies-nothing but goldies, in (eds R. Conrady and M. Buch) *Trends and Issues in Global Tourism 2007*. Berlin: Springer, pp. 3–17.

Vera Rebollo, J.F. (2006) Agua y modelos de desarrollo turístico: la necesidad de nuevos criterios para la gestión de los recursos. *Boletín de la Asociación de Geógrafos Españoles*, **42**, 55–178.

Vera Rebollo, J.F. and Ivars, J.A. (2003) Measuring sustainability in a mass tourist destination: pressures, perceptions and policy responses in Torrevieja, Spain. *Journal of Sustainable Tourism*, **11** (2/3), 181–203.

Vera Rebollo, J.F. and Ivars, J.A. (2009) Spread of low-cost carriers: tourism and regional policy effects in Spain. *Regional Studies*, **43** (4), 559–570.

Vera Rebollo, J.F. and Baños Castiñeira, C.J. (2010) Renovación y reestructuración de los destinos turísticos consolidados del litoral: las prácticas recreativas en la evolución del espacio turístico. *Boletín de la Asociación de Geógrafos Españoles*, **53**, 329–353.

Chapter 14

Sensitivity analyses of the ESPON Climate framework, on the basis of the case study on flooding in the Netherlands

Joost M. Knoop, Arno Bouwman and Hans Visser
PBL Netherlands Environmental Assessment Agency, A. van Leeuwenhoeklaan 9, 3721 MA Bilthoven, The Netherlands

Abstract

The ESPON Climate framework could play an important role in the EU's intended shift towards an adaptive climate policy. However, uncertainties about future developments as well as uncertainties caused by choices made in the framework's construction will make this application not as straightforward as one would hope. Often politicians and policymakers in particular find it difficult to deal with this phenomenon. Therefore, it is the task of scientists to provide tailored information, among other things, about the range of uncertainties and how to deal with them as responsibly as possible. This process starts by gaining insight into the potential sources of uncertainty. In this context, we analysed the sensitivity of one of the aggregated indicators – the potential impact of climate change on the Dutch NUTS3 regions – to several choices made in the construction of the framework, on the basis of a case study on flooding in the Netherlands.

The output of the framework was shown to be highly sensitive to the choice of flood map, hence to the choice of flood model used to calculate these maps. The basis of the framework development was the use of pan-European databases and models in order to avoid spatial differences due to variations in the quality of local information. The drawback of this approach is that very often more crude assumptions and simple models have to be used, such as two European-scale flood models (the DIVA coastal flood model and the LISFLOOD river flood model). A spatial comparison with calculated potential impacts, based on a more sophisticated Dutch flood modelling approach, showed a correlation of only 0.61, which became even less when results were classified (0.51–0.59, depending on the classification scheme used). The correlation was even worse (0.25) for the classes containing the estimated highest impacts. For the case of flooding in the Netherlands, the framework was not sensitive to choices with respect to the selection and weighting of the indicators reflecting the flood sensitivity, nor for the stage of normalisation in the framework. The choice of the risk concept (multiplicative or additive) had some impact, especially when the Dutch flood model was applied. This impact reduced to zero, however, if the results were classified according to an even distribution of the output instead of a classification based on equidistant classes. We also established a slight sensitivity of the output to extreme high-exposure values.

14.1   Introduction

The EU faces a challenging task to recover from a deep financial crisis, and at the same time switch to a low-carbon and adaptive climate policy, as stated in the main report of the ESPON Climate project (IRPUD, 2011). In 2009, the EU published a White Paper on *Adaptation to Climate Change*, with the aim of integrating adaptation into EU policies, such as the existing EU cohesion policy (Commission of the European Communities, 2009). The effect of climate change is likely to magnify geographical differences in competitiveness within the EU. The EU cohesion programmes are targeted to reduce these differences. The ESPON framework may play a role in the decision-making framework of this policy by showing expected regional differences in future vulnerability.

The ESPON framework is a science-based modelling framework, producing science-based results. Scientists and policymakers, however, use different mind frames (conceptual images, values, starting points, mental models), which easily lead to different interpretations (Wardekker *et al.*, 2009). Communication between scientists and policymakers, therefore, plays an essential role in projects such as ESPON Climate. Although misinterpretation or even misuse cannot be avoided, if only because of conflicting interests of politicians and policymakers, scientists have to provide the policy community with useful information; not only about what their research results may indicate, but also about what they do not. In particular, the so-called composite indicators, such as the vulnerability indicator estimated according to the ESPON Climate framework, have the advantage of providing the big picture at a glance, but they are difficult to produce in a sound and transparent way and bear the risk of misinterpretation (Saltelli *et al.*, 2004). The major part of the communication between scientists and policymakers, not least in the field of adapting to climate change, deals with uncertainties (as also mentioned in the ESPON Climate report (IRPUD, 2011). In its consecutive rounds, the IPCC has given this subject increasing attention, most visibly by natural scientists and to a lesser extent and in a very different way by social scientists (Swart, Bernstein, Ha-Duong and Petersen, 2009).

In order to facilitate communication between scientists and policymakers, one has to distinguish between the different aspects of uncertainty. Among others, Janssen *et al.* (2005) and Dessai and van der Sluijs (2007) distinguish between uncertainties induced by the judgement of experts and those by the model. The latter relates, for example, to model uncertainties caused by the structure, parameters or the inputs chosen. From a more operational point of view, these uncertainties can be divided according to three different levels: statistical uncertainties, scenario uncertainties and recognised ignorance (Dessai and Van de Sluijs, 2007). Statistical uncertainty concerns all sources that can be expressed statistically; for example, with ranges and probabilities. This category includes model uncertainty, among other things expressed in parameter estimation, but also with conceptual choices made in the way the real world is mirrored by the model itself. Scenario uncertainty concerns uncertainties about future developments, which almost by definition cannot be expressed adequately in statistical terms, for one thing because estimating probabilities is close to impossible. Recognised ignorance deals with known (theoretical) mechanisms that may cause or accelerate adverse effects, but cannot be established in any useful way; for instance, because there is no shred of evidence of them already occurring on any significant scale (e.g. the possible collapse of the West Antarctic Ice Sheet (Bindschadler, 1998)).

Because of their different limitations, these three levels are commonly dealt with in different ways. Additional research and monitoring is a common approach to reduce statistical uncertainties, yet even at this level uncertainties will always remain, which is by definition the case when looking into the future. For both the first and second class of uncertainties it is useful, therefore, to estimate their potential effect on the estimates of key impact variables, such as exposure and vulnerability. Ranges of possible outcomes estimated on the basis of possible future developments, such as the SRES scenarios used by the IPCC, are therefore the most commonly used way to express these uncertainties. Combined with physical models, these scenarios yield ranges of possible exposures and vulnerabilities as the most far-reaching way of expressing these uncertainties. Flooding, modelled according to this combination of scenarios and, in this case, global and regional circulation models, is shown to have a large impact (Dankers and Feyen, 2009; Veijalainen *et al.*, 2010). Yet both studies (Dankers and Feyen, 2009; Veijalainen *et al.*, 2010) used only one single flood model to estimate the exposure. Finally, the third class of uncertainties – recognised ignorance – is difficult to manage. These uncertainties have to be kept in mind all the time. Continuous monitoring, such as the size and movement of ice shields, is recommended by most.

The ESPON Climate framework is designed to estimate the differences between Europe's NUTS3 regions for the aggregated vulnerability increase

towards all possible impacts of all possible future changes in climate or climate induced changes, expressed according to a one to five classification. The above mentioned possible sources of uncertainties can play a role in nearly every stage within the ESPON Climate framework where specific choices are made with respect to the indicators, the structure and the input data. The input data often consist of other model outputs. This chapter analyses a number of possible sources of uncertainty of the ESPON Climate framework, on the basis of impact estimates for flooding in the Netherlands. Included in the analysis is the extent to which this geographical distribution of classes depends on choices made with respect to the inputs and to the framework calculation scheme for a selection of factors.

The first uncertainty to investigate in the field of impact from climate change would be the uncertainty in driving forces, such as rainfall and temperature induced by a combination of scenario uncertainties (caused by different views on future developments) and model uncertainties (caused by different views on relevant mechanisms, calibrated on sets of data which, although vast, are still limited). Within the framework of the ESPON Climate project, however, this information was not available. Therefore, we concentrated on other possible sources of errors. Still, a brief discussion is provided on ways to translate these types of uncertainty analyses into an adaptive strategy in the final section.

If the developed framework is applied to the case study of flooding in the Netherlands, the most specific potential sources of errors will be the choice of the flooding scenario used as model input. Therefore we will start with an extended analysis of the characteristics of the different approaches in this field. Also, choices regarding impact indicators and the way impact is calculated may influence the outcome. Finally, the potential impact of extreme values on the final distribution of the impact indicator is addressed.

Impact as a function of exposure and sensitivity, as defined within the ESPON Climate framework (IRPUD, 2011), can be regarded similar to risk; generally expressed as a function of hazard, exposure and vulnerability.[1] The term exposure used in our framework equals the hazard part of the risk concept and sensibility equals the part on vulnerability. Important, however, is the choice of whether risk is expressed as an added function of its different components, as is often done with risk mapping (De Bruijn, 2007), or as the product of hazard and vulnerability (or hazard, exposure and vulnerability, when regarded as different components of risk). Both approaches have advantages as well as disadvantages. The impact of this choice on the final results also was analysed.

As an introduction, a brief description of the Dutch situation is given with respect to flooding and flood protection policy, including the role of flood modelling.

14.2 Flood protection in the Netherlands

14.2.1 *The Netherlands*

Around 59% of the land area of the Netherlands, theoretically speaking, is susceptible to flooding by the sea, large lakes or rivers. The largest part of this

---

[1]Note that in this sense, vulnerability expresses the impact that exposure may have on assets or humans, and therefore has a different meaning than the vulnerability indicator of the ESPON Climate framework. It should also be noted that within the ESPON Climate framework the hazard is part of the exposure indicator, whereas in risk modelling these terms are usually separated.

fraction (55% of the 59%) is protected by embankments or dunes, including the 26% of the Dutch land area that lies below sea level. The remaining unprotected 4% of the flood-prone area is mainly above sea level. Although most parts of this unprotected area frequently become flooded, the effects remain very limited, as flooding of these areas always occurs very gradually. The protected land areas house around 70% of the total Dutch population and provide an equal percentage of the gross national product. Yet not all areas are equally vulnerable. The most densely populated areas and those with fixed assets that are located in the vicinity of large bodies of open water face the greatest risks.

The Netherlands has a long history of coastal and fluvial flooding; for example, the large coastal floods in the 17th and 20th centuries (Figure 14.1; Noordhoff, 2012). In 1953, the last major coastal flood disaster, caused by an unfortunate combination of spring tide and a devastating storm from the north-west, occurred in the south-western part of the country, with over 1700 casualties. After this disaster, the first Delta Committee was established to outline a new flood protection policy. Since then, the Dutch flood protection policy, based on the flood safety standards proposed by the first Delta Committee, has been geared primarily to the prevention of flooding and, therefore, is based essentially on managing the chances of flooding (Ten Brinke, Bannink and Ligtvoet, 2008).

In the decades following the 1953 flood disaster, many inlets in the South–West of the Netherlands

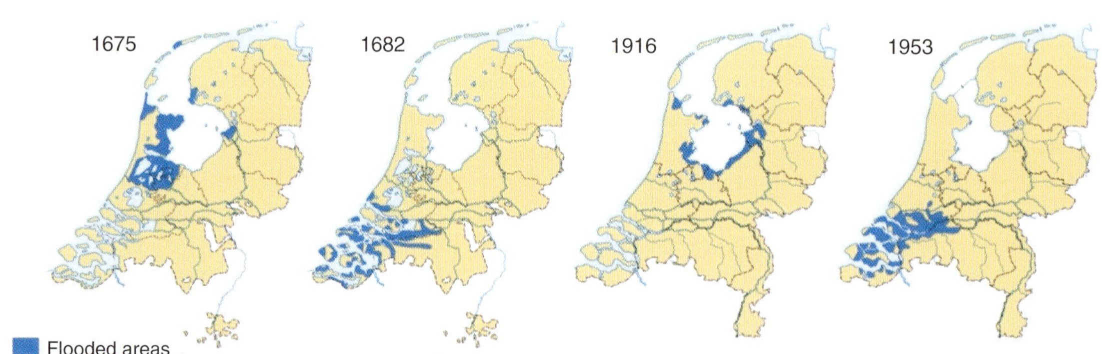

Figure 14.1   Four major historic coastal floods in the Netherlands. *Source*: Noordhoff Atlas productions (2012).

were also closed off by dams in order to protect the hinterland against flooding (as part of the Delta Works programme). Two inlets were kept open because they provide access to the ports of Antwerp and Rotterdam. Along the route to Antwerp, embankments were raised. The waterway connecting the port of Rotterdam to the North Sea can be closed off by a storm surge barrier.

Extreme river discharges in the Netherlands have also caused numerous floods. Embankment breaches in the past were often caused by ice dams. Once the thaw would set in, water could easily accumulate up to critical levels behind the ice dams. Because of river regulation, which began in the 19th century, ice dam formation no longer takes place. The adverse effect of upstream water management, however, is an acceleration of the upstream discharge, causing extreme water levels in areas located downstream during times of extreme rainfall, sometimes in combination with meltwater from the Alpine regions. This was the case in 1993 and 1995, when high water levels threatened Dutch embankments. In 1995, over 200 000 people were evacuated as a precautionary measure (however, the embankments held).

14.2.2 The Dutch flood defence system

The Netherlands has an extensive flood defence system, comprising of 3500 km of so-called primary defences (mostly embankments and along the coast around 300 km of dunes) protecting most of the Netherlands against flooding from the North Sea, the Wadden Sea, the rivers Rhine, Meuse and Westerschelde, the Oosterschelde and the IJsselmeer. In the Netherlands, certain areas are protected by an enclosed system of primary defences, called dyke rings. Each of these dyke ring areas has a safety standard that is related to exceeding a probability of a certain water level or wave height. These safety standards vary across the country, depending on the nature of the threat and the importance of the area, with high water level probabilities ranging from once in every 250 years to once in every 10 000 years.

14.2.3 Future flood sensitivity

For the future, flood risks are expected to increase further, as sea levels rise and river discharges increase (among others, Klijn, Kwadijk, de Bruijn and Hunink, 2010). If no adaptations are made, climate change will lead to a greater probability of flooding, which increases the flood hazard. The possible consequences of flooding are increased mainly by increases in investments in fixed assets and growing population concentrations in the most flood sensitive areas. This, in turn, raises the level of vulnerability.

14.2.4 Consequences of climate change for the Netherlands

14.2.4.1 Sea-level rise

Climate change in the Netherlands will result in a rise in sea level, and, during winter months, in an increase in peak drainage from rivers. Extreme wind speeds also may increase. Over the last century, the sea level along the Dutch coast rose by about 20 cm. Scenarios recently published by the Royal Netherlands Meteorological Institute (KNMI) predict a sea-level rise for the 21st century of between 35 and 85 cm (KNMI, 2006). The large difference between these two levels is due to the many factors that involve a large measure of uncertainty. These uncertainties relate to the sensitivity of the climate system as a whole, the way the Greenland and Antarctic ice sheets will melt, as well as future greenhouse gas emissions. Paleo research based estimates of maximal rates of sea-level rise vary from 5.6 m/millennium (Kopp *et al.*, 2009) up to between 0.7 and 1.7 m/century (Rohling *et al.*, 2008) for the interval within the last interglacials. The temporal resolution of the observations is too coarse to exclude higher rates for shorter periods, but does not provide information on the time lag between such extreme rates and more modest rates, such as the present 20 cm/century (Rohling *et al.*, 2008).

Extreme wind speeds are predicted to increase (Katsman *et al.*, 2011), but mainly from the southwest (Sterl *et al.*, 2009). However, because of the geometry of the North Sea, a North–Easterly wind direction is the most threatening to the Dutch coast (e.g. this was the wind direction during the 1953 flood disaster). Therefore, at present, no increase in the threat from that angle is expected.

14.2.4.2   Extreme discharges of the main rivers

Extreme discharges of the main rivers Rhine and Meuse also are expected to increase. However, for the Rhine basin, part of this increase will be reduced by upstream flooding, since the present exceeding probabilities of the upstream flood defences are so far not as high as they are in the Netherlands. For the Meuse basin, this effect of reducing extreme discharges by upstream flooding, has a much less substantial impact, since the room for flooding upstream of the Dutch border is naturally limited by the shape of the river valley. Estimated increases in the 1:1250 per year extreme discharge range from 15–25% for the river Rhine, for the year 2100, and are around 25% for the river Meuse, for the year 2100, based on the 2006 KNMI scenarios (Klijn *et al.*, 2010).

14.2.5 *Consequences of social economic changes for the Netherlands*

Population numbers, number of assets and GDP earned in flood prone areas in the Netherlands are expected to increase in the future, thereby increasing the vulnerability of the Netherlands to flooding. The PBL Netherlands Environmental Assessment Agency produced trend scenarios for 2040, for its study '*The Netherlands in the Future*' (Kuijpers-Linde *et al.*, 2007), based on two variants: one with moderate economic growth (1.7%) and moderate population growth (from the present 16 million to over 17 million by 2040), and one with higher economic growth (2.1%) and a population of almost 20 million by 2040. In both scenarios, the majority of any new urban development is assumed to take place in the urban areas of the Western Netherlands, hence in those parts of the Netherlands most sensitive to flooding.

This would continue the trend of recent decades and would mean that, in the period up to 2040, vulnerability to flooding of the Netherlands as a whole continues to increase, in terms of population numbers and economic value considered 'at risk' (Klijn, Kwadijk, de Bruijn and Hunink, 2010). The potential economic damage due to flooding is expected to increase in the period up to 2040 by a factor of between two and three, depending on economic and population growth. About 20–30% of this is due to new development (Figure 14.2), assuming that protection levels of the flood defence system are maintained in the period up to 2040 according to the standards defined in the Flood Defence Act.

In its National Water Plan (V&W, 2009), the Ministry of Infrastructure and the Environment announced a revision of the policy on flood risk management. The new policy will not only seek to prevent flooding, based on the chances of flooding, but also – more than before – seek to control the consequences if flooding does occur. The new strategy will explicitly address ways to reduce the risk of casualties and social disruption as well as economic damage.

14.3   Flood modelling

Adaptation of the Dutch flood defence system to changing conditions is based on a risk approach. To support this adaptation process, an extensive research project, Flood Risk in the Netherlands (VNK), is being carried out to estimate the probability of the different known failure mechanisms, as well as the consequences of a breach (V&A, 2005). Numerous potential breach locations have been identified and for every single location, flood scenarios have been calculated using advanced 2D hydrological models. In this way, flooding from the sea is treated equally to river flooding. Using a number of individual scenarios from this VNK

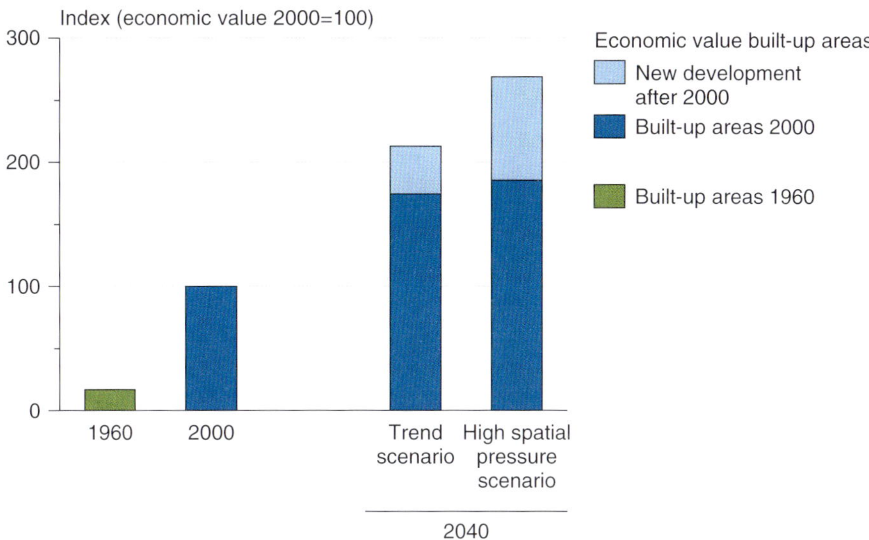

Figure 14.2   Potential economic damage in flood-sensitive areas (Kuijpers-Linde *et al.*, 2007).

project, a scenario was compiled for what would be considered the utmost possibility in flooding (called the Worst Credible Flood (WCF) scenario) (Kolen and Geerts, 2006) (Figure 14.3, right). This WCF scenario is to be used in the development of flood risk management plans. For the whole of Europe, however, this approach is far from feasible, since it requires detailed information on the condition of all local defence systems, in addition to enormous modelling efforts. Therefore within the ESPON Climate framework, simpler models were selected to estimate flood risks.

The assessment of the potential exposure to flooding, carried out within the ESPON Climate framework, was based on two models. For river flooding, a projection by the JRC LISFLOOD model was used (H12A2 scenario run. NB in ESPON Climate report: A1B scenario with CCLM climate model), and, for coastal flooding, the DIVA approach was taken (Dankers and Feyen, 2008; Barredo *et al.*, 2008) (Figure 14.3, left). To examine the sensitivity of the hazard assessment method, the ESPON Climate hazard map, based on these LISFLOOD/DIVA flood maps, was compared with a hazard map based on a worst credible flood map for the Netherlands

(Figure 14.3). The hazard maps for fluvial flooding show water depths that are expected to result from floods with a probability of 1:100 per year. The hazard maps for coastal flooding show water depths that are expected to result from a storm surge of 2 m above mean sea level (Barredo *et al.*, 2008).

14.3.1   *Differences in flood risk mapping between the JRC and the Netherlands*

Distinct differences between the maps by the Joint Research Centre of the European Commission (JRC) and the WCF map relate to flood extent and water depths. The combined LISFLOOD/DIVA map shows a much larger potentially flooded area along the coast than the Dutch maps do. In contrast, the areas along the main rivers in the eastern part of the Netherlands would stay dry according to the LISFLOOD/DIVA map, whereas the WCF map shows polders that potentially become flooded by the rivers Rhine and Meuse. In contrast, the LISFLOOD/DIVA map indicates flooding along some minor rivers, or in 'supposed catchments'.

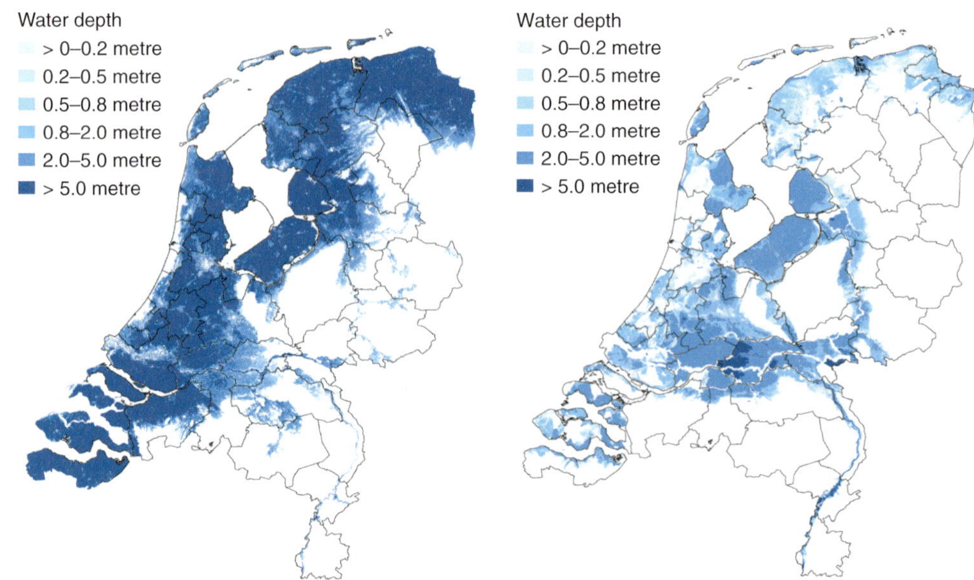

Figure 14.3   Flood depths (combined coastal and fluvial floods), estimated with the DIVA/LISFLOOD model (left) and the WCF model (right).

In general, the information in the hazard maps depends on the following assumptions and/or basic data.

1. **Choice of catchments**   The WCF flood map is based on the real water network, whereas LIS-FLOOD uses the modelled river network from the pan-European River and Catchment database (Vogt *et al.*, 2007). The algorithm that is used to derive slopes, flow direction and cumulated runoff from the digital elevation map, is difficult to apply to flat areas, such as the Netherlands.

Moreover, the method to base water courses on a catchment map cannot deal with river bifurcations. The result is that large rivers, such as the IJssel or Nederrijn, suddenly originate somewhere in the Netherlands instead of being mere branches of the Rhine, which originates in the Eastern Swiss Alps.

2. **Choice of the flood probability**   The coastal hazard map of the JRC (DIVA) shows water depths that result from a surge level of 2 m above mean sea level. Such a water level occurs almost yearly. However, because of the protection offered by dunes and embankments, this water level never causes any flooding. The WCF map is based on surge levels of at least 4.5–5 m above mean sea level.

The fluvial flood map of the JRC (LISFLOOD) shows flood water depths for floods of 1:100 per year in minor rivers. In the Netherlands, these are rather regarded as 'pluvial floods', floods caused by a drainage problem of minor rivers. Only less frequent floods in the major rivers (1:200–1:2000) are regarded to pose a 'risk'. The WCF map is based on these much lower, frequent floods.

3. **Coastal, estuarine fluvial and pluvial floods**   JRC produces separate maps for coastal hazards and fluvial hazards. The fluvial maps intend to show flooding from main rivers as well as from smaller canals. Because the catchment approach is used to identify water courses (point 1), the map produced is rather inadequate for the Netherlands. The Netherlands produces separate maps for floods resulting from breaches of the primary defences (along coasts and main rivers), and for flooding caused by failing embankments along minor inland water bodies (rivers and canals).

4. **Modelling flood water depths**   JRC uses a digital elevation model to obtain water depths,

by extrapolating water levels using a planar approximation of the flood level, and considering the local drainage direction. To obtain the 1:100 per year water levels in rivers, JRC applies the rainfall–runoff model LISFLOOD, for each 100 × 100 m grid cell. A Gumbel distribution is then fitted to the annual maximum values in every grid cell to estimate the probability of discharge levels with a 1:100 per year probability (Barredo *et al.*, 2008). Implicitly it is assumed that an unlimited amount of water is available for flooding, while, in reality, there is not enough water to fill up entire valleys. Therefore, the JRC method is likely to considerably overestimate water depths for wide river valleys and floodplain areas. The WCF map is also based on a rainfall–runoff model, but flooding is modelled by a 1D–2D hydrodynamic model assuming a limited amount of water (VNK, 2005).

5. **Flood defences** JRC uses digital elevation maps that do not take into account the existing flood defences and/or secondary embankments. A 1:100 per year flood will not cause significant flooding in the Netherlands, as embankments are high enough. Similarly, dunes and embankments will prevent coastal flooding up to a certain storm surge level. Secondary embankments and other elements in the area behind dykes may limit the extent of the floods. Two-dimensional simulations of the Netherlands do take account of this effect.

In summary, the JRC flood modelling approach is consistent and rigorous, which makes it applicable for the whole of Europe. This is very important information for the pan-European assessment within the ESPON Climate project.

Effects of sea-level rise and changes in extreme water levels due to climate and land-use changes also can be estimated. Yet, for the Netherlands, hazard maps based on this approach deviate considerably from hazard maps based on local conditions (and assumptions), among others due to differences in the assumed flooding probabilities (for the JRC maps 1:100 for fluvial flooding, whereas the coastal flooding approach may be practically considered as non-stochastical, while the Dutch maps assume at least a 1:1000 recurrence). Still, the impact of these

differences on the exposure assessment at the Dutch scale is also substantial but less pronounced, due to the fact that the ranking method is relative. On a European scale, however, the impact of these differences will be more pronounced, if the classification is based on non-Dutch extremes. This remains to be investigated. However, for the Netherlands, the impact of the exposure map on the overall vulnerability assessment will most likely be limited, due to its (in comparison with other European regions) large adaptive capacity with respect to flooding (although this depends on the construction of this indicator). Still, the point remains that the use of JRC maps may cause overestimation in the exposure assessment, especially in low-lying deltaic areas, such as the Po Valley in Italy, the Thames Estuary in the United Kingdom, the Elbe and Weser river mouths and the coastal plains along the North Sea in Germany. This impact of these potential differences on the local vulnerability assessment should be investigated.

14.4   Sensitivity analysis

The ESPON Climate framework estimates the overall vulnerability of every individual EU NUTS3 region to climate change. The objective of this section is to analyse the sensitivity of these estimates to uncertainties induced by choices made with respect to the construction of the framework and by the selection of inputs (constituting a potential contribution to the first class of uncertainties: the statistical uncertainties). For this purpose, flooding in the Netherlands will be used as an example.

Although the final vulnerability is a function of the impact as well as of the adaptive capacity, this last term is as good as the same for all Dutch NUTS3 regions (see map 56 of the final report (IRPUD, 2011)). Therefore, the final distribution of the vulnerability classes will not differ from the distribution of the impact classes. For this reason, we based our analyses on the results for the geographical distribution of the impact indicator in the Netherlands.

Along with the flood modelling, five other factors were selected which could have a significant impact

Table 14.1 Model factors, analysed for their possible contribution to errors in the exposure estimates.

Choice	Default	Alternative
1. Flood model	WCF model	LISFLOOD model
2. Risk concept	Additive	Multiplicative
3. Normalising	Only at the first and final stages	Throughout every stage
4. Indicators	All	Selection
5. Extreme values	All 40 NUTS3 areas included	The most extreme were left out
6. Weighting factor for the domains	Domains weighted by expert judgement	Domains weighted equally

on the final results: the flood exposure model, the risk concept, the stage of normalising, extreme values, and finally the factors by which the importance of the different domains, and their susceptibility to flooding, is weighted (Table 14.1).

The final estimates were classified into five different categories, ranging from no or very limited expected change (or even a positive change) in total vulnerability to an expected highly negative change. To explore the sensitivity, we analysed the similarities between all estimates, but also between the highest ones, since these will be the target of EU refunding.

14.4.1 Flood model

The previous section describes the differences between the available flood models, among others caused by the underlying river network – based on the real map or on a drainage model that cannot deal adequately with delta's and their bifurcating rivers. Other examples of model differences relate to the assumptions on the availability of water – whether considered infinite or based on hydrodynamic models – or the way the existing flood defence system is taken into account, varying from partially to completely ignoring it to using it as a starting point. The impact estimation was based on a combination of flood depth and flood extent.

The WCF map contains maximum water depths for flooding, irrespective of climate change. For the WCF map, the water depth classes are given. The LISFLOOD/DIVA maps were classified in the same way (see Figure 14.3). For each class of water depth, the percentage area per region was calculated. Weights were assigned, based on water depth.[2] The values (weight * percentage area) were subsequently aggregated. In addition to the score for this relative value, each region was scored for the absolute value of the flood prone area. The average of both of these scores constitutes the final score.

14.4.2 Risk concept

Within the risk society many ways are used to express the general concept of risk being a function of probability of an event and its consequences (e.g. FLOODsite, 2005; Marchand, 2009). Most commonly risk is defined as the product of probability and consequences, expressed as a function of its different components: hazard, exposure, receptors and stage damage (FLOODsite, 2005). Another way to operationalise the concept, however, would be to value the different components using classification schemes and finally adding all scores into one final risk score. This method is applied to risk mapping (De Bruijn, 2007; De Bruijn and Klijn, 2009). Although multiplying probability with (potential) effects seems to be the most obvious way to assess risk, adding both components has some advantages with respect to the previous mentioned third category of errors: the recognised ignorance. The differences emerge where either the probability is zero (or close to zero) or the effects are zero (or close to zero). In the case of multiplication, risk is estimated at zero (or close to zero) for both combinations, where addition still leaves some remaining risk that could reflect the potential impact of the as yet unforeseen (or not regarded as likely) but still imaginable possibility that things may change for the worse (with respect to both climate and effects).

---

[2] 0–0.2 m weight = 1; 0.2–0.5 m weight = 2; 0.5–0.8 m weight = 3; 0.8–2.0 m weight = 4; 2.0–5.0 m weight = 5; >5.0 m weight = 5.

14.4.3 Normalising

Within a framework as complex as the ESPON Climate framework, many different types of data are combined. To normalise all data (e.g. between 0 and 1 with the use of both extreme values) is a way of doing so. However, the stage at which this is performed may have an impact on the final result, since normalisation may change the mutual influence of different components. We compared a calculation where all factors were normalised at all stages with one where this was only done in the first and final stages.

14.4.4 Used indicators and weighting factors

In different stages of the ESPON Climate framework, individual indicators are aggregated into one overall indicator, using weight factors set by experts with the use of a Delphi method as an example of expert-judgement-induced uncertainty. With respect to the choice of indicators, the framework distinguishes five dimensions of susceptibility to flooding (physical, social, cultural, economic and environmental), each represented by the most important elements (Table 14.2) (IRPUD, 2011). An important constraint to the selection of the indicators was the availability of data.

The ESPON Climate method is more extensive than the method used in the Netherlands to estimate economic damage and fatalities. However, the latter method also takes into account indirect effects, albeit only to a limited extent, at present (for all flood prone areas, this is estimated at 1% of the total damage, a percentage that may be strongly underestimated). For all flood prone areas, the Dutch method estimates that private assets (e.g. houses, furniture) comprise 62% of all damage, trade and industry account for 18% and agriculture for 4% (Wouters, 2005). In the ESPON Climate method, the physical and, to a lesser extent, economic dimensions are most comparable with these Dutch estimations of potential damage. The ESPON Climate weighting factors of 19 for domestic housing and company housing and 24 for agriculture (out of a total of 100) are nowhere near the respective Dutch estimates of 80 and 4. In comparison with all EU Member States, the flood prone part of the

Table 14.2 Susceptibility to flooding indicators used in the ESPON Climate framework (IRPUD, 2011) and indicators used in the Netherlands (Wouters, 2005) for the estimation of economic damage (bold) and fatalities (italic).

Dimension	Physical	Social	Cultural	Economic	Environmental
Indicators	**Housing**	*Inhabitants*	Natural landscapes	Arable farming	Natura
	Housing area	Elderly (hh)	State monuments	Livestock	Natura 2000
	Working area	Low educated (hh)	Historic cities	Zero grazing livestock	
	Infrastructure (transport)		Archeological sites	Greenhouse culture jobs	
	High voltage distribution stations		UNESCO World Heritage sites		
	Power plants		Id. tentative		
	Hospitals				
	Schools				
Weighting factor	19	16	10	24	31

Netherlands is highly industrialised, with relatively many urban areas and few nature conservation sites.

14.4.5 Extreme values and classification scheme

Within the ESPON Climate framework, indicator values are divided into five classes. Although there are several ways of classification, the ESPON Climate method is most straightforward; it equally divides the parameter space between the highest and lowest values. This type of classification delivers results that are susceptible to relatively extreme values. In particular, extremely high values may influence the classification of the next highest category (this phenomenon may also occur in the classification of lower values, but these are not really of interest with respect to the framework target and, moreover, many of the data distributions are skewed towards higher values). Therefore, we analysed the impact of the highest value on the distribution of the impact indicator. Another type of classification would be to divide the results into evenly distributed classes (e.g. dividing the 40 Dutch NUTS3 regions into five

classes, each containing eight regions). Outliers have no impact on this method of distribution. We performed a sensitivity analysis using both classification schemes.

14.4.5.1 Method

A preliminary rough estimate showed that the choice of flood model was the dominant factor. We analysed the impact of the different factors by individually applying them to both exposure maps (LISFLOOD/DIVA and WCF) (see Table 14.1). The results were compared using a correlation between the continuous values, and a Spearman rank correlation between their classes, using an equidistant classification scheme and an even-distribution scheme. Comparison with the application of this last scheme was conducted for all five classes, as well as for the two highest classes separately (Tables 14.3 and 14.4).

14.4.5.2 Results

The estimates of the impact indicator, gradually scaled from 0 up to 1, were compared for different

Table 14.3 (Rank) Correlation coefficients between WCF model-based impact estimates, with the factors presented in Table 14.1, varying for continuous values and values classified with equidistant classes and with evenly distributed classes. Correlation coefficients between brackets represent only the two highest classes.

	Flood model	Risk concept	Normalising	Indicator selection	Extreme values	Weighting factors
Continuous	0.61	0.94	0.99	0.99	0.77	0.99
Equidistant classes	0.59	0.76	1.00	0.97	0.79	1.00
Evenly distributed classes	0.51	1.00	1.00	1.00	0.88	1.00
	[0.25]	[1.00]	[1.00]	[1.00]	[0.75]	[1.00]

Table 14.4 (Rank) Correlation coefficients between LISFLOOD/DIVA model-based impact estimates, with the factors presented in Table 14.1, varying for continuous values and values classified with equidistant classes and with evenly distributed classes. Correlation coefficients between brackets represent only the two highest classes.

	Flood model	Risk concept	Normalising	Indicator selection	Extreme values	Weighting factors
Continuous	0.61	0.97	0.99	0.99	0.86	0.99
Equidistant classes	0.59	0.92	0.99	0.98	0.88	1.00
Evenly distributed classes	0.51	1.00	1.00	0.99	0.88	0.99
	[0.25]	[1.00]	[1.00]	[1.00]	[0.75]	[1.00]

combinations of the six factors. Results clearly demonstrated that, in this specific case of impact of flooding in the Netherlands, the weighting of the different domains that are supposed to be sensitive to flooding hardly influenced the value of the impact indicator and had no influence at all on the classification. The same conclusion could be drawn with respect to the choice of indicators of sensitivity to flooding and the stage of normalisation within the framework. The choice of the risk concept itself did have minor influence on the ranking, which was highest in combination with the WCF model (with a rank correlation coefficient of 0.76).

The choice of flood model, however, was found to have a strong impact (Figure 14.4). Correlation was fairly low with $R = 0.61$, and even lower when values were classified. In particular the classification of the highest values showed a poor similarity, with a rank correlation coefficient of 0.25.

The analysis shows the sensitivity of the classification scheme to the highest value when applied to a combination of the WCF model with an equidistant classification scheme. According to the WCF model, this most extreme value implies that the second highest value would be at least 20% lower. When the LISFLOOD/DIVA exposure map was applied, seven NUTS3 areas had impact scores higher than 0.8, hence the impact of the highest value of the classification of the other NUTS3 areas would be far less significant (in this case, the impact was even as low as zero, since both – normalised – highest values were 1.0).

When the WCF exposure map was used, the region with the highest estimated impact consisted of both the IJsselmeer polders Flevopolder and Noordoostpolder (Figure 14.4, right). This highest value could represent a very likely scenario. The Noordoostpolder as well as the Flevopolder are expected to become completely flooded at high water levels, if the embankment is breached in at least one location. This expectation has not only been based on a physical model, but also on one of the rare incidences of relatively recent actual flooding. When, at the end of the Second World War, the embankments of the Wieringermeer, a comparable polder located at the opposite side of the IJsselmeer, were purposely breached by the German occupiers, the polder was completely flooded within 48 hours up to the water level of the IJsselmeer (Klingen, Hoes, Schuurmans and Strijker, 2011). The WCF exposure

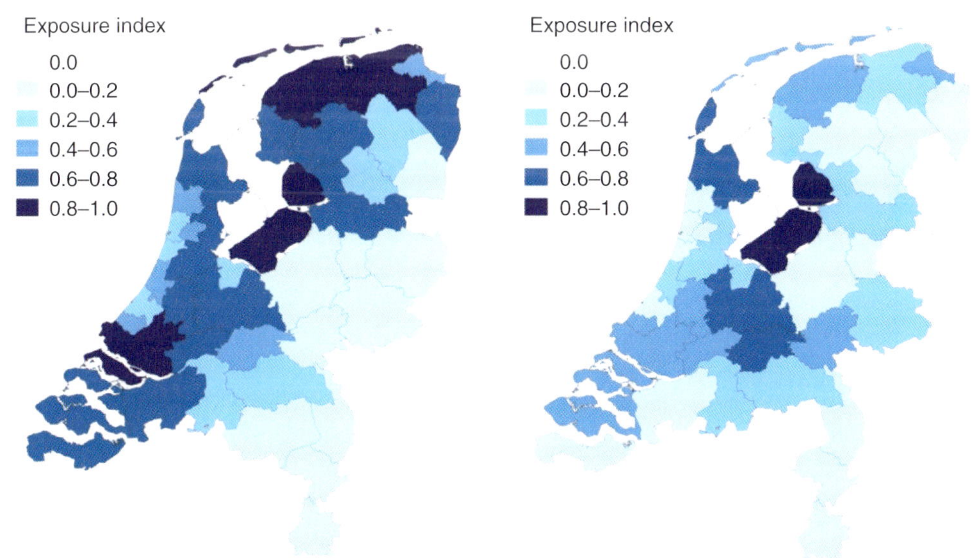

Figure 14.4   Exposure to flooding on the Dutch EU NUTS3 regions (classified from 1 to 5; 0 = no exposure), estimated with the DIVA/LISFLOOD flood map (left) and the WCF flood map (right).

map shows several other locations that could face even more extreme exposures when flooded than both IJsselmeer polders (Figure 14.3, right), yet these estimated flood plains encompass only certain parts of a NUTS3 region. This example clearly demonstrates the artefacts that may occur when the scale of the analysis does not match the scale of the potential problem, which thus can become more or less diluted; in this case yielding a regional impact value that does not directly reflect the estimated relative local risks.

14.5   Discussion and conclusions

The objective of our study was to analyse the sensitivity of the vulnerability indicator estimated by the ESPON Climate framework to choices made in the construction of the framework and to the selected inputs. For this purpose, the case of flooding in the Netherlands was taken as an example. Sensitivity was analysed for six different factors, bearing in mind that the focus for the ESPON Climate framework estimates was on the regions (NUTS3) with the highest ranking.

14.5.1   *Sensitivity to choices made with respect to the modelling concept*

Our analysis showed that, in this specific case of the impact of flooding in the Netherlands, the weighting of the different domains that are supposed to be sensitive to flooding had no influence on the impact indicator. This was also true for the choice of indicators of sensitivity to flooding or the stage of normalisation within the framework. Table 14.2 shows the difference between the elements that are sensitive to flooding and their mutual weighting, according to the ESPON Climate framework and a Dutch method of estimating the potential impact. Although, for the Dutch regions, these differences are insufficiently discriminative to have a significant

impact on the final impact indicator, this could be the case in a comparison of pan-European regions. The method's emphasis on agriculture within the economic dimension and the high weighting of the environmental dimension bears the risk of a structural underestimation of the sensitivity of more urban and industrialised regions.

However, whether the risk concept is additive or multiplicative has an effect on the estimated exposure indicator, both on value and ranking. The latter is more important, since the primary goal of the ESPON Climate framework is that of ranking, while in the case of the 20% most important NUTS3 areas (the highest ranked) the effect on the ranking was zero, independent of the use of the flood exposure map. The choice of risk concept is more conceptual (should we, up to some level, take unforeseen but feasible phenomena into account or not) rather than scientific. Still, the possible impact of such a choice on the final results is something that must be considered.

Depending on the choice of flood exposure model, the most extreme value may be of importance. Leaving out this value would have a substantial impact on ranking when using the WCF flood model, but not so with the use of the LIS-FLOOD/DIVA model. Furthermore, the impact only occurred when the classification scheme was based on the highest (and lowest) values. In case of the WCF model, this most extreme value had a substantial impact, since the second-highest value was at least 20% lower. When the LISFLOOD/DIVA model was applied, seven NUTS3 regions had impact scores higher than 0.8, and so the impact of the highest value of the classification of the other NUTS3 regions would be far less significant (in this case, the impact was even zero, since both – normalised – highest values were 1.0). This example shows the potential impact of choices with respect to a combination of a flood exposure model and classification scheme, given the purpose of the ESPON Climate framework as a guidance tool for EU adaptive policy, which will be focused on the highest ranked regions. Using the WCF exposure map, the region consisting of two large IJsselmeer polders would suffer the highest potential impact. However, the WCF exposure map shows several other locations that may face even more extreme exposures when flooded (Figure 14.3, right), while these estimated flood plains encompass only certain parts of a NUTS3 region. This example clearly demonstrates the artefacts that may occur when the scale of the analyses does not match the scale of the potential problem, which thus could become more or less diluted; in this case, yielding a regional impact value that does not directly reflect the estimated relative local risks.

14.5.2 Sensitivity to the choice of flood exposure model

The estimates of the potential impact are highly sensitive to the choice of flood model (Figure 14.5). Although the LISFLOOD/DIVA approach is less sensitive to extreme values, the method has some major disadvantages. Both models (the LIS-FLOOD model for river flooding and the DIVA model for coastal flooding) assume an unlimited amount of water involved in the flooding, whereas, in actual practice, this is not the case. Especially

in flat areas, such as in the Netherlands, this approach therefore tends to overestimate the potential impact, which is clearly demonstrated by the distribution according to the impact indicator, compared with the distribution based on the WCF model (Figure 14.6).

The question of how sensitive the framework estimates are to the choices made within the framework (including choices with respect to the inputs) is answered by focusing on the highest values, which are the most important estimates, since they play a major role in the use of the framework. The final estimates were classified into five different categories, ranging from no or very limited expected change in total vulnerability to an expected highly negative change. In these highest classes, the correspondence between estimates made by the LISFLOOD/DIVA and WCF models was even worse with a Spearman rank coefficient of 0.25.

The ESPON Climate framework, however, is targeted at indicating areas with the largest increase in vulnerability, rather than areas with highest vulnerability. This increase (or, more generally, change) is based on the difference between exposure estimates

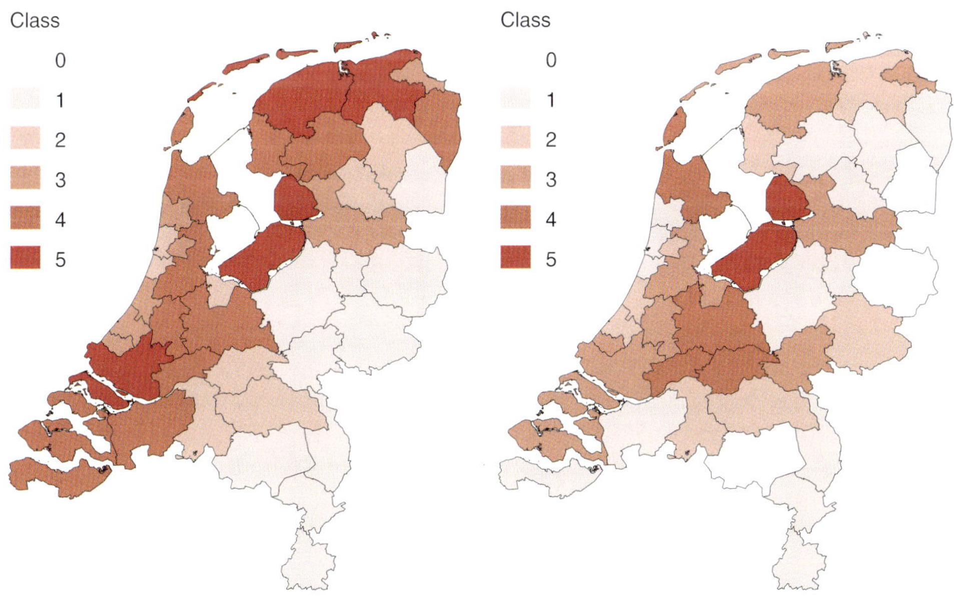

Figure 14.5 Potential impact of flooding on the Dutch EU NUTS3 regions (classified from 1 to 5; 0 = no impact), estimated with the DIVA/LISFLOOD flood map (left) and the WCF flood map (right).

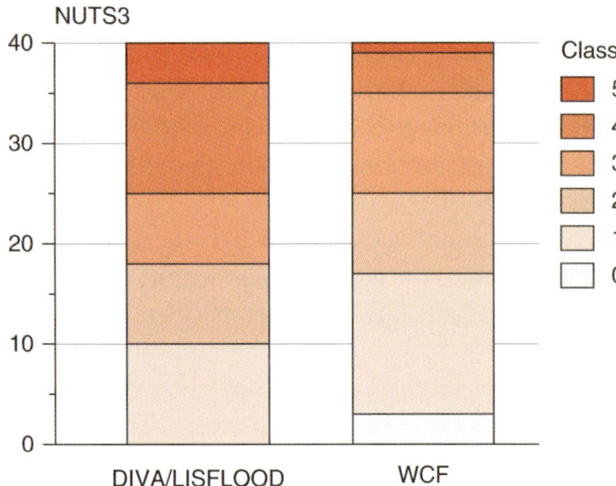

Figure 14.6 Distribution of the total impact scores (count) for the Dutch EU NUTS3 regions, estimated with the DIVA/LISFLOOD flood map (left) and the WCF flood map (right); 0 = no impact; 1 = lowest potential impact; 5 = highest potential impact.

for the present and the future situation. Unfortunately, these maps were not available during our analyses. However, this did not alter our conclusions about the sensitivity of the impact indicator to several factors. The calculated impact showed a high sensitivity to the choice of flood exposure model used, hence the difference between estimates of future and present exposure will most certainly also be sensitive to this factor. The final ESPON Climate vulnerability map shows high values in the NUTS3 regions bordering on areas exposed to coastal floods as estimated with the DIVA model (see maps 49 and 24 of the final report (IRPUD, 2011)). Flood depth is also taken into account in the final version of the ESPON Climate framework (which, according to the straightforward DIVA method, will increase over the entire estimated exposure area with the exact same value as assumed for sea-level rise), whereas the combined effect of extension and increase in depth is expected to emerge along the edges of the estimated flood extension. The Dutch WCF model is based on assumed breaches at the weakest spots

in the existing flood defence system. Several of these locations are situated along the coastal line. Sea-level rise will locally imply a flood plain extension as well as an increase in flood depth, but only in regions directly located along the coast (Figure 14.7).

We compared the results from the WCF exposure map with those from the exposure map estimated by JRC, based on a combination of the DIVA coastal flood modelling approach and JRC's LISFLOOD model for river flooding. The indicated strong differences in the spatial estimates of potential impact were mainly caused by the different expectations about coastal flooding. Although we have good reason to assume that the LISFLOOD model also tends to overestimate river flooding in flat areas, as indicated in the above section on flood modelling, this phenomena could not be demonstrated very clearly for the Netherlands. This was mainly due to the fact that the LISFLOOD exposure estimates were based on a 1:100 per year extreme discharge; whereas the Dutch flood defence system deals with crest heights that are far above these values (presumably at least partly recognised by the digital elevation model that was used to estimate the extent of the flood plain).

Furthermore, the final ESPON Climate results indicated no increase in potential impact caused by river flooding, simply because the combination of a CO_2 emission scenario with either a global or a regional climate model predicted no significant increase in the 1:100 per year extreme river discharge of the rivers Rhine and Meuse (see map 23 of the final report (IRPUD, 2011)). Dankers and Feyen (2009) estimated the same 1:100 per year extreme discharges for eight possible combinations of two emissions scenarios, two global climate models and two regional climate models, which showed increases as well as decreases. In contrast, Klijn, Kwadijk, De Bruijn and Hunink (2010) estimated only increases, using the KNMI scenarios, which can be considered interpretations of the complete set of IPCC AR4 climate model experiments (KNMI, 2006). These increases range from around 15 to 25%

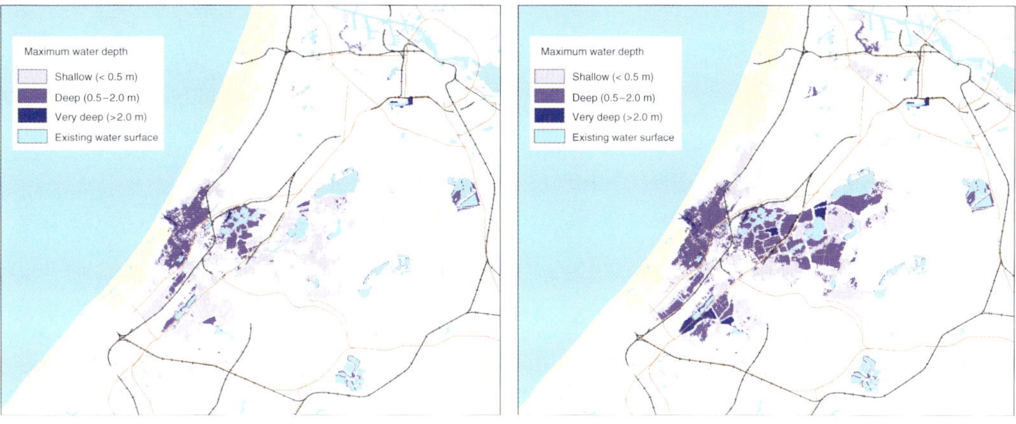

Figure 14.7   Estimated exposure to coastal flood caused by a breach at Katwijk, at the present sea level (left) and with an assumed seal level rise of 1.3 m (right).

for a 1:1250 flood event (discharges corresponding with the safety standard for the embankments along the major Dutch rivers), to even over 30% for a 1:100 per year flood event.

In particular, the use of the DIVA coastal flood model within the ESPON Climate framework also showed a potential disadvantage related to the use of more general indicators to estimate the adaptive capacity of NUTS3 regions. The positive contribution of the Dutch flood defence system, which in comparison with all other European nations is highly protective and adaptive, does not become apparent, if the modelling is based on indicators such as educational commitment, number of patents and income per capita. Up to a certain level, this component of the total Dutch adaptive capacity can be mirrored with the more general indicators used in the ESPON Climate framework, which reflects prosperity and welfare as the driving force behind it. However, even compared with other similarly prosperous nations such as Germany and the United Kingdom, the Dutch flood protection system induces significantly lower flood risks, whereas flooding, together with freshwater availability, is considered a great if not the most important challenge of climate change that the Netherlands is facing (Ligtvoet, 2009a, b; V&W *et al.*, 2009).

14.5.3   Some recommendations with respect to the use of the ESPON Climate framework results

In the introduction, we referred to numerous authors indicating the huge differences in framing between the people who make these types of analyses (scientists) and the people who would like to use these analyses (stakeholders, policymakers). How to deal with statistical and scenario uncertainty is often very difficult if not impossible for policymakers to comprehend. Based on work by, for example, Dessai and van der Sluijs (2007), we therefore would like to end this chapter with some recommendations with respect to the use of the ESPON Climate framework in policy-making.

The (modelling) framework must always be run several times, for several combinations of likely greenhouse gas emissions scenarios and (global as well as regional) climate models. The output of these combined runs may then be classified into groups, for example: one where most of the estimates point to an increase in vulnerability; one where most of the estimates indicate a limited or even negative change in vulnerability; and a third group containing regions for which both positive and limited or even negative increases are predicted. For the first group, taking actions is most likely a good strategy, whereas for the third group, the group with the highest uncertainty, an adaptive strategy is more obvious (a strategy that is mostly based on monitoring, in order to obtain more certainty about the direction and rate of change, in combination with preparing to take action if necessary). In case no (or almost no) negative impacts are expected for the near future, close monitoring would still be a good strategy, especially to anticipate the third class of uncertainties of the thinkable but at the present time not likely mechanisms which could change things for the worse. This strategy should be combined with one that leaves enough room (physically, but also with respect to, for example, government) to act if observations start deviating from predictions (which is essentially the same adaptive strategy as is advised for the third group).

References

Barredo, J., Salamon, P. and Bodis, K. (2008) *Towards an Assessment of Coastal Flood Damage Potential in Europe*, JRC, Ispra.

Bindschadler, R. (1998) Future of the West Antarctic Ice Sheet. *Science*, **282**, 428–429.

Commission of the European Communities (2009) *Impact Assessment. Commission Staff Working Document accompanying the White Paper Adapting to Climate Change: Towards a European Framework for Action*, Commission of the European Communities, Brussels.

Dankers, R. and Feyen, L. (2008) Climate change impact on flood hazard in Europe: An assessment based on high-resolution climate simulations. *Journal of Geophysical Research*, **113**, 1–17.

Dankers, R. and Feyen, L. (2009) Flood hazard in Europe in an ensemble of regional climate scenarios. *Journal of Geophysical Research*, **114**.

De Bruijn, K.M. (2007) *Risky places in the Netherlands: a first approximation for floods*. WL Delft Hydraulics.

De Bruijn, K.M. and Klijn, F. (2009) Risky places in the Netherlands: a first approximation for floods. *Journal of Flood Risk Management*, **2**, 58–67.

Dessai, S. and van de Sluijs, J. (2007) *Uncertainty and Climate Change Adaptation - a Scoping Study*, Copernicus Institute of the University of Utrecht, Utrecht.

FLOODsite (2005) *Language of risk; project definitions*. (online) Available at: <http://www.floodsite.net/html/publications.asp> (accessed 13 September 2012).

IRPUD (2011) *ESPON Climate - final report*. TU Dortmund University, Dortmund.

Janssen, H.M., Petersen, A.C., van der Sluijs, J.P., *et al.* (2005) Towards Guidance in Assessing and Communicating Uncertainties. *Water Science & Technology*, **52**, 125–131.

Katsman, C., Sterl, A., Beersma, J., *et al.* (2011) Exploring high-end scenarios for local sea level rise to develop flood protection strategies for a low-lying delta – the Netherlands as an example. *Climatic Change*, **109**, 617–645.

Klijn, F., Kwadijk, J., de Bruijn, K. and Hunink, J. (2010) *Overstromingsrisico's en droogterisico's in een veranderend klimaat; verkenning van wegen naar een klimaatveranderingsbestendig Nederland*, Deltares, Delft.

Klingen, L., Hoes, O., Schuurmans, W. and Strijker, J. (2011) Historich onderzoek Wieringermeer met moderne technieken. *H2O*, **22**, 14–16.

KNMI (2006) *Klimaat in de 21e eeuw - vier scenario's voor Nederland*, KNMI, De Bilt.

Kolen, B. and Geerts, R. (2006) *Als het toch misgaat: Overstromingsscenario's voor rampenplannen*, Betooglijn. HKV lijn in water.

Kopp, R.E., Simons, F.J., Mitrovica, J.X., *et al.* (2009) Probabilistic assessment of sea level during the last interglacial stage. *Nature*, **462**, 863–867.

Kuijpers-Linde, M.A.J., Geurs, K.T., Knoop, J.M., *et al.* (2007) *Nederland Later - Tweede Duurzaamheidsverkenning, deel Fysieke leefomgeving Nederland*, MNP, Bilthoven.

Ligtvoet, W. and van Minnen, J. (2009a) *Roadmap to a climate-proof Netherlands*, PBL, Bilthoven.

Ligtvoet, W., Knoop, J.M., Strengers, B. and Bouwman, A.F. (2009b) *Flood protection in the Netherlands: framing long-term challenges and options for a climate-resilient delta*. PBL, Bilthoven.

Marchand, M. (2009) *Modelling coastal vulnerability: design and evaluation of a vulnerability model for tropical storms and floods*. Technical University Delft, Ios Press.

Noordhoff Atlas productions (2012) Water atlas of the Netherlands, Groningen.

Rohling, E.J., Grant, K., Hemleben, C., *et al.* (2008) High rates of sea-level rise during the last interglacial period. *Nature Geoscience*, **1**, 38–42.

Saltelli, A., Nardo, M., Saisana, M., *et al.* (2004) Composite indicators - the controversy and the way forward. *OECD World Forum on Key Indicators*. Palermo.

Sterl, A., van den Brink, H., de Vries, H., *et al.* (2009) An ensemble study of extreme storm surge related water levels in the North Sea in a changing climate. *Ocean Science*, **5**, 369–378.

Swart, R., Bernstein, L., Ha-Duong, M. and Petersen, A. (2009) Agreeing to disagree: uncertainty management in assessing climate change, impacts and responses by the IPCC. *Climatic Change*, **92**, 1–29.

Ten Brinke, W.B.M., Bannink, B.A. and Ligtvoet, W. (2008) The evaluation of flood risk policy in the Netherlands. *Proceedings of the ICE - Water Management*, **161**, 181–188.

V&W (2005) *Veiligheid Nederland in Kaart - Hoofdrapport onderzoek overstromingsrisico's*, Rijkswaterstaat, Den Haag.

V&W, VROM and LNV (2009) Nationaal Waterplan, in *Ministerie van Verkeer en Waterstaat, M.V.V., Ruimtelijke Ordening en Milieu*, (eds Ministerie van Landbouw, Natuur en Voedsel) V&W, VROM, LNV.

Veijalainen, N., Lotsari, E., Alho, P., *et al.* (2010) National scale assessment of climate change impacts on flooding in Finland. *Journal of Hydrology*, **391**, 333–350.

VNK (2005) *Veiligheid Nederland in Kaart - Hoofdrapport onderzoek overstromingsrisico's*. V&W, Rijkswaterstaat.

Vogt, J., Soille, P., de Jager, A., *et al.* (2007) *A pan-European River and Catchment Database*. JRC.

Wardekker, J.A., Boer, J.D., Kolkman, M.J., *et al.* (2009) *Tool catalogue frame-based information tools*. Utrecht University, Copernicus Institute, Utrecht.

Wouters, K. (2005) Veiligheid Nederland in Kaart - Globale schadeberekening. HKV lijn in water, RWS/Dienst Wegen Waterbouwkunde, Delft.

Chapter 15

Vulnerability and adaptation to climate change in the Alpine space: a case study on the adaptive capacity of the tourism sector

Sylvia Kruse, Manuela Stiffler, Daniel Baumgartner and Marco Pütz

Swiss Federal Institute for Forest, Snow and Landscape Research WSL, Zürcherstrasse 111, 8903 Birmensdorf, Switzerland

Abstract

The objectives of this case study are to analyse the existing capacities of the tourism sector in the European Alps and to investigate opportunities for improving adaptation enacted by tourism stakeholders. Alpine tourism is expected to be among the economic sectors most affected by climate change effects. Because tourism development often requires mid- and long-term investments in infrastructure and services, early action is needed to build capacities to adapt to climate change and prevent economic losses to the regional economies. Up to now, only a few studies have considered both summer and winter tourism when evaluating the adaptive capacity and possible adaptations of the tourism sector. Two research questions are investigated: (i) what characterises the adaptive capacity of the tourism sector in the European Alps? and (ii) which adaptation activities have the potential to reduce the vulnerability of the tourism sector? By investigating these questions, the study endeavoured to characterise the adaptive capacity and identify current and potential adaptation activities for both summer and winter

European Climate Vulnerabilities and Adaptation: A Spatial Planning Perspective, First Edition.
Edited by Philipp Schmidt-Thomé and Stefan Greiving.
© 2013 John Wiley & Sons, Ltd. Published 2013 by John Wiley & Sons, Ltd.

tourism in the European Alps. To measure the adaptive capacity of the tourism sector a semi-standardised questionnaire survey was administered to tourism stakeholders, that is, representatives of tourism organisations as well as public authorities for economic development, environmental agencies and spatial planning authorities at both the national and regional levels. The area of the study comprises all countries of the European Alps and is limited to the perimeter defined by the Alpine Convention. The results manifest a broad range of adaptation strategies and measures that are already being realised in Alpine tourism today. Concerning the currently realised adaptation strategies and the willingness to adapt in the future, the results show that adaptation activities within the tourism sector are primarily realised in an autonomous, private and local manner. In particular a broad range of corporate measures (e.g. providing snowmaking equipment) are actualised on a private basis without necessarily following a comprehensive and long-term strategy. Mostly short- and medium-term adaptation options are realised, which can be seen as no-regret or low-regret measures with benefits even without climate change. Further, the results show that, on the one hand, climate adaptation in the tourism sector is prompted by economic incentives. In general, tourism stakeholders are not motivated to take adaptive action to address climate change, but rather, strive to prepare for increasing losses in winter tourism and to improve tourism infrastructure and supply to meet future tourism demand. On the other hand, adaptation decisions are very complex and uncertain concerning their effects if they aim at long-term development and take into account a variety of social, institutional and political barriers.

15.1 Introduction

Mountain areas number among the world's regions that are most affected by climate change (Kohler *et al.*, 2010). Current and future climate changes affect not just mountain ecosystems but societies and economies as well. The tourism sector is one of the most important for the economy of mountain regions. In its White Paper on Climate Change Adaptation, the European Commission (2009a) declared the European Alps as one of the areas that is most vulnerable to climate change impacts. Moreover, Alpine tourism is mentioned as one of the sectors that is most likely to be affected by a changing climate (Commission of the European Communities, 2009a). Because tourism development often requires mid- and long-term investments in infrastructure and services, the impact assessment of the European Commission concludes that early action is needed to 'incorporate as soon as possible climate proofing and long term planning and to build capacity to adapt in the tourism sector' (Commission of the European Communities, 2009b).

Recently, many scholars have conducted research specifically on the exposure of the Alpine space to climate change stimuli (Beniston, 2005; Bavay, Lehning, Jonas and Lowe, 2009; EEA, 2009). Few studies have also assessed the impacts of climate change on tourism in general (Amelung and Moreno, 2009). Research on mountain tourism has mainly focused on winter sports and snow reliability (Steiger and Mayer, 2008; Hoffmann *et al.*, 2009; Pütz *et al.*, 2011). To date, few studies have considered both summer and winter tourism when evaluating the adaptive capacity and possible adaptations of the tourism sector in the European Alps (for an explorative approach, cf. Müller and Weber, 2008; Hill, Wallner and Furtado, 2010).

Therefore, the objectives of the case study 'Alpine Space' within the ESPON Climate project were to analyse the existing capacities of the tourism sector in the European Alps and to investigate opportunities for improving adaptation enacted by tourism stakeholders. Accordingly, we raised the following research questions.

- What characterises the adaptive capacity of the tourism sector in the European Alps?
- Which adaptation activities have the potential to reduce the vulnerability of the tourism sector in the European Alps?

By investigating these questions, the ESPON Climate case study 'Alpine Space' endeavoured to characterise the adaptive capacity and identify current and potential adaptation activities for the tourism sector in the European Alps. The remainder of this chapter is structured as follows: in Section 15.2, we review existing studies of climate change impacts on Alpine tourism, including adaptation strategies on various levels; in Section 15.3, we elaborate on the design of the adaptive capacity assessment and the methods used; Section 15.4 summarises the results of the regional adaptive capacity assessment in the European Alps and the investigation of current and potential adaptation strategies; and, finally, we identify specific policy implications for enhancing climate change adaptation in Alpine tourism (see Section 15.5).

15.2 Climate change impacts and the adaptation of Alpine tourism

15.2.1 *Climate change impacts on Alpine tourism*

Over the last two centuries, both summer and winter tourism have emerged as core economic sectors in the Alps. Today, the Alps are the second most-favoured holiday destination in Europe (EEA, 2007), and with 370 million overnight guests (Mayer, Kraus and Job, 2011), tourism is the most important economic sector in many rural regions within the European Alps. At the same time, mountain tourism in the European Alps is one of the economic sectors that is most vulnerable to climate change (Elsasser and Bürki, 2002).[1] Within the last 150 years, the climate in the European Alps has changed significantly (EEA, 2009); the mean temperature has increased by approximately

2 °C, which is more than twice the average rate of warming in the northern hemisphere. Increasing temperatures have resulted in the retreat of glaciers, changes in seasonal mean temperatures and precipitation patterns and a decline in snow cover (EEA, 2009). This development is expected to be reinforced by future climate change and its impacts; recent climate scenarios expect a further increase in temperature, a higher variability of precipitation and an increasing occurrence and intensity of natural hazards (Beniston, 2005; ClimChAlp, 2008a; CH2011, 2011).

Owing to the seasonality of tourism, climatic stimuli that are expected to change due to climate alterations will have different effects on summer and winter tourism. The eight exposure indicators used for the pan-European exposure analysis within the ESPON Climate project (Chapter 4 in this book) and their corresponding effects on Alpine summer and winter tourism are listed in Table 15.1.

The anticipated effects of changing climate stimuli on the tourism sector can also be differentiated according to the altitude of different regions within the European Alps (EEA, 2007; OECD, 2007; Steiger and Mayer, 2008; Rixen *et al.*, 2011). For high Alpine tourism (above 2000 m a.s.l. (above sea level)), increases in mean temperatures and the number of summer days are expected to have a positive seasonal effect because of the freshness of summer resorts (Abegg and Steiger, 2011; Serquet and Rebetez, 2011). In contrast, a decline in the attractiveness of snow sport activities is likely for high Alpine winter tourism due to changes in precipitation patterns, which would result not only in a decrease in the number of days with snow cover but also an increase in the occurrence of natural hazards. The former applies especially for mid- and low-lying ski resorts (1200–2000 m a.s.l.) (OECD, 2007). In high Alpine resorts the amount of snow might increase as some climate scenarios predict more precipitation in the winter season, especially for the Northern Alps (EEA, 2009). Hence, destinations at high altitude will have an advantage compared with mid- and low-lying regions. In lower mountain areas, rural tourism is expected to benefit in the summer because of the increased attractiveness of the

---

[1] For example for Switzerland, Meier (1998) calculated the potential annual costs of climate change at CHF 2.3–3.2 billion by the year 2050. This is 0.6 to 0.8% of the Swiss gross national product for 1995. CHF 1.8–2.3 billion would be accounted for by tourism.

Table 15.1 Economic sensitivity of Alpine tourism related to climatic stimuli and triggered climate effects.

Economic sensitivity	Climatic stimuli									Triggered climate effects
	Change in annual mean temperature	Decrease in the number of frost days	Change in the number of summer days	Change in mean winter precipitation	Change in mean summer precipitation	Change in the number of heavy rainfall days	Change in annual Mean evaporation	Change in the number of days with snow cover	Change in the occurrence of river flooding	Change in mean sea level
Summer tourism	X		X		X				X	
Winter tourism	X	X		X				X		

lake regions. In winter, mid-altitude and low-lying tourist destinations are expected to experience a significant decrease in snow reliability and a corresponding change in the length of the season. In the lowlands of the European Alps, urban tourism in particular is expected to gain attractiveness because of the prolonged summer season and an increased number of summer days, while heat days might have a negative influence on city tourism in summer (SECO, 2011).

15.2.2 Climate change adaptation strategies for the tourism sector in the European Alps

Only recently has climate change adaptation found its way into international, national and regional policies, and adaptation strategies for the tourism sector have now been implemented. On the supranational level, the Alpine Convention developed the Action Plan on Climate Change in the Alps, which was adopted in 2009 by the Xth Alpine Conference. The territorial dimension of this Action Plan is delimited by the perimeter of the Alpine Convention. The Action Plan will be implemented by the Alpine Convention, the Working Groups that are set for a two-year term, the Platforms of the Alpine Convention and the contracting parties, which include the eight Alpine States (Switzerland, Austria, Germany, Slovenia, Monaco, France, Italy and Liechtenstein) (Permanent Secretary of the Alpine Convention, 2009). The Action Plan on Climate Change in the Alps addresses a broad range of sectors, including tourism. The strategies for the tourism sector mainly focus on mitigating greenhouse gas emissions by tourism activities and tourism-related transportation. Adaptation to climate change impacts is addressed via the goal to adapt winter tourism and diversify tourism offers (Permanent Secretary of the Alpine Convention, 2009).

Tourism issues have also been addressed by several climate-related projects of the Alpine Space Programme within the European Territorial Cooperation 2007–2011 framework and its predecessor INTERREG B programmes, for example, ClimAlpTour, 'Climate Change and its Impact on Tourism in the Alpine Space' (2008–2011, ClimAlpTour, 2009) and ClimChAlp, 'Climate Change, Impacts and Adaptation Strategies in the Alpine Space' (2005–2008, ClimChAlp, 2008b).

The Alpine countries are currently in different stages of preparing, developing and implementing national adaptation strategies (NASs): Austria (NAS adopted in 2012), France (NAS adopted in 2006; National Adaptation Plan 2011–2015), Germany (NAS adopted in 2008), Switzerland (NAS adopted in 2012), Italy (National Conference on Climate Change 2007), Slovenia (N.A. (not available)),

Table 15.2 Potential adaptation options for Alpine tourism.

Potential adaptation strategies and measures for Alpine tourism

Developing and securing snow sport activities	● Use of snowmaking equipment ● Cooperation or merger of cableway companies ● Construction of water supply reservoirs for artificial snow-making ● Extension of existing ski areas to higher elevations ● Building of new high-altitude ski areas ● Promotion of glacier skiing
Promoting year-round tourism	● Creation of new summer attractions ● Development of spa programs and the promotion of health-specific activities ● Development of year-round tourism offers that are independent of climate and weather ● Increase the attractiveness of the region by emphasising regional specialities ● Improvement of learning opportunities and cultural offers
Other adaptation activities	● Withdrawal from tourism in ski areas at lower elevations ● Informing tourists of climate change impacts ● Promotion of research and development projects ● Monitoring the impacts of climate change on the tourism sector ● Improving natural hazards management and risk reduction ● Use of insurance instruments ● Others

Sources: Bürki, 2000; Kämpf and Hunziker, 2008; Steiger and Mayer, 2008; Hoffmann *et al.*, 2009; Hill, Wallner and Furtado, 2010; Rixen *et al.*, 2011.

Liechtenstein (N.A.) and Monaco (N.A.) (EEA, 2011; also see Biesbroek *et al.*, 2010). In the existing NASs of France and Germany, tourism has been listed as a relevant sector for climate change adaptation without mentioning detailed strategies for adapting the tourism sector in the Alpine region. In Switzerland, the State Secretariat for Economic Affairs SECO investigated the ecological and economic impacts of climate change on tourism and developed a number of strategies for tourism adaptation as a contribution to the NAS (UVEK, 2011). Three major fields of actions were identified: supply development, risk minimisation and communication (SECO, 2011).

Current studies discuss a broad range of adaptation strategies for mountain tourism in the Alpine space at the regional and local level and vary from short-term reactions in the form of technical adaptation measures (e.g. artificial snow production), economic risk reduction via organisational and financial instruments (e.g. insurances) to long-term adaptation strategies, such as the diversification of tourism offers and a withdrawal from ski tourism at lower elevations (Table 15.2). Nevertheless, comparative studies of the implementation or willingness of stakeholders to implement these strategies in the European Alps are lacking.

15.3 Assessing adaptive capacity: methods and measures

To characterise the adaptive capacity and to identify adaptation activities that have the potential to reduce the vulnerability of the tourism sector in the European Alps the methodology for generic adaptive capacity assessment developed within the ESPON Climate project was chosen. For this assessment, indicators to measure the adaptive capacity of mountain tourism in the European Alps at the regional scale were specified. Values of these indicators were obtained from a questionnaire-based survey of tourism experts.

15.3.1 Assessing the adaptive capacity of sectors and regions

Recently, many vulnerability studies have been conducted with an emphasis on developing and refining a common methodology (Brooks, Adger and Kelly, 2005; Füssel, 2007; Greiving *et al.*, 2011). According to the research framework developed by Füssel and Klein (2006), which is widely used for vulnerability assessments, the vulnerability of a system (e.g. a region or an economic sector) is not only determined by the climatic stimuli and the sensitivity of the system but also by its adaptive capacity. Adaptive capacity is defined as the ability or potential of a system to respond successfully to climate variability and change and includes adjustments in behaviour, resources and technologies (Adger *et al.*, 2007). Accordingly, the identification of the adaptive capacity is a crucial prerequisite for the design and implementation of adaptation strategies. The Fourth Assessment Report of the Intergovernmental Panel on Climate Change (IPCC) outlines two required dimensions of the adaptive capacity: a generic capacity and an impact-specific capacity (Adger *et al.*, 2007). A generic adaptive capacity refers to the general ability of a system to respond to climate change. The impact-specific capacity relates to the adaptation requirements, that is, a specific climate change impact, such as a drought or a flood that poses a threat to the system. Schneiderbauer, Pedoth, Zhang and Zebisch (2011) have added a third dimension, the sector-specific adaptive capacity to describe the capacity of a certain economic sector.

15.3.2 Specifying adaptive capacity indicators for Alpine tourism

Research so far has identified determinants that assess adaptive capacity at the national, regional and local levels (Yohe and Tol, 2002; Keskitalo *et al.*, 2011; Westerhoff, Keskitalo and Juhola, 2011). Based on the IPCC, most studies used six generic determinants to describe adaptive capacity: knowledge and awareness, technology, infrastructure, institutions, economic resources and equity (Adger *et al.*, 2007). For the ESPON Climate project, equity was not considered as a separate determinant but is considered within economic resources and institutional dimensions (see Chapter 7).

To analyse the regional and sectoral adaptive capacity (AC) of tourism in the Alpine space we adopted and modified these generic adaptive capacity determinants (Figure 15.1). Each of these five determinants was measured by a specific set of indicators as is now outlined.

The first determinant, 'knowledge and awareness' for the tourism sector in the Alpine space, was measured by four indicators: problem awareness (AC1.1), access to information (AC1.2), knowledge gaps (AC1.3) and demand for further training (AC1.4). The underlying assumptions are that: (a) regions with a high problem awareness concerning climate change impacts in their own region have a higher adaptive capacity, (b) regions where information regarding climate change impacts is available and (c) are aware of knowledge gaps have a higher adaptive capacity and (d) the demand for further education/professional training regarding possible courses of action concerning adaptation to climate change is a first step in adaptation activities (Hoffmann *et al.*, 2009).

The second determinant 'infrastructure' was operationalised by quantifying the green (AC2.1) and cultural (AC4.2) infrastructure of a region. The underlying assumption is that regions with a high proportion of natural assets (AC2.1) as well as cultural assets (AC4.2) have a higher potential for diversifying their tourism offers in summer and

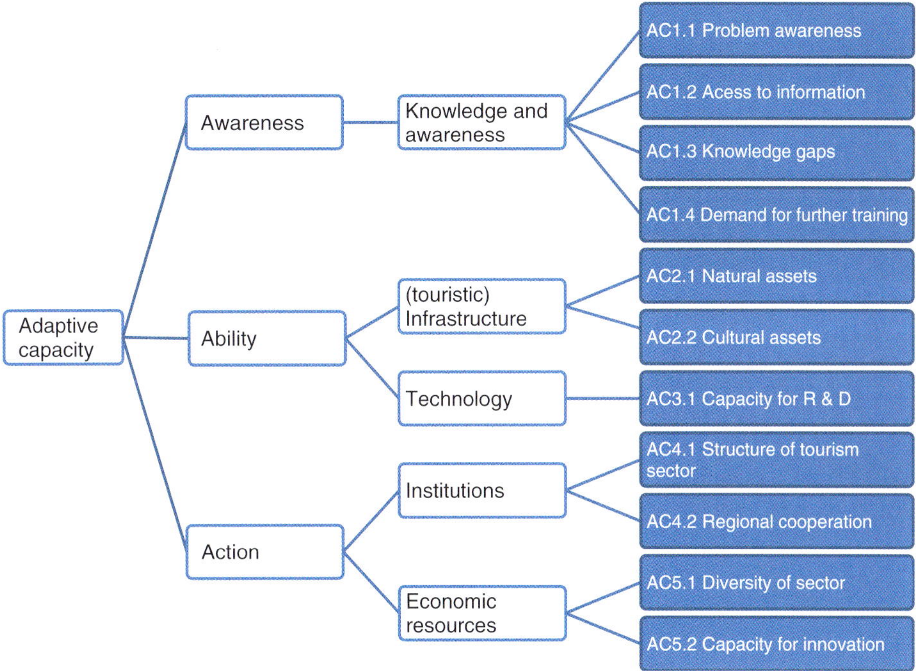

Figure 15.1 Specified adaptive capacity (AC) for Alpine tourism. *Source*: own draft, modified from Chapter 7, in this book.

throughout the year complementing winter sport activities (Loibl and Walz, 2010).[2]

The third determinant "technology" was operationalised by previous investments for research and development (AC3.1) as we assume that regions that have already invested in research and development in the field of adaptation tourism to climate change have a higher technological capacity than regions that have not (yet) invested in such research and development (Brooks, Adger and Kelly, 2005).

The fourth determinant 'institutions' was measured by two indicators: vertical (AC4.1) and horizontal integration (AC4.2) of the tourism sector. The underlying assumptions are that: (a) tourism organisations, which are divided into a few large sections, have a higher adaptive capacity than tourism organisations, which are divided into small sections, and (b) tourism organisations that

have intensive cooperation with local, regional and national authorities have a higher adaptive capacity than tourism organisation with few cooperation activities (Kämpf and Hunziker, 2008).

The determinant 'economic resources', finally, was operationalised by two indicators: diversity of sector (AC5.1) and capacity for innovation (AC5.2). The assumptions are that: (a) a more diverse tourism sector corresponds with a higher adaptive capacity, (b) if winter tourism is contributing to the added value of a region at a high level, the adaptive capacity is diminished and (c) a higher level of innovative tourism options is associated with a higher adaptive capacity (Kämpf and Hunziker, 2008).

15.3.3 *Measuring adaptive capacity of the tourism sector in Alpine regions*

To measure the adaptive capacity of the tourism sector in the European Alps a semi-standardised

---

[2] Note that the term 'infrastructure' is used consistent with the ESPON Climate terminology (Chapter 7, in this book) and thus must not be confused with technical infrastructure of the tourism industry, such as cable cars, ski lifts and so on.

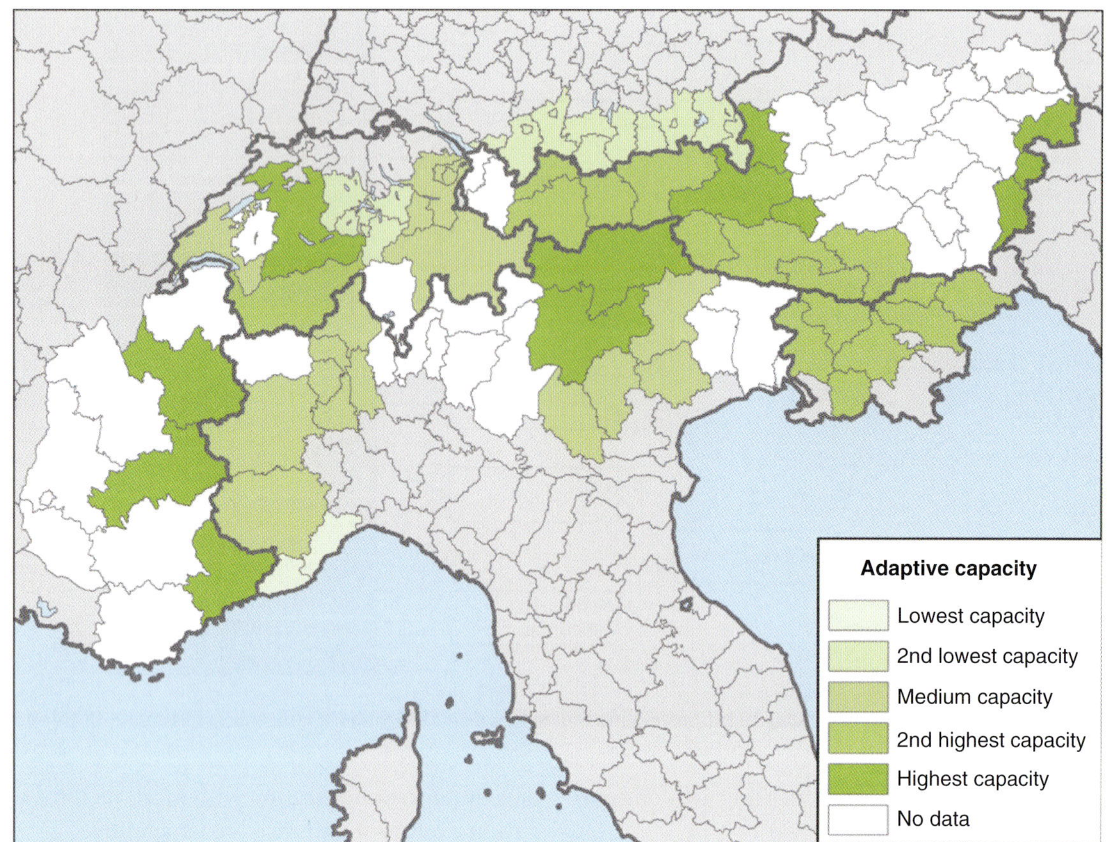

Figure 15.2   Adaptive capacity in the Alps: results of the aggregated adaptive capacity scores for the Alpine regions. *Data source*: EuroGeographics for the administrative boundaries; *cartography*: Felix Neff. Note: The research area is delimited by the perimeter of the Alpine Convention using the boundaries of NUTS3 regions.

questionnaire survey was administered to tourism stakeholders, that is, representatives of tourism organisations as well as public authorities for economic development, environmental agencies and spatial planning authorities at both the national and regional levels. The area of the study was limited to the perimeter defined by the Alpine Convention (Figure 15.2). The survey was conducted with stakeholders at the level of the Länder/Cantons/Provinces/Departments. In Liechtenstein, Monaco and Slovenia, the questionnaire was administered to tourism stakeholders at the national level only.[3] For the analysis of the

survey data, certain cantons in Switzerland were grouped together because a number of tourism organisations cover several smaller cantons (e.g. Eastern Switzerland covers Appenzell Outher Rhoden, Appenzell Inner Rhoden, Glarus and St. Gallen).

Three to five tourism stakeholders per administrative region who were the contact persons for climate adaptation and tourism in their organisation were contacted. These stakeholders were identified through web searches or personal recommendations. The survey was conducted online with the alternative of answering a print version of

---

[3]For more details on the administrative and statistical perimeter of the regions as well as on the participating organisations and the number of responses per region, see Kruse, Pütz, Stiffler and Baumgartner, 2011.

the questionnaire. A total of 193 stakeholders were initially contacted by mail to introduce the project and to explain the goals and procedures of the survey; 94 stakeholders participated, yielding a response rate of 48.7%. No responses were received from five out of 36 regions (i.e. Vorarlberg (Austria), Alpes-des-Hautes-Provences (France), Isère (France), Var (France), Vaucluse (France)). Determinants and indicators measuring adaptive capacity per region were retrieved using different types of survey questions (closed/open questions, single/multiple choice questions and a three- or five-level Likert scale). All scores were normalised on a scale from zero (lowest level) to ten (highest level) to enable an aggregation of the various items. The aggregation of the different indicators within each determinant was achieved by equal weighting (for details regarding the description, calculation and aggregation of the indicators used here, refer to Kruse, Pütz, Stiffler and Baumgartner, 2011). If questions were not answered by any expert in a particular region, they were marked as missing (N.A.) and could not be included in the calculation of the indicators. As a result, the aggregated score (ACagg) contained missing data and was only available for 20 regions. For indicators AC1, AC2 and AC5, results for 20, 25 and 29 regions, respectively, were at hand.

To investigate the previously realised and potential adaptation options, the survey included questions where the expert could choose between a selection of adaptation options (Table 15.2) and add additional options. The experts were also asked to rank the realised and potential adaptation options according to their effectiveness. The hypotheses behind these items are as follows: (a) regions where adaptation activities already have been implemented have a higher adaptive capacity; (b) regions that are implementing long-term adaptation strategies have a higher adaptive capacity than regions that are implementing short-term adaptation strategies; and (c) a greater number of adaptation options in a region corresponds with a higher capacity for adaptation to climate change. Finally, we integrated two control questions to assess the self-estimation of the adaptive capacity (control question 1) and the willingness to adapt (control question 2). These questions are

based on the assumption that a higher awareness of one's own adaptive capacity and a higher willingness to implement adaptation measures in a region is directly correlated with a higher probability to realise adaptation options.

15.4 Results

15.4.1 *Adaptive capacity of Alpine tourism: regional spotlights*

The results show a differentiated picture of strengths and weaknesses concerning the five adaptive capacity determinants. The results are based on the questionnaire survey and focus on the main determinants 'knowledge and awareness' (AC1), 'institutions' (AC4) and 'economic resources' (AC5). In addition, the aggregated score (ACagg) is examined (Figure 15.2).

In the following presentation of the results, we discuss how the regions of each country were assessed relative to the other regions according to the highest seven and lowest seven scores of all Alpine regions. For Switzerland, 15 cantons were contacted for the survey, and at least one questionnaire was returned from all cantons. Berne was among the top seven Alpine regions with respect to the overall score for adaptive capacity as well as the determinant 'knowledge and awareness' (AC1). Additionally, in the self-assessment of adaptive capacity (control question 1), Berne achieved one of the highest scores among all regions. Valais and Ticino were also among the top seven with respect to the 'institutions' determinant (AC4). Eastern Switzerland, Central Switzerland, Valais and Vaud were among the seven lowest-scoring areas with respect to the 'knowledge and awareness' (AC1) determinant.

In the Austrian part of the Alps, eight Bundesländers were contacted. No reply was received from Voralberg. Among the other regions, Burgenland and Salzburg were within the top seven among all regions of the Alpine environment for the aggregated score (ACagg). With regard to the determinant 'knowledge and awareness' (AC1), Salzburg and

Tyrol had the highest scores due to a relatively high awareness of climate change impacts on tourism in the respective regions, which is a reflection of existing knowledge gaps, the demand for further information and training for individuals in the tourism sector. With regard to the determinant 'institutions' (AC4), the regions Carinthia, Tyrol and Upper Austria were in the top seven regions, whereas Styria was among the bottom seven. For the determinant 'economic resources' (AC5), Burgenland, Lower Austria and Upper Austria had the highest scores, mainly due to their capacity for innovation and their low dependency on winter tourism, whereas Tyrol and Styria were in the lowest range because winter tourism in both regions is economically most important while the tourism sector is not considered to be innovative. Both of these factors combine to yield a low adaptive capacity regarding economic resources. At the same time, the respondents of both Tyrol and Styria considered their own capacity to adapt tourism to climate change impacts as high (control question 1).

For Germany, the state of Bavaria was assessed, and all experts who were contacted participated in the survey. The results showed a relatively high problem awareness compared with other regions in the Alpine space. In particular, the access to information was positive due to the availability of regional climate change scenarios and vulnerability assessments. Tourism offers in this area are diverse, and winter tourism is secondary for the regional tourism economy.

In the French part of the Alps, all nine departments were contacted for the survey. We did not receive any reply from experts from Alpes-de-Haute-Provence, Isère, Var or Vaucluse. The departments Alpes-Maritimes and Savoie were among the top seven of all Alpine regions for the aggregated score. Alpes-Maritimes and Hautes-Alpes achieved high scores in the determinant 'knowledge and awareness' (AC1), especially due to a sufficient knowledge and information base at the regional level (e.g. regional climate change scenarios are available). Additionally, various measures to overcome knowledge gaps and to enhance professional training on climate issues within the tourism sector have been realised. The department Alpes-Maritimes and Drôme received high scores in the determinant 'economic resources' (AC5). Their tourism offers are very diverse, and winter tourism is not among the main economic drivers for the tourism sector.

Seven provinces were contacted in the Italian part of the Alps. Answers were not received from the Friulia Venezia Giulia province. Trentino Alto Adige was in the top seven of all Alpine regions in the overall adaptive capacity score (ACagg) as well as the scores for 'knowledge and awareness' (AC1) and 'institutions' (AC4). The current information available is considered to be sufficient for taking action and ranges from national and regional climate change scenarios to climate impact studies and vulnerability assessments. The experts state that no additional information is needed, although a higher level of acceptance of climate change activities by tourism stakeholders is necessary. Therefore, the main objective of further training in Italy should be to increase problem awareness within the tourism sector. The tourism sector in Trentino Alto Adige is sufficiently diverse, although winter tourism contributes the largest share of the economic value within the tourism sector.

In Slovenia, experts from the national level participated in the survey. The overall adaptive capacity scores from Slovenia were in the upper middle range of all Alpine regions. In the field of 'knowledge and awareness' (AC1), Slovenia was among the highest scores, especially due to a good informational basis and a high awareness of the knowledge gaps. The main challenges for enhancing adaptive capacity are considered to be an increase in the awareness and education of tourism stakeholders and the identification of financial support for realising adaptation activities in the tourism sector.

We received only one reply from Monaco, and this response was missing certain information. Therefore, an aggregated score could not be calculated. In comparison with the other Alpine regions, Monaco had a high score concerning the self-assessment of adaptive capacity but had rather low scores in 'problem awareness' (AC1.1), 'access to information' (AC1.2) and 'institutions' (AC4). These results should be considered in the context of the fact that Monaco is a relatively small territory. Thus, Monaco is dependent on regional climate data and scenarios from the surrounding countries and regions (especially France). Nevertheless, Monaco has a diversified tourism sector that is primarily dependent on forms of tourism that are not dependent on climate, such as congress tourism, cultural tourism and event trips. Together with a rather innovative tourism sector, these factors led to a high score for the determinant 'economic resources'.

For Liechtenstein, which is a small country, the informational basis for implementing adaptation options is a major challenge and is only considered partially sufficient. At the same time, the tourism economy relies mainly on revenues from winter tourism, and the tourism sector is not considered to be innovative. Therefore, the aggregated score for the adaptive capacity of Liechtenstein is relatively low compared with other Alpine regions.

15.4.2 *Current and potential climate change adaptation strategies*

As outlined in Section 15.2.2, comparative studies of the implementation or the willingness of tourism stakeholders to implement climate change adaptation strategies have not been previously performed. The results of the questionnaire survey show that almost 78% of tourism stakeholders in the Alpine space who participated in the survey agree or strongly agree that it would be sensible to immediately enact measures to prevent the negative consequences of climate change on tourism in their region. For example, adaptation strategies have already been implemented in all regions that were surveyed in the European Alps. One of the areas of adaptation that is currently widely realised by tourism stakeholders in the Swiss Alps involves developing and securing snow sport activities (Figure 15.3). There has been a major effort to improve artificial snowmaking, that is, investing in additional snowmaking equipment or constructing water reservoirs for snowmaking. In addition

to securing snow sport activities, stakeholders also engage in many adaptation activities geared towards year-round tourism. This includes the innovation and diversification of tourism offers, for example, the creation of new summer attractions, the development of spa programmes, the promotion of health-related activities and other climate- and weather-independent activities. Other widespread adaptation strategies are dedicated to increasing the attractiveness of different regions by emphasising their regional specialties (e.g. local food) and improving natural hazard management.

Of the adaptation strategies and measures that stakeholders consider to have the potential to be realised by 2020, the most important strategy will be the creation of new summer attractions that compensate for predicted losses during the winter season due to climate change effects. In addition, the closure of ski areas at lower elevations and the dissemination of climate change information to tourists will become more important.

A comparison of current and potential future adaptation strategies indicates that developing and securing snow sport activities will become considerably less important by 2020 from the point of view of tourism stakeholders. In particular, adaptation options that use additional snowmaking equipment and the construction of water reservoirs for artificial snowmaking are expected to become less feasible in the future. At the same time, the promotion of year-round tourism is expected to increase until 2020. According to tourism stakeholders, the monitoring of climate change impacts on the tourism sector and promoting research and development projects that actively improve the adaptation of tourism in response to climate change will become much more important in the future.

When asked what additional activities should be realised to increase the adaptive capacity of the tourism sector, the most common answer from many tourism stakeholders was the raising of awareness of climate change impacts among tourism participants. Other activities for fostering adaptive capacities include the creation of financial support for innovative tourism offerings and the initiation of long-term strategies in the tourism sector.

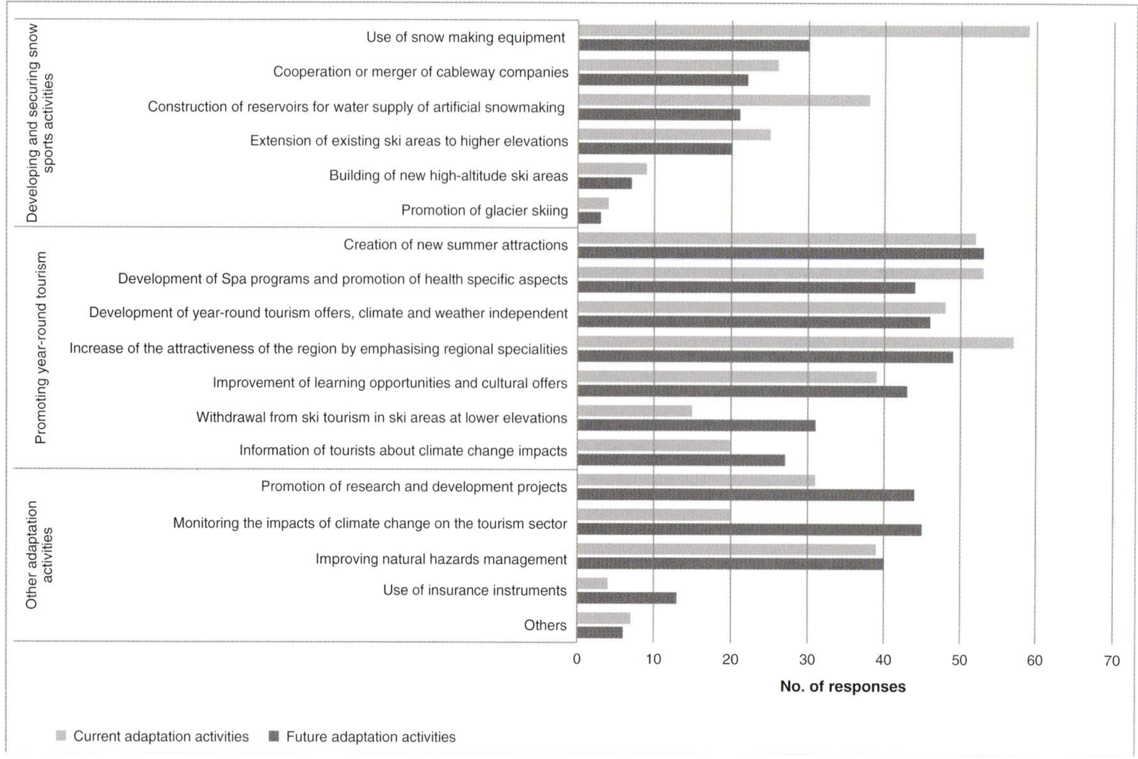

Figure 15.3 Current and future adaptation activities. The results represent the rate of positive answers of all respondents (N = 79). Survey questions: 'Which adaptation activities have been implemented in the tourism sector in your region so far?' (current adaptation activities) and 'Which adaptation activities are feasible within the tourism sector in your region by 2020?' (future adaptation activities). Multiple answers were possible.

15.5 Discussion and conclusions

The results manifest a broad range of adaptation strategies and measures that are already being realised today. These findings corroborate the European Commission's assumption that early action and capacity development is needed within the tourism sector, which is likely to be affected by climate-related impacts (Commission of the European Communities, 2009b).

Concerning the currently realised adaptation strategies and the willingness to adapt in the future, the results show that adaptation activities within the tourism sector are primarily realised in an autonomous, private and local manner. In particular,

the tourism stakeholders that we interviewed listed a broad range of corporate measures (e.g. providing snowmaking equipment) that are actualised on a private basis without a comprehensive and long-term strategy. They mention mostly short- and medium-term adaptation options, which can be seen as no-regret or low-regret measures with benefits even without climate change. However, a number of these autonomously and locally realised strategies and measures (e.g., the use of snow making equipment, extension of existing ski areas to higher elevations, construction of reservoirs for water supply of artificial snowmaking and the promotion of glacier skiing) that strive to prevent the negative economic impacts of climate change may produce negative side effects on the environment. Several studies have shown, for example, that artificial snowmaking might have negative impacts on the seasonal and regional water balance and on the natural environment (Vanham, Fleischhacker and Rauch, 2009; Rixen *et al.*, 2011; Roux-Fouillet, Wipf and Rixen, 2011). Our results show that artificial snowmaking equipment and infrastructure are currently among the most widely used autonomous adaptation measures. Nevertheless, not all private activities focus on securing snow sport activities for future development. Many private and local activities emphasise the development and promotion of year-round and weather-independent tourism and attempt to increase the attractiveness of regional specialties. According to tourism stakeholders, these adaptation strategies will become even more important in the future, whereas securing snow sport activities and additional snowmaking equipment will lose importance by 2020. These results can be contextualised in two ways: on the one hand, snowmaking is expected to become less efficient and too expensive in lower lying regions with fewer number of frost days (OECD, 2007; Rixen *et al.*, 2011); on the other hand, today some skiing resorts, especially in Austria and Italy, have already equipped a high proportion of slopes with artificial snowmaking and, thus, have only slight potential for additional snowmaking (OECD, 2007; Steiger and Mayer, 2008).

The chosen methodology involves some limitations of the scope of results. Because the assessment is based on expert opinion rather than quantifiable statistical data, the results are not comparable in absolute terms but reflect the relative self-evaluation of the participating tourism stakeholders. Additionally, the number of responses per region is very small (1–10) and not representative. Further, there might be a bias in the responses, which can also be seen in other expert-based surveys: those who reply to the survey tend to see climate change as a problem and moderate a positive image of their own region and activities. Nevertheless, the assessment demonstrates how tourism stakeholders characterise the adaptive capacity of their own region and where they see opportunities for improvement. Even if the comparative design of the study levels out regional distinctions and, in the face of the limitations which

go along with qualitative and expert-based surveys, the comparative assessment of adaptive capacity identifies the performance of the regions according to the different determinants of adaptive capacity. The results can thus lead to starting points for discussion on climate change adaptation within each region, involving more data and stakeholders.

We believe that the results and methodology presented here are partly transferable to other mountain regions in Europe where the regional economy is also highly dependent on the tourism sector and where winter tourism in particular is important (e.g. the Pyrenees). However, certain indicators may need to be modified depending on the specific sensitivity of the regional tourism offers to climate change. To transfer the methodology to other regions that are dependent on tourism but generate most of their yearly revenues in summer (e.g. Mediterranean regions), specific indicators would have to be revised (Perch-Nielsen, 2008). Additionally, further research is needed on the interconnections between different determinants and indicators as well as hindering and supporting factors for adaptive capacity, which have not yet been fully understood (e.g. Haddad, 2005).

Finally, it is necessary to note that, on the one hand, climate adaptation in the tourism sector is prompted by economic incentives. In general, tourism stakeholders are not motivated to take adaptive action to address climate change but rather strive to prepare for increasing losses in winter tourism and to improve tourism infrastructure and supply to meet future tourism demand. On the other hand, adaptation decisions are very complex and uncertain concerning their effects if they aim at long-term development and take into account a variety of social, institutional and political barriers (Smith *et al.*, 2011).

References

Abegg, B. and Steiger, R. (2011) Will Alpine summer tourism benefit from climate change? A review, in: (eds A. Borsdorf, E. Stötter and E. Veulliet) *Managing Alpine Future II – Inspire and Drive Sustainable Mountain*

Regions – Proceedings of the Innsbruck Conference November 21–23, 2011, Verlag der Österreichischen Akademie der Wissenschaften, Wien, Austria, pp. 268–277.

Adger, W.N., Agrawala, S., Mirza, M.M.Q., *et al.* (2007) Assessment of adaptation practices, options, constraints and capacity, in: (eds M.L. Parry, O.F. Canziani, J.P. Palutikof, P.J. van der Linden, C.E. Hanson) *Climate Change 2007: Impacts, Adaptation and Vulnerability – Contribution of Working Group II to the Fourth Assessment Report of the Intergovernmental Panel on Climate Change*, Cambridge University Press, Cambridge, pp. 717–743.

Amelung, B. and Moreno, A. (2009) *Impacts of climate change in tourism in Europe. PESETA-Tourism study.* (pdf) Office for Official Publications of the European Communities, Luxembourg. Available at: <http://ftp.jrc.es/EURdoc/JRC55392.pdf> (accessed 30 November 2011).

Bavay, M., Lehning, M., Jonas, T. and Lowe, H. (2009) Simulations of future snow cover and discharge in Alpine headwater catchments. *Hydrological Processes*, **23** (1), 95–108.

Beniston, M. (2005) Mountain climates and climatic change: An overview of processes focusing on the European Alps. *Pure and Applied Geophysics*, **162** (8–9), 1587–1606.

Biesbroek, G.R., Swart, R.J., Carter, T.R., *et al.* (2010) Europe adapts to climate change: Comparing National Adaptation Strategies. *Global Environmental Change*, **20** (3), 440–450.

Brooks, N., Adger, W.N. and Kelly, P.M. (2005) The determinants of vulnerability and adaptive capacity at the national level and the implications for adaptation. *Global Environmental Change*, **15** (2), 151–163.

Bürki, R. (2000) Klimaänderung und Anpassungsprozesse im Wintertourismus. Ph.D. thesis, Zürich University.

CH2011 (2011) *Swiss Climate Change Scenarios CH2011.* C2SM, MeteoSwiss, ETH, NCCR Climate and OcCC, Zurich, Switzerland.

ClimAlpTour (2009) *Climate change and its impact on Tourism in the Alpine Space.* (online) Available at: <www.climalptour.eu> (accessed 27 June 2012).

ClimChAlp (2008a) *Impacts of Climate Change on Spatial Development and Economy: Synthesis and Model Region Studies. Extended Scientific Report of WP7 of the Interreg III B Alpine Space Project ClimChAlp.* (pdf) Available at: <http://www.climchalp.org/index.php?option=com _docman&task=doc_download&gid=196&&Itemid =125> (accessed 30 November 2011).

ClimChAlp (2008b) *ClimChAlp.org – Home.* (online) Available at: <www.climchalp.org> (accessed 27 June 2012).

Commission of the European Communities (2009a) *White paper – Adapting to climate change: Towards a European framework for action.* (pdf) COM, Brussels. Available at: <http://eur-lex.europa.eu/LexUriServ/LexUriServ.do?uri =COM:2009:0147:FIN:EN:PDF> (accessed 30 November 2011).

Commission of the European Communities (2009b) Impact Assessment – Commission staff working document. (pdf) SEC, Brussels. Available at: <http://eur-lex.europa.eu /LexUriServ/LexUriServ.do?uri=SEC:2009:0387:FIN:EN :PDF> (accessed 30 November 2011).

EEA (European Environment Agency) (2007) *Europe's Environment: The fourth assessment.* (pdf) Copenhagen: EEA. Available at: <http://www.eea.europa.eu/publications /state_of_environment_report_2007_1/Belgrade_EN_all _chapters_incl_cover.pdf> (accessed 30 November 2011).

EEA (2009) *Regional Climate Change and Adaptation – The Alps facing the challenge of changing water resources.* (pdf) EEA, Copenhagen. Available at: <http://www .eea.europa.eu/publications/alps-climate-change-and-adaptation-2009> (accessed 30 November 2011).

EEA (2011) *National adaptation strategies.* (online) Available at: <http://www.eea.europa.eu/themes/climate/national-adaptation-strategies> (accessed 29 November 2011).

Elsasser, H. and Bürki, R. (2002) Climate change as a threat to tourism in the Alps. *Climate Research*, **20** (3), 253–257.

Füssel, H.M. (2007) Vulnerability: A generally applicable conceptual framework for climate change research. *Global Environmental Change*, **17** (2), 155–167.

Füssel, H.M. and Klein, R.J.T. (2006) Climate change vulnerability assessments: An evolution of conceptual thinking. *Climatic Change*, **75** (3), 301–329.

Greiving, S., Flex, F., Lindner, C., *et al.* (2011) *ESPON CLIMATE: Climate Change and Territorial Effects on Regions and Local Economies – Final Report.* (pdf) ESPON and IRPUD, Dortmund. Available at: <http://www.espon.eu/export/sites/default/Documents/ Projects/AppliedResearch/CLIMATE/ESPON_Climate _Final_Report-Part_B-MainReport.pdf> (accessed 30 November 2011).

Haddad, B.M. (2005) Ranking the adaptive capacity of nations to climate change when socio-political goals are explicit. *Global Environmental Change*, **15** (2), 165–176.

Hill, M., Wallner, A. and Furtado, J. (2010) Reducing vulnerability to climate change in the Swiss Alps: a study of adaptive planning. *Climate Policy*, **10** (1), 70–86.

Hoffmann, V.H., Sprengel, D.C., Ziegler, A., *et al.* (2009) Determinants of corporate adaptation to climate change in winter tourism: An econometric analysis. *Global Environmental Change*, **19** (2), 256–264.

Kämpf, R. and Hunziker, C. (2008) *Erfolg und Wettbewerbsfähigkeit im alpinen Tourismus.* BAK (Basel Economics), Basel.

Keskitalo, E.C.H., Dannevig, H., Hovelsrud, G.K., *et al.* (2011) Adaptive capacity determinants in developed states:

examples from the Nordic countries and Russia. *Regional Environmental Change*, **11** (3), 579–592.

Kohler, T., Giger, M., Hurni, H., *et al.* (2010) Mountains and climate change: A global concern. *Mountain Research and Development*, **30**, 53–55.

Kruse, S., Pütz, M., Stiffler, M. and Baumgartner, D. (2011) *ESPON CLIMATE: Climate Change and Territorial Effects on Regions and Local Economies. Applied Research Project 2013/1/4. Final Report. Annex 1: Case Study Alpine Space*, ESPON and WSL, Birmensdorf.

Loibl, W. and Walz, A. (2010) Generic regional development strategies from local stakeholders' scenarios – an Alpine village experience. *Ecology and Society*, **15** (3), 3.

Mayer, M., Kraus, F. and Job, H. (2011) Tourismus – Treiber des Wandels oder Bewahrer alpiner Kultur und Landschaft? *Mitteilungen der Österreichischen Geographischen Gesellschaft*, **153**, 31–74.

Meier, R. (1998) *Sozioökonomische Aspekte von Klimaänderungen und Naturkatastrophen in der Schweiz (Projektschlussbericht NFP 31)*, Vdf Hochschulverlag, Zurich.

Müller, H. and Weber, F. (2008) Climate change and tourism – scenario analysis for the Bernese Oberland in 2030. *Tourism Review*, **63** (3), 57–71.

OECD (Organisation for Economic Co-operation and Development) (2007) *Climate Change in the European Alps: Adapting Winter Tourism and Natural Hazard Management*. OECD, Paris.

Perch-Nielsen, S.L. (2008) Climate change and tourism intertwined. Ph.D. thesis, ETH Zurich University.

Permanent Secretary of the Alpine Convention (2009) The Action Plan on Climate Change in the Alps. (pdf) Available at: <http://www.alpconv.org/NR/rdonlyres/193D7A9E-0F5E-475D-A48D-E3276F11D292/0/AC_X_B6_en_new_fin.pdf> (accessed 30 November 2011).

Pütz, M., Gallati, D., Kytzia, S., *et al.* (2011) Winter tourism and climate change: attitudes towards artificial snow and regional economic impacts of snowmaking. *Mountain Research and Development*, **31**, 357–362.

Rixen, C., Teich, M., Lardelli, C., *et al.* (2011) Winter tourism and climate change in the Alps. An assessment of resource consumption, snow reliability and future snowmaking potential. *Mountain Research and Development*, **31** (3), 229–136.

Roux-Fouillet, P., Wipf, S. and Rixen, C. (2011) Long-term impacts of ski piste management on alpine vegetation and soils. *Journal of Applied Ecology*, **48** (4), 906–915.

Schneiderbauer, S., Pedoth, L., Zhang, D. and Zebisch, M. (2011) Assessing adaptive capacity within regional climate change vulnerability studies – an Alpine example. *Natural Hazards*, 1–15.

SECO (Staatssekretariat für Wirtschaft) (2011) *Der Schweizer Tourismus im Klimawandel Auswirkungen und Anpassungsoptionen*, Staatssekretariat für Wirtschaft, Bern.

Serquet, G. and Rebetez, M. (2011) Relationship between tourism demand in the Swiss Alps and hot summer air temperatures associated with climate change. *Climatic Change*, **108** (1), 291–300.

Smith, M.S., Horrocks, L., Harvey, A. and Hamilton, C. (2011) Rethinking adaptation for a 4° world. *Philosophical Transactions of the Royal Society A*, **369** (1934), 196–216.

Steiger, R. and Mayer, M. (2008) Snowmaking and climate change future options for snow production in Tyrolean ski resorts. *Mountain Research and Development*, **28** (3/4), 292–298.

UVEK (Eidgenössisches Departement für Umwelt, Verkehr, Energie und Kommunikation) (2011) *Anpassung an die Klimaänderung in der Schweiz – die Strategie des Bundes. Entwurf*, UVEK, Bern.

Vanham, D., Fleischhacker, E. and Rauch, W. (2009) Impact of snowmaking on alpine water resources management under present and climate change conditions. *Water Science & Technology*, **59** (9), 1793–1801.

Westerhoff, L., Keskitalo, E.C.H. and Juhola, S. (2011) Capacities across scales: local to national adaptation policy in four European countries. *Climate Policy*, **11** (4), 1071–1085.

Yohe, G. and Tol, R.S.J. (2002) Indicators for social and economic coping capacity – moving toward a working definition of adaptive capacity. *Global Environmental Change*, **12** (1), 25–40.

Chapter 16
Comparative analysis of the case studies

Stefan Greiving

*TU Dortmund University, Institute for Spatial Planning (IRPUD),
August-Schmidt-Strasse 10, 44227 Dortmund, Germany*

Abstract

The main features of the seven case studies are summarised, and prove the applicability of the common methodological framework that had been agreed to ensure compatibility with the pan-European analyses and to enable comparisons across the case studies. It was shown that this framework is flexible in terms of spatial scales and indicators for exposure, sensitivity and adaptive capacity. These studies are very good examples that the new comprehensive ESPON approach meets the demands of spatial planning: a new, more complex picture of the patterns of vulnerability became visible and can therefore be seen as a step forward from pure sector-based studies towards a more comprehensive view on vulnerability.

16.1 Comparative analysis

In order to ensure compatibility with the pan-European analyses of the other research actions of the project and to enable comparisons across the case studies, a common methodological framework was agreed and was followed throughout the performance of the seven case studies. This framework covered the following general aspects:

1. General characterisation of the region.
2. Vulnerability assessment
   a. main effects of climate change on case study region so far;
   b. exposure to climate change;
   c. sensitivity to climate change;
   d. impacts of climate change;

e. adaptive capacity in regard to climate change;
f. vulnerability to climate change.
3. Response strategies and policy development in regard to mitigation and adaptation.
4. Further aspects specific to the respective case study area.
5. Discussion of validity of European-wide analysis from a regional perspective.
6. Transferability of results to other regions.

Table 16.1 summarises the main features of each case study and is structured along the common framework, which was explained above.

All in all, the case studies proved the applicability of the conceptual framework. It was shown that this framework is flexible in terms of spatial scales and indicators for exposure, sensitivity and adaptive capacity. The seven case studies are very good examples that the new comprehensive ESPON approach meets the demands of spatial planning: a new, more complex picture of the patterns of vulnerability became visible and can therefore be seen as a step forward from pure sector-based studies towards a more comprehensive view on vulnerability.

The spatial patterns between the pan-European assessment and the case study assessments are quite similar when comparing, for example, the pan-European cluster analysis with the analysis conducted for the NRW study: its case study area is divided into the same three different climate change types although slightly different exposure indicators were chosen. However, particularly the more fine-grained case study on North Rhine–Westphalia, but also the Tisza River case study, show a more differentiated picture in terms of impact, adaptive capacity and vulnerability than the results of the pan-European assessment for these areas. This is mainly due to the normalisation of data: the existing relative differences between the municipalities of the case study area are fairly small compared with the differences across the whole continent; even those municipalities that are marked in red on the case study map are only moderately vulnerable from a pan-European perspective. Thus, the pan-European vulnerability map shows a more homogenous picture for North Rhine–Westphalia.

This clearly underlines the scale-dependency of any vulnerability assessment. The Tisza River case study shows what an uncertainty analysis could look like. Each exposure indicator provided by the pan-European assessment was intensively validated by comparison with available results from other studies and scientific literature that cover the case study area. This approach is principally useful for any vulnerability assessment on the regional and local level in order to reduce the inherent uncertainty in the models and indicators.

Institutional and cultural issues were only partly covered by the case studies mostly due to the lack of adequate data, but also available resources. There was a particular focus on these topics in the Alpine case study, which was based on an extensive questionnaire survey. To conclude, a more qualitative approach is needed in order to understand the driving forces for institutional settings and related response strategies. All the case studies pointed out that adaptation has to be addressed in a more comprehensive way by spatial planning on the different spatial scales.

However, there is no visible connection between the attention paid to spatial planning and the type of administrative or planning system of the respective country. Consequently, other factors such as political priorities of a national government might determine the relevance of adaptation on the political agenda. This observation was also demonstrated by other studies (Greiving and Fleischhauer, 2012).

The results of the economic sensitivity assessment on tourism correspond almost completely with the results of the case study on coastal Mediterranean Spain: there is a gradient from the North to the South where both studies calculated the greatest potential impact and vulnerability. However, the case study results are much more fine-grained (LAU2) and reflect possible situations of 'maladaptation' and therefore possible conflicts between mitigation and adaptation measures on the very local level to which national and regional strategies on climate change, at least for the case of Spain, have not responded adequately yet. Here, the added value of the case study approach becomes clearly visible, which is also underlined

Table 16.1　Main results of the case studies.

Case study	Exposure	Sensitivity	Impact	Adaptive capacity	Vulnerability	Policy response strategies	Applicability of European method	Transferability
Bergen	Is based on different climatic models. All variables for temperature were used + precipitation + sea-level rise and flooding	Physical, cultural and economic dimensions were considered for a quantitative assessment. Study area (LAU2) was divided into city districts	No quantitative assessment	Pan-European approach (awareness, ability, action) was implemented on a city district level	No quantitative assessment	Bergen was the first municipality in Norway to work out a climate plan in year 2000. Response strategies focus on dealing with extreme events	No general validation, but its applicability to the local level was successfully proved	Test of modified CBA makes clear that use is difficult for assessing vulnerability to climate change
Alpine space (focus on tourism)	All climatic stimuli were considered + inundation	Focus on institutional and cultural dimensions. Study is based on NUTS 3	No quantitative assessment	Semi standardised survey. All three dimensions were covered. Indicators are partly the same, but some adjustments for local needs	No quantitative impact/vulnerability assessment	Several options discussed which are of relevance for mountain areas. Adaptation is seen as a multi-level governance issue	No general validation, but adaptive capacity was proved as relevant factor for the vulnerability assessment	Generally speaking possible to transfer to other mountain areas which are dependent on tourism. Semi-quantified survey as option for regionalised Delphi weighting
Netherlands (focus on flooding)	LISFLOOD map for fluvial flooding + coastal flooding + worst credible flood map = focus on indirect climate change	All five dimensions are covered. Each indicator has a relative and absolute value = final value as average of relative and absolute values. Study is based on LAU2	Impact is calculated by adding exposure and sensitivity score. Exposure and sensitivity are weighted for each dimension (0.44 + 0.56)	In line with the ESPON approach, the estimation of the adaptive capacity is based on generic features: percentage of inhabitants with tertiary degree, computer use, highway density, GDP and age distribution on the municipal level	The final merging of the adaptive capacity and impact into a vulnerability map on the municipal level resembles the impact map, but with a more smoothed pattern due to the almost uniform distribution of the adaptive capacity over the Dutch municipalities	Focus on national level is on water (sea-level rise, flooding) while heat is understood as an issue for some major cities only	The pan-European concept has been successfully applied. However, it became clear that a full vulnerability assessment is only partly suitable for homogenous regions. Moreover, the applicability of the LISFLOOD model was proved.	The approach is transferable to the others since it follows the pan-European assessment

(continued overleaf)

Table 16.1 (*continued*)

Case study	Exposure	Sensitivity	Impact	Adaptive capacity	Vulnerability	Policy response strategies	Applicability of European method	Transferability
Tisza River	All climatic stimuli were considered and extensively validated by using reference runs with other models and literature sources Weighting of stimuli according to results of validation process	Focus on physical, social, environmental and economic sensitivity Particular attention is paid to agriculture Weighting of dimensions according to regional expert's opinion Study is based on NUTS 3	Matrix with five classes (impact on human health and agriculture)	Approach follows the pan-European assessment Indicators used: technology, infrastructure and economic resources Weighting of the three dimensions according to regional expert's opinion	Vulnerability is calculated on the basis of potential impact and aggregated adaptive capacity Additional regional socio-economic assessments for Lake Tisza and Romanian Tisza River based on questionnaires	Seen as a multi-level governance. Particular focus on the transnational Tisza River Programme Extensively discussed for national and regional levels of Hungary, Romania, Slovakia	The agricultural vulnerability assessment can be used in regions where the role of the agriculture and the changes in the climate parameters affecting agriculture change similarly to that in the Tisza region. Such are, e.g., the countries of South-Eastern Europe	Advanced validation concept of exposure indicators is principally transferable depending on the availability of other studies
North Rhine–Westphalia	All climatic stimuli were used, but instead of summer days heat days (30 °C) + storm days as new variable (better represented in study area) Owing to the more similar climate in NRW, absolute changes were used for hydrologic variables Own cluster analysis (3 cl) Inundation as indirect climate change	Four dimensions were considered (culture is missing without a justification) Used indicators are almost the same as for the pan-European assessment Study is based on LAU2	Multiplication of exposure and sensitivity values	Consistent with pan-European assessment, but some adjustments concerning indicators due to small size of study areas Economic resources of municipalities and households Knowledge and awareness (education level + participation in climate change initiatives)	No use of regionalised Delphi weighting Results show higher potential impacts in urban areas which are characterised by a low adaptive capacity A large gradient is apparent between NRW (return period as basis for design of dikes 1 in 100 year event) and the Netherlands, which apply 1250 years	Widely discussed. Need for cross-sectoral as well as vertical integration of response strategies in the political context Existing political focus on urban areas is in line with the results of the case study Given new scientific finding (projected increase in flood risk) current adaptation to flooding should be re-evaluated in NRW	The pan-European concept has been jointly developed together with this case study concept. The overall methodological frame could thus be well applied to the regional scale of NRW in a quantified way	This methodology is in general transferable to other regions The selection of impact chains should be adapted to the specific regional relevance Given a better data source, for some sectors, absolute vulnerabilities or impacts can be determined

Spanish coast	All climatic stimuli were used, but no indirect climate change	All five dimensions are covered, but culture was excluded due to a lack of data Reduced set of indicators (one indicator per dimension), tailor-made for case study, which is based on LAU2 level	Impact is calculated by adding exposure and sensitivity score Exposure and sensitivity are weighted for each dimension (0.44 + 0.56)	Reduced set of indicators: economic (regional income) + technological (desalinisation capacity + water re-use)	ESPON Climate Standard procedure. Case study made use of pan-European weighting scores. The results show hot spots in the south (i.e. Costa del Sol)	Access to resources that are climate independent (desalinisation) is currently regarded as important Need for new forms of governance in order to avoid dependency on costly water technologies Relevance of urban patterns	The pan-European concept has been successfully applied Exposure to climate change tends to be high, sensitivity tends to be high as well; as a consequence, impacts are high and adaptive capacity tends to be low at least in comparison with other European regions	Methods are transferable to other European tourist areas where adaptive capacity actions may produce unintended negative effects (?)
Coastal aquifers	Particular focus on coastal aquifers = sectoral study Owing to data shortcomings, the study focused on Finland mainly All stimuli + indirect effects were considered, eight were proved as important	All five dimensions were considered, but culture excluded due to a lack of data and its irrelevance to coastal aquifer case study. Only three indicators developed for the pan-European study were used, but complemented by four tailor-made indicators Study is based on NUTS 3 level	Quantitative assessment for the Finnish case study came to the conclusion that the physical, environmental, social and economic impacts are marginal	Use of pan-European concept and indicators. Complemented by two tailor-made indicators (availability of alternative water sources + national, regional, local adaptation strategies)	ESPON Climate model was used to assess vulnerability. Vulnerability of coastal aquifers towards potential climate change impacts in Finnish case study areas showed to be marginal at pan-European scale	Discussed by the example of Finland. No need for response actions due to the marginal impact of climate change on coastal aquifers	The pan-European assessment was principally proved and completely applied in Finnish part of the case study areas	The transferability of the case study approach was shown by the choice of different coastal aquifers

by the in-depth study on coastal aquifers: each cause–effect chain from exposure to sensitivity, impact, adaptive capacity and vulnerability has to be studied in detail in order to create an evidence base for adaptation strategies. This was simply not possible on the pan-European level within the given time frame and budget restrictions. However, it clearly shows further research needs.

Reference

Greiving, S. and Fleischhauer, M. (2012) National climate change adaptation strategies of European states from a spatial planning and development perspective. *European Planning Studies*, **20** (1), 27–47.

Chapter 17

Implications for territorial development and challenges for the territorial cohesion of the European Union

Stefan Greiving[1], Philipp Schmidt-Thomé[2], Simin Davoudi[3],
Lasse Peltonen[4] and Teresa Sprague[2]

[1]*TU Dortmund University, Institute for Spatial Planning (IRPUD), August-Schmidt-Strasse 10, 44227 Dortmund, Germany*
[2]*Geological Survey of Finland (GTK), P.O. Box 96, 02151 Espoo, Finland*
[3]*Newcastle University, Claremont Tower, School of Architecture, Planning and Landscape, Newcastle upon Tyne NE1 7RU, United Kingdom*
[4]*Finnish Environment Institute (SYKE), P.O. Box 140, 00251 Helsinki, Finland*

Abstract

At the level of the EU as whole, compared with other major economic regions in the world, Europe will be less affected by climate change. This is particularly the case for the economic core of Europe, which also has, as shown in the ESPON Climate project, a high level of mitigative and adaptive capacity. If this capacity is capitalised, it will certainly enhance the competitiveness of the EU in the global market. Another important point is that the diversity of climatic regions in Europe allows for a degree of economic adjustment. For the competitiveness of the EU as a whole, this implies that a potential loss of tourism in one part of Europe may be compensated by a potential gain in another part. Furthermore, climate mitigation and energy efficiency policies are one of the four key priorities of the renewed Lisbon

Strategy. This means that through the development of knowledge base and support for research and innovation, EU action on climate change can converge with the Lisbon Strategy. Nevertheless, without effective adaptation measures, such transformations may lead to increased disparities in Europe.

17.1   Introduction

This chapter is comprised of two sections. Section 1, 'Climate change and its implications for existing European policies', provides a current EU level policy base with respect to territorial cohesion and other relevant EU policies and programmes. Section 2, 'Policy options for climate change adaptation and mitigation', elaborates on what strategies are recommended within the baseline established in the previous section and the output of the ESPON Climate project. Particular attention is paid in the latter section to the regional level of governance systems, with a focus on implications for spatial planning and development. The chapter concludes with a number of recommendations for further research.

17.2   Climate change and its implications for existing European policies

Understanding the implications of climate change for current European policies is of utmost importance, considering the current challenges faced by European countries, including, most significantly, the financial crisis, unemployment, social exclusion, demographic change, the transition to a low-carbon economy and the need for adaptation to climate change impacts. Responding to these challenges requires effective and urgent policy initiatives and actions at all levels of governance including the European, national and regional levels as well as across different policy sectors. This section firstly outlines some of the key implications of climate change for the EU competiveness and cohesion policy and, secondly, considers the relevance of other EU policies and programmes to climate change.

17.2.1   *Competitiveness and cohesion policy*

Climate change will have significant economic, social and environmental impacts across the EU with some European regions, economic sectors and social groups being affected more than others. In responding to the challenges, the EU as a whole, along with individual member states, needs to take action toward both mitigation and adaptation. The latter in particular requires a place-based approach due to the spatial variation of climatic impacts. As regards mitigation, the EU has already set up a number of energy goals aimed at reducing greenhouse gas (GHG) emissions while increasing energy security.[1]

Since the adoption of the Lisbon Strategy, the EU's overarching competitiveness agenda has been to make the EU into the world's most competitive knowledge-based economy. In light of the recent economic and financial crisis and the realisation of the impacts of climate change, the European Council adopted the *Europe 2020 Strategy* in 2010 to provide a route map for recovery (COM, 2010). The Strategy recognises that '(s)trong dependence on fossil fuels such as oil and inefficient use of raw materials expose (. . .) consumers and businesses to harmful and costly price shocks' and threaten Europe's 'economic security (. . .) (while) contributing to climate change' (COM, 2010, p. 6). Thus, competitiveness of Europe as a whole is discussed within an internal perspective in the following sections, in addition to the implications of climate change on territorial cohesion. In an effort toward resolution of the challenges posed, the Strategy puts forward the following priorities (COM, 2010, p. 3):

- **Smart growth:** developing an economy based on knowledge and innovation.

---

[1]To reduce GHG emissions to 20% below 1990 levels by 2020; to increase the share of renewable energy to 20% by 2020; and to achieve 20% energy efficiency by 2020 (COM, 2010).

- **Sustainable growth:** promoting a more resource efficient, greener and more competitive economy.
- **Inclusive growth:** fostering a high-employment economy delivering social and territorial cohesion.

The Strategy also identifies seven flagship initiatives to which the EU and the Member States are committed. One of these is 'Resource efficient Europe', which implies: decoupling of economic growth from the use of resources; shifting towards a low carbon economy; increasing the use of renewable energy sources; modernising the transport sector: and promoting energy efficiency (COM, 2010, p. 4). All these will contribute not only to climate change mitigation but also to the future competitiveness of the EU. As part of its 'smart, sustainable and inclusive growth' agenda, the *Europe 2020 Strategy* emphasises the need for improving resource efficiency to limit emissions as well as to 'save money and boost economic growth' (COM, 2010, p. 13). Overall, the future competitiveness of the EU depends on an adequate supply of energy and resources, which in the current climate is uncertain given the increasing level of energy insecurity and the international obligations to reduce the use of fossil fuels.

This is of particular concern because, according to the European Environment Agency, pledges to reduce emissions to the level of the 2009 Copenhagen Accord will not prevent an increase of over $2\,^{\circ}\mathrm{C}$ beyond pre-industrial levels (EEA, 2010).[2] With respect to this fact, even if pledges are met there will still be far-reaching consequences for the EU's economic competiveness because of the changes in climate that are already underway. This means that adaptation actions are imperative in enhancing the resilience of the EU and these actions, as is widely acknowledged, must be taken now to prevent substantially higher costs of inaction (OECD, 2009). An important step taken by the EU is the adoption of the EU *White Paper on Adaptation to Climate Change* in 2009, which proposes a framework for

action based on developing the knowledge base and integrating adaptation into EU policies through increasing the resilience of health and social policies, agriculture and forests, biodiversity, ecosystems and water, coastal and marine areas and production systems and physical infrastructure (CEC, 2009). The White Paper also discusses the regional variability of climate change impacts. The Commission has acknowledged the variability of impacts and its crucial role in the future competitiveness of Europe. However, initiatives proposed by the Commission do not seem to take into account the significant differences in the mitigative capacity of different European regions and their ability to meet the EU-wide targets. Disparities between different regions of Europe with respect to both mitigation and adaptation capacities is a crucial factor for both the future competitiveness and cohesion of the European Community.

The White Paper and the outputs of the ESPON Climate project elaborated within the previous chapters discuss the differences in the adaptation and mitigation capacities between Europe's regions. It is important to acknowledge disparities, for example, South and East Europe both have less mitigative capacity than North Europe and tend to demonstrate a weaker performance in EU competitiveness indicators. Another example is found with the gradient between East and South Europe with respect to adaptive capacity, which is seen as much lower at the peripheral regions than at the core. The *Impact Assessment of the EU White Paper* offers a potential explanation for capacity inequalities in stating that '(a)daptive capacity is often positively correlated with economic development, (where) access to efficient adaptation is greater for high-income groups and richer areas, and less for the poor' (CEC, 2009a, p. 16). The document further elaborates that these effects are significantly influenced by awareness, information availability and insurance (CEC, 2009a). Aside from the White Paper, the Commission also adopted a communication on reinforcing disaster response capacity in 2008 (CEC, 2008b) and a second one on disaster risk prevention in 2009 (CEC, 2009b), which aims to integrate policies and instruments related to disaster risk assessment, forecasting,

---

[2]For direct content access, please consult the European Environment Agency website at http://www.eea.europa.eu/soer/policy-makers/climate-change-mitigation

prevention, preparedness and recovery. Extreme events addressed within the communication as well as creeping changes involve multiple levels of complex variability where the severity of climate change impacts varies in various regions and for different economic sectors and social groups.

Economic sectors most directly affected, as revealed through the results of the ESPON Climate project, include the primary, tourism and energy sectors with the primary sector in the peripheral regions demonstrating the highest vulnerability to climate change. Taking into account regional imbalances, the estimated impact inequalities are important when considering the future of cohesion policy and particularly in relation to solidarity. Attention to this matter is imperative due to the fact that those regions which are most impacted, as estimated by the ESPON Climate project, appear to also have less adaptive capacity. Differentiated climate change impacts will in turn affect European territorial cohesion and may lead to the widening and deepening of territorial disparities. Therefore, the compounding inequalities, which may result due to climate change, need to be addressed when considering the future of cohesion policy and particularly in relation to solidarity.

In order to address these issues, there needs to be a mainstreaming of climate issues into the rural development policy in the interest of a balanced territorial development of European rural areas. Such mainstreaming is also required under the Renewed Social Agenda, which is based on a holistic approach to social policy (CEC, 2008). Furthermore, as the frequency and intensity of natural hazards increases, leading to more flooding, drought, heat waves and forest fires, there will be significant impact to the EU physical capital (such as infrastructures, roads and utilities) and human capital (such as loss of working days and even lives), both of which will greatly affect EU economies and competiveness. These, too, are spatially differentiated and can therefore potentially exacerbate the current territorial disparities. It is important to note, however, that some climate change impacts can provide opportunities, which, if capitalised, can reduce such disparities. Measures to remedy inequities and to pursue potential opportunities require a degree of

oversight and responsibility at the EU level to complement the actions at national level and to ensure cohesion.

Cohesion Policy is the EU's primary instrument for seeking harmonious development throughout and across the Union (EU, 2010). Its vision is to pursue both economic development and environmental sustainability particularly for regions that lag in these respects. The vision pays specific attention to the need for supporting vulnerable social groups and for appreciating the EU's territorial and cultural diversity (EU, 2010). The Fifth Cohesion Report (5CR), published in November 2010 for comments, is the first report which is to be adopted under the Lisbon Treaty and focuses more on sustainable development than previous reports, stressing also its territorial specificity (COM, 2010a). In a wider sense, the Lisbon Treaty brought about what is considered as a 'third dimension' to cohesion policy and is seen explicitly within the 5CR. This dimension primarily encompasses that 'the EU "shall promote economic, social and territorial cohesion"' (EU, 2010, p. 24). Economic and social cohesion are distinguishable from territorial cohesion in that their focus lies with regions and their respective competitiveness and well-being. The territorial dimension of cohesion, however, stresses importance on 'access to services, sustainable development, functional geographies and territorial analysis' (EU, 2010, p. 24). The 5CR further elaborates within the sustainable development element of the territorial dimension by stating that both climate change and current renewable energy goals will necessitate coordination between all levels of governance to prevent counteracting policies. There is a need for this coordination especially at the regional and local levels because of the regional variation of impacts, which significantly influences mitigation and adaptation strategies and the strain on currently existing capacities.

As demonstrated through the outputs of the ESPON Climate project, the degree of adaptive capacity for tackling climate change is a significant driver of potential future disparities. Attention, therefore, should be directed to the different level of efforts and investments needed to mitigate and adapt to climate change in different parts of Europe. Though the 5CR addresses this issue through its chapter on 'Enhancing environmental sustainability', which acknowledges that climate change will hit Southern and Eastern Europe hardest, there is little mention of how these varied climate change impacts will be reflected in future cohesion policy. There appears to be a continuous emphasis on economic indicators for providing financial support for the regions (EU, 2010). The 5CR states that this financial support is differentiated based on GDP per capita measurement between what are considered ' "less" or "more" developed regions' (COM, 2010a, p. 10). The ESPON Climate project assists with regard to this measurement issue by providing a means to identify regions that should be targeted for financial assistance through regional analysis of social and economic impacts of climate change. The project's findings, for instance, should inform the allocation of EU funds in a way that regions targeted are those which are most adversely impacted and have the lowest capacity to adapt or to mitigate such impacts. This regional allocation of funds is also linked to Cohesion Policy as well as the *Europe 2020 Strategy*, which stresses the need for, and links between, supporting research and development, innovation, increased resource efficiency, improved competitiveness and job creation (COM, 2010). This linkage is stated within the 5CR as follows (EU, 2010, XXIII):

> The explicit linkage of Cohesion Policy and Europe 2020 provides a real opportunity: to continue helping the poorer regions of the EU catch up, to facilitate coordination between EU policies, and to develop Cohesion Policy into a leading enabler of growth, also in qualitative terms, for the whole of the EU, while addressing societal challenges such as ageing and climate change.

This quote demonstrates that Cohesion Policy and future growth strategies are inevitably linked in the effort to achieve harmonious development. Impacts of climate change, as estimated by the ESPON Climate project outputs, encourage the territorial dimension of Cohesion Policy by stressing the spatial differentiation of impacts through provision of a territorial analysis. It must be additionally stated that, to achieve the EU cohesion goals, it is necessary to work toward a convergence in climate resilience,

which will require coordination at all governance levels with respect to policy implementation. With respect to implementation, it is also important, as highlighted in the next sub-section, to briefly consider other relevant policies and programmes at the EU level that are considered to play a role in future action toward territorial cohesion.

17.2.2   Other relevant EU policies and programmes

The purpose of this section is to provide a wider picture of the other relevant policies and programmes. Particular emphasis is placed within the INTERREG IV B & C programmes analysed by the ESPON Climate project.[3] This analysis serves as an example of how the ESPON Climate project can be applied to other relevant EU policies and programmes particularly with a regional focus. The policies selected do not constitute an exhaustive list.

Between 2007 and 2013, four inter-regional, 13 trans-national and 52 trans-boundary programmes were launched within the framework of the European Territorial Cooperation. One such programme, INTERREG, is the focus of part of the analysis provided by the ESPON Climate project. As indicated by the summary given in Table 17.1, regional analysis is differentiated based on the INTERREG IV B & C programmes' respective ten European trans-national regions covering the entire territory of the EU. Through this analysis, the following events were found to be reiterated: increase in sea level, floods, forest fires, droughts, extreme weather conditions and events, and an increase in the frequency of natural damage. It was found that adaptation to unfavourable climate change impacts was not shown as an explicit priority in any of these programmes. However, methods to address these impacts appear indirectly in the majority of these programmes as a result of efforts to achieve other priority goals such as environmental protection, sustainable development or avoidance of natural risk. Actions toward achieving these goals

---

[3] Please see http://www.interreg4c.eu/about_programme.html for further details.

Table 17.1 The table provides each trans-national region specified by the INTERREG IV B & C programmes. With the inclusion of options for further programme development and the climate change issues generated from the output of the ESPON Climate project, this provides an example of implications and applications of the project on other EU policies and programmes.

Region identified	Climate change issues identified by the ESPON Climate project	Options for further programme development
Northern Periphery	Flood Sea-level rise	Risk management for settlements potentially affected by river floods related to climate change
Baltic Sea	Storm surges Sea-level rise Floods and flash floods Changing frost Changing precipitation	Further development of Regional Adaptation Strategies related to climate change impacts on forestry Climate change impact assessments on coastal and island areas, including tourism and water quality (algae blooming)
North West Europe	Flood (river and flash) Sea-level rise Storm surges	Combination of flood and storm surge prevention and spatial planning as cross-border and trans-national initiatives
North Sea	Flood (river and flash) Sea-level rise Storms (and surges)	Combination of flood and storm surge prevention and spatial planning as cross-border and trans-national initiatives
Atlantic Coast	Flood (river and flash) Sea-level rise Storms (and surges)	Development of regional strategies to anticipate the impact of river floods Development of regional strategies to anticipate the impact of storms and storm surges
Alpine Space	Floods and flash floods Changing precipitation	Diversification of tourism, also interlinked with water scarcity Integration of sustainable cross-border adaptation and mitigation concepts Options of enhancing synergies to avoid conflicts (especially on adaptation measures) Over regional and trans-national water management approaches, especially focusing on the Alps as a 'water tower'
Central Europe	Floods & flash floods Changing frost Changing precipitation + summer days + summer temp. Sea-level rise	Development of regional climate change adaptation strategies on floods, heat waves, forest fires Development of regional climate change adaptation strategies on water scarcity Development of regional climate change adaptation strategies on tourism Development of regional climate change adaptation strategies for agriculture and forestry
South West Europe	Agriculture Forestry Flood Sea-level rise	Development of regional transnational climate change adaptation strategies on heat waves, water shortage and forest fires

Table 17.1   *(continued)*

Region identified	Climate change issues identified by the ESPON Climate project	Options for further programme development
Mediterranean	storm surges	Management of public (including tourism) water demand
	droughts	Identification of possibilities to save water instead of relying on current
	floods	water management schemes and further development of desalinisation
	forest fires	plants
	changing precipitation	Avoidance of mal-adaptation, e.g. transferring costs and risks from water
	+ summer days	sector to energy sector
	Sea level rise	Management of land take (urban sprawl)
South East Europe	flood	Emphasize analysis and management concepts on impacts of climate
	sea level rise	change on forestry and agriculture
	changing precipitation	Development of common (crossborder) methodology for land use
	changing evaporation	restructuring, including integrated water management planning
	+ summer days	
	Sea level rise	

mitigate unfavourable impacts most commonly by the development of water management and the use of various means of risk prevention.

According to the ESPON Climate project, when applying the climatic stimuli, the most prominent impacts in all regions are floods, storms, changes in precipitation and then sea-level rise, in this order. ESPON Climate has also stressed the need for more holistic approaches to be taken in identifying climate change impacts. The approach, it is stressed, should take a particular focus depending on the most pronounced changes in a given region, such as demographic change, tourism and structural development. There is also the need for development plans and emphasis on the necessity to consider already existing structures, as this is a significant challenge for spatial planning particularly with respect to adaptation strategies. Planning is emphasised most notably within the recommendation toward implementation of a Regional Adaptation Strategy (RAS), or regional adaptation plan, for each respective region. Further focus is additionally placed on the need for more social measurements, such as demographic changes. Furthermore, there is a need to concentrate on water management and specifically integrated water management. In

addition, in the South East region there is a need for trans-national flood risk prevention.

Further recommendations for future projects consider the need for additional research and development particularly for activities requiring international cooperation, including development of models and forecasting systems, knowledge transfer, planning methods (including both spatial and regional planning practice), coping strategies for impacts of climate change and specifically for natural risks. Emphasis, in general, is placed on the theme of water management. Regional planning to fulfil these recommendations can benefit from the results of the ESPON Climate project. Owing to the project's regional analysis of expected climate change impacts, severity of impact can be identified for each trans-national region and can assist in determining regional goals and priorities for future programmes. Based on the results of the vulnerability assessment of the ESPON Climate project, the ten assessed trans-national regions can be classified into the three major groups, identified and elaborated within Table 17.2.

There are a number of additional policies that should be considered and analysed. The following provides a few examples.

Table 17.2 The three major groups based on the ESPON Climate project vulnerability assessment and INTERREG's ten trans-national regions.

1. Regions where vulnerability is expected, as a rule, to increase at a high or medium rate

Regions: Mediterranean Region, South-Western and South-Eastern Europe

Description: impacts accrue primarily in the environmental and economic dimension, overall growth of vulnerability related to poor adaptation capacity

Proposed solutions: programme measures addressing, for example, water management, preservation of water, forest fire forecasts, preparation for heat waves and regulation of land use would potentially have the greatest importance

2. Regions where vulnerability is expected, as a rule, to grow at a low rate

Regions: Northern Sea region, the North-Western European and the Atlantic coastal regions (includes also some Alpine regions)

Description: increase of climate change impacts is projected primarily for the physical and social dimensions

Proposed solutions: measures addressing natural disasters such as floods and coastal storms would have the greatest positive effects

3. Regions, where vulnerability is expected, as a rule, not to change significantly

Regions: Baltic Sea region, the Northern periphery and Central Europe (the lattermost can also be in part of the second group)

Description: climate impacts are diverse but not extreme

Proposed solutions: adaptation measures should stress the importance of the role of water management and prevention of natural disasters

17.2.2.1 EU Nature and Biodiversity Policy

The EU will develop its own plan to assess and reduce adverse impacts to biodiversity as a result of climate change adaptation and mitigation strategies (or activities) for the period 2011–2020 (EC, 2011). This is in accord with the international agreement on biodiversity as the EU is party to the Convention on Biological Diversity.

17.2.2.2 EU Directive on the Assessment and Management of Flood Risks

The main purpose of this directive is to reduce flood risk and adverse consequences thereof for the following: health and life, environment, cultural heritage, economy and infrastructure (EC, 2007). Coordination is encouraged using the river basin unit through river basin management plans. The directive states that the focus of flood risk management plans should be: prevention, protection and preparedness.

17.2.2.3 EU Integrated Maritime Policy

The policy encourages implementation of integrated management and governance for oceans, seas and coasts (CEC, 2007). The policy further harbours interaction amongst all the EU's sea-related policies.

17.2.2.4 Marine Strategy Framework Directive

The purpose of this directive is to protect the EU's marine environment and to achieve good environmental status for marine waters by the year 2020 (EC, 2008). The directive additionally aims to protect marine resources and their dependent economic and social activities.

17.2.2.5 EU Common Transport Policy

Within this ten year policy plan, the EU strives to develop cleaner vehicles, fuel and emission-reducing technologies (EC, 2001). Spatial planning methods are particularly encouraged for land use and transportation planning.

17.2.2.6 Directive on Environmental Impact Assessment

This directive is focused toward achieving a high level of environmental protection particularly for the implementation of any projects or programmes in order to reduce adverse environmental impact (CEC, 1985).[4]

17.2.2.7   Directive on Strategic Environmental Impact Assessment

The directive, similarly to the Directive on Environmental Impact Assessment, also concentrates attention on the requirement of environmental considerations in projects and planning but ensures that this occurs prior to decision-making (EC, 2001a).

17.3   Policy options for climate change adaptation and mitigation

This sub-section elaborates on the foundations set by the previous sub-chapter with respect to policy options for both adaptation and mitigation strategies. The EU, national and regional level are considered with particular attention paid to the regional, or more local, level.

17.3.1   *Adaptation to climate change*

Adaptation within the context of the ESPON Climate research refers to the IPCC definition stating that adaptation is an 'adjustment in natural or human systems in response to actual or expected climatic stimuli or their effects, which moderates harm or exploits beneficial opportunities' (Klein *et al.*, p. 750). Different forms of adaptation include both autonomous (through individuals or businesses) and planned adaptation (through government initiated public policy measures) (Leary, 1999; Smit and Pilifosova, 2001). Adaptation policy development can be reactive or proactive, especially with respect to disaster risk management for increasing intensity or frequency of hazards where one can see either pre- or post-event action toward

adaptation measures and implementation thereof. Development of this policy, however, is met with challenges, such as sectoral coordination and policy integration, as well as the potential for maladaptation where unintentional, adverse consequences occur as a result of adaptation efforts (Barnett and O'Neill, 2010). Concentrating on the vertical dimension of governance, the focus of adaptation policy discussion within this sub-chapter discusses planned adaptation and requires the consideration of multiple levels of governance. This is due to the nature of policy implementation, where implementation at one level is not independent of implementation at other levels.

17.3.2   *Adaptation at the EU level*

Although the national and regional levels are where most adaptation measures will be implemented, it is necessary to consider the EU level because climatic impacts are not confined neatly into political or administrative boundaries. This is why trans-national approaches are needed for adaptation policy, particularly with respect to policy coordination and the need to ensure adaptation capabilities for the most disadvantaged regions. EU level legislation provides the description of the EU's Adaptation Framework through The White Paper '*Adapting to climate change: Towards a European framework for action*' with the purpose of building resilience to climate change impacts (CEC, 2009). The framework presents a two-phase approach where the first phase, from 2009 to 2012, is comprised of preparation of comprehensive EU adaptation strategies. This requires coordination between all governance levels and is constructed of four pillars, which are briefly highlighted in Table 17.3.

Within the third pillar, it is important to note that there are EU level funding mechanisms that serve as existing tools to assist RAS development. The funds can be used for developing knowledge, testing and validating this knowledge, monitoring RAS development, RAS implementation and general stakeholder and public awareness. Some of the specific available funding mechanisms include: the

---

[4]This document has been amended as of 2009.

Table 17.3 The pillars represent the details comprising Phase 1 of the EU Adaptation Framework (CEC, 2009). Deadline for actions are indicated by parentheses. Actions provided within the brief descriptions are not an exhaustive list. For further reference, consult pages 7–16 of the White Paper on Adaptation.

Pillars of action	Brief description of actions
Development of a knowledge base	Establish a Clearing House Mechanism (2011)
	Development of methods, models, data sets and prediction tools (2011)
	Development of indicators to improve monitoring of climate impacts (2011)
	Conduct assessment of costs and benefits for adaption options (2011)
Integration into EU policies	Development of guidelines and surveillance for health impacts (2011)
	Assess impact of climate change and adaptation strategies on employment and vulnerable groups
	Ensure implementation of the Floods Directive accounts for climate change
	Development of guidelines for ESI and SEA Directives to account for climate change (2011)
Financial instruments	Estimation of adaptation costs to be taken into account for future policy
	Examination of innovative funding measures for use in adaptation
	Exploration of insurance, among other products, to enable risk sharing and to complement adaptation strategies
	Encouragement of Member States to use ETS (Emissions Trading Scheme) revenues for adaptation efforts
Member State partnership and cooperation	Establish an Impact and Adaptation Steering Group (IASG) to improve and increase coordination (2009)
	Encourage development of NAS and RAS toward mandatory strategies (2012)
	Improve efforts in mainstreaming adaptation within EU external policy
	Strengthen dialogue on adaptation issues with partner countries
	Take Framework for Action on Adaptation to the UNFCCC

European Regional Development Fund (ERDF), the European Social Fund (ESF), LIFE + and INTER-REG and European Territorial Cooperation (ETC) funding.

17.3.3 *Adaptation at the national level*

Adaptation policy at the national level is primarily evaluated on the basis of National Adaptation Strategies (NAS). It is of utmost importance to acknowledge, when discussing NAS, that the adoption and development of these strategies in a given country is determined by a number of factors, including the country's vulnerabilities, severity of impacts from climate change and capacities to adapt. Table 17.4 provides a brief picture of current NAS development. However, the reader is encouraged to consult the EEA website for most recent updates.

Table 17.4 Overview of national adaptation strategies. Dates of NAS adoption indicated in parentheses.

NAS adopted	NAS in progress	Countries without NAS
Belgium (2010)	Austria	Iceland
Denmark (2008)	Czech Republic	Liechtenstein
Finland (2005)	Estonia	Lithuania
France (2006, updated 2011[a])	Ireland	Luxembourg
Germany (2008)	Latvia	Poland
Hungary (2008)	Romania	Slovak Republic
Netherlands (2008)	Switzerland	Turkey
Norway (2008)		
Portugal (2010)		
Spain (2006)		
Sweden (2009)		
United Kingdom (2008)		

[a]Update provided by the République Française Ministry of Ecology, Sustainable Development, Transport and Housing French National Climate Change Impact Adaptation Plan, 2011.
Sources: The information is derived from the EEA website and from the Partnership for European Environmental Research (PEER) Report (EEA, 2010; Swart *et al.*, 2009). Please consult the EEA website for the most recent update to NAS development.

The Partnership for European Environmental Research (PEER) Report provides a comparison of ten NAS structured around six key themes (Swart *et al.*, 2009). The report analyses NAS development, the role of research in NAS development, the role of communication, aspects of multilevel governance, adaptation integration into sectoral policies and the role of end processes including monitoring, evaluation and enforcement. The report also identified a number of motivating factors that tend to encourage NAS development, which include climate negotiations at the international level, experience with extreme events and existing climate change research. The last factor is related to the additional finding that a significant role is played within a given country by climate information availability. In general, it was determined that the further developed a country's climate change research is, the more likely it is that the county has taken adaptation into consideration. Another important factor, more related however to the success of an NAS, is communication. This is seen as a crucial element to establishing a successful NAS, but evidence remains to be found to determine the most effective means of relaying information across sectors and to relevant stakeholders. As mentioned previously, multilevel governance is also recognised as having an important role, although there is an issue with regard to the lack of clarity of roles and responsibilities between levels and there have been few studies conducted that provide multiple scale analyses.

A few other important points not elaborated within the PEER Report are found within the EUR-Adapt project *Organising adaptation to climate change in Europe*, which focuses on analysis of adaptation policy in Finland, Italy, Sweden and the United Kingdom (Keskitalo, 2010). The findings of this project also follow the need for attention to multilevel governance that triggers the development of adaptation strategies through both a top-down and a bottom-up component. The top-down stems from international dialogue influencing national level policy while bottom-up action is rooted in local level, or sub-national, experiences with various weather impacts (Keskitalo, Westerhoff and Juhola, 2012).

Studies analysing national scale adaptation strategies are not limited solely to consideration of NAS. A study by Gagnon-Labrun and Agarwala (2006; 2007), for example, examined national communications (NCs) from the United Nations Framework Convention on Climate Change (UNFCCC) Annex I countries.[5] NCs assessment enabled the study of different national adaptation measures and strategies based on three criteria: impacts and vulnerability, adaptation options and measure implementation, and supporting institutional mechanisms. The results of the study indicated that adaptation issues are limited and that there is an emphasis on vulnerability to impacts. The study identifies the difference between what is intended and what is translated into action within institutional mechanisms, existing policies and measures. The authors also identify three categories of countries based on their impact assessment development. These are as follows: early to advanced impact assessment where adaptation is not discussed in the NCs, very advanced with limited introduction to adaptation and advanced assessment toward adaptation strategy implementation.

Another study by Massey and Bergsma (2008) concentrates on focus areas comprised of concerns, recommendations and measures by country by analysing the NCs. The results of this study indicate that Western Europe is most advanced regarding implementation of adaptation measures followed by Southern Europe. Northern and Central Europe, however, are more advanced with respect to policy recommendations. The study also provides four categories of adaptation objectives, which consist of general adaptation capacity building, risk and sensitivity reduction, coping capacity for extreme events and climate change capitalisation (Massey and Bergsma, 2008). The main objective in all European regions, according to the study, is the reduction of risk and sensitivity. The study also frames analysis around physiographic regions including the

---

[5]National Communications are reports that must be written in a standard format. This enables cross-national comparison, reflecting the general perspective of a given government (Gagnon-Lebrun and Agarwala, 2006).

Alpine region, the Tatra and Carpathian region, the Atlantic, North Sea, Baltic Sea, Mediterranean and the Black Sea, emphasising the need for identifying adaptation goals within a spatial or regional specificity. The importance of a spatial focus has been revealed through reviews of the spatial planning perspectives within European NASs where spatial planning is seen as specifically related to land use planning and general adaptation strategies (BMVBS, 2010). It is underlined that, though the role of spatial planning is growing in importance, it is not well recognised (BMVBS, 2010). However, several states analysed within the reviews have granted spatial planning at least an assigned role (Germany, France and Hungary) if not a central role (Netherlands and United Kingdom) though some states have yet to ascribe any particular role (Finland and Spain) (BMVBS, 2010). The notion of a spatial focus is growing in importance within policy development with respect to adaptation policy and is particularly applicable to development at the regional level.

17.3.4 *Adaptation at the regional level*

As at the national level, the regional level of adaptation policy in Europe has seen the emergence of what is termed Regional Adaptation Strategies (RAS). The RAS is, generally speaking, a relatively new concept, which is featured in only a few studies. This is partly as a consequence of the fact that few have actually been implemented. Regardless of implementation maturity, analysis of these strategies must be considered because of the nature of impacts to climate change. Impacts will be felt at the more local levels because the severity of impact is dependent upon the regional context influenced by physical as well as social-economic factors. The role of the region is also important due to the responsibilities to provide utilities and health and emergency services. These are vital to disaster risk management, including pre- and post-disaster actions, and to overall adaptation due to the potential increase in intensity and frequency of hydro-meteorological hazards. The development

of RAS tends to be supported by two processes. As previously mentioned, there is a top-down and a bottom-up process, where the top-down consists of national responses to the need for adaptation while the bottom-up is comprised of local self-initiative dependent upon local resources.

Information regarding regional strategies, however, faces specific hindrances. Of important mention is the fact that most information at the regional level is only available in the local language (Ribeiro *et al.*, 2009). Other informational challenges include lack of climate change knowledge at the more localised levels, including higher uncertainty that accompanies higher resolution and the variation in quality of information with which to base adaptation strategies. However, local knowledge of the physical, social and cultural elements of given regions can often be more context informative than that of the national level. To evaluate both of these levels, a study by Ribeiro *et al.* (2009) considers both the local and national levels with respect to adaptation policy development. A two-phase approach is taken to analyse 31 RAS in six countries (France, Germany, Netherlands, Sweden, Spain and the United Kingdom). The first phase consists of a holistic analysis of strategies, preparation and information thereof, while the second phase provides a categorisation of terms and actions within RAS. The study reveals two types of regional processes involved in the formulation of adaptation strategies at more localised levels: (i) processes involving sub-national governments and (ii) processes involving the city level or Local Adaptation Strategies (LAS). Policies analysed within the study tend to develop through the interaction of both central and regional governments. This occurs because countries typically first have an NAS and then begin to implement an RAS, which is sometimes encouraged through the NAS framework. However, the NASs offer limited guidance for Regional Adaptation Strategies and none of the countries in the study provide any requirement for RAS development. Other findings include the reiteration of the lack of clarity within roles and responsibilities between different levels of governance and the identification of most popular methods of participation, which include consultation workshops and sectoral and cross-sectoral working groups. However, it is important to note that, in general, preferred methods will vary by regional context and level of locality. Another way in which regional strategies differ is in the allocation of priority. Whatever is particularly vulnerable, such as a specific sector, will commonly be put at the highest priority within the regional strategy. Popular priorities include sectors pertaining to health, landscape (including flooding, sea-level rise and drought), agriculture and water supply, among others.

Variations exist not only between regional strategies but also between their levels of implementation. Existence of an RAS does not necessarily equate to action taken. To evaluate the degree of implementation, Ribeiro *et al.* (2009) further developed the previously mentioned classification of divisions established by Massey and Bergsma (2008), which differentiates between concerns, recommendations and measures. One of the outputs from this evaluation was that implementation of actual policy measures only took place in less than 20% of strategies evaluated (Ribeiro *et al.*, 2009).

The ESPON Climate project takes analysis of strategies a step further by utilising an approach that identifies region specific options for adaptation policy. Impact maps derived from climate model outputs within the ESPON Climate project enabled this identification process. Within the analysis, impact is taken to be the combination of exposure and sensitivity. For recommendation of policy options, the exposure and sensitivity components of impact are highlighted with respect to five impact dimensions: social, cultural, economic, environmental and physical impact. However, the suggestions produced from this analysis should not be considered as the only possible options. The project considers policy options additionally through the aforementioned classification of adaptation objectives from Massey and Bergsma (2008); namely, enhancing adaptive capacity, risk and sensitivity reduction, coping capacity and capitalisation.

This system of impact dimensions and adaptation objectives classification forms part of the methodological framework of the ESPON Climate project.

The social dimension focuses on populations that are adversely affected by estimated changes, such as the elderly, poor and urban populations. Exposure within this dimension primarily pertains to an increase in heat and flood events and sea-level rise. The adaptation objectives here relate mainly to policy measures that reduce risk and sensitivity by, for example, preventing loss of human life through planning policies that reduce the heat island effect. Increasing adaptive capacity through establishment of emergency services and early warning systems is another focus.

With respect to the cultural dimension, attention is focused on historical sites and landscapes that are susceptible to climate change impacts. The primary exposure is that of flooding. Adaptation measures to cope with flooding impact within this dimension are closely related to those that apply to both the social and physical dimensions. However, a limitation is placed, with measures to increase coping capacity and reduction of risk, because relocation is not an option for many historical sites. The potential for these measures is highly dependent upon the regional capacity to adapt to climate change. This also holds true for measures that focus on capitalisation on impacts, particularly in relation to the tourism sector.

The economic dimension is granted a variety of mixed effects where the impact, positive or negative, is dependent upon the sector affected. For the reduction of risks, policy recommendations include the need for flexibility and general support for autonomous adaptation for businesses. Coping capacity is the objective targeted within policies geared toward prevention of disruption to production processes, particularly with regard to coping with extreme events. To build adaptive capacity, it is suggested that concentration should be placed on research and development and general education toward innovation in adaptation. Exposures such as changing duration of summer days, winter precipitation and summer precipitation might provide opportunities for capitalisation. Examples of this are found within the tourism and agricultural sectors with regard to new tourist location and new crop varieties.

The environmental impact dimension with the project analysis focuses mainly on indicators related to soil and the ecosystem. Building adaptive capacity can be found through encouragement of policies and measures that work toward the maintenance of ecosystem services. This requires more scientific information of climate change impacts to ecosystems in given regions. Other objectives such as coping capacity and risk reduction can be targeted through policies to prevent and handle forest fires. The capitalisation objective, like most of the other dimensions, has seen very little development within the policy addressing climate change impacts.

Policy related to the physical impact dimension can reduce risk through review of current planning regulations and building codes for areas prone to flooding. Coping capacity can also be addressed through an improvement in the capabilities of emergency services particularly with respect to storms and flooding. The objective of adaptive capacity building is supported through policies encouraging the provision of information, relocation of settlements currently located in flood prone areas and an increase in the technological capacity to manage flooding.

Generally speaking, adaptation policy for all dimensions is needed for both the short and long term. Trends in the analysis indicate particular characteristics of the categories of adaptation objectives. Reduction of risk and sensitivity typically equates to the revision of current and specific policies and regulations. Increasing coping capacity is normally found within the response of emergency services while increasing adaptive capacity can commonly be a result of early warning systems and education and information sharing. Capitalisation has been fairly tourism centric, with very few examples in practice. There is also a lack of clarity in terms of what options exist for capitalisation of, for example, floods and coastal surges. The ability to address objectives within these dimensions is highly dependent on the spatial context and the capabilities of a given region.

The above analysis invokes the need to consider spatial planning within the place-based characteristics of regional development. Owing to the regional specificity of impacts, spatial planning provides tools for approaches to reduce negative impacts and can potentially play an important role because of its cross-sectoral character (Davoudi *et al.*, 2009). This role, however, is not currently emphasised or well developed with respect to mitigation and adaptation strategies (Mickwitz *et al.*, 2009). Spatial planning is more relevant for regions that are still developing, since one of the main challenges for spatial planning is areas that are already built-up. This is because conflicts between private property rights and the need for adaptation enforcement arise. For this reason, regulatory (or conforming) planning is not the preferred method for enforcing adaptation in pre-existing settlements. If pursued, compensation for private property owners will be necessary for legitimising new regulation and the use of land for a public purpose. Though this can be seen as a potentially negative aspect with regard to the role of spatial planning, it is important to note that vulnerability assessment and addressing policy recommendations thereof can benefit through this spatial focus and via spatial typologies.

The ESPON Climate project provides a pan-European vulnerability assessment that identifies specific types of regions. Variation in climate change impacts occurs within and between countries and even between like regions where, for example, one mountain region will face a different impact than that of another mountain region. However, because of the difficulty in identifying regional vulnerability given a pan-European assessment, spatial typologies are used as a medium to provide some general policy recommendations. The typologies are listed as follows: metropolitan/urban, rural, mountain, coastal, sparsely populated, islands, border regions. An overall category is also analysed. Analysis of these regions considers impact, adaptive capacity, vulnerability and recommendations.

Findings from the metropolitan/urban regions indicated that high impact was likely for the highly populated coastal regions, Alpine areas and particularly Southern Europe. Much of this is dependent upon the population factor where there is a concentration of people but also a concentration of infrastructure. These regions are characterised by high impact and low vulnerability and there is a need for collaboration between planners and civil society, specifically to establish a common 'spatial vision' or shared planning effort. There are a few essential parts to creating this spatial vision. The first of these is efficiency, which requires the management of less space and fewer resources. The second is diversity, which necessitates the need to differentiate the sensitivity of various structures. The latter two parts to the vision are redundancy and robustness, where redundancy relates to the need of an urban system to have redundant or interchangeable main elements and robustness refers to the strength of, for example, infrastructure and vegetation to resist adverse impacts.

The rural regions feature different characteristics where there is a range between low to high potential impact with problematic regions highlighted in Southeast Europe. These areas typically have low adaptive capacity, as they are not as economically robust as other regions to enable capacity building. There is therefore a problem within these regions, particularly those that are economically weak. Rural areas within Southeast Europe will need to remedy capacity deficits through improvement of economic development. Of significant importance is the attention that must be paid to regions that face continuous water scarcity.

Mountains regions face a medium to high impact particularly in Southeast Europe, Greece, Spain and the southern side of the Alps. In addition to adverse changes in demographic conditions, these areas are not as accessible via transport and are typically more vulnerable to hazards triggered by climate change. What is needed within these regions is an assessment of hazards and risk mapping with attention stressed on the necessity of trans-national coordination.

The coastal regions are also estimated to face medium to high levels of impact. This is in large part due to the potential for sea-level rise. Integrated Coastal Zone Management is a concept that should be used to improve coastal defence systems and to adjust settlement restrictions. There are also potential opportunities for tourism development in Northern Europe due to the potential increase in tourist comfort.

Sparsely populated regions include some interior areas of Spain, Scandinavia and Scotland. Here the impact is measured more in relative terms than in absolute. In regard to absolute terms, there is not much impact because of the few assets and very low population within these regions. Relative change, however, is of more significance as these regions are typically classified as periphery regions and improving accessibility may also improve adaptive capacity.

Islands analysed are concentrated in the Mediterranean and the Atlantic and face severe impact. The island regions that tend to have lower adaptive capacity and a high dependency on the tourism and agricultural sectors are particularly at risk. There must be a diversification of the economy in order to increase resilience. More specifically, there is a need in these regions to encourage diversification in activities that are less climate sensitive in order to reduce stress to the natural environment, especially with regard to conservation of freshwater resources.

The border regions, similarly to the rural regions, face a wide range of impacts. Here, the variation is due to different sensitivities including adaptive capacity, population density, settlement patterns and economic development. There is a need in these regions for development strategies that balance climate change with issues such as economic development and demographic change as well as environmental issues. Examples such as the Tisza Catchment Area Development (TICAD) project underline what is recommended for these regions, a spatial plan and joint policy particularly for water management, sustainable economic development, efficient use of common resources and relations within settlement systems.

Within the overall analysis, the highest vulnerability is found in the Mediterranean and Southeast Europe. The explanation for this is the potential for high impact combined with low adaptive capacity. Particularly vulnerable areas include the Alps and Mediterranean especially in regard to the tourism sector. It is necessary to focus attention on the different starting points and levels of effort and investment needed in the most susceptible regions to mitigate and adapt to climate change. This variation and identification of most vulnerable regions are acknowledged in the 5CR but, as previously stated, there is no further explanation provided for future policy development, especially within the continuing dialogue of cohesion policy. In relation to specific vulnerability via water shortages, which play an important role in the Mediterranean, it is useful to consider the Regional Mitigation and Adaptation Strategies currently developing in several countries. These are of importance because of their relation to territorial development and explicitly to the local and regional scales that form a key component of overall vulnerability reduction. It is further recommended that formation of joint water management bodies or the establishment of water markets at regional or local levels be encouraged for the development of new governance strategies.

What is also needed is a cross-sectoral approach for adaptation through multi-level governance. This approach is needed because of the nature of climate change impacts where various sectors are differently affected. This calls for cooperation across sectors, particularly due to the existing strong interdependences. This connectedness and the need for cooperation may prove to be a sound reasoning for what Mickwitz *et al.* (2009) address as a potential role to be played by spatial planning and the following of a comprehensive planning approach. One way to go about implementation can be found in the example of the Netherlands, where the role of the ministries in spatial planning was strengthened, especially with respect to adaptation strategies. However, the institutional implementation alone is not sufficient as there is a need to establish the legitimacy of actions. This additionally requires the involvement of multiple levels of social groups as well as relevant stakeholders, as measures and strategies implemented often tend to quantify what characteristically has a normative value. This gives support to the need for climate governance, which features incentives and a more inclusive discourse-based approach. This is particularly cited as an important factor for successful adaptation strategies in the European level legislation as well as recent literature (Ribeiro *et al.*, 2009; Swart *et al.*, 2009; Meister *et al.*, 2009).

The local level is seen as the most important level for establishing much needed trust to legitimise strategy implementation and as a crucial element to territorial development within the context of climate change adaptation especially with respect to planning law (Greiving and Fleischhauer, 2012). Elaboration of the strengths and weaknesses of territorial development and climate change adaptation is provided through a study by Greiving and Fleischhauer (2012). The study reviews the potential for territorial development to cope with climate change through the following three areas: assessment of long-term consequences, climate proofing and disaster prevention. Table 17.5 summarises the main findings.

There is particular attention given to a spatial planning focus that provides important insight when considering the regional, and more local, context for implementation of adaptation strategies. The regional level is also considered with increasing importance in the discussion surrounding mitigation strategies, which are addressed within the next sub-section.

17.3.5 *Climate change mitigation*

The main aim of mitigation strategies is to achieve stabilisation of emission levels as is stated in Article 2 of the UNFCCC (Rogner *et al.*, 2007). The ESPON Climate project uses the term mitigation to follow the definitions provided from the IPCC (2007) and Füssel and Klein (2002) where mitigation is the attempt to reduce climate change by enhancing carbon sinks and reducing greenhouse gas (GHG) emissions. More specifically, the aim of stabilisation through mitigation hopes to be achieved at a pace that allows ecosystems to naturally adjust to climate change, to ensure the continuation of food production and economic development. The process of determining and implementing strategies to prevent emission levels that equate to a dangerous alteration of the climatic system is difficult both scientifically and politically, due to the complication of

Table 17.5 Potentials for territorial development to cope with climate change.

Area of analysis	Brief description and potential for territorial development
Assessment of long-term consequences	*(Fair)* Current assessment and appraisal of impacts is seen as the milestone in this area where continued emphasis should be placed on comprehensive planning. This approach takes an integrative view of multiple processes including demography, economy, environment and climate.
Climate proofing	*(Good)* Land use identification, new guiding principles, adaptation of new developments are all seen with positive potential especially for the ease in which they can be included into existing planning practices and strategic environmental assessments. *(Poor)* Adaptation of already existing spatial structures is considered to have poor potential. What is needed to remedy this are incentives and a change of perception for current property owners. This is sometimes very difficult but an attempt to establish a latent demand for specific changes could be pursued.
Disaster prevention	*(Poor)* Assessment of frequency and magnitude of extreme events (exposure) has poor potential for territorial development because responsibility is actually held by specialised authorities (i.e. water boards). However, there are opportunities for communication with the responsible authorities where collaboration would be helpful to all parties involved. *(Poor)* Relocation and retreat from threatened areas is also poor and is again related to conflicts with existing private property rights and built-up areas. *(Fair)* Differentiated land-use according to the given risk is fair, but is not effective with regard to the impediment of existing structures. *(Fair)* Adaptation of existing building structures is also considered fair but must overcome difficulties induced by property rights issues. However, this can be remediated in part through incentives and stakeholder communication. *(Good)* Keeping disaster prone areas free of further development has good potential, particularly within conforming planning systems via regulatory zoning instruments.

Source: The table is adapted from the findings from Greiving and Fleischhauer's analysis of the potential for territorial development to cope with climate change.

projection estimations and the normative dialogue surrounding consequential actions. Difficulty is expressly pervasive in light of conflict with economic growth, especially of developing nations, and the need for more sustainable energy use. The conflict is evident in the continuous rise in emissions, which, according to estimates from the International Energy Agency (IEA), in 2010 were the highest on record (IEA, 2011). Within the European context, the IPCC Fourth Assessment Report states that efforts within the European Union have exhibited effective climate policy development but have seen great difficulty in the full implementation and coordination toward an agreed set of objectives (Rogner *et al.*, 2007). The focus of the ESPON Climate project provides an elaboration into the European context within this dialogue. Similarly to the adaptation analysis, the ESPON Climate project first provides a brief outline of the EU level mitigation policy, followed by national or country level approaches, and concludes with a brief review of regional examples.

17.3.6   *Mitigation at the EU level*

In 2007, the EU Parliament re-established the position of the European Union with respect to climate change mitigation by adopting a resolution and setting legally binding targets. This was also to signal the position of the EU as a leading example in the international arena of climate change mitigation policy. The European Commission in 2008 followed this resolution with the EU Climate Change and Energy Package 2020, consisting of measures to achieve the previously set targets (CEC, 2008a).

This document includes three principle measures, the expected contribution of each of the Member States and proposed measures to achieve targets. The first target is the reduction of at least 20% of GHGs by the year 2020. The second target states that renewable energy must supply 20% of energy usage. The third and final target requires a 20% reduction in energy consumption by 2020. The final target, the document states, is to be achieved through enhancing energy efficiency in sectors such as transport, building, power generation, transmission and distribution. The Package also recognises the need to simultaneously consider and minimise the cost of strategy implementation and further acknowledges that the burden on Member States (MS) must be equitable. This acknowledgement takes into account the different starting points and variations in capabilities of each MS to meet the requirements.

The Emissions Trading Scheme (ETS) is used as a market-based system serving as the main tool for achieving these targets. Within this system, companies receive allowances from governments and can trade these amongst each other if they have maintained the proper emission level. The Package recognises the ETS needs to be strengthened and updated in order to meet its objectives and states that, as it is currently, the system faces a potential risk of distorting the functionality of the market and general competition. The ETS is not limited to CO_2 but includes other GHGs in addition to all other common industrial emissions. According to the Package, there is still a need for substantial investment into renewable energy to reach current goals. Incentive for this investment, however, is found in the document's observation that renewable energy will improve energy security within member states. Renewables currently make up 8.5% of energy consumption but this will need to increase by 11.5% in order to meet targets (CEC, 2008a). There is therefore a need for substantial investment in renewable energies throughout the Union.

Under current legislation, Member States are required to develop a national action plan identifying how they will meet their climate mitigation targets. However, as stated within the Package, the targets will need to be realistic, based on the capabilities of each MS. This applies within the ETS framework, where the structure enables the first half of the efforts to reach renewable energy targets to be equally shared amongst the MS and the latter half to be divided on a GDP per capita basis. Each MS has the ability to decide for itself how to go about meeting these targets and what shall constitute the primary focus such as, for example, a concentration of efforts toward solar, wind or biomass, or whether to pursue options beyond their state borders. The Package additionally highly encourages the pursuit of biofuels as the most practical alternative to fossil fuel consumption. However, the document also acknowledges that development of these energy sources must not lead to negative environmental externalities, especially with respect to land use change and biodiversity. Technological solutions such as carbon capture and storage (CCS) are also considered, with emphasis placed on the construction of plants by 2015 (CEC, 2008a).

EU level instruments and general policies influencing national level development of climate change mitigation strategies were first explored within the first European Climate Change Programme (ECCP I). This was initiated in 2000 by the European Commission as an effort to identify potential policies and measures toward achieving the targets in the Kyoto Protocol. In 2005, the second ECCP (II) was launched providing a review of its predecessor and investigating the availability of new policy options (EEA, 2009). Progress to date demonstrates that the measures identified within the ECCP are at least implemented if not already in an advanced preparatory stage. This progress is also in accordance with the 2009 Climate Change and Energy Package, which serves as an important baseline influencing development of mitigation strategies at the national level.

17.3.7 Mitigation at the national level

At the national level, mitigation policy across Europe can rely on a wide range of policy options reflecting the different modes of climate governance. The IPCC lists many of these options for country-wide implementation in its Fourth Assessment Report and are presented for the purpose of achieving national targets (Gupta *et al.*, 2007). The most common instruments are regulatory and economic in nature followed by voluntary and informational instruments. Regulatory instruments consist primarily of regulations and standards, which, for example, require the use of specific technologies or an emission threshold. The economic instruments include taxes or fees such as a carbon tax or flat fee for GHGs emitted. Other economic instruments include subsidies and incentives, which can, for example, consist of tax credits. Another instrument is an emission credit trading systems such as the EU ETS. Voluntary instruments are those where an agreement is struck between a government and another party toward emission mitigation. Informational instruments are found with research and development and public information campaigns to encourage innovation to reduce fossil fuel dependency and to increase awareness and apply public pressure. It is also important to note the non-climate policies that impact GHG emissions in a given country. These include policies in agriculture, energy, transport, trade and land use related sectors. It is likely that any policy that encourages an increased usage of natural resources simultaneously will increase the production of emissions (Gupta *et al.*, 2007).

In general, a combination of the above categories of instruments will be necessary to achieve mitigation targets and encourage sustainable development. How policy instruments are selected on the national level is based on certain criterion such as 'environmental effectiveness, cost-effectiveness, distributional considerations and institutional feasibility' (Gupta *et al.*, 2007, p. 789). A study by Geeraerts *et al.* (2007), commissioned by the European Parliament's Temporary Committee on Climate Change, analysed national level legislation, initiatives and programmes. The study provided vast and detailed knowledge of each country examined. Though the details of the study are not discussed in depth within this chapter, the reader is encouraged to view the original publication.

Another useful reference is the compilation of policies and measures for climate change mitigation in European countries provided by the European Environment Agency entitled *Climate change policies and measures in Europe* (PAM). The list includes policies that have been amassed from UNFCCC National Communications and categorises them based on the Member State, the policy type, its applicable sector, its implementation status and the GHG it mitigates. The policy types follow the distinction between regulatory, economic, voluntary and educational (or informational) and are collected from the EU 27 Member States. The findings of the database output indicate that the majority of implemented policies are regulatory and include examples such as directives for energy efficiency and savings. The second most popular type of policy is economic, which includes, for example, the ETS and development plans for sectoral climate mitigation. The least used policy types are informational including education and research measures. The rationale for the regulatory and economic emphasis can be explained in part through the fact that energy and transport sectors tend to dominate mitigation efforts.

Every Member State (as of the EU 27) has mitigation policies. However, there is great variation of the number of policies adopted within each country. One could imply that countries in which there are a greater number of adopted policies tend to consider climate change with greater importance, though the quantity of policies adopted is not an adequate indicator by itself for the quality of implementation of these policies. Further studies should evaluate the validity of this potential correlation, particularly with respect to other factors including size and socio-economic development of a given country, governance levels of policy implementation and political cultures.

17.3.8 *Mitigation at the regional level*

It must be reiterated that the regional level is influenced by other governance levels, with direction from both the national level and initiative from the local level. Each region's actions toward climate change mitigation are dependent upon these influences as well as the regional context. Case studies of strategy implementation within the regional level have been examined by Monni and Raes (2008), Smith (2007), the Assembly of the European Regions (AER), and the Environmental Conference of the European Regions (ENCORE) and provide examples of implementation within the European context.

Monni and Raes (2008) provide a study of regional implementation of EU directives at both the national and regional levels in the Helsinki Metropolitan Area Council in Finland. The study analyses the implementation of four directives for emission reduction. The results of the analysis indicate that decision-making power exists at the more local government levels, particularly with respect to land use and regulations for transportation and building construction. This is revealed to be an important characteristic of the regional system even though the local levels are not granted legislative powers. Other results determine that the regional level free rides with regard to the country-wide emission reduction targets and that one of the explanations for this lies within the inadequate financial incentives for encouraging behavioural change at the local level, especially for investment in renewable energies.

Another example by Smith (2007) demonstrates similar issues with multilevel implementation processes within the UK context. Like the study by Monni and Raes (2008), Smith (2007) focuses on implementation of strategies toward renewable energies. The system within the United Kingdom is such that encouragement of particular strategies first goes through the national, regional and then local business and non-state actor networks. The horizontal governance level is also in operation within each level of these networks and through co-ordination between public and private actors. From the study observations, Smith states that process of implementation has both ' "ordered" and "messy" forms of governance' (Hooghe and Marks, 2003; Smith, 2007). The 'ordered' form of governance occurs through the regional level's use of authority granted by the national level. However, 'messy'

forms occur through the functioning of governance with policy networks at the regional level, which are allotted no particular powers of authority. The study further reveals that there is a hindrance to strategy implementation whereby the national level does not empower the regional level; thus producing the consequential regional policy networks.

In general, there are not many examples of guidelines for best practices at the regional level for Europe. The Assembly of European Regions (AER) initiated a Working Group on climate Change in an attempt to remedy this problem. The objective of the Working Group was to emphasise the role of the regions and to amalgamate regional best practices for both mitigation and adaptation. Another entity, the Environmental Conference of the European Regions, considered 19 European regions and created a virtual Climate Change Working Group.

In understanding regional climate change mitigation policies within the EU, it is important to also consider the linkages to territorial development and cohesion. The *2007 Territorial Agenda* stresses the need for sustainable development while the *Green Paper on Territorial Cohesion* addresses how development of the territorial agenda should ensure that any given region within the EU should not be disadvantaged (CEC, 2008c). The Green Paper further places sustainable development at the centre of policy development connecting the three dimensions of cohesion policy, economic, social and territorial cohesion. The Territorial Agenda strengthens linkages to regional mitigation strategies through identification of Six Territorial Priorities. Please see Table 17.6 for the list of priorities.

A study by Sykes and Fischer (2009) analyses linkages between the Territorial Agenda and mitigation. Important points from the study reveal that attention should be paid to transport policy and urban sprawl. The rationale for this, they argue, is that there must be an additional transport policy produced to ease the difficulty of reducing GHG emissions. Urban sprawl must also be addressed due to its relation with transportation. The authors conclude that economic and social dimensions are predominant without recognition of their impact on

Table 17.6 Brief view of the Six Territorial Priorities within the Territorial Agenda of the European Union.

Six Territorial Priorities of the Territorial Agenda of the European Union

1	Polycentric development
2	New forms of partnerships between rural and urban
3	Regional clusters to promote competition
4	Strengthen trans-European networks
5	Risks need to be managed on a trans-European scale
6	Strengthen ecological structures and cultural resources

Source: TA, 2007.

the objectives of climate change and environmental policy.

Davoudi (2009) also addresses the issues of energy demand in connection with Territorial Agenda policies. Davoudi highlights the importance of specific land use policy in respect of managing energy demand within the topic of territorial policies. This is especially relevant for policies that target reduction of car travel time and enhancing energy efficiency within the built-up environment. The author states that territorial policies can positively influence climate change mitigation through regulatory efforts to guide development of land use policies toward mitigative behaviour. In addressing the need to create a more efficient built-up environment, the author states that territorial policies can focus on the location and layout as well as landscaping and new development sites (Davoudi, 2009).

Within the ESPON Climate project outputs, mitigative capacity is analysed on a regional scale identifying, as termed by the IPPC, 'a country's (or region's) ability to reduce anthropogenic greenhouse gas emissions or enhance natural sinks' (IPCC, 2007, p. 818). The project takes into consideration the social factors when estimating mitigative capacity for any particular region. Factors for analysis include educational commitment and attitudes towards climate change as well as the availability of technology for reducing emissions. Additional factors include the availability of non-carbon energy sources, types of land use, policies for mitigation, government effectiveness

and income per capita as a means of indicating economic resources. The study produces four types of regions as follows:

1. regions with high capacity and low emissions;
2. regions with high capacity and high emissions;
3. regions with low capacity and low emissions;
4. regions with low capacity and high emissions.

The two most important of the above, are number (2) regions with high capacity and high emissions and number (4) regions with low capacity and high emissions, as these indicate the most obvious areas where measures to reduce emissions should be taken.

17.3.9　*Policy development opportunities through adaptation and mitigation in Europe*

Though it is difficult to estimate sectoral developments specifically due to the uncertainty in climate change impacts, a few generally applicable findings can be discussed. Within overall policy development, capitalisation is a new concept that has seen relatively little development within adaptation strategies. The predominant focus has rather been toward risk assessment and avoidance of damage. However, capitalisation also applies to mitigation strategies with respect to production and consumption of energy and can provide new market opportunities through carbon neutral technology development.

Development opportunities within a transnational regional focus can be determined through a spatial distribution. Results from the aforementioned Massey and Bergsma (2008) study reveal that, within the Western, Northern and Southern European regions, policies pertaining to capitalisation are given the lowest priority. Within the Central regions, however, capitalisation is more greatly emphasised and surpasses priority given to measures to increase coping capacity. At least half of all measures analysed within the study concentrate on risk reduction. Consideration for physiographic regions produced findings identifying potential policy development opportunities. Within the Alpine region it was identified, similarly to the Black Sea region, that risk and sensitivity reduction is the main policy focus followed by capitalisation. More specifically, Switzerland was identified as having the most capitalisation measures while Italy had the least. Within the Tatra and Carpathian region, there is again a main focus on risk and sensitivity reduction. However, Poland and Slovakia appear to be leading the region in relation to implementation of capitalisation measures. Identification of leaders in capitalisation for other physiographic regions indicated that the Atlantic regions sees Ireland as the leader, Denmark leads for the North Sea, and in the Baltic Sea region Sweden leads with 40% of measures having a capitalisation focus. The Mediterranean, in comparison with all other regions, has the least measures for capitalisation.

Another study by Iglesias *et al.* (2007), produced for the European Commission, considers a more sectoral focus, concentrating on primarily the agricultural sector and recommendations for adaptation strategies. The results of the study demonstrate that adverse impacts for agriculture will likely be experienced as a consequence of the increased frequency of extreme events (Iglesias *et al.*, 2007). Another finding indicates that climatic changes will likely shift the location of optimum production areas for specific crops and will additionally change carbon and nitrogen cycles, equating to alterations in the soil erosion and water quality in given areas (Iglesias *et al.*, 2007). Water availability and changes in the hydrological cycle were indicated as particularly important with respect to potential reduction in crop yields and especially in consideration of possible loss of indigenous varieties as well as conditions for new pests and diseases. The stakeholder survey conducted within this study provided evidence that the Southern regions are typically more drought aware than regions in the North (Iglesias *et al.*, 2007). The study also provides further details of risks and opportunities within agro-climatic zones. The reader is therefore recommended to consult the original publication for further information.

Another particular sector in which recommendations for future policy development can be made is the tourism sector. Hydro-meteorological changes related to climate change impacts will alter water availability, providing significant implications for both tourism and agriculture. This is of importance particularly because the peak water demand for both sectors occurs during the summer months, simultaneously as the seasonal reduction in water supply (Amelung and Moreno, 2009). Within a regional focus, tourism plays an important role as an economic revenue provider for European regions such as the Mediterranean and the Alps. The susceptibility of these regions is acknowledged in the EU White Paper on adaptation, which states that tourism sectors within coastal and mountainous areas are particularly susceptible to climate change impacts (CEC, 2009). Adaptation to these impacts, at least for tourism, occurs mostly autonomously as there is no clear role set for the EU level action toward adaptation specifically within this sector (CEC, 2009). However, linkages exist in relation to infrastructure and development, which are supported by EU structural funds. Within the context of future development of policy concerning tourism as well as agriculture, it is important to reiterate the commonly accepted fact that preventative and precautionary measures are less expensive than that of reactive measures. This, thus, implies the need for more immediate pursuit of development opportunities.

The ESPON Climate project highlights opportunities particularly within the tourism sector. With respect to mitigation, opportunities lie in further development and implementation of measures to ensure climate friendly facilities. In regard to adaptation, there is a general need for new development in diversifying the tourism industry with respect to adaptation strategies. Within this context the project especially identifies the Alpine region. As demonstrated in the Tisza River case study, there is also a need for development of joint adaptation plans specifically targeting flooding and land use policies. The study also reveals a need for further development of policies for flood protection, spatial and rural development, and water resources protection.

Another policy development option discussed by the ESPON Climate project is the need for further exploration of linkages between mitigation and adaptation strategies. The IPCC reports, especially the Fourth Assessment Report, recognise and discuss these linkages (Klein *et al.*, 2007). Three types of relationships are identified within the Fourth Assessment Report. The first is a direct relationship, which involves the use of the same resources such as land and stakeholders. The second is an indirect relationship, which connects adaptation and mitigation, for example, through budgetary allocations. The final relationship identified is a remote relationship. This can be found within the example of currency exchanges. Understanding these relationships is necessary for the effectiveness of future policy development for both climate mitigation and adaptation strategies.

Aside from these types of relationships, the IPCC also identifies four types of inter-relationships of which the ESPON Climate project discusses two: (1) adaptation actions that have consequences for mitigation, and (2) mitigation activities that have impacts on adaptation. With respect to the first inter-relationship, many adaptation options require additional energy inputs, whether this is measured as one large input through major infrastructural construction or as incremental inputs for goods and services. The Fourth Assessment Report estimates that the construction of infrastructure for adaptation will occur primarily for the water sector and coastal zones. As for the second inter-relationship, an important issue lies with land use and land cover, such as deforestation due to agriculture and its impact on GHG emissions. A reversal of, for example, deforestation also influences adaptation efforts through its impact on water resources and biodiversity in a given region.

In general, the importance of synergies and trade-offs between adaptation and mitigation will increase at the regional and local levels. Though there has previously been a concentration on mitigation efforts at these levels, a more recent shift has been identified, which points to the need for further policy development with respect to both climate change strategies (Ribeiro *et al.*, 2009). In addition to this observation, there is also a need for research that looks at policies and institutions and the role they play in inter-relationships between adaptation and mitigation (Klein *et al.*, 2007).

17.3.10 *Challenges for territorial cohesion*

As is evident from the preceding sub-sections of this chapter, there are several considerations to be made for the present challenges to territorial cohesion policy with respect to climate change. This sub-section continues the dialogue with additional observations from discussion within the ESPON Internal Seminar in 2011. One of the challenges is found in how to use the Europe 2020 Strategy to implement the regional, and perhaps more spatial, focus of territorial development in light of anticipated climate impacts. Within the Polish Presidency, there has been emphasis placed on the need to stress the importance of the potential role of spatial planning. The message from the Polish Presidency is that there is a need to improve institutional elements to create regionally, and especially spatially, specific strategies with respect to implementation through networks such as ESPON. To improve the implementation process there is also a need for more regional and local input throughout strategy development. Another consideration within the current dialogue is the requirement of flexibility in the translation of the objectives and targets of the 2020 Strategy to the local levels. A reason for this is the notion that some of these targets are unrealistic or undesirable for particular regions. An example of this can be found at the very local level in terms of adaptation policies with respect to property rights. As previously mentioned within this chapter, the potential conflict specifically with private property rights and the need for change in land use planning strategies poses a major challenge. In general, however, development of strategies for implementation should be based on selection of priorities at the regional level. This is considered highly important for the successful implementation of climate change strategies within the context of territorial development.

Territorial development will likely be impacted by climate change in a way that increases regional inequalities; thus contributing to the challenges to overall Cohesion Policy and especially to the solidarity principle. More attention must be paid to how future Cohesion Policy will address the variations and potential disparities induced by climate change impacts. Another consideration is the combination of climate change effects and demographic changes. There is a potential for an increase in imbalances, particularly with the addition of demographic changes. This may produce an increase in the wealth gap in areas with greater expected changes and should be considered for financial resources particularly with respect to regional development (ESPON, 2010). Emphasis should be made especially to future challenges of the peripheral regions in regard to rising energy prices, in particular those areas with extensive commuting. Additionally, depending on carbon tax costs, there may be a further issue concerning 'carbon leakage' in some areas where high emitter industries will relocate to prevent increased production costs (ESPON, 2010). A potential avenue for addressing some of these issues is the encouragement of integration of climate change issues into policy for rural development, particularly reflecting the goal of balanced territorial development within the rural areas of Europe's regions.

Additional challenges can be seen for the progress toward realisation of the sustainability pillar within the 2020 Strategy. As stated within the Fifth Cohesion Report, there must be coordination at and between all governance levels with respect to both renewable energy and climate change goals. This coordination is required for the prevention of counteracting policies. It must be noted that the cohesion policy cannot do much for ETS but it can help with sectors that are not covered within ETS through sharing of the burden where the share of renewable energy implementation varies. Regional characteristics determine the extent to which a given EU region can produce renewable energy. Thus, with respect to related infrastructure development, attention needs to be directed to the local context for implementation. Another challenge lies in the lack of data at the sub-national levels, which, as mentioned previously, is also limited in availability due to information only being in the local language. Within a general context, the 2020 Strategy is ambitious and requires flexibility for implementation at the regional level. Actions need to be tailored to the local context and need time to direct development efforts toward the 2020 objectives. In order to reach the goals set by the 2020 Strategy, the EU needs to move toward 'a more resource efficient, greener and more competitive economy' where the EU reduces its current 16% share of the global ecological footprint (EC, 2010, p. 3; ESPON, 2010). Owing to estimated climate change impacts, challenges and development opportunities should be identified, especially at the local level for regions that concentrate on seasonally dependent sectors, such as tourism, agriculture and forestry (ESPON, 2010). With regard to mitigation and renewable energies, there is also a need to connect the Northern regions, which have favourable wind power conditions, to areas of high energy consumption (ESPON, 2010). Each of these challenges and suggestions for improvement toward achieving the goals of the 2020 Strategy reflect the current strategy for Cohesion Policy, which is centred on the need for place-based integrative policy.

17.4 Conclusions

This chapter first provided a basis within the broader European policy context, elaborating on the implications of climate change impacts to current competitiveness and cohesion policy. This baseline was necessary to establish the core of the EU level territorial development dialogue. The following sections focused specifically on the climate change strategies at three governance levels: the EU, national and regional levels. The structure of the discussion based on multilevel governance provided a crucial starting point, as policy development and implementation at one level of the vertical dimension are influenced by all other levels. The amount of influence, however, differs in each country in terms of the weight of the top-down or bottom-up influence. Adaptation, as seen through the discussion, is still being developed by both processes in territorially specific ways, while mitigation strategies have mainly seen more influence from the top-down. The emerging trend, however, is the increase of regional and more local climate strategies for both mitigation and adaptation. The implementation of these strategies, are region specific just as the climate change impacts are also region specific. Projects, such as the ESPON Climate project, assist the making of the tailor-made strategies with respect to a more local context. Understanding and implementing context appropriate strategies will be crucial to overcoming the present challenges to territorial cohesion and to establish harmonisation of overall territorial development.

References

Amelung, B. and Moreno, A. (2009) Impacts of Climate Change in Tourism and Europe. PESETA Tourism Study, European Commission, Joint Research Centre, Institute for Prospective Technological Studies, Seville.

Barnett, J. and O'Neill, S. (2010) Maladaptation. *Global Environmental Change*, **20** (2), 211–213.

BMVBS (ed.) (2010) National strategies of European countries for climate change adaptation: A review from a spatial planning and territorial development perspective. *BMVBS-Online Publikation*, 21/2010.

Commission of the European Communities (CEC) (1985) Council Directive of 27 June 1985 on the assessment of the effects of certain public and private projects on the environment. 85/337/EEC. *Official Journal of the European Union*, NO. L 175, 05/07/1985, p. 0040–0048.

Commission of the European Communities (CEC) (2007) Communication from the Commission to the Parliament, the Council, the European Economic and Social Committee and the Committee of the Regions. An Integrated Maritime Policy for the European Union. 10 October 2007, COM(2007) 575 final, Brussels.

Commission of the European Communities (CEC) (2008) Communication from the Commission of the European Parliament, the Council, the European Economic and Social Committee and the Committee of the Regions. Renewed social agenda: Opportunities, access and solidarity in 21st Century Europe. 2 July 2008, COM(2008) 412 final, Brussels.

Commission of the European Communities (CEC) (2008a) Communication from the Commission of the European Parliament, the Council, the European Economic and Social Committee and the Committee of the Regions. 20 20 by 2020. Europe's climate change opportunity. 23 January 2008, COM(2008) 30 final, Brussels.

Commission of the European Communities (CEC) (2008b) Communication from the Commission to the European Parliament and the Council on Reinforcing the Union's Disaster Response Capacity. 5 March 2008, COM(2008) 130 final, Brussels.

Commission of the European Communities (CEC) (2008c) Communication from the Commission to the Council, the European Parliament, the Committee of the Regions and the European Economic and Social Committee. Green Paper on Territorial Cohesion. Turning Territorial Diversity into Strength. 6 October 2008, COM(2008) 616 final, Brussels.

Commission of the European Communities (CEC) (2009) White Paper – Adapting to Climate Change: Towards a European Framework for Action. 1 April 2009, COM(2009) 147 final, Brussels.

Commission of the European Communities (CEC) (2009a) Commission Staff Working Document accompanying the White Paper – Adapting to Climate Change: Towards a European Framework for Action. Impact Assessment. 1 April 2009, SEC(2009) 387, Brussels.

Commission of the European Communities (CEC) (2009b) Communication from the Commission of the European Parliament, the Council, the European Economic and Social Committee and the Committee of the Regions. A Community Approach on the Prevention of Natural and Man-made Disasters. 23 February 2009, COM(2009) 82 final, Brussels.

Davoudi, S., Crawford, J. and Mehmood, A. (eds) (2009) *Planning for Climate Change: Strategies for Mitigation and Adaptation for Spatial Planners*, Earthscan, London.

European Commission (EC) (2001) White Paper – European transport policy for 2010: time to decide. Office for Official Publications of the European Communities, Luxembourg.

European Commission (EC) (2001a) Directive 2001/42/EC of the European Parliament and of the Council. On assessment of the effects of certain plans and programmes on the environment. *Official Journal of the European Communities*, NO. L 197, 21/07/2001, p. 0030–0037.

European Commission (EC) (2007) Directive 2007/60/EC of the European Parliament and of the Council. On the assessment and management of flood risks. *Official Journal of the European Union*, NO. L288/27, 6/11/2007.

European Commission (EC) (2008) Directive 2008/56/EC of the European Parliament and of the Council. Establishing a framework for community action in the field of marine environmental policy (Marine Strategy Framework

Directive). *Official Journal of the European Union*, NO. L 164/19, 25/6/2008.

European Commission (EC) (2010) Communication from the Commission. Europe 2020: A strategy for smart, sustainable and inclusive growth. 3 March 2010, COM(2010) 2020 final, Brussels.

European Commission (EC) (2010a) Communication from the Commission to the European Parliament, the Council, the European Economic and Social Committee, the Committee of the Regions and the European Investment Bank. Conclusions of the fifth report on economic, social and territorial cohesion: the future of cohesion policy. 9 November 2010, COM(2010) 642 final, Brussels.

European Commission (EC) (2011) Communication from the Commission to the European Parliament, the Council, the Economic and Social Committee and the Committee of the Regions. Our life insurance, our natural capital: an EU biodiversity strategy 2020. 3 May 2011, COM(2011) 244 final, Brussels.

European Environment Agency (EEA) (2009) Greenhouse gas emission trends and projections in Europe 2009. Tracking progress towards Kyoto targets, European Environment Agency, Copenhagen.

European Environment Agency (EEA) (2010) Climate change mitigation. Available online at http://www.eea .europa.eu/soer/policy-makers/climate-change-mitigation, updated on 2010 (accessed 18 November 2011).

European Spatial Planning Observation Network (ESPON) (2010) First ESPON 2013 Synthesis Report. ESPON Results by summer 2010. New Evidence on Smart, Sustainable and Inclusive Territories, Luxembourg.

European Union (EU) (2007) Treaty of Lisbon Amending the Treaty on European Union and the Treaty Establishing the European Community, 13 December 2007, 2007/C 306/01.

European Union (EU) (2010) Fifth report on economic, social and territorial cohesion. Publications Office of the European Union, Luxembourg.

Füssel, H-M. and Klein, R.J.T. (2002) Assessing vulnerability and adaptation to climate change: An evolution of conceptual thinking, in *A Climate Risk Management Approach to Disaster Reduction and Adaptation to Climate Change. Proceedings of the UNDP Expert Group Meeting on "Integrating Disaster Reduction and Adaptation to Climate Change"*. Havana, Cuba, 17–19 June 2002, pp. 45–59.

Gagnon-Lebrun, F. and Agarwala, S. (2006) Progress on adaptation to climate change in developed countries. An analysis of broad trends, OECD, Paris.

Gagnon-Lebrun, F. and Agarwala, S. (2007) Implementing adaptation in developed countries: an analysis of progress and trends. *Climate Policy*, **7**, 392–408.

Geeraerts, K., Bassi, S., Chiavari, J., *et al.* (2007) National Legislation and national initiatives and programmes (since 2005) on topics related to climate change, Economic and scientific policy, Policy Department, European Parliament, Brussels.

Greiving, S. and Fleischhauer, M. (2012) National climate change adaptation strategies of European states from a spatial planning and development perspective. *European Planning Studies*, **20** (1), 27–47.

Gupta, S., Tirpak, D.A., Burger, N., *et al.* (2007) Policies, Instruments and Co-operative Arrangements, in *Climate Change 2007: Mitigation*. Contribution of Working Group III to the Fourth Assessment Report of the Intergovernmental Panel on Climate Change.

Hooghe, L. and Marks, G. (2003) Unravelling the central state, but how? Types of multi-level governance. *American Political Science Review*, **97** (2) 233–243.

Iglesias, A., Avis, K., Benzie, M., *et al.* (2007) Adaptation to climate change in the agricultural sector. AEA Energy & Environment and Universidad de Politecnica de Madrid, Didcot, UK.

Intergovernmental Panel on Climate Change (IPCC)(2007) *Climate Change 2007: Mitigation. Contribution of Working Group III to the Fourth Assessment Report of the Intergovernmental Panel on Climate Change* (eds B. Metz, O.R. Davidson, P.R. Bosch, R. Dave and L.A. Meyer), Cambridge University Press, Cambridge and New York.

International Energy Agency (IEA) (2011) Latest Information. Prospect of limiting the global increase in temperature to 2 °C is getting bleaker. Available online at http://www.iea.org/index_info.asp?id=1959 (accessed 9 November 2011).

Keskitalo, E.C.H. (ed.) (2010) *The Development of Adaptation Policy and Practice in Europe: Multilevel Governance of Climate Change*, Springer Verlag, Berlin.

Keskitalo, E.C.H., Westehoff, L. and Juhola, S. (2012). Agenda-setting on the environment: the development of climate change adaptation as an issue in European States. *Environmental Policy and* Governance, **22** (6), 381–394.

Klein, R.J.T., Huq, S., Denton, F., *et al.* (2007) Interrelationships between adaptation and mitigation, in (eds M.L. Parry, O.F. Canziani, J.P. Palutikof, P.J. van der Linden and C.E. Hanson) *Climate Change 2007: Impacts, Adaptation and Vulnerability. Contribution of Working Group II to the Fourth Assessment Report of the Intergovernmental Panel on Climate Change*, Cambridge University Press, Cambridge, pp. 745–777.

Leary, N.A. (1999) A framework for benefit-cost analysis of adaptation to climate change and climate variability.

Mitigation and Adaptation Strategies for Global Change, **4** (3-4), 307–318.

Massey, E. and Bergsma, E. (2008) *Assessing adaptation in 29 European countries,* IVM Institute of Environmental Studies, Vrije University, Amsterdam.

Meister, H.P., Kröger, I., Richwien, M., *et al.* (2009) Floating Houses and Mosquito Nets: Emerging Climate Change Adaptation Strategies around the World, IFOK Study, Meister Consultants Group, Inc., Boston.

Mickwitz, P., Aix, F., Beck, S., *et al.* (2009) Climate Policy Integration, Coherence and Governance, PEER Report No. 2. Helsinki.

Ministry of Ecology, Sustainable Development, Transport and Housing (2011) French National Climate Change Impact Adaptation Plan 2011–2015. République Française.

Monni, S. and F. Raes (2008) Multilevel climate policy: the case of the European Union, Finland and Helsinki. *Environmental Science & Policy*, **11** (8), 743–755.

OECD (2009) The Economics of Climate Change Mitigation. Policies and Options for Global Action Beyond 2012.

Ribeiro, M., Losenno, C., Dworak, T., *et al.* (2009) Design of guidelines for the elaboration of Regional Climate Change Adaptation Strategies, Ecologic Institute, Vienna.

Rogner, H., Zhou, D., Bradley, R., *et al.* (2007) Introduction, in *Climate Change 2007: Mitigation,* Cambridge University Press, Cambridge.

Smit, B. and Pilifosova, O. (2001) Adaptation to climate change in the context of sustainable development and equity, in: *IPCC 2001: Climate Change 2001 – Impacts, Adaptation, and Vulnerability*. Contribution of the Working Group II to the Third Assessment Report of the Intergovernmental Panel on Climate Change, Cambridge University Press, Cambridge, pp. 877–912.

Smith, A. (2007) Emerging in between: The multi-level governance of renewable energy in the English regions. *Energy Policy*, **35** (12), 6266–6280.

Swart, R., Biesbroek, G.R., Binnerup, S., *et al.* (2009) Europe Adapts to Climate Change. Comparing National Adaptation Strategies, Vammalan Kirjapaino Oy. Sastamala.

Sykes, O. and Fischer, T. (2009) The territorial agenda of the European Union – progress for climate mitigation and adaptation? in (eds S. Davoudi, J. Crawford and A. Mehmood) *Planning for Climate Change. Strategies for Mitigation and Adaptation for Spatial Planners*. Earthscan, London, pp. 111–125.

Territorial Agenda of the European Union (TA) (2007) Towards a More Competitive and Sustainable Europe of Diverse Regions. Informal Ministerial Meeting on Urban Development and Territorial Cohesion, Leipzig.

Westerhoff, L., Keskitalo, E.C.H. and Juhola, S. (2010) Capacities across scales: local to national adaptation policy in four European countries. *Climate Policy*, **11** (4), 1071–1085.

Index